T0236121

AC Motor Control and Electrical Vehicle Applications

Second Edition

AC Motor Control and Electrical Vehicle Applications

Second Edition

Kwang Hee Nam

CRC Press
Taylor & Francis Group
Boca Raton London New York

CRC Press is an imprint of the
Taylor & Francis Group, an **informa** business

CRC Press
Taylor & Francis Group
6000 Broken Sound Parkway NW, Suite 300
Boca Raton, FL 33487-2742

First issued in paperback 2020

© 2019 by Taylor & Francis Group, LLC
CRC Press is an imprint of Taylor & Francis Group, an Informa business

No claim to original U.S. Government works

ISBN-13: 978-1-138-71249-2 (hbk)
ISBN-13: 978-0-367-73286-8 (pbk)

Visit the Taylor & Francis Web site at
http://www.taylorandfrancis.com

and the CRC Press Web site at
http://www.crcpress.com

Contents

Preface

The importance of motor control technology has resurfaced recently, since electrification of various power sources reduces green house gas. Autonomous vehicle technology opens a new world of unmanned delivery with the expansion of drone applications. Sooner or later, electric planes and flying cars are popularly used for passenger transport. The motor application is more accelerated as the battery costs are reduced.

Control engineers need to understand many motor design issues to meet the challenging design specifications. From this point of view, various aspects including motor control, motor design, practical manufacturing, testing, and programming are considered in this book. This book was written as a textbook for a graduate level course on AC motor control and electric drive. Not only the motor control, but also some motor design basics are covered to give a comprehensive view in the multidisciplinary age. Theoretical integrity in the modeling and control of AC motors is pursued throughout the book.

In the second edition, many EV projects and teaching experiences at POSTECH and industrial sites are reflected. The contents become richer by adding more exercises and problems that utilize Excel spreadsheet and MATLAB Simulink.

There is a little barrier for the beginners to understand the principles of the AC rotating machine, because many physical phenomena are interpreted in the moving frame. The essential machinery is the ability to understand voltage to current dynamics in the rotating frame. Firstly, this book is focused on illustrating how the rotating field is synthesized with the three phase winding. Also, the benefits of coordinate transformation are stressed in the dynamic modeling of AC motors. For example, many mathematical tools are utilized to show how the voltage and current limits affect the torque maximization. Loss minimizing and sensorless controls are also covered.

In the second part of this book, many issues regarding EV motor design and fabrication are expressed. In Chapter 11 and 12, a motor design method is suggested based on the requirements of power, torque, power density, etc. under voltage and current limits. In addition, experimental procedure and inverter programming technique are introduced that provide an optimal current control strategy under varying (battery) voltage conditions. In the last part, the basics of vehicle dynamics and EV power trains are shown including calculation methods of driving range and efficiency of the vehicle.

This book is intended to bring combined knowledge and problems to the students who wish to learn the electric power train. So a fusing approach is attempted while covering control, signal processing, electro-magnetics, power electronics, material properties, vehicle dynamics, etc. Many control issues that can lead to on-going research are discussed.

Finally, the authors would like to say a word of thanks to the family who supported me and encouraged me. Also, many thanks are given to Jongwon Choi, Yoonjae Kim, Bonkil Koo, Jeonghun Lee, Minhyeok Lee, Taeyeon Lee, Pooreum Jang, and Heekwang Lee who helped me by providing many solutions, simulation results, and interesting reflections.

An MATLAB® files found in this book are available for download from the publisher's Web site. MATLAB® is a registered trade mark of The MathWorks, Inc. For product information, please contact:

The MathWorks, Inc.
3 Apple Hill Drive
Natick, MA 01760-2098 USA
Tel: 508-647-7000
Fax: 508-647-7001
E-mail: info@mathworks.com
Web: www.mathworks.com

About the Author

Dr. Kwang Hee Nam received his B.S. in chemical technology and his M.S. in control and instrumentation from Seoul National University in 1980 and 1982, respectively. He also earned an M.A. in mathematics and a Ph.D. in electrical engineering from the University of Texas at Austin in 1986. Since 1987, he has been at POSTECH as a professor of department of electrical engineering. From 1987 to 1992, he participated in constructing a 2.5 GeV synchrotron light source as a beam dynamics group leader. He also served as the director of POSTECH Information Research Laboratories from 1998 to 1999.

He published more than 150 publications in motor drives and power converters and received a best paper award from the Korean Institute of Electrical Engineers in 1992 and a best transaction paper award from the Industrial Electronics Society of IEEE in 2000. Dr. Nam has worked on numerous industrial projects for major Korean industries such as POSCO, Hyundai Motor Company, LG Electronics, and Hyundai Mobis. He served as a president of Korean Institute of Power Electronics in 2016. Presently, his research areas include sensorless control, EV propulsion systems, motor design, and EV chargers.

Chapter 1

Preliminaries for Motor Control

1.1 Basics of Electromagnetics

A vector can be assigned to the tangent space defined at a point. Similarly, another vector may be assigned to the tangent space of a neighboring point. In this way, we can assign vectors to all point of the manifold. In mathematics, a vector field is a mapping rule which assigns a vector to each point in a subspace. Using a local coordinate system of a manifold, continuity and differentiability of vector fields are defined. Vector fields are often used to model the speed and direction of a moving fluid or the magnitude and direction of force, such as magnetic or gravitational force.

1.1.1 Tensors

Tensors are coordinate-free geometric operators that describe linear relations between vectors, scalars, and other operators. Gradient, divergence, curl, Laplacian operations are typical differential operators used frequently in electromagnetics.

Table 1.1: Differential operators in electromagnetics.

Operator	Symbol	Operand	Result	Example
Gradient	∇	scalar	vector	$\nabla V = -\mathbf{E}$
Divergence	$\nabla\cdot$	vector	scalar	$\nabla \cdot \mathbf{E} = \frac{\rho}{\epsilon}$
Curl	$\nabla\times$	vector	vector	$\nabla \times \mathbf{H} = \mathbf{J}$
Laplacian	∇^2	scalar	scalar	$\nabla^2 V = -\rho/\epsilon$

Gradient is an operator acting on a scalar function to give gradients to the x, y, and z directions.

$$\nabla f = \frac{\partial f}{\partial x}\mathbf{i}_x + \frac{\partial f}{\partial y}\mathbf{i}_y + \frac{\partial f}{\partial z}\mathbf{i}_z \quad : \quad \text{scalar} \quad \rightarrow \quad \text{vector}.$$

It gives an orthogonal vector at each point to the equi-potential line, $f(x, y, c) = c$, where c is a constant. A typical example of gradient flow is the electric field:

$$\mathbf{E} = -\nabla V.$$

1

Divergence acts on a vector to give a scalar. It appears as the inner-product operation between the differential operator and a vector field.

$$\nabla \cdot \mathbf{A} = \begin{bmatrix} \dfrac{\partial}{\partial x}, & \dfrac{\partial}{\partial y}, & \dfrac{\partial}{\partial z} \end{bmatrix} \begin{bmatrix} A_x \\ A_y \\ A_z \end{bmatrix} = \frac{\partial A_x}{\partial x} + \frac{\partial A_y}{\partial y} + \frac{\partial A_z}{\partial z} \quad : \quad \text{vector} \rightarrow \text{scalar.}$$

The divergence of a vector $\mathbf{A}$ at a point x can be interpreted as the net outgoing flux per volume of an infinitesimally small polyhedron or sphere that encloses x. If more flux flows out of the volume, the divergence of $\mathbf{A}$ is positive about the point x. On the other hand, if the net flux flows into the volume, the divergence of $\mathbf{A}$ is negative. If the divergence is equal to zero at a point, it means that the point is not a source or sink. Physically, a magnet monopole does not exist. Thus, $\nabla \cdot \mathbf{B} = 0$.

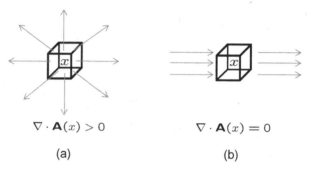

$$\nabla \cdot \mathbf{A}(x) > 0 \qquad\qquad \nabla \cdot \mathbf{A}(x) = 0$$

(a)   (b)

Figure 1.1: Applications of divergence to example flows.

Curl is a special form of differentiation which acts on a vector to give another vector. It can be computed as a cross product of differential operator and an operand vector:

$$\nabla \times \mathbf{A} = \begin{vmatrix} \mathbf{i}_x & \mathbf{i}_y & \mathbf{i}_z \\ \dfrac{\partial}{\partial x} & \dfrac{\partial}{\partial y} & \dfrac{\partial}{\partial z} \\ A_x & A_y & A_z \end{vmatrix}$$

$$= \mathbf{i}_x \left(\frac{\partial A_z}{\partial y} - \frac{\partial A_y}{\partial z} \right) - \mathbf{i}_y \left(\frac{\partial A_z}{\partial x} - \frac{\partial A_x}{\partial z} \right) + \mathbf{i}_z \left(\frac{\partial A_y}{\partial x} - \frac{\partial A_x}{\partial y} \right).$$

The curl of vector $\mathbf{A}$ at a point x can be visualized as a small paddle wheel in a force field, $\mathbf{A}$. If the forces on both sides are balanced, the paddle will not rotate as shown in Fig. 1.2 (a). In that situation the curl operation yields zero. Otherwise, $\nabla \times \mathbf{A} \neq \mathbf{0}$. Then, the paddle wheel will be rotated by the field. The vector direction is determined according to the right hand rule. Fig. 1.2 (b) shows a case of rotating field where $\nabla \times \mathbf{A} \neq \mathbf{0}$. The rotation velocity is represented by the magnitude of z-component.

The curl is a tool of determining integrability. For example, if $\frac{\partial A_y}{\partial x} - \frac{\partial A_x}{\partial y} = 0$ and the region of integration is simply-connected (i.e., there is no hole), then there exists a scalar function $\varphi : \mathbb{R}^2 \rightarrow \mathbb{R}$, $\varphi(x, y) = c$ for some c such that $\frac{\partial \varphi}{\partial x} = A_y$ and $\frac{\partial \varphi}{\partial y} = A_x$. Note that $\varphi(x, y) = c$ is an integral manifold, i.e., if $\frac{\partial A_y}{\partial x} = \frac{\partial A_x}{\partial y}$, the

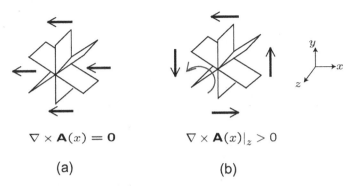

$$\nabla \times \mathbf{A}(x) = \mathbf{0} \qquad\qquad \nabla \times \mathbf{A}(x)|_z > 0$$

(a)   (b)

Figure 1.2: Curl applications.

vector, (A_x, A_y) can be integrated to a manifold, and it follows sufficiently that

$$\frac{\partial^2 \varphi}{\partial x \partial y} = \frac{\partial^2 \varphi}{\partial y \partial x}.$$

1.1.2   Riemann Integral and Fundamental Theorem of Calculus

Consider the upper sum and the lower sum of a function $f : \mathbb{R} \to \mathbb{R}$ on an interval $[a, b]$ defined as:

$$\text{Upper Sum :} \quad U_a^b(n) \;=\; \sum_{i=0}^{n-1} \sup_{\xi_i \le x < \xi_{i+1}} \{f(x)\} \, (\xi_{i+1} - \xi_i) \tag{1.1}$$

$$\text{Lower Sum :} \quad L_a^b(n) \;=\; \sum_{i=0}^{n-1} \inf_{\xi_i \le x < \xi_{i+1}} \{f(x)\} \, (\xi_{i+1} - \xi_i), \tag{1.2}$$

for a given partition

$$a = \xi_0 < \xi_1 < \cdots < \xi_n = b.$$

Note that $\sup_{\xi_i \le x < \xi_{i+1}} \{f(x)\}$ and $\inf_{\xi_i \le x < \xi_{i+1}} \{f(x)\}$ denote the supremum and the infimum of $f(x)$ over the subinterval $\xi_i \le x < \xi_{i+1}$, respectively. The partition does not need to be uniform. Consider the limiting case as $n \to \infty$ while letting the maximum interval approach to zero, i.e., $\max_{0 \le i \le n-1}(\xi_{i+1} - \xi_i) \to 0$. If the limits of $U_a^b(n)$ and $L_a^b(n)$ exist and converge to the same value, i.e.,

$$\lim_{n \to \infty} U_a^b(n) = \lim_{n \to \infty} L_a^b(n) = S, \tag{1.3}$$

then we say that the function is Riemann integrable on $[a, b]$, and denote the limit value by

$$S = \int_a^b f(x) dx.$$

Roughly speaking, the Riemann integral is the limit of the Riemann sums as the partition gets fine.

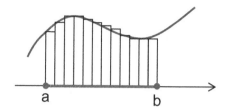

Figure 1.3: Riemann integral and fundamental theorem of calculus.

Exercise 1.1   Consider a function $f : [0, 1] \rightarrow [0, 1]$:

$$f(x) = \begin{cases} 1, & x = \text{rational number,} \\ 0, & \text{otherwise.} \end{cases} \tag{1.4}$$

Check whether f is Riemann integrable or not.

Solution. It is not Riemann integrable, since $U_a^b(n) = 1$ whereas $L_a^b(n) = 0$ for all feasible partitions. Note that there is a rational number between any two different numbers.   ■

As we all know, the integration is not taken by using the Reimann sums. Instead, we use the anti-derivative and evaluate at the boundary. Such an evaluation is based on the fundamental theorem of calculus:

Theorem (Fundamental theorem of calculus): Let f and F be real-valued functions defined on a closed interval $[a, b]$ such that

$$\frac{d}{dx}F(x) = f(x), \qquad \forall x \in [a, b].$$

If f is Riemann integrable on $[a, b]$ then

$$\int_a^b f(x)dx = F(b) - F(a).$$

Note that F is called *anti-derivative* of f. The integral is defined as the limit of the Riemann sum. However, the fundamental theorem of calculus gives us the relation between integration and differentiation. It states that the integral is simply an evaluation of anti-derivative of the original function at the boundary of an interval. Note that the boundary of the interval, $[a, b]$ is a and b, and that the evaluation of F at the boundary is denoted by $\partial_{a,b} F \equiv F(b) - F(a)$. For example, when we evaluate $\int_0^1 xdx$, we do $\frac{x^2}{2}\Big|_0^1 = 1/2$ using the anti-derivative. Now, consider a scalar function, $g : \mathbb{R}^2 \rightarrow \mathbb{R}$ defined on the real plane: It is desired to integrate g over a closed region, $S \in \mathbb{R}^2$:

$$\iint_S g(x, y)dxdy$$

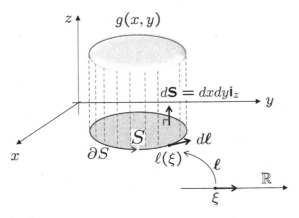

Figure 1.4: Green's theorem.

The fundamental theorem of calculus can be extended to the higher dimensional cases:

Theorem (Green's Theorem): Suppose that there exist C^1-functions, G_1, G_2 : $\mathbb{R}^2 \to \mathbb{R}$ such that

$$\iint_S g(x,y)dxdy = \oint_{\partial S} G_1(x,y)dx + G_2(x,y)dy, \tag{1.5}$$

where ∂S is the boundary of a closed area, S. Then, it should follow that

$$\frac{\partial G_2}{\partial x} - \frac{\partial G_1}{\partial y} = g(x,y). \tag{1.6}$$

The anti-derivative of a scalar function, $g : \mathbb{R}^2 \to \mathbb{R}$ is given as the two-dimensional function, $[G_1, G_2] : \mathbb{R}^2 \to \mathbb{R}^2$ that satisfies a specialized form (1.6) of differentiation. The Green's theorem tells us that the Riemann integral of g over S should be equal to an evaluation of some function, $[G_1, G_2]$ along the boundary, ∂S. The line integral on the right side of (1.5) is written equivalently as

$$\oint_{\partial S} [G_1(x,y), G_2(x,y)] \cdot d\boldsymbol{\ell}(x,y),$$

where $\boldsymbol{\ell} : \mathbb{R} \to \mathbb{R}^2$ is a line imbedding into the plane, i.e., a (one-parameter group) function onto the boundary ∂S.

Theorem (Stokes' Theorem): Consider a function $\mathbf{F} \equiv [F_x, F_y, F_z]^T : \mathbb{R}^3 \to \mathbb{R}^3$ on a closed region, $S \in \mathbb{R}^2$. Then,

$$\iint_S (\nabla \times \mathbf{F}) \cdot d\mathbf{S} = \oint_{\partial S} \mathbf{F} \cdot d\boldsymbol{\ell}, \tag{1.7}$$

where $d\mathbf{S} = dydz\mathbf{i}_x + dxdz\mathbf{i}_y + dxdy\mathbf{i}_z$ and $[\ell_x(\xi), \ell_y(\xi), \ell_z(\xi)]^T \in \partial S$ for $-\infty \leq \xi \leq \infty$.

Note that $d\boldsymbol{\ell}$ is a tangent vector to ∂S and that (1.7) is expressed equivalently as

$$\iint_S \left\{ \left(\frac{\partial F_z}{\partial y} - \frac{\partial F_y}{\partial z} \right) dydz + \left(\frac{\partial F_x}{\partial z} - \frac{\partial F_z}{\partial x} \right) dxdz + \left(\frac{\partial F_y}{\partial x} - \frac{\partial F_x}{\partial y} \right) dxdy \right\}$$

$$= \oint_{\partial S} \left\{ F_x \frac{\partial \ell_x}{\partial \xi} + F_y \frac{\partial \ell_y}{\partial \xi} + F_z \frac{\partial \ell_z}{\partial \xi} \right\} d\xi \tag{1.8}$$

Exercise 1.2   Using the Stokes' theorem, derive the Green's theorem.

Solution. Let $\mathbf{G} = G_1(x,y)\mathbf{i}_x + G_2(x,y)\mathbf{i}_y$. Then,

$$\nabla \times \mathbf{G} = \begin{vmatrix} \mathbf{i}_x & \mathbf{i}_y & \mathbf{i}_z \\ \frac{\partial}{\partial x} & \frac{\partial}{\partial y} & \frac{\partial}{\partial z} \\ G_1 & G_2 & 0 \end{vmatrix} = \left(\frac{\partial G_2}{\partial x} - \frac{\partial G_1}{\partial y} \right) \mathbf{i}_z.$$

Note that $d\mathbf{S} = dydz\,\mathbf{i}_x + dxdz\,\mathbf{i}_y + dxdy\,\mathbf{i}_z$. Thus,

$$(\nabla \times \mathbf{G}) \cdot d\mathbf{S} = \left(\frac{\partial G_2}{\partial x} - \frac{\partial G_1}{\partial y} \right) dxdy.$$

On the other hand, it follows from the right hand side of (1.8) that

$$\oint_{\partial S} \mathbf{G} \cdot d\boldsymbol{\ell} = \oint_{\partial S} \left\{ G_1 dx + G_2 dy \right\}. \qquad\qquad \blacksquare$$

1.1.3   Ampere's Law

Current flow is always accompanied by magnetic field, and the Ampere's law states the relation between currents and magnetic fields. Neglecting the displacement current, the Ampere's law in the free space is

$$\nabla \times \mathbf{B} = \mu_0 \mathbf{J}, \tag{1.9}$$

where $\mathbf{B}$ is the magnetic flux density in tesla (T), $\mu_0 = 4\pi \times 10^{-7}$ (Henries/m) is the permeability of free space, and $\mathbf{J}$ is the current density in ampere per square meter, A/m^2. Applying the double integral to both sides of (1.9) over a closed region, S we obtain

$$\iint_S (\nabla \times \mathbf{B}) \cdot d\mathbf{S} = \mu_0 \iint_S \mathbf{J} \cdot d\mathbf{S}. \tag{1.10}$$

Then by the Stokes' theorem, the left hand side of (1.10) is equal to

$$\iint_S (\nabla \times \mathbf{B}) \cdot d\mathbf{S} = \oint_{\partial S} \mathbf{B} \cdot d\boldsymbol{\ell}. \tag{1.11}$$

Further, the integral of the right hand side of (1.10)

$$\iint_S \mathbf{J} \cdot d\mathbf{S} = I \tag{1.12}$$

yields the current flowing transversally to S. Therefore, the integral form of the Ampere's law is equal to

$$\oint_{\partial S} \mathbf{B} \cdot d\ell = \mu I. \tag{1.13}$$

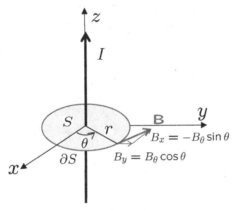

Figure 1.5: B-field around the line current, I.

Consider a specific case shown in Fig. 1.5, in which a line current, I penetrates the xy-plane at the origin. It is desired to evaluate field density, B_θ around a circle of radius, r. We set a region S to be a disk of radius, r located at the origin of the xy-plane. Then, the current density is described mathematically by a two-dimensional delta function:

$$J_z = I\delta(x, y), \tag{1.14}$$

where

$$\delta(x, y) \begin{cases} \neq 0, & (x, y) = (0, 0) \\ = 0, & \text{otherwise.} \end{cases}$$

and

$$\iint_S \delta(x, y) dS = 1.$$

Since the magnetic field is constant, $|\mathbf{B}| = B_\theta > 0$ along ∂S, we have $(B_x, \dot{B}_y) = (-B_\theta \sin\theta, B_\theta \cos\theta)$. Note that ∂S can be parameterized, for example, by $(\ell_x, \ell_y) =$

$(r\cos\theta, r\sin\theta)$ for $0 \le \theta < 2\pi$. Therefore,

$$\iint_S (\nabla \times \mathbf{B})_z \, dx dy = \int_0^{2\pi} B_\theta r(\sin^2\theta + \cos^2\theta)d\theta = 2\pi r B_\theta$$

$$\iint_S J_z = I$$

Finally, it follows that

$$B_\theta = \frac{\mu_0 I}{2\pi r}. \tag{1.15}$$

Exercise 1.3   Consider the toroidal core of permeability $\mu \gg \mu_0$ shown in Fig. 1.6. Current I flows through the N-turn coil wound on the core. Determine the magnetic flux.

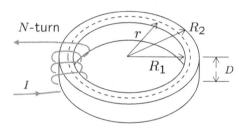

Figure 1.6: Magnet flux density of a toroidal wound core.

Solution. Consider a band of radius $R_1 < r < R_2$. Applying the Ampere's law, we obtain $B_\theta(r)$:

$$\oint_{\partial S} \mathbf{B} \cdot d\boldsymbol{\ell} = B_\theta(r)2\pi r = \mu NI, \qquad R_1 < r < R_2.$$

Therefore, the flux over the core section is

$$\Phi = D \int_{R_1}^{R_2} B_\theta(r)dr = \frac{\mu DNI}{2\pi} \int_{R_1}^{R_2} \frac{dr}{r} = \frac{\mu DNI}{2\pi} \ln\frac{R_2}{R_1}. \qquad \blacksquare$$

Exercise 1.4   Consider a ferrite C-core with the permeability, $\mu \approx 2000\mu_0$. The air gap height is $g = 2$ mm, and the number of coil turns are $N = 8$ as shown in Fig. 1.7. Assume that the flux density is uniform in the core, as well as in the air gap, i.e., neglect the fringe field effect. Calculate the air gap field density when 50 A current flows.

Solution. Applying the Ampere's law, we obtain

$$\oint_{\partial S} \frac{B}{\mu}d\ell = \frac{B}{\mu}\ell_{core} + \frac{B}{\mu_0}g = NI, \tag{1.16}$$

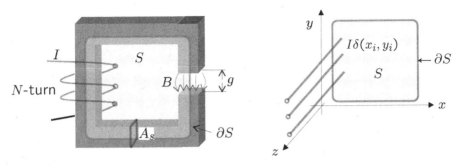

Figure 1.7: C-core with an air gap, g.

where ℓ_{core} is the mean length of the core. The core permeability is very large compared with the air gap. Thus, it is assumed that $\frac{B}{\mu}\ell_{core} \ll \frac{B}{\mu_0}g$. Therefore, we have an approximate air gap field density

$$B \approx \mu_0 \frac{NI}{g} = 4\pi \times 10^{-7}\frac{8 \times 50}{0.002} = 0.25 \text{ T.}$$

Fig. 1.8 shows FEM simulation results showing flux line and flux density. Note from Fig. 1.8 (c) that the air gap field density is slightly lower than the calculated value. The difference was caused by neglecting the line integral along the core in (1.16). ■

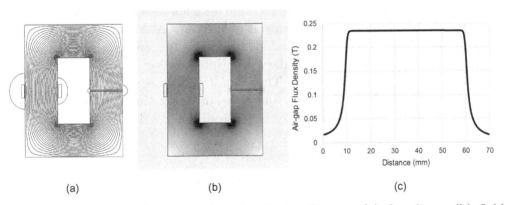

(a)   (b)   (c)

Figure 1.8: FEM simulation results of a ferrite C-core: (a) flux lines, (b) field density by grey level, and (c) air gap field density.

1.1.4 Faraday's Law

An electromagnetic circuit refers to a circuit coupled with magnetic flux. Consider a simple electromagnetic circuit shown in Fig. 1.9. The Maxwell-Faraday equation states that a time-varying magnetic field induces an electric field, and vice versa. It is written as

$$\nabla \times \mathbf{E} = -\frac{\partial \mathbf{B}}{\partial t}. \tag{1.17}$$

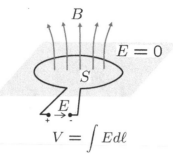

Figure 1.9: Open loop circuit linked by flux, Φ.

Let S be the loop area through which the flux passes. Integrating both sides of (1.17) over S, it follows that

$$\text{Left Hand Side} \qquad \iint_S (\nabla \times \mathbf{E}) \cdot d\mathbf{S} = \oint_{\partial S} \mathbf{E} \cdot d\ell = V \qquad (1.18)$$

$$\text{Right Hand Side} \qquad -\iint_S \frac{\partial \mathbf{B}}{\partial t} \cdot d\mathbf{S} = -\frac{\partial}{\partial t} \iint_S B\, dx dy = -\frac{\partial \Phi}{\partial t}, \qquad (1.19)$$

where Φ is the amount of flux that passes through the loop. Of the total flux, it is Φ that matters. Flux flowing outside the loop does not affect induction. If Φ passes through a N-turn loop, the flux linkage is defined as $\lambda \equiv N\Phi$.

Applying Stokes' theorem to the left hand side, the surface integral turns out to be a line integral along the boundary ∂S. Note however that the electric field is equal to zero inside the conductor. Thus, non-trivial integration is taken between the terminals. On the other hand, the order of integration can be interchanged between the space and time variables in the right hand side. As a result, we have Lenz's law:

$$V = -\frac{d\lambda}{dt} = -N\frac{d\Phi}{dt}. \qquad (1.20)$$

It says that the voltage is induced to counteract the field variation.

Let us consider how we can create a time varying field. One may place another coil nearby and supply an alternating current. Then an alternating field is generated by the Ampere law, and it may pass through the original coil. It states the induction principle of transformer. The second case is to shake a permanent magnet over the coil. It also induces voltage that corresponds to the principle of generator, which converts mechanical power into electrical power. One has to be careful with the meaning of d/dt. It is interpreted that no electrical power is generated without mechanical work. This explains why a huge steam turbine is necessary for electrical power generation in a power plant.

Exercise 1.5 Consider a two pole cylindrical machine shown in Fig. 1.10. The rotor has two permanent magnets (PMs) on its surface. The stator inner radius is $R_m = 5$ cm, and the PM coverage angle is $\varphi = 160°$. Assume that the core

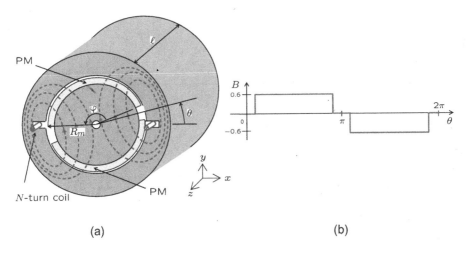

Figure 1.10: Two pole cylindrical PM machine (Exercise 1.5).

permeability is infinity. Then the air gap field has only radial component. Assume further that the air gap flux density is $B = \pm 0.6$ T at the top of the PMs, and zero elsewhere as shown in Fig. 1.10 (b). The axial (stacking) length is $\ell = 7$ cm. It has two slots along the z-axis, and the number of coil turns are $N = 12$. Ignoring the slot opening, answer to the following questions:

    a) Sketch the flux linkage of the coil as a function of rotor angle, θ.
    b) Sketch the terminal voltage when the rotor rotates at 6000 rpm.
    c) Find the RMS value of the fundamental component of the voltage obtained from b).

Solution. The flux linkage has a trapezoidal waveform as shown in Fig. 1.11 (a), and the maximum flux linkage is equal to

$$\lambda = N \cdot R_m \cdot \varphi \cdot \ell \cdot B = 12 \times 0.05 \times \frac{160}{180}\pi \times 0.07 \times 0.6 = 0.07 \text{ Wb.}$$

Note that one period of the rotating field is equal to $T = 10$ ms, since the rotor rotates at the speed of 6000 rpm. Thus, the interval of a transition is $\Delta T = 5 \times \frac{160}{180} = 4.444$ ms, and the flat interval takes 1.111 ms. Therefore, the peak of induced voltage is equal to

$$V = \frac{2N \cdot R_m \cdot \ell \cdot B_r}{\Delta T} = \frac{2 \times 0.07}{4.444 \times 10^{-3}} = 31.5 \text{ V.}$$

The induced voltage is shown in Fig. 1.11 (b). The fundamental component is equal to

$$V_{[1]} = 31.5 \times \frac{4}{\pi} \int_0^{\frac{4\pi}{9}} \cos(\omega t) d(\omega t) = 31.5 \times \frac{4}{\pi} \cdot \sin \frac{4\pi}{9} = 39.5. \qquad \blacksquare$$

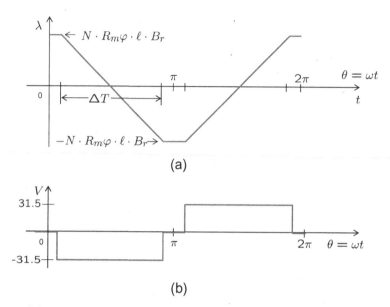

Figure 1.11: (a) Flux linkage as a function of rotor angle and (b) induced voltage (Solution to Exercise 1.5).

1.1.5   Inductance

The flux linkage is developed in a coil by conducting current in the coil, i.e., self-induction. Further, the strength of flux linkage is proportional to the current, as shown in Fig. 1.12 (b). The flux increasing rate is termed as the inductance. Then, the flux linkage is expressed two ways:

$$\lambda = N\Phi = Li. \tag{1.21}$$

Therefore, the inductance of a magnetic circuit is often defined as

$$L = \frac{\lambda}{i} = \frac{N\Phi}{i}. \tag{1.22}$$

That is, the inductance is flux linkage per ampere. However, saturation develops in most of the core materials. If a field exceeds a certain level, the field increasing rate drops as shown in Fig. 1.12 (b). Before the saturation, the $\lambda - i$ relation is linear. But in the saturation region, a local inductance, sometimes called incremental inductance, is far less than the average inductance.

The inductance of a core shown in Fig. 1.12 (a) is equal to

$$L = \frac{NBA_s}{i} = \mu\frac{A_s}{g}N^2, \tag{1.23}$$

where A_s is the core sectional area. From (1.23), it needs to be emphasized that the inductance consists of three elements:

1) material property: μ,
2) geometrical dimensions of the core: A_s and g,
3) number of coil turns: N^2.

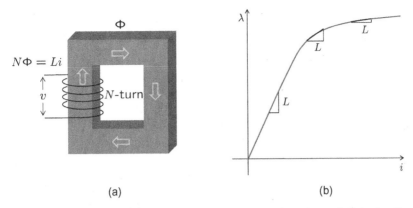

(a)   (b)

Figure 1.12: Flux linkage: (a) an electro-magnetic circuit and (b) the flux-current relation.

Also, note that the inductance of the C-core shown in Fig. 1.7 is given by

$$L = \frac{NBA_s}{i} = \mu_0 \frac{A_s}{g} N^2. \tag{1.24}$$

Exercise 1.6 Derive a formula for the inductance of the toroidal core shown in Fig. 1.6.

Solution.

$$L = \frac{N\Phi}{I} = \frac{\mu D N^2}{2\pi} \ln \frac{R_2}{R_1}. \qquad\blacksquare$$

Exercise 1.7 Consider the construction of an inductor with no air gap using a commercial ferrite core (PC90). The core permeability is $\mu = 2200 \ \mu_0$, and the maximum flux density is $B_s = 0.45$ T at $100°C$. It is desired to have 1 mH inductance when 10 A current flows. Determine the minimum core cross sectional area, A_s and number of coil turns, N when the central core periphery is $\ell = 18$ cm.

Solution. $L = \mu_r \mu_0 \frac{A_s N^2}{\ell} = 0.01535 \times A_s N^2$. On the other hand, note that $L = \frac{N B_s A_s}{I} = 0.045 \times N A_s$. Therefore, $N = 2.93 \approx 3$ turns. Correspondingly, $A_s = 0.0074$ m^2 = 74 cm^2. $\blacksquare$

1.1.6   Analogy of Ohm's Law

Note that NI is called magneto-motor force (MMF). As the name implies, NI is a driving force to generate the magnetic flux. The resistance of the magnetic circuit is called *reluctance*, and is defined

$$\mathcal{R} = \frac{NI}{\Phi}. \tag{1.25}$$

In other words, the magnetic flux, Φ is determined by dividing MMF by its reluctance. That relation is identical to the Ohm's law. Note also that the inductance is expressed using $\mathcal{R}$ as

$$L = \frac{N^2}{\mathcal{R}}. \tag{1.26}$$

The magnetic circuit terms and the Ohm's law analogy are summarized in Tables 1.2 and 1.3, respectively.

Table 1.2: Terms for magnetic circuits.

Name	Symbol (Units)	Equation
Magneto Motive Force	MMF (At)	$F = Ni$
Field Intensity	H (At/m)	$H = Ni/\ell$, ℓ = contour length.
Flux Density	B (T)	$B = \mu H$
Permeability	μ (Vs/Am)	$\mu_0 = 4\pi \times 10^{-7}$: free space
Flux	Φ (Wb)	$\Phi = BA_s$, A_s = cross-sectional area
Flux Linkage	λ (Wbt)	$\lambda = N\Phi$
Inductance	L (H)	$L = \lambda/i = N^2/\mathcal{R}$
Reluctance	$\mathcal{R}$ (At/Wb)	$\mathcal{R} = \ell/\mu A_s$.

Exercise 1.8 Consider a toroidal-core inductor shown in Fig. 1.6. Show that

$$\mathcal{R} = \frac{\pi(R_1 + R_2)}{\mu(R_2 - R_1)d}.$$

Solution. From the Ampere's law, $\oint H \cdot dl = \frac{B}{\mu} \times \pi \cdot (R_1 + R_2) = NI$. Therefore the flux is equal to $\Phi = \mu \frac{NI}{\pi(R_1+R_2)} \times A_s$. ∎

Table 1.3: Ohm's law analogy.

Electrical		Magnetic	
Current	$I = V/R$ (A)	Flux	$\Phi = \text{MMF}/\mathcal{R}$ (Wb)
EMF	V (V)	MMF	MMF(At)
Resistance	R (Ω)	Reluctance	$\mathcal{R}$ (At/Wb)
Conductivity	σ (S/m)	Permeability	μ (Vs/Am)

1.1.7 Transformer

Consider a rectangular core made with a highly permeable material around which two sets of coils are wound tightly, as shown in Fig. 1.13. Thereby, all the flux is assumed to be confined in the core. The numbers of primary and secondary coil turns are N_1 and N_2, and their positions are denoted by (x_k^a, y_k^a) and (x_m^a, y_m^a) in a rectangular plane, S whose periphery is aligned with the centerline of the core. The mean length and the sectional area of core are denoted by ℓ and A_s, respectively. Then the current density, $\mathbf{J}$ on S is described as $N_1 + N_2$ delta functions such that

$$\sum_{k=1}^{N_1} i_1 \delta(x - x_k^a, y - y_k^a) + \sum_{m=1}^{N_2} i_2 \delta(x - x_m^b, y - y_m^b), \tag{1.27}$$

where i_1 and i_2 are the currents of primary and secondary coils.

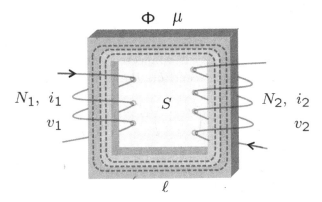

Figure 1.13: Transformer.

Applying the Ampere's law over S, we obtain

$$\oint_{\partial S} \mathbf{B} \cdot d\ell = B\ell = \mu \iint_S \mathbf{J} \cdot d\mathbf{S} = \mu(N_1 i_1 + N_2 i_2) \tag{1.28}$$

The flux is equal to

$$\Phi = B \cdot A_s = \frac{\mu A_s}{\ell}(N_1 i_1 + N_2 i_2) \tag{1.29}$$

Let

$$L_1 \equiv \frac{\mu A_s}{\ell} N_1^2,$$

$$L_2 \equiv \frac{\mu A_s}{\ell} N_2^2,$$

$$M \equiv \frac{\mu A_s}{\ell} N_1 N_2.$$

Then the flux linkages of the primary and secondary coils are

$$\lambda_1 \equiv N_1 \Phi = \frac{\mu A_s}{\ell} N_1(N_1 i_1 + N_2 i_2) = L_1 i_1 + M i_2, \tag{1.30}$$

$$\lambda_2 \equiv N_2 \Phi = \frac{\mu A_s}{\ell} N_2(N_1 i_1 + N_2 i_2) = M i_1 + L_2 i_2. \tag{1.31}$$

The terminal voltages of coils are

$$v_1 = \frac{d\lambda_1}{dt} = L_1 \frac{di_1}{dt} + M \frac{di_2}{dt}, \tag{1.32}$$

$$v_2 = \frac{d\lambda_2}{dt} = M \frac{di_1}{dt} + L_2 \frac{di_2}{dt}. \tag{1.33}$$

Since $M = \frac{N_2}{N_1} L_1 = \frac{N_1}{N_2} L_2$, the voltage ratio turns out to be

$$\frac{v_2}{v_1} = \frac{N_2}{N_1}. \tag{1.34}$$

From the perspective of power balance, it follows that $v_1 i_1 = v_2 i_2$. Therefore, the current ratio is the inverse of the voltage ratio:

$$\frac{i_2}{i_1} = \frac{N_1}{N_2}. \tag{1.35}$$

Suppose that a load resistance, R_2 is connected across the secondary terminal. Then, $i_2 = v_2 / R_2$. The impedance seen from the primary side is equal to

$$R_2' = \frac{v_1}{i_1} = \frac{N_1}{N_2} v_2 \left(\frac{N_2}{N_1} i_2 \right)^{-1} = \left(\frac{N_1}{N_2} \right)^2 R_2. \tag{1.36}$$

The square of a turn ratio must be multiplied if a load impedance is viewed through a transformer.

T-Equivalent Circuit

Let κ be an arbitrary number and modify (1.30) and (1.31) such that

$$\lambda_1 = L_1 i_1 + \kappa M \frac{i_2}{\kappa} = L_1 i_1 + L_m i_2'$$

$$\lambda_2' = \kappa \lambda_2 = \kappa M i_1 + \kappa^2 L_2 \frac{i_2}{\kappa} = L_m i_1 + L_2' i_2',$$

where $L_m \equiv \kappa M$, $i_2' = \frac{i_2}{\kappa}$, $\lambda_2' \equiv \kappa \lambda_2$, and $L_2' \equiv \kappa^2 L_2$. Then it follows that

$$\lambda_1 = (L_1 - L_m) i_1 + L_m (i_1 + i_2') = L_{1\sigma} i_1 + L_m i_m \tag{1.37}$$

$$\lambda_2' = L_m (i_1 + i_2') + (L_2' - L_m) i_2' = L_m i_m + L_{2\sigma}' i_2', \tag{1.38}$$

where $i_m = i_s + i_2'$, $L_{1\sigma} = L_1 - L_m$, and $L_{2\sigma}' = L_2' - L_m$. Based on (1.37) and (1.38), we have a $T-$equivalent circuit as shown in Fig. 1.14. Note that L_m is the magnetizing inductance, and i_m is the magnetizing current. When $\kappa = \frac{N_1}{N_2}$, the secondary part variables are transferred to the primary side along with the leakage inductance. Note that the dotted box in the right part represents an ideal transformer. By the ideal transformer, we mean a converter that changes simply voltage and current according to the turn ratio. There is neither power loss nor creation. As $\mu \rightarrow \infty$, it follows that $L_m \rightarrow \infty$ and $L_{1\sigma}, L_{2\sigma} \rightarrow 0$. Then only the dotted box remains in Fig. 1.14.

Practical Transformers

In practice, the permeability of the core is not sufficiently high and the coil is not wrapped tightly around the core. A part of flux remains in the free space and does not link with the other coil. In Fig. 1.15 the leakage fluxes around the primary and

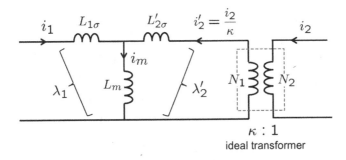

Figure 1.14: T-equivalent circuit of a transformer.

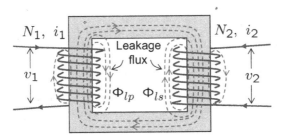

Figure 1.15: Practical transformer carrying leakage flux.

secondary windings are denoted by Φ_{lp} and Φ_{ls}, respectively. Circuit model of a physical transformer is shown in Fig. 1.16. It contains series resistors, r_1 and r_2 which represent the coil losses, and a parallel resistor, r_c that represents the core loss. The coupling coefficient is defined to be

$$k \equiv \frac{L_m}{\sqrt{L_1 L_2}} = \frac{L_m}{\sqrt{(L_m + L_{l1})(L_m + L_{l2}\kappa^2)}}. \tag{1.39}$$

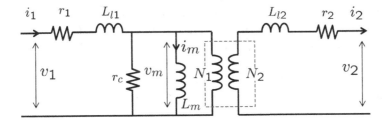

Figure 1.16: Circuit model of a physical transformer.

Figs. 1.17 (a) and (b) show phasor diagrams of a transformer under light and rated load conditions, respectively. In the diagram, the core loss is not considered. The magnetizing current is almost the same in both cases, since it mainly depends on the terminal voltage. Hence, the proportion of reactive component is high in the light load case. As a result, the power factor improves as the load current increases.

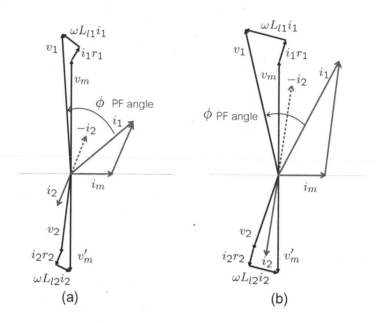

Figure 1.17: Phasor diagrams for transformer: (a) light load condition and (b) rated load condition.

1.1.8 Three Phase System

A polyphase system is advantageous in delivering power and particularly useful for transmitting power to electric motors. Therefore, most power grids consist of the three phase power system. A three phase system is called *balanced* when the phase magnitudes are the same, but the phase angles differ by 120° with one another. That is, $|v_a| = |v_b| = |v_c|$ and $\angle v_a - \angle v_b = \angle v_b - \angle v_c = \angle v_c - \angle v_a = 120°$. The phase voltages are written as $v_a = V\cos(\omega t - \phi)$, $v_b = V\cos(\omega t - \frac{2\pi}{3} - \phi)$, and $v_c = V\cos(\omega t - \frac{4\pi}{3} - \phi)$.

Fig. 1.18 shows the equivalence in a balanced three phase system. Fig. 1.18 (a) shows three independent single phase systems connected to an equal load, R. In Fig. 1.18 (b), three return paths are merged into one. However, it does not alter any current in Fig. 1.18 (a). An important observation is that it is possible to eliminate the common line if its current is equal to zero. It is feasible when the voltage sources are balanced and the loads are equal. In such a case,

$$i_a + i_b + i_c = \frac{V}{R}\left(\cos(\omega t - \phi) + \cos(\omega t - \frac{2\pi}{3} - \phi) + \cos(\omega t - \frac{4\pi}{3} - \phi)\right) = 0.$$

The common point is called the *neutral point*, and no connection is made between the two neutral points as shown in Fig. 1.18 (c). Yet, the entire power delivered to load is not reduced, i.e., $3 \times \frac{V^2}{R}$. Thereby, 50% of wire can be saved with the three phase system. Considering the fact that the transmission line length is several hundred kilometers, the copper saving is enormous.

Fig. 1.19 shows the equivalence between the ideal core and a practical three phase core. As the current sum is zero, the flux sum of a balanced three phase system is

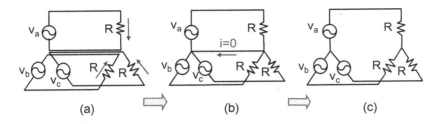

Figure 1.18: Three phase transformer: (a) three single phase systems, (b) with neutral line connected, and (c) three phase source and load.

also equal to zero, i.e., $\Phi_a(t) + \Phi_b(t) + \Phi_c(t) = 0$ for all t. Correspondingly, it is not necessary to give a common leg to the three phase core. In addition, the core permeance is so high that it is not necessary to make a core as a symmetrical object as shown in Fig. 1.18 (b). Instead, the three phase core is usually made by stacking flat steel sheets as shown in Fig. 1.18 (c).

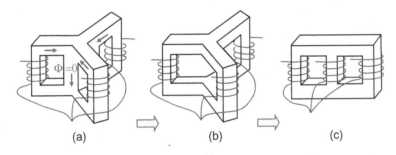

Figure 1.19: Three phase transformer core: (a) three phase core with a common leg, (b) three phase core without a common leg, and (c) practical three phase transformer core.

1.2 Basics of DC Machines

An electric motor is a machine that converts electrical oscillation into the mechanical oscillation. The armature winding is wound on the rotor. Although a DC current is supplied to the machine, an alternating current is developed in the armature coil through a mechanical commutation made by brush and commutator. Fig. 1.20 shows a photo of brush and commutator. DC motors are popularly used since the torque or speed controller (chopper) is simple, and its cost is lower than the inverter cost. They are still widely used in numerous areas such as in traction systems, mill drives, robots, printers, wipers, power windows, etc. However, DC motors are inferior to AC motors in power density, efficiency, and reliability.

DC motors have two major coil sets: field winding and armature winding. The DC field is generated by either field winding or permanent magnets (PMs). Fig. 1.21 shows a schematic diagram of a DC motor with PMs. The basic principle of a DC motor operation is illustrated in Fig. 1.22. Fig. 1.22 (a) shows a moment of torque

Figure 1.20: Brush and commutator of a DC machine.

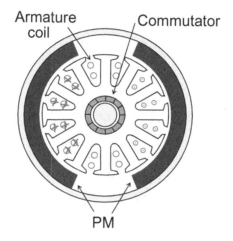

Figure 1.21: Cross section of a typical PM DC motor.

production, where the armature coil lies in the middle of the field magnet. Applying the Flemming left hand rule, we can see that an upward force is generated in the left conductor. Similarly, downward force takes place in the right conductor. As a result, torque is developed on the rotor assembly clockwise. At this moment, the induced voltage in the coil is also the largest as the rate of change of the flux linkage is the highest. Note however that the magnitude of flux linkage is equal to zero in the horizontal position. Fig. 1.22 (b) shows a disconnected state in which the armature winding is separated from the voltage source. Correspondingly, no force is generated. However, another set of armature coil is engaged to produce torque during this period. In Fig. 1.22 (c), the armature coil is re-engaged to the circuit, generating torque in the same direction. This state is the same as that in Fig. 1.22 (a) though the coil positions are switched. Note however that the current direction is reversed when it is seen from the coil.

Since most brushes are made of carbon, they wear out continuously like a pencil lead. Thus, DC motors require regular maintenance. Further, the carbon and the mechanical contact cause the voltage drop, causing an additional loss. As the motor power increases, the commutator dimension also needs to be increased to provide a larger current and a higher voltage isolation between neighboring commutator

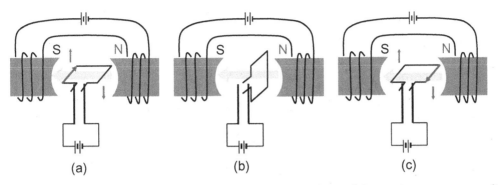

Figure 1.22: DC motor commutation and current flow: (a) maximum torque, (b) disengaged, and (c) maximum torque.

segments. Then it increases the commutator surface scrubbing speed. It limits the maximum speed and the power rating of DC motor.

1.2.1 DC Machine Dynamics

In the DC machine, two electromagnetic phenomena take place concurrently:

EMF generation: When the armature coil rotates in a magnetic field, the flux linkage changes, i.e., the amount of flux passing through the coil loops varies. Then the EMF is developed in the coil according to Faraday's law, and is proportional to the speed. It is called *back EMF* and described as $e_b = K_b \omega_r$, where K_b is the back EMF constant and ω_r is the rotor angular speed. The back EMF constant depends on the field current, but has nothing to do with the armature current.

Torque generation: When a current-carrying conductor is placed in a magnetic field, Lorentz force is generated on the conductor. The electromagnetic torque is expressed as $T_e = K_t i_a$, where K_t is the torque constant and i_a is armature current.

An equivalent circuit of a separately wound DC motor is shown in Fig. 1.23. Applying the Kirchhoff's voltage law to the equivalent circuit, we obtain

$$v_a = r_a i_a + L_a \frac{di_a}{dt} + e_b, \tag{1.40}$$

$$e_b = K_b \omega_r, \tag{1.41}$$

$$T_e = K_t i_a \tag{1.42}$$

where v_a, i_a, r_a, and L_a are the armature voltage, current, resistance, and inductance, respectively. Fig. 1.23 also shows a field winding circuit, and its flux is denoted by ψ. The back EMF constant, K_b, and torque constant, K_t depend on ψ. Below a base speed, ψ is usually controlled to be constant. But it is weakened in the high speed region.

Note that the electrical power over the EMF is equal to $e_b i_a$, whereas the mechanical power is $T_e \omega_r$. Neglecting power loss due to the armature coil resistance, the electrical power and mechanical power should be the same, $T_e\omega_r = e_b i_a$. Therefore, it follows that $K_t i_a \cdot \omega_r = K_b \omega_r \cdot i_a$. As a result, $K_t = K_b$.

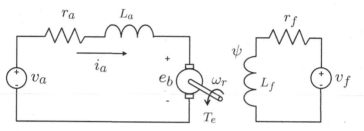

Figure 1.23: Equivalent circuit for a DC motor.

The motor can act as a generator and vice versa. There are two voltage sources, v_a and e_b that compete each other over r_a and L_a. In the motoring action, $v_a > e_b$ and current ($i_a > 0$) is supplied to e_b from the external source, v_a. Then the motor receives power, $e_b i_a$ and it is converted into mechanical power. In contrast, an external force rotates the machine shaft in the generation mode. Suppose that a back EMF is larger than the voltage source, i.e., $v_a < e_b$. Then, the current flows out from the machine to the source ($i_a < 0$). At this time, a load torque is developed by the motor, consuming the mechanical power.

Exercise 1.9   Calculate K_t and K_b for a DC motor with the parameters listed below:

Table 1.4: Example DC motor parameters.

Power (rated)	3.73kW
Voltage (rated)	240V
Armature current (rated)	16A
Rotor speed (rated)	1220 rpm (127.8 rad/s.)
Torque (rated)	28.8 Nm
Resistance, r_a	0.6Ω

Solution. The back EMF is equal to $e_b = v_a - r_a i_a = 240 - 16 \times 0.6 = 230.4$ V. Hence, $K_b = e_b/\omega_r = 230.4/127.8 = 1.8$Vs/rad. Since $K_t = T_e/i_a = 28.8/16 = 1.8$ Nm/A, one can check $K_t = K_b$. ∎

Exercise 1.10   Consider a DC motor with armature voltage 125V and armature resistance $r_a = 0.4Ω$. It is running at 1800 rpm under no load condition.

    a) Calculate the back EMF constant K_b.
    b) When the rated armature current is 30A, calculate the rated torque.
    c) Calculate the rated speed.

Solution. a) With no load condition, $i_a = 0$. Thus,

$$K_b = \frac{125 \text{ V}}{1800 \text{ rpm}} \times \frac{60 \text{ rpm}}{2 \pi \text{rad/s.}} = 0.663 \text{ Vs/rad.}$$

b) Since $K_t = K_b$, $T_e = K_t i_a = 0.663 \times 30 = 19.89$ Nm.

c) In the steady state, $\frac{di}{dt} = 0$. Therefore, $T_e \cdot \omega_r = (v_a - r_a i_a) i_a$. Hence,

$$\omega_r = \frac{(125 - 0.4 \times 30) \times 30}{19.89} = 170.44 \text{ rad/s} = 1628 \text{ rpm.} \qquad \blacksquare$$

1.2.2 Field Weakening Control

The back EMF increases as the motor speed increases. In general, motors are designed such that the back EMF reaches the maximum armature voltage v_a^{max} at a rated speed ω_r^{rated}. That is, the back EMF reaches the maximum allowable source voltage at the rated speed, i.e., $v_a^{max} \approx K_b \omega_r^{rated}$. Then the question is "How can we increase the speed above the rated speed?" The answer is to use *the field weakening control* which is illustrated in the following:

Note that torque constant K_t is proportional to the flux ψ. Therefore, we can let $K_t = k\psi_0$ for some $k > 0$, where ψ_0 is a nominal (rated) flux. In the steady state, the armature current is constant. Thus, $L_a \frac{di_a}{dt} \approx 0$. Thereby,

$$T_e = K_t i_a = K_t \frac{v_a^{max} - K_b \omega_r}{r_a} = \frac{k\psi_0 v_a^{max}}{r_a} - \frac{k^2 \psi_0^2}{r_a}\omega_r \qquad (1.43)$$

Fig. 1.24 shows three torque curves for different k's. Note that as k decreases, the

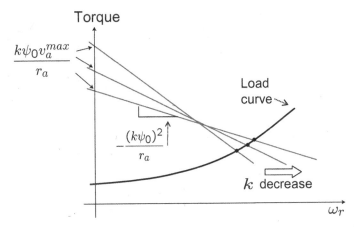

Figure 1.24: Torque curves for three different $k\psi_0$'s: Speed increase by field weakening.

y-intercept, $k\psi_0 v_a^{max}/r_a$ decreases while the slope, $-(k\psi_0)^2/r_a$ approaches zero. It means that the torque function is flattened as the field is weakened. Since the speed is determined at the intersection with a load curve, we can see that the speed increase as $k\psi_0$ reduces. It states that a higher speed can be secured by decreasing the field.

The necessity for field weakening is seen clearly from the power relation. Power, $P_e = T_e \omega_r = k\psi_0 i_a^{rated}\omega_r$ is kept constant above the rated speed. The current is assumed to be constant in all speed range. To keep the power constant, it is necessary to decrease $k\psi_0$ as ω_r increases. Specifically, it is necessary to let $k \propto \frac{1}{\omega_r}$.

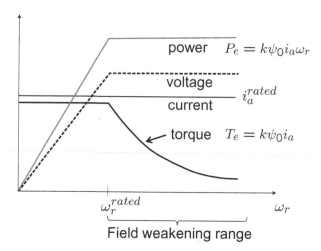

Figure 1.25: Power and torque versus speed in the field weakening region.

Correspondingly, $T_e = k\psi_0 i_a^{rated}$ also decreases as shown in Fig. 1.25. Note also that both mechanical power, $T_e\omega_r = k\psi_0 i_a\omega_r$ and electrical power, $v_a i_a$ remain constant in the field weakening region.

Exercise 1.11   Consider a DC motor with $r_a = 0.5\ \Omega$, $K_t = 0.8$ Nm/A, and $K_b = 0.8$Vs/rad.

    a) Assume that the motor terminal voltage reaches $v_a^{max} = 120\ V$ when the motor runs at a rated speed, $\omega_r = 140$ rad/s. Calculate the rated current and rated torque.

    b) Assume that a load torque, $T_L = 6$ Nm, is applied and that the field is weakened to a half, i.e., $K_t = 0.4$ Vs/rad. Determine the speed.

Solution. a) $i_a^{rated} = (120 - 0.8 \times 140)/0.5 = 16$ A. $T_e = 0.8 \times 16 = 12$ Nm.
b) Using (1.43), $T_e = 6 = \frac{0.4 \times 120}{0.5} - \frac{0.4^2}{0.5}\omega_r$. Thus, $\omega_r = 281$ rad/s.   ∎

Exercise 1.12   Consider a DC motor with $r_a = 0.2\ \Omega$ and $L_a = 500\ \mu$H. Assume that the torque constant is proportional to the field current, i.e, $K_t = K_f i_f$, where $K_f = 0.02$ Nm/A^2. The inertia and damping coefficient are $J = 10^{-3}$ kg · m^2 and $B = 10^{-3}$ Nm · s. The system is under a constant load, $T_L = 10$ Nm. When the armature voltage is fixed at $V_{max} = 100\ V$, compute the following:

    a) Draw torque curves for $i_f = 4,\ 5$ A. Find the intersection with the load line and determine the operation speed in each case.

    b) Using MATLAB Simulink, construct a DC motor model. Apply step voltage $v_a = 100u_s(t)$ when $i_f = 5$ A. Then change the field current to 4A after 1 sec, i.e., set $i_f(t) = 5[u_s(t) - u_s(t-1)]$. Plot the shaft speed, ω_r and compare the speeds with the previous analytic results.

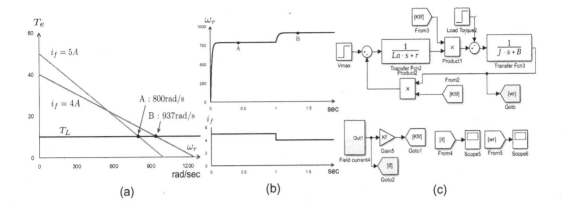

Figure 1.26: (a) Analytic method to find operation speeds, (b) MATLAB simulation result, and (c) Simulink model for a DC motor (Solution to Exercise 1.12).

1.2.3 Four Quadrant Operation

Depending on the polarities of torque and speed, there are four operation modes:

Motoring:
Supplying positive current into the motor terminal, positive torque is developed, which results in a forward movement.

Regeneration:
The motor shaft is rotated in the reverse direction by an external force, while the armature current generates a positive torque. The motor generates electric power with a braking torque. In electric vehicle, this mode is referred to as regenerative braking.

Motoring in the reverse direction:
If the armature current is negative, the motor will turn in the reverse direction.

Regeneration in the forward direction:
External torque is positive and greater than the negative torque generated by negative armature current. The motor generates electrical power as the motor rotates in the forward direction.

Quadrants 1 and 2 correspond to the crane lifting and lowering of a load. Quadrants 3 and 4 are the symmetries of Quadrants 1 and 2 in the reverse direction.

1.2.4 DC Motor Dynamics and Control

The dynamics of the mechanical part is described as

$$J\frac{d\omega_r}{dt} + B\omega_r + T_L = T_e, \tag{1.44}$$

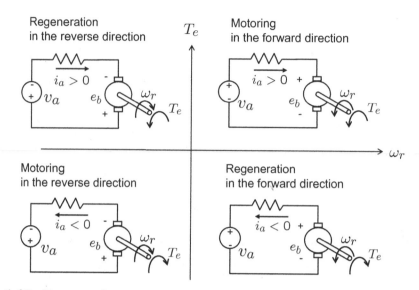

Figure 1.27: Four quadrant operation characteristics.

where J is the inertia of the rotating body, B is the damping coefficient, and T_L is a load torque. Combined with the electrical dynamics (1.40)–(1.42), the DC motor block diagram is shown in Fig. 1.28. Note that a load torque T_L acts as a disturbance to the DC motor system and that the back EMF $K_b\omega_r$ forms a negative feedback loop.

A DC motor controller normally consists of two loops: current control loop and speed control loop. Generally both controllers utilize proportional integral (PI) controllers. Since the current loop lies inside the speed loop, it is called *the cascaded control structure*.

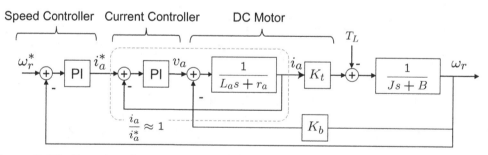

Figure 1.28: Speed and current control block diagram for DC motor.

Current Control Loop

Let the current proportional and integral gains be denoted by K_{pc} and K_{ic}, respectively. With the PI controller $K_{pc} + K_{ic}/s$, the closed loop transfer function of the current loop is given by

$$\frac{i_a(s)}{i_a^*(s)} = \frac{K_{pc}s + K_{ic}}{L_a s^2 + (r_a + K_{pc})s + K_{ic}} = \frac{(K_{pc}/L_a)s + K_{ic}/L_a}{s^2 + 2\zeta\omega_n s + \omega_n^2}, \tag{1.45}$$

where i_a^* is a current command, $\omega_n = \sqrt{K_{ic}/L_a}$ and $\zeta = (r_a + K_{pc})/(2\omega_n L_a)$. The control bandwidth is $[0, \omega_n]$ when $\zeta = 1$. A bandwidth requirement determines the I-gain. Similarly, the P-gain is selected to meet a system specification for the overshoot. In the case of the current control, a small overshoot is allowed normally to shorten the rise time. The current control bandwidth is usually $5 \sim 10$ times wider than the speed bandwidth.

Speed Control Loop

The current controller should react much faster than a mechanical system. Accordingly, the current control bandwidth is greater than the speed control bandwidth. For example in the normal drive system below 10 kW, the current control bandwidth is in the range of $80\sim150$ Hz, whereas the speed control bandwidth is less than 30 Hz. Since there is a big difference, the whole current block can be treated as unity in determining the speed PI gains, (K_{ps}, K_{is}). In particular, we let $i_a(s)/i_a^*(s) = 1$ in the speed loop. Then the whole speed loop is reduced to a second-order system:

$$\frac{\omega_r(s)}{\omega_r^*(s)} = \frac{K_t(K_{ps}s + K_{is})}{Js^2 + (B + K_t K_{ps})s + K_t K_{is}} = \frac{(K_t K_{ps}/J)s + \omega_{sn}^2}{s^2 + 2\zeta_s \omega_{sn} s + \omega_{sn}^2}, \tag{1.46}$$

where $\omega_{sn} = \sqrt{K_t K_{is}/J}$ is a corner frequency and $\zeta_s = (B + K_t K_{ps})/(2\sqrt{JK_t K_{is}})$ is a damping coefficient. The corner frequency (or the natural frequency) ω_{sn} is determined by the I-gain, K_{is}, whereas the damping coefficient, ζ_s is a function of P-gain, K_{ps}.

Consider the motor parameter in Table 1.4. Suppose the rotor inertia is $J = 0.01$ kg $\cdot$ m^2 and the friction coefficient is $B = 0.0002$ Nm $\cdot$ s. The torque constant is $K_t = 28.8/16 = 1.8$ (Nm/A). Then, the transfer function is

$$\frac{\omega_r(s)}{\omega_r^*(s)} = \frac{180(K_{ps}s + K_{is})}{s^2 + (0.02 + 180K_{ps})s + 180K_{is}}. \tag{1.47}$$

Note that the damping coefficient is $\zeta_s = (0.02 + 180K_{ps})/(2\sqrt{180K_{is}})$. The step responses of (1.47) for different P-gains are shown in Fig. 1.29. Note that overshoot increases as the damping coefficient reduces. Due to the presence of a zero, the response has an overshoot even when $\zeta_s = 1$.

Exercise 1.13   Consider a DC motor with $r_a = 0.2$ Ω, $L_a = 500$ μH, $K_t = 0.03$ Nm/A, $J = 10^{-4}$ kg $\cdot$ m^2, and $B = 2 \times 10^{-5}$ Nm $\cdot$ s.

a) Determine current PI gains, K_{pc} and K_{ic} so that the current control bandwidth is 150 Hz and the damping coefficient is 0.7.

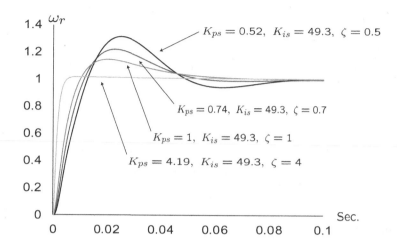

Figure 1.29: Speed step responses of a DC motor for different damping coefficients.

b) Treating the current block as unity, determine speed loop PI gains, K_{ps} and K_{is} so that the speed loop bandwidth is 15 Hz and the step response does not have overshoot. Use MATLAB to choose a proper P-gain, K_{ps}.

Solution.
a) For simplicity, let $[0, \, w_n]$ be the control bandwidth. With the given parameters, $w_n = 2\pi \times 150 = 942$. Therefore, $K_{ic} = L_a w_n^2 = 444$ and $\zeta = 0.7 = (0.2 + K_{pc})/(2 \times 5 \times 10^{-4} \times 942)$. Thus, $K_{pc} = 0.46$.
b) With the given parameters, $w_{sn} = 2\pi \times 15 = 94$. Therefore, $K_{is} = \frac{10^{-4} \times 94^2}{0.03} = 29.4$. Fig. 1.30 (a) shows the speed step responses for different K_{ps}'s. Note that the overshoot disappears with $K_{ps} = 6$. ∎

1.3   Dynamical System Control

Consider a plant, $G(s)$ with a controller, $C(s)$ shown in Fig. 1.31. One may wish a perfect tracking control via an open loop. The control objective may be achieved via an open loop by letting $C(s) = G(s)^{-1}$. In theory, the input to output transfer function will be unity for all frequencies. But, it cannot be realized practically for the following reasons:

i)   The plant may be in nonminimum phase, i.e., the plant has zeros in the right half plane. Then, its inverse will have poles in the right half plane.

ii) Normally, the plant is strictly proper, thereby its inverse contains a differentiator. Then, the controller may require an unrealistically large input under a step command.

iii) When there is a disturbance, d or parameter uncertainty in the plant, an output error cannot be repelled or attenuated.

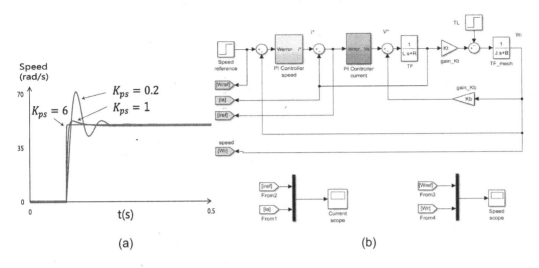

Figure 1.30: (a) Speed step responses for different proportional gains while $K_{is} =$ 29.4 and (b) Simulink model for a DC motor with current and speed controllers (Solution to Exercise 1.29).

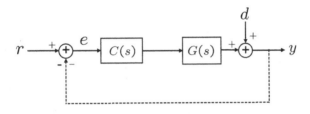

Figure 1.31: Plant with unity feedback controller.

In most control systems, the design objectives are classified as *set point tracking* and *disturbance rejection*. These goals can be achieved via a proper feedback control though the solution is bandlimited. For the closed loop system, the sensitivity function is defined as

$$S \equiv \frac{y(s)}{d(s)} = \frac{1}{1 + CG},$$

representing the effect of disturbance, d on output, y. Therefore, a smaller S implies a better disturbance rejection. On the other hand, the complementary sensitivity function is defined as

$$T(s) = \frac{CG}{1 + CG} = \frac{y(s)}{r(s)}.$$

Note that $T \equiv 1 - S$. in this example. The complementary sensitivity function reflects the tracking performance. Ideally, it is desired to be unity in the whole frequency range. But, it cannot be realized practically. The goals can be satisfied partially in a low frequency region. Bode plots of the typical sensitivity and complementary sensitivity functions are shown in Fig. 1.32. Note that $T(j\omega) \approx 0$dB and $S(j\omega)$ is far lower than 0 dB in a low frequency region.

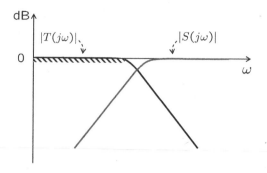

Figure 1.32: Typical sensitivity and complementary sensitivity functions.

1.3.1 Gain and Phase Margins

The phase delay is an intrinsic nature of a dynamic system, and the delay sometimes causes instability in a closed loop system. Instability occurs in the feedback loop when two events take place at the same time: unity loop gain and 180° phase delay. Consider a pathological example:

$$|C(j\omega)G(j\omega)| = 1 \quad \text{and} \quad \angle C(j\omega)G(j\omega) = -180°.$$

Assume that the reference command is sinusoidal as shown in Fig. 1.33. Then, $-y(t)$ is in phase with $r(t)$, i.e., the error will be $e(t) = r(t) - y(t) = 2r(t)$ after proceeding one loop. Repeating the same, $e(t)$ grows indefinitely.

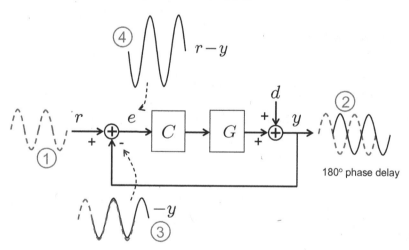

Figure 1.33: Instability mechanism when loop gain is unity and phase delay is 180°.

Phase margin and gain margin refer to the magnitudes of buffers from the unstable points. The gain margin is defined by

$$
\begin{aligned}
A_{mg} &= 20\log_{10} 1 - 20\log_{10} |C(j\omega_p)G(j\omega_p)| \\
&= 20\log_{10} \frac{1}{|C(j\omega_p)G(j\omega_p)|},
\end{aligned}
\tag{1.48}
$$

where ω_p is the frequency at which $\arg[C(j\omega_p)G(j\omega_p)] = -180°$. That is, the gain margin is a difference from 0 dB at the frequency ω_p, where the phase delay is 180°. On the other hand, the phase margin is defined by

$$\phi_{mg} = \arg[C(j\omega_c)G(j\omega_c)] - (-180) \tag{1.49}$$

at ω_c where $|C(j\omega_c)G(j\omega_c)| = 1$. In other words, the phase margin is an angle difference from $-180°$ at the frequency when the loop gain has unity. Gain and phase margins are marked by arrows in the Bode plot shown in Fig. 1.34.

If a system is set to have a high gain margin, it can be stable even under a large disturbance. The response, however, is slow and therefore it cannot meet certain control objectives. In short, the gain margin is usually traded with the control bandwidth. But in the other extreme, the system will be unstable. Therefore, the controller must be well-tuned in the sense that agility is balanced with stability, i.e., the system has to be stable, but it should not sacrifice the performance. Proper selection ranges are: Gain margin $= 12\sim20$ dB and Phase margin $= 40° \sim 60°$.

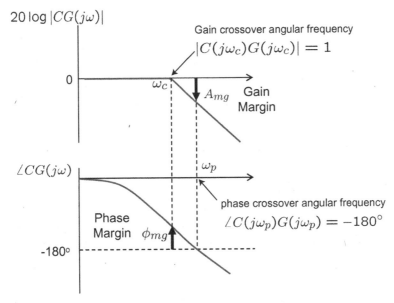

Figure 1.34: Gain and phase margins in a Bode plot.

Exercise 1.14 Suppose that $C(s) = K_p + \frac{1}{s}$ and $G(s) = \frac{1}{s}$ in Fig. 1.31. Determine K_p such that the phase margin is equal to 30°.

1.3.2 PD Controller

We consider a closed loop system by unity feedback shown in Fig. 1.31. Let ω_c be a crossover frequency of the system, $G(s)$, i.e., $|G(j\omega_c)| = 1$ and assume that $C(s)$ is a PD controller. The proportional-derivative (PD) controller adds a zero to the loop, thereby it improves the system performance by reducing the phase delay.

That is, it acts like a leading compensator. But it requires the use of a low pass filter (lagging compensator) to prevent the unnecessary gain increase in the high frequency region. Thus, it turns out to be a *lead-lag* compensator.

Denote by ω_z and ω_p zero and pole of a lead-lag compensator, and let $\omega_z \ll \omega_p$. Then it has the form

$$C_{pd}(s) = K_p \frac{1 + s/\omega_z}{1 + s/\omega_p}, \tag{1.50}$$

and its typical Bode plot is shown in Fig. 1.35.

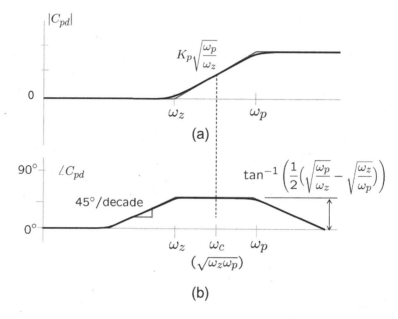

Figure 1.35: Bode plot of a typical PD (lead-lag) controller.

The maximum phase lead is achieved at the center frequency defined by $\log \omega_m = \frac{1}{2}(\log \omega_z + \log \omega_p)$, or $\omega_m = \sqrt{\omega_z \omega_p}$. Substituting $s = j\omega_m$ into (1.50) we have an angle advance such that

$$\phi_{pd} = \max(\angle C_{pd}) = \tan^{-1}\left(\frac{1}{2}\left(\sqrt{\frac{\omega_p}{\omega_z}} - \sqrt{\frac{\omega_z}{\omega_p}}\right)\right). \tag{1.51}$$

It is equivalently expressed as

$$\frac{\omega_z}{\omega_p} = \frac{1 - \sin \phi_{pd}}{1 + \sin \phi_{pd}}. \tag{1.52}$$

It is often required to make the controller gain unity $|C_{pd}(\omega_c)| = 1$ at the crossover frequency, ω_c of a target plant:

$$\omega_c = \sqrt{\omega_z \omega_p}. \tag{1.53}$$

Then the frequency of maximum phase lead is obtained at ω_c. It follows from (1.52) and (1.53) that

$$\omega_z = \omega_c \sqrt{\frac{1 - \sin\phi_{pd}}{1 + \sin\phi_{pd}}}, \tag{1.54}$$

$$\omega_p = \omega_c \sqrt{\frac{1 + \sin\phi_{pd}}{1 - \sin\phi_{pd}}}. \tag{1.55}$$

It is a rule of setting zero and pole of a lead-lag compensator. The maximum phase lead is plotted against $\frac{\omega_p}{\omega_z}$ in Fig. 1.36.

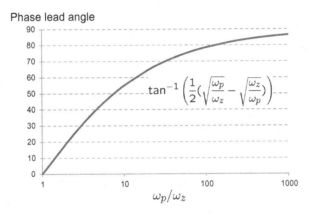

Figure 1.36: Phase lead versus ω_p/ω_z of a PD controller.

Exercise 1.15 Derive (1.51) and (1.52).

Exercise 1.16 Show that the gain satisfying $|C_{pd}(j\omega_c)| = 1$ is

$$K_p = \sqrt{\frac{1 - \sin\phi_d}{1 + \sin\phi_d}}. \tag{1.56}$$

Exercise 1.17 Consider a system

$$G(s) = \frac{1}{1 + s/30 + s^2/900}.$$

a) Determine ω_c of $G(s)$.

b) It is required to have a phase advance, $\phi_d = 30°$.
   Select zero and pole of a PD (lead-lag) controller.
   Determine K_p so that the $|C_{pd}(j\omega_c)G(j\omega_c)| = 1$.

c) Draw Bode plots of $G(s)$ and $C_{pd}(s)G(s)$.

d) Find the gain margin improvement in degree resulted from the PD controller.

Solution. a) The denominator of current transfer function is

$$\left|1 + j\frac{\omega}{30} - \left(\frac{\omega}{30}\right)^2\right| = \sqrt{\left(1 - \left(\frac{\omega}{30}\right)^2\right)^2 + \left(\frac{\omega}{30}\right)^2} = 1. \tag{1.57}$$

Thus, $\omega_c = 30$.

b) Since $\sin\phi_d = 0.5$, it follows from (1.54) and (1.55) that $\omega_z = 30/\sqrt{3}$ and $\omega_p = 30\sqrt{3}$. From (1.56), we have $K_p = 1/\sqrt{3}$. ∎

1.3.3 PI Controller

We consider a case in which a PI controller is installed in the unity feedback system as shown in Fig. 1.31. The PI controller adds a pole, ω_i to the loop:

$$C_{pi}(s) = K_p\left(1 + \frac{\omega_i}{s}\right). \tag{1.58}$$

It consists of two subcontrollers: The proportional controller applies the same gain, K_p to all frequency components. On the other hand, the integral controller differentiates its gain depending on the frequency: Infinite gain is applied to the DC component. However, a lower value is multiplied to a higher frequency component. Thus, the integral controller is effective rejecting a DC offset error, or getting a good tracking performance in a low frequency band. Due to the ability of rejecting DC disturbance, the PI controller is an indispensable component to most control systems. However, it causes a phase lag, reducing the gain margin. Thus, a downside is that it increases the possibility of making the system unstable. Therefore, there should be a compromise between a good tracking performance and a wide control bandwidth. The Bode plot of a PI controller (1.58) is shown in Fig. 1.37.

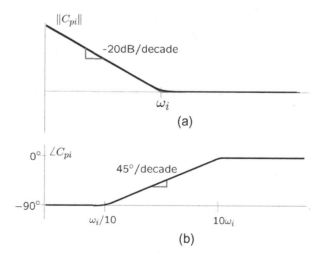

Figure 1.37: Bode plot for a PI controller.

Exercise 1.18 Consider a first order system, $G(s) = G_0/(1+s/\omega_0)$ with the PI controller (1.58). Determine K_p so that the open loop system $T_{cg}(s) = C_{pi}(s)G(s)$ has the unity gain at ω_c. Plot $|T_{cg}(j\omega)|$ and $1/|1+T_{cg}(j\omega)|$.

Solution. For a sufficiently large ω, the open loop transfer function is approximated as $T_{cg}(j\omega) \approx K_p G_0/(\omega/\omega_0)$. Therefore, the desired proportional gain is $K_p = \omega_c/(G_0\omega_0)$. ∎

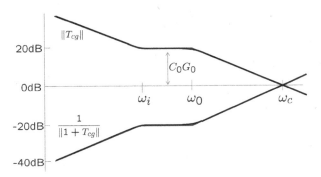

Figure 1.38: Bode plots for open and closed loop transfer functions with a PI controller, where $T_{cg}(s) = C_{pi}(s)G(s)$.

Note from Fig. 1.31 that the influence of a disturbance, d to the system output is described by $1/(1+T_{cg})$. Fig. 1.38 shows that $20\log_{10}(1/|1+T_{cg}|) = -40 \ dB$ at a low frequency. Theoretically, $1/|1+T_{cg}|$ is zero for a DC disturbance. As a result, a feasible DC offset error due to the DC disturbance is rejected from the output.

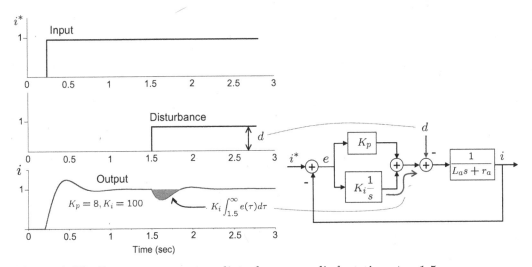

Figure 1.39: Response to a step disturbance applied at time $t = 1.5$ s.

Fig. 1.39 shows an integral action in a time-domain simulation. A step disturbance $d(t) = d \cdot u_s(t - t_1)$ is applied at $t_1 = 1.5$ s. Applying the final value theorem [2], it follows that

$$\lim_{t\to\infty} e(t) = \lim_{s\to 0} sE(s)\frac{de^{-st_1}}{s} = \lim_{s\to 0} \frac{de^{-st_1}}{1 + \left(\frac{K_p s + K_i}{s}\right)\left(\frac{1}{s+\alpha}\right)} = 0.$$

It is interpreted that the steady state error caused by a DC disturbance can be eliminated completely. Note further that the amount of cumulated error (error integral) is equal to the level of the DC disturbance, i.e.,

$$K_i \int_{t_1}^{\infty} e(\tau)d\tau = d.$$

That is, the shaded area is equal to d. As far as $e(t) \neq 0$, the integral action takes place until the error sum is equal to the magnitude of the disturbance. However, a very high integral gain makes the system unstable, producing a large overshoot.

Selecting a PI Gain for Speed Control System

In this part, a guideline for selecting PI gains is presented. Fig. 1.40 (a) shows a typical motor speed control loop, in which the whole current control block of an inverter is represented as a low pass filter, $1/(1 + s/\omega_{ic})$. Thus, the current control bandwidth is $[0, \omega_{ic}]$. The motor with a load is simply represented by a rotating object that has an inertia, J.

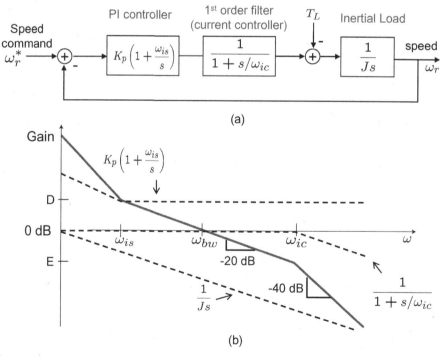

(a)

(b)

Figure 1.40: (a) Speed control loop and (b) Bode plot for the open loop.

The PI control block has a zero at $s = \omega_{is}$. It needs to be sufficiently small such that $\omega_{is} \ll \omega_{ic}$. Then the whole frequency range can be divided into three parts:

[0, ω_{is}), [ω_{is}, ω_{ic}), and [ω_{ic}, ∞). Note that $1 + \frac{\omega_{is}}{s} \approx \frac{\omega_{is}}{s}$ and $1/(1 + s/\omega_{ic}) \approx 1$ in [0, ω_{ic}]. Thus, the (open) loop gain is approximated as $\frac{K_p \omega_i}{J s^2}$ in the low frequency region, [0, ω_{is}). However, it is close to $\frac{K_p}{Js}$ in the middle frequency region, [ω_i, ω_{ic}) since $1 + \frac{\omega_{is}}{s} \approx 1$. Finally, the loop gain is approximated as $\frac{K_p \omega_{ic}}{J s^2}$ in [ω_{ic}, ∞). Summarizing the above, we have the following asymptotic lines:

$$\frac{K_p}{Js}\left(1 + \frac{1}{s/\omega_{is}}\right)\left(\frac{1}{1 + s/\omega_{ic}}\right) \approx \begin{cases} \frac{K_p \omega_{is}}{J s^2} & (-40\text{dB}), \quad [0,\ \omega_{is}) \\ \frac{K_p}{Js} & (-20\text{dB}), \quad [\omega_{is},\ \omega_{ic}) \\ \frac{K_p \omega_{ic}}{J s^2} & (-40\text{dB}), \quad [\omega_{ic},\ \infty) \end{cases}$$

Denote by ω_{bw} the frequency at which the open loop gain plot passes through the 0 dB line. Then [0, ω_{bw}) will be the speed bandwidth. Therefore, it is necessary to make $\frac{K_p}{J\omega} \approx 1$ for $\omega < \omega_{bw}$. Hence, the proportional gain should be selected as

$$K_p = J\omega_{bw}. \tag{1.59}$$

It is also necessary to make the zero, ω_{is} of the PI controller much smaller than ω_{bw}. A common rule is to choose

$$\omega_{is} = \frac{\omega_{bw}}{5}. \tag{1.60}$$

Then with the gain $K_p = J\omega_{bw}$ and $\omega_{is} = \omega_{bw}/5$, the closed loop transfer function is

$$\frac{K_p\left(1 + \frac{\omega_{is}}{s}\right)\frac{1}{Js}}{1 + K_p\left(1 + \frac{\omega_{is}}{s}\right)\frac{1}{Js}} = \frac{\omega_{bw}s + \frac{\omega_{bw}^2}{5}}{s^2 + \omega_{bw}s + \frac{\omega_{bw}^2}{5}}. \tag{1.61}$$

Thus, the damping coefficient is $\zeta = \sqrt{5}/2$.

Exercise 1.19   Consider the speed control loop shown in Fig. 1.40 (a) with $J = 0.01$ kg·m^2 and $\omega_{ic} = 1000$ rad/s. It is desired to make the speed control bandwidth [0, 120] rad/s. Determine the coefficients, K_p and ω_{is} of the PI controller according to the design rules (1.59) and (1.60). Plot the step response using MATLAB Simulink.

Solution. $K_p = 1.2$ and $\omega_{is} = 24$.   ∎

1.3.4   IP Controller

Different types of controllers are compared in Fig. 1.42. Note that the block of 'K_t' represents a current control system and that the speed controller output is fed to the current block as the current command, i^*. The integral-proportional controller (IP controller) is a variation of PI controller which has the proportional gain in

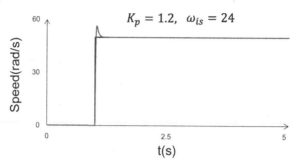

Figure 1.41: Speed step response (Solution to Exercise 1.19).

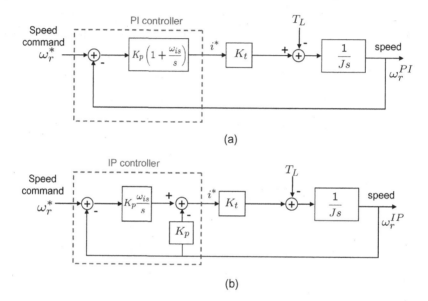

Figure 1.42: Speed control loops with (a) PI and (b) IP controllers.

the feedback path, as shown in Fig. 1.42 (b). However, the location of the integral controller is the same. The transfer functions are equal to

$$\text{PI Controller} \Rightarrow \frac{\omega_r^{PI}(s)}{\omega_r^*(s)} = \frac{\frac{K_t K_p}{J}(s + \omega_{is})}{s^2 + \frac{K_t K_p}{J}s + \frac{K_t K_p \omega_{is}}{J}} \tag{1.62}$$

$$\text{IP Controller} \Rightarrow \frac{\omega_r^{IP}(s)}{\omega_r^*(s)} = \frac{\frac{K_p K_t \omega_{is}}{J}}{s^2 + \frac{K_t K_p}{J}s + \frac{K_t K_p \omega_{is}}{J}}. \tag{1.63}$$

Note that both functions have the same denominator, but the numerators are different: The IP controller does not have differential operator 's', whereas the PI controller does. Let $\omega_r^{PI}(t) = \mathcal{L}^{-1}\{\omega_r^{PI}(s)\}$ and $\omega_r^{IP}(t) = \mathcal{L}^{-1}\{\omega_r^{IP}(s)\}$, where $\mathcal{L}^{-1}$ is the inverse Laplace operator. Then,

$$\omega_r^{PI}(t) = \omega_r^{IP}(t) + \frac{1}{\omega_{is}}\frac{d}{dt}\omega_r^{IP}(t).$$

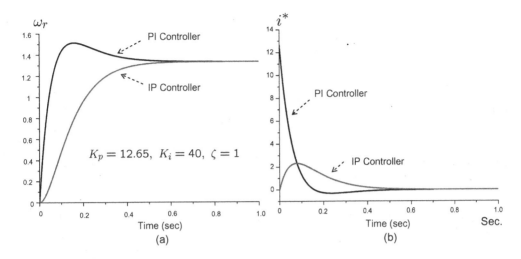

Figure 1.43: (a) Speed responses and (b) current commands for the PI and IP controllers shown in Fig. 1.42.

That is, in the PI controller output the derivative, $\dot{\omega}_r^{IP}$ is added additionally to the output of IP controller. Therefore, the PI controller yields faster response. However, it tends to make a larger overshoot or undershoot. Fig. 1.43 shows the speed responses and the magnitudes of current commands for the same K_p and ω_{is}. Note that the PI controller produces a higher current command, i^* (about 5 times), than the IP controller for the step input. Observe also that ω_r^{PI} has an overshoot, whereas ω_r^{IP} does not. Thus, IP controllers may be used in the speed loop to avoid a possible current peaking.

Exercise 1.20 Obtain the transfer functions from the disturbance to the output $\frac{\omega(s)}{T_L(s)}$ for the PI and IP cases shown in Fig. 1.42.

1.3.5 PI Controller with Reference Model

In the previous section, it is shown that the integral action is required to eliminate a DC offset. But, the integral controller increases the system order. The order increase causes more phase delay, narrowing down the phase margin. In this section, we consider a modified form, named *PI controller with reference model*. It provides a disturbance rejection capability without increasing the system order between the input and output.

Fig. 1.44 shows a speed control block diagram of the PI controller with reference model [3]. Note that the current loop is modeled simply as $1/(1 + \tau_\sigma s) \approx 1$. The block diagram consists of two parts: The bottom part (dotted line) shows the tracking control loop, whereas the upper part shows the disturbance rejection part. It should be noted that the integral controller appears only in the disturbance rejection part, not in the tracking part. As a result, the closed loop transfer function

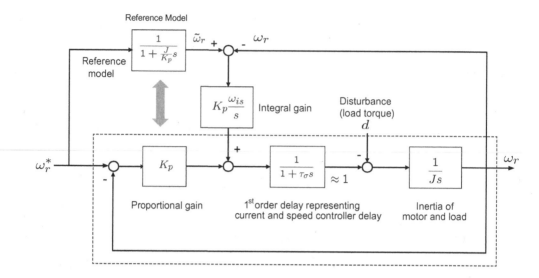

Figure 1.44: PI controller with reference model.

of $\omega_r(s)/\omega_r^*(s)$ appears as a second order system:

$$\frac{\omega_r(s)}{\omega_r^*(s)} = \frac{1}{\frac{J\tau_\sigma}{K_p}s^2 + \frac{J}{K_p}s + 1} \approx \frac{1}{\frac{J}{K_p}s + 1}. \tag{1.64}$$

A first order reduced model, $1/(\frac{J}{K_p}s + 1)$ is used as a reference model, and is employed to detect an unknown disturbance. The same command, ω_r^* is also applied to the model to produce a virtual output. But if a disturbance, d is present in the plant, the two outputs will not be the same. However, the difference, $\tilde{\omega}_r - \omega_r$ carries information on d. So, it is integrated to produce a compensation term. In the steady state, it follows that

$$K_p \omega_{is} \int_0^t (\tilde{\omega}_r(\tau) - \omega_r(\tau))d\tau = d.$$

The transfer functions for tracking and disturbance rejection are compared in Table 1.5. For disturbance rejection the PI with reference model is the same as the conventional PI controller. But the difference lies in the tracking part: The tracking part remains the second order, since the I-controller is not involved. In contrast, the conventional PI controller has a third order transfer function.

Therefore, the PI with reference model is expected to be more stable with a larger gain margin. Fig. 1.45 shows step responses when $J = 0.015$, $\tau_\sigma = 0.000548$, $K_p = 13.69$, and $\omega_{is} = 454.5$. The conventional PI controller creates a higher overshoot for the same gain. It is due to the presence of a zero which does not exist in the PI with reference model. Fig. 1.46 shows the Bode plots of $\omega_r(s)/\omega_r^*(s)$. One can see that the PI controller with reference model has less phase delay than the conventional PI, meaning that the former has a larger stability margin.

Table 1.5: Comparision of transfer functions

	PI with reference model	Conventional PI
Tracking $\dfrac{\omega_r(s)}{\omega_r^*(s)}$	$\dfrac{1}{\dfrac{J\tau_\sigma}{K_p}s^2 + \dfrac{J}{K_p}s + 1}$	$\dfrac{s/\omega_{is} + 1}{\dfrac{\tau_\sigma J}{K_p\omega_{is}}s^3 + \dfrac{J}{K_p\omega_{is}}s^2 + \dfrac{s}{\omega_{is}} + 1}$
Dist. rejection $\dfrac{\omega_r(s)}{d(s)}$	$\dfrac{\dfrac{1}{K_p\omega_{is}}s(\tau_\sigma + 1)}{\dfrac{\tau_\sigma J}{K_p\omega_{is}}s^3 + \dfrac{J}{K_p\omega_{is}}s^2 + \dfrac{s}{\omega_{is}} + 1}$	$\dfrac{\dfrac{1}{K_p\omega_{is}}s(\tau_\sigma + 1)}{\dfrac{\tau_\sigma J}{K_p\omega_{is}}s^3 + \dfrac{J}{K_p\omega_{is}}s^2 + \dfrac{s}{\omega_{is}} + 1}$

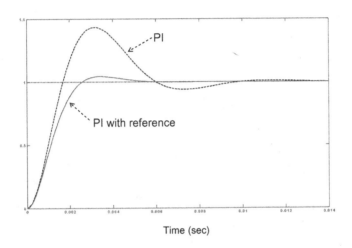

Time (sec)

Figure 1.45: Step responses with the PI controller with reference model and the conventional PI controller for the same gains.

Double Ratio Rule

The double ratio rule give a systematic guide to select proper gains for higher dimensional systems. It was developed based on a damping optimum of the closed control loop [3]. Consider a closed loop system

$$H(s) = \frac{y(s)}{u(s)} = \frac{b_n s^n + b_{n-1}s^{n-1} + \cdots + b_1 s + b_0}{a_n s^n + a_{n-1}s^{n-1} + \cdots + a_1 s + a_0}$$

Make a sequence of coefficient ratios

$$\frac{a_n}{a_{n-1}}, \ \frac{a_{n-1}}{a_{n-2}}, \ \frac{a_{n-2}}{a_{n-3}}, \ \cdots , \ \frac{a_1}{a_0},$$

and assume that

$$\left|\frac{a_n}{a_{n-1}}\right| \ll \left|\frac{a_{n-1}}{a_{n-2}}\right| \ll \cdots \ll \left|\frac{a_2}{a_1}\right| \ll \left|\frac{a_1}{a_0}\right|.$$

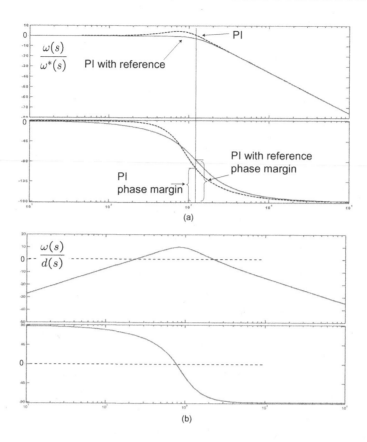

Figure 1.46: Bode plots of the systems with the PI controller with reference model and the conventional PI controller: (a) for tracking and (b) for disturbance rejection.

Then the poles of $H(s)$ are approximated as [4]

$$-\frac{a_0}{a_1}, \quad -\frac{a_1}{a_2}, \quad \cdots, \quad -\frac{a_{n-1}}{a_n}.$$

Now, we choose the controller coefficients such that the ratios between adjacent components are less than or equal to $1/2$:

$$\frac{\frac{a_k}{a_{k-1}}}{\frac{a_{k-1}}{a_{k-2}}} = \frac{a_k a_{k-2}}{a_{k-1}^2} \leq \frac{1}{2}, \quad \text{for} \quad 2 \leq k \leq n. \tag{1.65}$$

This pole allocation method leads to a good command and disturbance response without extensive calculation. This method results in a robust response, i.e., it makes the system less susceptible to parameter variation.

For example, consider a second order system, $1/(a_2 s^2 + a_1 s + a_0)$. Applying the double ratio rule, (1.65), we obtain $a_2 = \frac{a_1^2}{2a_0}$. Then the denominator is equal to

$$\frac{a_1^2}{2a_0}\left[s^2 + 2\left(\frac{a_0}{a_1}\right)s + 2\left(\frac{a_0}{a_1}\right)^2\right] = \frac{a_1^2}{2a_0}\left[s^2 + 2 \cdot \frac{1}{\sqrt{2}}\left(\frac{\sqrt{2}a_0}{a_1}\right)s + \left(\frac{\sqrt{2}a_0}{a_1}\right)^2\right]$$

Therefore, the double ratio rule leads to a damping coefficient, $\zeta = \frac{1}{\sqrt{2}} = 0.707$, which is desirable.

Now we apply the double ratio rule to the third order system in Table 1.5. Then

$$\frac{\frac{\tau_\sigma J}{K_p \omega_{is}^2}}{\left(\frac{J}{K_p \omega_{is}}\right)^2} = \frac{K_p \tau_\sigma}{J} \leq \frac{1}{2}$$

$$\frac{J}{K_p \omega_{is}} \cdot \omega_{is}^2 = \frac{J \omega_{is}}{K_p} \leq \frac{1}{2}$$

These lead to gain selection:

$$K_p = \frac{J}{2\tau_\sigma} \tag{1.66}$$

$$\frac{1}{\omega_{is}} = \frac{2J}{K_p} = 4\tau_\sigma . \tag{1.67}$$

Note that the integral gain is equal to

$$K_i = K_p \omega_{is} = \frac{J}{8\tau_\sigma^2}. \tag{1.68}$$

This tells us that as the delay increases, both the proportional and integral gains should be reduced. On the other hand, both gains can be increased as the inertia increases. After applying the optimal gains (1.66) and (1.67), the transfer functions turn out to be

$$\frac{\omega_r(s)}{\omega_r^*(s)} = \frac{1}{2\tau_\sigma^2 s^2 + 2\tau_\sigma s + 1} \tag{1.69}$$

$$\frac{\omega_r(s)}{d(s)} = \frac{8\tau_\sigma s(\tau_\sigma s + 1)}{8\tau_\sigma^3 s^3 + 8\tau_\sigma^2 s^2 + 4\tau_\sigma s + 1} \tag{1.70}$$

Exercise 1.21 Consider the closed loop transfer function (1.69). Show that the damping coefficient is $\zeta = \frac{1}{\sqrt{2}} = 0.707$ and the phase margin is $65.5°$.

This section can be summarized as

1) The PI controller with reference model does not increase the order of the transfer function from the command to the output. Thereby, it has lower phase lag and larger phase margin. On the other hand, it has the same characteristics with the conventional PI controller in the disturbance rejection.

2) The double ratio rule is a convenient rule that yields an optimum damping. It can be used for determining the PI gains. The double ratio rule provides us a good reference for gain selection. It gives us some insights between the magnitude of gain and a system delay. As the delay increases, the gain needs to be lowered. On the other hand, it tells that a high gain can be used for a speed control system with a large inertia.

1.3.6   2-DOF Controller

In general, PI controllers do not utilize the structural information of the plant. The internal model control (IMC), as the name stands for, employs the model of the plant inside the controller. The IMC combined with two-degree-of-freedom (2-DOF) controller is shown in Fig. 1.47 [5]. The IMC utilizes a model, $\hat{G}(s)$ in the control loop, which is an estimate of the plant, $G(s)$. Both the plant and the model receive the same input, and then the outputs are compared. For the purpose of illustration, let $\hat{G}(s) = G(s)$. Due to the presence of disturbance, the outputs are not the same. The output error is fed back to the input through the feedback compensator, $Q_d(s)$. On the other hand, the IMC also has a feedforward compensator, $Q_r(s)$.

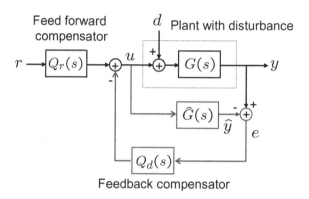

Figure 1.47: IMC structure for plant $G(s)$ with disturbance d.

The transfer function of the whole system is

$$y = \left[\frac{G(s)Q_r(s)}{1 + Q_d(s)(G(s) - \hat{G}(s))} \right] r + \left[\frac{G(s)(1 - Q_d(s)\hat{G}(s))}{1 + Q_d(s)(G(s) - \hat{G}(s))} \right] d. \qquad (1.71)$$

If there is no plant model error, i.e., $G = \hat{G}$, then

$$y = G(s)Q_r(s)r + G(s)(1 - Q_d(s)G(s))d.$$

The sensitivity function S and the complementary sensitivity function T are

$$S(s) \equiv \left. \frac{y(s)}{d(s)} \right|_{r(s)=0} \quad = \quad G(s)(1 - Q_d(s)\hat{G}(s)), \qquad (1.72)$$

$$T(s) \equiv \left. \frac{y(s)}{r(s)} \right|_{d(s)=0} \quad = \quad G(s)Q_r(s). \qquad (1.73)$$

Note that $S(s)$ is affected by the feedback compensator $Q_d(s)$, whereas $T(s)$ is affected by the feedforward compensator $Q_r(s)$. That is, $Q_d(s)$ serves for disturbance rejection, whereas $Q_r(s)$ functions to enhance the tracking performance. Since the sensitivity function and the complementary sensitivity function can be designed independently, it is called the *two-degree-of-freedom* controller.

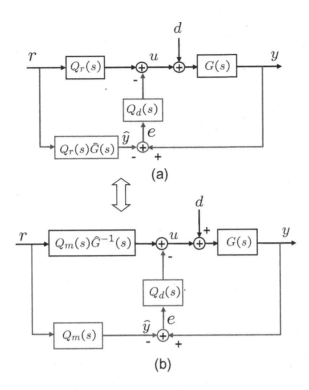

Figure 1.48: Equivalent 2-DOF structures when $Q_r(s) = Q_m(s)G(s)^{-1}$.

1.3.7 Variations of 2-DOF Structures

Ideal performances, $S = 0$ and $T = 1$ are obtained when $Q_d(s) = G(s)^{-1}$ and $Q_r(s) = G(s)^{-1}$. In this case, the controllers will be improper, i.e., they will contain differentiators. Therefore, the ideal controller cannot be realized practically. Despite not being perfect, it is desired to let $S \approx 0$ and $T \approx 1$ in a low frequency range. To resolve the problem of differentiation, we put a low pass filter $Q_m(s)$ in front of the inverse dynamics $G(s)^{-1}$, i.e.,

$$Q_r(s) = Q_m(s)G(s)^{-1}, \qquad (1.74)$$

where $Q_m(s)$ is a filter that prevents improperness of $Q_r(s)$. Thus, $Q_m(s)$ is desired to have a unity gain in a low frequency region. A practical choice is

$$Q_m(s) = \frac{1}{(\tau s + 1)^n}, \qquad (1.75)$$

where $1/\tau > 0$ represents a cut-off frequency and $n > 0$ is an integer. As τ gets smaller, the wider (frequency) range of unity gain is obtained. The order of the filter, n needs to be the same as the relative degree of $G(s)$. Then $Q_m(s)G(s)^{-1}$ turns out to be proper.

Variations of the IMC block diagram are shown in Fig. 1.48 [6]. Note that the IMC in Fig. 1.47 is equivalent to the one in Fig. 1.48 (a). With the feedforward compensator, (1.74), the block diagram turns out to be Fig. 1.48 (b). The controller shown in Fig. 1.48 (b) is the 2-DOF controller containing inverse dynamics. The role of $Q_d(s)$ is to nullify the effects of the disturbance, $d(s)$ in a low frequency region. In [7], a PI controller was selected for $Q_d(s)$.

1.3.8   Load Torque Observer

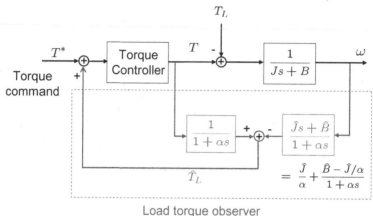

Load torque observer

Figure 1.49: A load torque observer.

A load torque observer is often called a disturbance observer. The load torque is different from the state observer, since the disturbance is not a state variable. It is quite often used in the speed loop to reject the disturbance torque. Fig. 1.49 shows a typical structure of the load torque observer, where the load torque, T_L, is treated as an external disturbance. The objective is to estimate the unknown load torque and feedback it to the input of the torque controller, so that the unknown load is compensated before the speed (PI) controller is activated.

In the following analysis, the load torque is assumed to be a constant. The load torque is filtered out by the plant dynamics, so that the observer has, in nature, a structure of the inverse dynamics. To circumvent the use of the differentiator, $\hat{J}s + \hat{B}$, it utilizes a low pass filter with unity gain, $1/(1 + \alpha s)$ [8]. Note that the impulse response is equal to $\frac{1}{\alpha}e^{-\frac{1}{\alpha}t}u_s(t)$, where $u_s(t)$ is the unit step function. Since $\int_0^\infty \frac{1}{\alpha}e^{-\frac{1}{\alpha}t}dt = 1$ independently of α, it approaches the delta function as α decreases to zero. Therefore, practically $\frac{\hat{J}s+\hat{B}}{1+\alpha s}w_r(s) \approx (\hat{J}s + \hat{B})w_r(s)$ for a large $\alpha > 0$ while a direct differentiation is avoided:

$$\frac{\hat{J}s + \hat{B}}{1 + \alpha s}w_r(s) = \frac{\hat{J}}{\alpha} + \frac{\hat{B} - \hat{J}/\alpha}{1 + \alpha s}w_r(s).$$

The same low pass filter is applied to the input, $T(s)$. Then, a torque estimate $\hat{T}_L$ is obtained as

$$\hat{T}_L = \frac{1}{1 + \alpha s}T(s) - \frac{\hat{J}s + \hat{B}}{1 + \alpha s}w(s). \tag{1.76}$$

Hong and Nam[8] put an artificial delay in the input path of the load torque (disturbance) observer for a system with measurement delay as shown in Fig. 1.50. It was shown that the observer with the artificial delay showed a better performance like a Smith predictor [9].

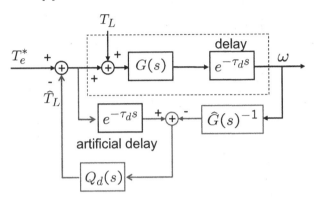

Figure 1.50: A load torque observer with an artificial delay in the input path.

1.3.9   Feedback Linearization

Consider a single input, single output nonlinear system

$$\dot{x} = f(x) + ug(x), \tag{1.77}$$
$$y = h(x), \tag{1.78}$$

where $x \in \mathbb{R}^n$, $f : \mathbb{R}^n \to \mathbb{R}^n$, $g : \mathbb{R}^n \to \mathbb{R}^n$, $h : \mathbb{R}^n \to \mathbb{R}$, $u : [0, \infty) \to \mathbb{R}^n$. Assume that f, g, and h are differentiable infinitely. Let

$$L_f h \equiv < dh, f > = \sum_{i=1}^{n} \frac{\partial h}{\partial x_i} f_i .$$

Correspondingly, define for $k \geq 1$

$$L_f^k h \equiv < dL_f^{k-1} h, f > .$$

We construct a new map $T : \mathbb{R}^n \to \mathbb{R}^n$ defined by

$$T(x) = \begin{bmatrix} h \\ L_f h \\ \vdots \\ L_f^{n-1} h \end{bmatrix} . \tag{1.79}$$

Let x_e be an equilibrium point of $f(x)$, i.e., $f(x_e) = 0$. Suppose further that the Jacobian $DT(x_e)$ has rank n, i.e.,

$$\text{rank}\,[DT(x_e)] = \text{rank} \begin{bmatrix} dh \\ dL_f h \\ \vdots \\ L_f^{n-1} h \end{bmatrix} (x_e) = n. \tag{1.80}$$

Then T is a local diffeomorphism in a neighborhood of x_e. In other words, T is a differentiable homeomorphism. When a function is one-to-one and onto, and its inverse is continuous, then the function is called homeomorphism.

We define a new coordinate system $z = T(x)$. Assume further that

$$\begin{cases} L_g L_f^k h = 0, & 0 \le k \le n - 2. \\ L_g L_f^{n-1} h \ne 0. \end{cases} \tag{1.81}$$

Then, it follows from (1.81) that

$$\dot{z} = \begin{bmatrix} L_{f+ug} h \\ L_{f+ug} L_f h \\ \vdots \\ L_{f+ug} L_f^{n-1} h \end{bmatrix} = \begin{bmatrix} L_f h \\ L_f^2 h \\ \vdots \\ L_f^n h + u L_g L_f^{n-1} h \end{bmatrix}. \tag{1.82}$$

Now, we choose

$$u = \frac{1}{L_g L_f^{n-1} h}(v - L_f^n h). \tag{1.83}$$

Then it follows from (1.82) that

$$\dot{z} = \begin{bmatrix} z_2 \\ z_3 \\ \vdots \\ z_n \\ v \end{bmatrix} = \begin{bmatrix} 0 & 1 & 0 & \cdots & 0 \\ 0 & 0 & 1 & \cdots & 0 \\ & & & \ddots & \\ 0 & 0 & 0 & \cdots & 1 \\ 0 & 0 & 0 & \cdots & 0 \end{bmatrix} z + \begin{bmatrix} 0 \\ 0 \\ \vdots \\ 0 \\ 1 \end{bmatrix} v. \tag{1.84}$$

That is, through coordinate transformation (1.79) and feedback (1.83), a linear system is obtained [10]. Let $ad_f g = [f, g]$ and $ad_f^k g = [f, ad_f^{k-1} g]$, where $[f, g]$ is the Lie bracket of vector fields f and g. A necessary and sufficient condition for the existence of a function that satisfies (1.81) is

$$\{g, ad_f g, \cdots, ad_f^{n-1} g\} \quad \text{are linearly independent;}$$
$$\{g, ad_f g, \cdots, ad_f^{n-2} g\} \quad \text{are involutive,}$$

where involutive means a distribution is closed under Lie bracket operation[10]. An application of feedback linearization theory to a PWM converter-inverter system is found in [11].

Simplified Modeling of Practical Current Loop

Nowadays, most control algorithms are implemented using microcontrollers, and thereby associated delay elements are introduced in the control loop:

i) command holder,
ii) computing cycle time,
iii) pulse width modulation (PWM) execution delay,
iv) value detection

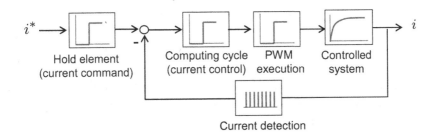

Figure 1.51: Practical current loop based on a microcontroller.

The command is refreshed every sampling period. Since the command value is held until the next sampling period, a holding delay takes place, which is estimated about half of the sampling period, $T_{sa}/2$, where T_{sa} is the current sampling period. A computing cycle is required for current control and calculation of the PWM intervals. Also, there is a delay in executing the PWM. Finally, a delay is required for current sensing and A/D conversion. Fig. 1.51 shows the delay elements in the current control loop. We sum up all the delay elements in the current loop, and denote it by τ_σ. The total delay of the current loop is estimated as $\tau_\sigma = 1.5T_{sa} \sim 2T_{sa}$ [3]. Note further that

$$e^{-\tau_\sigma s} = \frac{1}{e^{\tau_\sigma s}} \approx \frac{1}{1 + \tau_\sigma s}. \tag{1.85}$$

Hence, the total delay can be represented as a a first order filter. In the speed loop, the current block is often treated as a first order system.

References

[1] R. Ordenburger, Frequency response data presentation, Standards and Design Criteria, *IEEE Trans. ASME*, no. 53-A-11, Nov. pp. 1155–1954.

[2] B. C. Kuo and F Golnaraghi, *Automatic Control Systems*, 8th Ed., Wiley, 2003.

[3] H. Gross, J. Hamann and G. Wiegartner, *Electrical Feed Drives in Automation, Basics, Computation, Dimensioning*, Publics MCD Corporate Publishing, 2001.

[4] R. W. Erickson and Maksimovic, *Fundamentals of Power Electronics*, 2nd Ed., Springer, 2001.

[5] M. Morari and E. Zafirious, *Robust Process Control*, Prentice Hall, 1989.

[6] T. Sugie and T. Yoshikawa, General solution of robust tracking problem in two-degree-of-freedom control systems, *IEEE Trans. Automat. Contr.*, vol. AC-31, pp. 552–554, June 1986.

[7] N. Hur, K. Nam, and S. Won, A two degrees of freedom current control scheme for dead-time compensation, *IEEE Trans. Ind. Electron.*, vol. 47, pp. 557–564, June 2000.

[8] K. Hong and K. Nam, A load torque compensation scheme under the speed measurement delay, *IEEE Trans. Ind. Electron.*, vol. 45, no. 2, pp. 283–290, Apr. 1998.

[9] K. J. Astrom, C. C. Hang, and B. C. Lim, A new Smith predictor for controlling a process with an integrator and long dead-time, *IEEE Trans. Automat. Contr.*, vol. 39, pp. 343–345, Feb. 1994.

[10] A. Isidori, *Nonlinear Control Systems II*, Springer, 1999.

[11] J. Jung, S. Lim, and K. Nam, A feedback linearizing control scheme for a PWM converter-inverter having a very small DC-link capacitor, *IEEE Trans. Ind. Appl.*, vol. 35, no. 5, Sep./Oct., pp. 1124–1131, 1999.

Problems

1.1   Verify the vector identity.

    a) $\mathbf{B} \times (\nabla \times \mathbf{A}) = \nabla(\mathbf{A} \cdot \mathbf{B}) - (\mathbf{B} \cdot \nabla)\mathbf{A}$
    b) $\nabla \cdot (\nabla \times \mathbf{A}) = 0$
    c) $\nabla \times (\nabla f) = \mathbf{0}$
    d) $\nabla \times (\nabla \times \mathbf{A}) = \nabla(\nabla \cdot \mathbf{A}) - \nabla^2\mathbf{A}.$

1.2   Draw the vector fields, $\mathbf{F}$ in the (x, y)-plane, and obtain $\nabla \cdot \mathbf{F}$ and $\nabla \times \mathbf{F}$.

    a) $\mathbf{F} = x\mathbf{i}_x + y\mathbf{i}_y$
    b) $\mathbf{F} = -y\mathbf{i}_x + x\mathbf{i}_y$
    c) $\mathbf{F} = (-x + 2y)\mathbf{i}_x + (-2x - y)\mathbf{i}_y$

1.3   Compute $T = \frac{3P}{4}(\boldsymbol{\lambda} \times \mathbf{i})_z$. Let $\boldsymbol{\lambda} = \lambda_d\mathbf{i}_x + \lambda_q\mathbf{i}_y$ and $\mathbf{i} = i_d\mathbf{i}_x + i_q\mathbf{i}_y$, where $\lambda_d = L_d i_d + L_{dq}i_q + \lambda_m$, $\lambda_q = L_q i_q + L_{qd}i_d$, and λ_m is a constant.

1.4   Let $f(x, y, x) = xy^2 - yz$. Find ∇f. Show that $\nabla \times (\nabla f) = \mathbf{0}$.

1.5   Verify Stokes theorem for

$$\mathbf{F} = (x^2 - 3z)\mathbf{i}(y^2 - x)\mathbf{j}xy\mathbf{k}$$

over the surface S on the plane $z = 0$. The boundary, ∂S is defined by $y = x$, $y = 0$, and $x = 1$, and oriented counterclockwise as shown below:

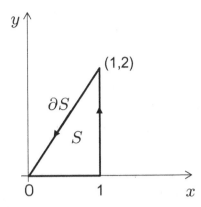

1.6 Derive the following for a vector, **A** and a scalar function, f in cylindrical coordinate:

$$\nabla \cdot \mathbf{A} = \frac{1}{r}\frac{\partial(r A_r)}{\partial r} + \frac{1}{r}\frac{\partial A_\phi}{\partial \phi} + \frac{\partial A_z}{\partial z}$$

$$\nabla^2 f = \frac{1}{r}\frac{\partial}{\partial r}\left(r\frac{\partial f}{\partial r}\right) + \frac{1}{r^2}\frac{\partial^2 f}{\partial \phi^2} + \frac{\partial^2 f}{\partial z^2}$$

1.7 Consider an infinitely long cylindrical conductor of radius R. Assume that the axial center is aligned with the z-axis and that current I flows with current distribution $J(r) = \alpha r$ for $0 \le r \le R$, where α is a constant, and r is a radius of an orthogonal disk centered at the z-axis.

a) Calculate the magnetic field B_r for $0 \le r \le R$ (inside the conductor).

b) Calculate the magnetic field B_r for $r > R$ (outside the conductor).

c) Plot B_r as a function of r.

1.8 Consider a coaxial cable shown below: The radii of the outer coil are a and b. In the inner coil, current I flows from the left to the right, while the same current flows in the opposite direction in the outer wire. Using Ampere's law, find the H field for each region below:

a) $r < a$
b) $a < r < b$
c) $b < r$
d) Assume that a material with permeability μ is filled in the region of $r < a$. Derive the inductance per unit length.

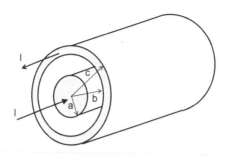

1.9 Consider N−turn circular conductor loop with radius 1 centered at the origin in the $x - y$ plane. Suppose that magnetic field is given by
$\mathbf{B} = B_0(\exp(x^2 + y^2) - e)\sin(\omega t)\mathbf{i}_z$, where ω is the angular frequency.

a) Derive the flux linkage of the circular loop.
b) Find the EMF induced in the loop.

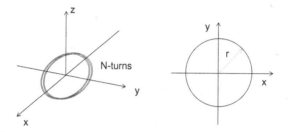

1.10 Let $c(t) = (3t - 2, t + 1)$ for $1 \le t \le 3$ be a parameterization of a wire. The density of the wire at point (x, y) is given by $f(x, y) = x + 2y$. Compute the mass of the wire.

1.11 Consider a single phase transformer that has the turn ratio of $N_1 : N_2 = 2400 : 240$. It operates at $f = 60$ Hz. When $R_1 = 0.75 \, \Omega$, $R_2 = 0.0075 \, \Omega$, $X_{l1} = 1 \, \Omega$, $X_{l2} = 0.01 \, \Omega$, $X_m = 50 \, \Omega$, draw an equivalent circuit referred to the primary side. Also draw the phase diagram when the primary voltage is 440 V and a 0.5 Ω load is connected to the secondary terminal.

1.12 An infinitely straight conductor carrying current I is placed at $x = s$ to the left of a rectangular coil with width a and length b. Assume that the straight conductor moves to the right over the loop. Find the mutual inductance of the system for $0 \le s < c$, $c \le s < c + a$, and $s \ge c + a$. If the conductor traveling speed is 10 m/s, find the induced voltage on the loop.

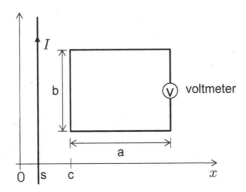

1.13 Consider a DC motor with $r_a = 0.5\ \Omega$, $K_t = 0.4\ \text{Nm/A}$, and $K_b = 0.4\ \text{Vs/rad}$. Assume that the maximum armature voltage is $v_a^{max} = 120\ V$.

a) Determine the motor speed when the load torque is $T_L = 8\ \text{Nm}$.

b) Suppose the back EMF constant is reduced to $K_b = 0.3\ \text{Vs/rad}$ under the same load. Recalculate the speed.

1.14 Consider a system

$$G(s) = \frac{10000}{s^2 + 28s + 10000}.$$

a) Determine ω_c of $G(s)$.
b) It is desired to advance a phase by $20°$, i.e., $\phi_d = 20°$.
   Select zero and pole of a PD (lead-lag) controller.
   Determine K_p so that the $|C_{pd}(j\omega_c)G(j\omega_c)| = 1$.
c) Draw Bode plots of $G(s)$ and $C_{pd}(s)G(s)$.
d) Find the gain margin improvement brought by the PD controller.

1.15 Consider a DC motor with the following parameters: $r_a = 0.5\ \Omega$, $L_a = 12\ \text{mH}$, $K_t = 1.8\ \text{Nm/A}$, $K_b = 1.8\ \text{Vs/rad}$, and $J = 0.1\ \text{kg} \cdot \text{m}^2$. Then the current loop transfer function with a PI controller, $K_p + K_i/s$ is

$$\frac{I_a(s)}{I_a^*(s)} = \frac{(K_p + \frac{K_i}{s})\frac{1}{L_a s + r_a}}{1 + (K_p + \frac{K_i}{s})\frac{1}{L_a s + r_a}} = \frac{\frac{K_p}{L_a}s + \frac{K_i}{L_a}}{s^2 + \frac{r_a + K_p}{L_a}s + \frac{K_i}{L_a}} = \frac{\frac{K_p}{L_a}s + \omega_n^2}{s^2 + 2\zeta\omega_n s + \omega_n^2}$$

a) Determine the proportional gain, K_p such that $\zeta\omega_n = 1000$. Also, choose K_i such that overshoot of the step response is about 10%.

b) The current loop is approximated as

$$\frac{I_a(s)}{I_a^*(s)} = \frac{\frac{K_p}{L_a\omega_n^2}s + 1}{\frac{1}{\omega_n^2}s^2 + 2\frac{\zeta}{\omega_n}s + 1} \approx \frac{1}{2\frac{\zeta}{\omega_n}s + 1}.$$

Refer to the speed loop shown in Fig. 1.40 (a) and use the approximate current model. Given $\omega_2 = 1000$ rad/s, determine the speed loop gain by utilizing (1.59), (1.60) and $\omega_{bw} - \omega_1 = \frac{1}{2}(\omega_2 - \omega_{bw})$.

c) Draw the Bode plot of the speed loop with the gains obtained in b) and determine the phase margin.

1.16 Consider a DC motor with $r_a = 0.2 \ \Omega$, $L_a = 500 \ \mu H$, $K_t = 0.1$ Nm/A, $J = 0.01$ kg $\cdot$ m^2, and $B = 0.01$ Nm $\cdot$ s.

a) Determine current I-gain, K_{ic} so that the current control bandwidth is 200 Hz.
b) Using MATLAB Simulink, construct a current control loop as shown below. Plot the current response to current command and adjust current P-gain, K_{ip} so that the overshoot is less than 10%.
c) Add the speed controller and determine K_{ps} and K_{is} so that the speed loop bandwidth is 20 Hz and the step response does not have an overshoot. Plot the speed response with the step speed command.
d) Show disturbance rejection performance via speed response when a step load torque, $T_L = 30 \cdot u_s(t - 0.2)$ is applied.

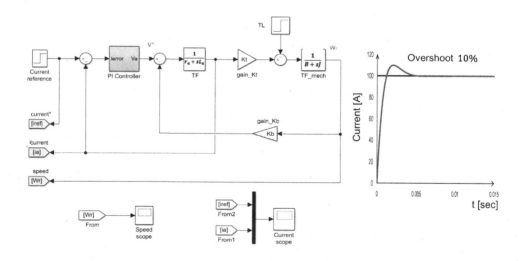

1.17 Consider the speed loop transfer function, (1.64). Suppose that the proportional gain is selected two times higher than the one in (1.66), i.e., $K_p = J/\tau_\sigma$. At this time, ω_{is} changes correspondingly. Determine the damping coefficient and the phase margin.

1.18 Consider a standard second order system:

$$\frac{1}{1 + \frac{s}{\omega_0} + \frac{s^2}{\omega_0 \omega_2}} = \frac{1}{1 + 2\zeta \frac{s}{\omega_c} + \frac{s^2}{\omega_c^2}} = \frac{1}{1 + \frac{s}{Q\omega_c} + \frac{s^2}{\omega_c^2}},$$

where $\omega_c = \sqrt{\omega_0 \omega_2}$ and $Q = \frac{1}{2\zeta} = \frac{\omega_0}{\omega_c} = \sqrt{\frac{\omega_0}{\omega_2}}$. Derive the following relation between Q and the phase margin ϕ_m:

$$Q = \frac{\sqrt{\cos \phi_m}}{\sin \phi_m} \qquad \text{or} \qquad \phi_m = \tan^{-1} \sqrt{\frac{1 + \sqrt{1 + 4Q^2}}{2Q^4}}.$$

1.19 Consider Fig. 1.48 (a). Let

$$\begin{bmatrix} y \\ u \end{bmatrix} = A(s) \begin{bmatrix} r \\ d \end{bmatrix}.$$

Determine $A(s) \in \mathbb{R}^{2 \times 2}$.

1.20 Consider a load torque observer for the system with measurement delay shown in Fig. 1.50. Let $G(s) = \hat{G}(s) = 1/(Js + B)$ and $Q_d(s) = 1/(1 + \alpha s)$.

a) Determine the closed loop transfer function, $\hat{T}_L(s)/T_L(s)$.

b) Repeat the same when the artificial delay is not present in the input side.

c) Approximate $e^{-\tau_d s} \approx 1 - \tau_d s$. Discuss from the stability viewpoint why the observer with the artificial delay is better compared with the ordinary one.

Chapter 2

Rotating Magnetic Field

Most rotating electrical machines have cylindrical air gaps. A traveling magneto motive force (MMF) can be generated in the air gap when multi-phase sinusoidal currents flow through the multi-phase coils. Since the coils are confined in the slots, the MMFs have quantized levels. The MMF harmonics are analyzed by Fourier series. The voltage, the current, and the flux of the three phase system can be identified with some rotating vectors in the complex plane. Interestingly, the direction of the flux vector coincides with that of real flux. Coordinate transformation maps are defined, and illustrated with examples.

Figure 2.1: Stator coil winding of a small induction motor.

2.1 Magneto Motive Force and Inductance

A photo of a stator winding is shown in Fig. 2.1. The stator is constructed by stacking punched steel sheets and then inserting coils into the slots on the stator core. Figs. 2.2 (a) and (b) show a section of a cylindrical machine and its a cut and stretched view, respectively. It is assumed that the number of coil turns per phase is N_{ph}. Though Fig. 2.2 (a) depicts a two pole machine, the theory can be extended to p pole pair machines. When the air gap diameter is D, the pole pitch is defined as $\tau_p = \pi D/(2p)$. Note that the core stack length is l_{st}, the gap height is g, and the angle, $\theta = \frac{\pi}{\tau_p}x$ is taken counterclockwise from the horizontal axis, where x is a space variable.

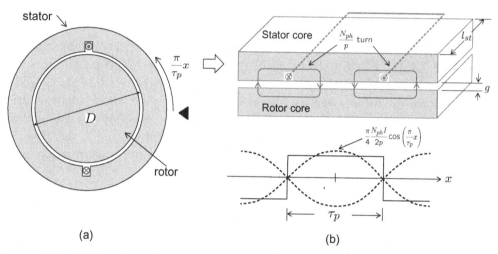

(a)   (b)

Figure 2.2: MMF of a cylindrical uniform air gap machine: (a) sectional view and (b) cut and stretched view.

Note here that the number of turns per pole pair is N_{ph}/p. Then the MMF is given as a periodic square function with amplitude $\pm\frac{N_{ph}}{p}\frac{I}{2}$. It is described mathematically as

$$F(x) = \sum_{\nu=1}^{\infty} a_\nu \cos\left(\nu\frac{\pi}{\tau_p}x\right), \tag{2.1}$$

where the Fourier coefficient is

$$a_\nu = \frac{1}{\tau_p}\int_{-\tau_p}^{\tau_p} F(x)\cos\left(\nu\frac{\pi}{\tau_p}x\right)dx \tag{2.2}$$

$$= \frac{4}{\tau_p}\int_{0}^{\tau_p/2} F(x)\cos\left(\nu\frac{\pi}{\tau_p}x\right)dx.$$

The square MMF is expressed as

$$F(x) = \frac{4}{\pi}\frac{N_{ph}I}{2p}\sum_{\nu=1}\frac{1}{\nu}\left(\sin\nu\frac{\pi}{2}\right)\cos\nu\frac{\pi}{\tau_p}x, \tag{2.3}$$

$$= \frac{4}{\pi}\frac{N_{ph}I}{2p}\left(\cos\frac{\pi}{\tau_p}x - \frac{1}{3}\cos 3\frac{\pi}{\tau_p}x + \frac{1}{5}\cos 5\frac{\pi}{\tau_p}x - \frac{1}{7}\cos 7\frac{\pi}{\tau_p}x + \cdots\right).\tag{2.4}$$

Obviously, the square function contains infinitely many harmonics. It will be shown later that only the fundamental component is used for torque production. The other high order harmonics act as the source for various loss and noise-vibration. Fig. 2.3 shows the fundamental component and a partial sum up to 7th order. Here it is worth remembering that the scaling factor, $\frac{4}{\pi}$ is common to all components, and the rest harmonic coefficients decrease after $1, -\frac{1}{3}, \frac{1}{5}, -\frac{1}{7}, \cdots$.

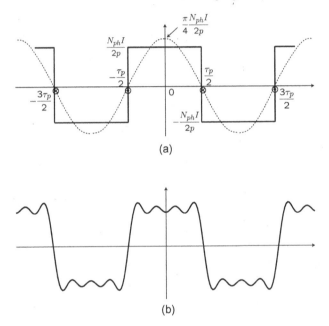

(a)

(b)

Figure 2.3: (a) Square MMF and (b) Fourier series approximation up to 7th order.

2.1.1 Single Phase Inductance

Now we assume that the current is alternating such that $I(t) = \sqrt{2}I_{rms}\cos\omega t$. The gap field, B_g is calculated using the Ampere's law. Consider the integral contour shown in Fig. 2.2 (b). Since the core permeability is so high compared with μ_0, it is treated as infinity. Then the integral is evaluated only in the air gap, g. For the convenience of derivation, only the fundamental component is considered in (2.4). Thus, we have

$$B_g(x,t) = \mu_0\frac{F(x,t)}{g} = B_{mg}\cos(\frac{\pi}{\tau_p}x)\cos(\omega t), \tag{2.5}$$

where

$$B_{mg} = \frac{4}{\pi}\frac{\mu_0 N_{ph}}{2pg}k_w\sqrt{2}I_{rms}. \tag{2.6}$$

Note here that the winding factor k_w is used to reflect the coil pitch and distribution factors. The winding factor is treated in Chapter 11.

Since

$$\int_{-\frac{\tau_p}{2\nu}}^{\frac{\tau_p}{2\nu}} \cos(\frac{\pi}{\tau_p}x)dx = \frac{2}{\pi}\tau_p. \tag{2.7}$$

the flux per pole is equal to

$$\Phi(t) = l_{st}\int_{-\frac{\tau_p}{2\nu}}^{\frac{\tau_p}{2\nu}} B(x,t)dx = \frac{2}{\pi}\tau_p l_{st}\frac{\mu_0}{2g}\frac{4}{\pi}k_w\frac{N_{ph}}{p}\sqrt{2}I_{rms}\cos\omega t. \tag{2.8}$$

Since the flux linkage per pole pair per phase is equal to $\lambda = \frac{N_{ph}}{p}k_w\Phi(t)$, the fundamental component of phase voltage is

$$V_{ph} = -\frac{d}{dt}p\lambda = \omega\frac{2}{\pi}\tau_p l_{st}\frac{\mu_0}{2g}\frac{4}{\pi}\frac{k_w^2 N_{ph}^2}{p}\sqrt{2}I_{rms}\sin\omega t \tag{2.9}$$

Therefore, the rms value is $V_{ph(rms)} = \omega\mu_0\frac{4}{\pi^2}\frac{\tau_p l_{st}}{pg}k_w^2 N_{ph}^2 I_{rms}$.

The fundamental component of phase inductance is obtained from $V_{ph(rms)} = \omega L_{ms}I_{rms}$. Therefore, the phase inductance is

$$L_{ms} = \mu_0\frac{4}{\pi^2}\frac{\tau_p l_{st}}{pg}k_w^2 N_{ph}^2 = \mu_0\frac{2}{\pi}\frac{Dl_{st}}{p^2g}k_w^2 N_{ph}^2. \tag{2.10}$$

It is useful to note that the inductance decreases as the pole number increase when N_{ph} is fixed.

To calculate high order inductances, consider the full expression of (2.3). Then,

$$B_g(x,t) = \mu_0\frac{F(x,t)}{g} = \sum_\nu B_{mg(\nu)}\cos(\nu\frac{\pi}{\tau_p}x)\cos\omega t, \tag{2.11}$$

where

$$B_{mg(\nu)}(x) = \frac{4}{\pi}\frac{\mu_0 N_{ph}}{2pg}\frac{k_{w(\nu)}}{\nu}\sqrt{2}I_{rms}. \tag{2.12}$$

Note also that

$$\int_{-\frac{\tau_p}{2\nu}}^{\frac{\tau_p}{2\nu}} \cos\left(\nu\frac{\pi}{\tau_p}x\right)dx = \frac{2\tau_p}{\pi\nu} \tag{2.13}$$

The flux corresponding to ν^{th} harmonic is

$$\Phi_\nu(t) = l_{st}\int_{-\frac{\tau_p}{2}}^{\frac{\tau_p}{2}} B_{mg(\nu)}(x)dx = \frac{2}{\pi}\tau_p l_{st}\frac{\mu_0}{2g}\frac{4}{\pi}\frac{k_{w(\nu)}}{\nu^2}\frac{N_{ph}}{p}\sqrt{2}I_{rms}\cos\omega t. \tag{2.14}$$

Therefore, the νth order harmonic inductance for a single phase is given by

$$L_{ms(\nu)} = \mu_0\frac{4}{\pi^2}\frac{\tau_p l_{st}}{pg}\frac{k_{w(\nu)}^2}{\nu^2}N_{ph}^2. \tag{2.15}$$

Exercise 2.1   Estimate the phase inductance of a uniform air gap machine with the following specifications: $p = 4$, $N_{ph} = 24$ turns, $D_r = 172$ mm, $k_w = 0.95$, $g = 0.8$ mm, $l_{st} = 120$ mm.

Solution.

$$L_{ms} = 4\pi \times 10^{-7} \times \frac{4}{\pi^2} \frac{0.172 \times \pi}{8} \frac{0.12}{4 \times 0.0008} \times 0.95^2 \times 24^2 = 671 \ (\mu H). \quad \blacksquare$$

2.1.2   Inductance of Three Phase Uniform Gap Machine

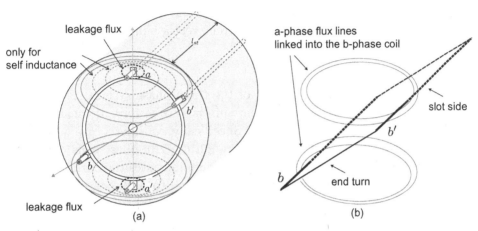

Figure 2.4: *a*-phase flux linking into the *b*-phase coil: (a) flux lines excited by the *a*-phase coil and (b) linking into the *b*-phase coil loop.

Consider a cylindrical machine that has a uniform air gap. Fig. 2.4 (a) shows flux lines that are excited by the *a*-phase current. Note that the innermost lines do not cross the air gap. They depict the leakage flux and its amount is expressed as $L_{ls}i_a$, where L_{ls} is the leakage inductance. The gap crossing flux is denoted by $L_{ms}i_a$. Then the total flux produced by the *a*-phase current is equals to $\lambda_a = (L_{ms} + L_{ls})i_a$.

In the three phase system, different phase windings are linked by flux. Note that *a* and *b*-phase windings are displaced angularly by 120°. A part of *a*-phase flux passes through the loop of $b - b'$ as shown in Fig. 2.4. Such crossing flux makes a linking with the *b*-phase coil. It describes the mutual flux between *b* and *a* phase coils. Specifically, the *b*-phase flux linkage excited by the *a*-phase current is $\lambda_b = -\frac{1}{2}L_{ms}i_a$, i.e., about a half of the flux is linked. It can be understood by considering 120° angular difference, i.e., $\cos\frac{2\pi}{3} = -\frac{1}{2}$. Due to the symmetry among the phases, the whole flux linkage is expressed as

$$\boldsymbol{\lambda}_{abc} = \mathbf{L}_{abcs}\mathbf{i}_{abc}, \tag{2.16}$$

where $\boldsymbol{\lambda}_{abc} = [\lambda_a, \ \lambda_b, \ \lambda_c]^T$, $\mathbf{i}_{abc} = [i_a, \ i_b, \ i_c]^T$, and

$$\mathbf{L}_{abcs} = \begin{bmatrix} L_{ms} + L_{ls} & -\frac{1}{2}L_{ms} & -\frac{1}{2}L_{ms} \\ -\frac{1}{2}L_{ms} & L_{ms} + L_{ls} & -\frac{1}{2}L_{ms} \\ -\frac{1}{2}L_{ms} & -\frac{1}{2}L_{ms} & L_{ms} + L_{ls} \end{bmatrix}. \tag{2.17}$$

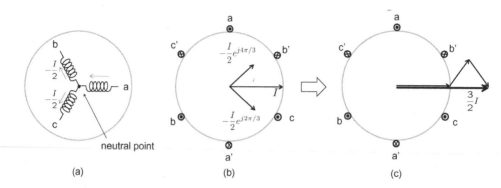

Figure 2.5: Vector sum in three phase machine: (a) wye connection, (b) current vectors when $[I, -\frac{1}{2}I, -\frac{1}{2}I]$, and (c) the current vector sum.

Fig. 2.5 (a) shows a wye-connection of the three phase system. Note that the neutral point is floated so that the current sum has to be zero. It means that it is impossible to conduct only one phase coil. Specifically, consider a case when $i_a(=I)$ splits at the neutral point and conducts through b and c phase coils equally, i.e., $[i_a, i_b, i_c] = [I, -\frac{1}{2}I, -\frac{1}{2}I]$, where the negative sign indicates the current that flows out from a coil. Obviously, the algebraic sum is equal to zero, i.e., $i_a + i_b + i_c = 0$. However, the current vector sum is not. Instead, it is equal to

$$I - \frac{I}{2}e^{j\frac{2\pi}{3}} - \frac{I}{2}e^{j\frac{4\pi}{3}} = \frac{3}{2}I \tag{2.18}$$

as depicted in Fig. 2.5 (b) and (c). Likewise, the three phase flux linkage is equal to

$$L_{ms}I - \frac{1}{2}L_{ms}(-\frac{1}{2}Ie^{j\frac{2\pi}{3}}) - \frac{1}{2}L_{ms}(-\frac{1}{2}Ie^{j\frac{4\pi}{3}}) = \frac{3}{2}L_{ms}I. \tag{2.19}$$

That is, the inductance is increased by a factor $\frac{3}{2}$ when it is compared with the single phase case. The three phase inductance called *synchronous inductance*, is derived by multiplying (2.10) by $\frac{3}{2}$:

$$\frac{3}{2}L_{ms} = \mu_0 \frac{\tau_p l_{st}}{pg} \frac{6}{\pi^2} k_w^2 N_{ph}^2. \tag{2.20}$$

Correspondingly, the ν-th order synchronous inductance is equal to

$$L_{ms(\nu)} = \mu_0 \frac{6}{\pi^2} \frac{\tau_p l_{st}}{pg} \frac{k_{w(\nu)}^2}{\nu^2} N_{ph}^2. \tag{2.21}$$

2.2 Rotating Field

The phase coils are distributed along the air gap periphery to shape the MMF as close as possible to a sinusoidal wave. Fig. 2.6 (a) shows an ideal coil distribution of a two pole machine, in which the coil density is described as a sine function of θ, i.e., $N_a(\theta) = \frac{N_{ph}}{2}\sin\theta$. Note that the sum over a half period is equal to the number of phase winding, $\int_0^\pi \frac{N_{ph}}{2}\sin\theta d\theta = N_{ph}$.

Consider an MMF, $F_a(\theta)$ at an angular position, θ when the a-phase winding conducts. It is an a-phase MMF to the direction of $e^{j\theta}$. Then

$$F_a(\theta) = \frac{1}{2}\left(\int_\theta^{\pi+\theta} \frac{N_{ph}}{2}\sin\theta d\theta\right)\cdot i_a = \frac{1}{2}N_{ph}i_a\cos\theta. \tag{2.22}$$

Note that the sinusoidal coil distribution yields a perfect sinusoidal MMF, as shown in Fig. 2.6 (b) [1]. In the three phase machine, coils are displaced by 120° from one another. The MMFs to the direction of $e^{j\theta}$ excited by i_b and i_c are

$$F_b(\theta) = \frac{1}{2}N_{ph}i_b\cos(\theta - \frac{2\pi}{3}) \tag{2.23}$$

$$F_c(\theta) = \frac{1}{2}N_{ph}i_c\cos(\theta - \frac{4\pi}{3}). \tag{2.24}$$

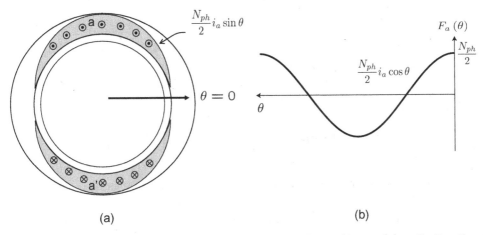

(a)   (b)

Figure 2.6: MMF made by sinusoidally distributed windings: (a) coil distribution and (b) MMF plot.

On the other hand, we have a liberty of choosing phase currents arbitrarily. Suppose that

$$i_a = I\cos(\omega t), \tag{2.25}$$

$$i_b = I\cos(\omega t - \frac{2\pi}{3}), \tag{2.26}$$

$$i_c = I\cos(\omega t - \frac{4\pi}{3}). \tag{2.27}$$

Then, the total MMF is equal to

$$F_{sum}(\theta) = F_a(\theta) + F_b(\theta) + F_c(\theta) = \frac{1}{2}N_{ph}I\Big[\cos\theta\cos\omega t$$

$$+ \cos(\theta - \frac{2\pi}{3})\cos(\omega t - \frac{2\pi}{3}) + \cos(\theta - \frac{4\pi}{3})\cos(\omega t - \frac{4\pi}{3})\Big]. \tag{2.28}$$

Utilizing the cosine law, it follows that

$$\cos\theta\cos\omega t = \frac{1}{2}\left(\cos(\theta+\omega t)+\cos(\theta-\omega t)\right),$$

$$\cos(\theta-\frac{2\pi}{3})\cos(\omega t-\frac{2\pi}{3}) = \frac{1}{2}\left(\cos(\theta+\omega t-\frac{4\pi}{3})+\cos(\theta-\omega t)\right),$$

$$\cos(\theta-\frac{4\pi}{3})\cos(\omega t-\frac{4\pi}{3}) = \frac{1}{2}\left(\cos(\theta+\omega t-\frac{8\pi}{3})+\cos(\theta-\omega t)\right).$$

Note the identity

$$\cos(\theta+\omega t)+\cos(\theta+\omega t-\frac{4\pi}{3})+\cos(\theta+\omega t-\frac{2\pi}{3}) = 0 \qquad \text{for all t.}$$

This implies that the component sum that rotates clockwise (negative sequence) is equal to zero. Therefore, it follows that

$$F_{sum}(\theta, t) = \frac{3}{4}N_{ph}I\cos(\theta-\omega t). \tag{2.29}$$

This is a traveling wave equation. For example, consider a case of observing the MMF at a moving position, $\theta = \omega t$. Then, the MMF looks constant. It is interpreted as an MMF that rotates counterclockwise (positive sequence) at the electrical angular speed, ω. In summary, a traveling MMF can be obtained by applying balanced three phase current to sinusoidally distributed three phase windings [1].

2.2.1　Rotating Field Generation by Inverter

Fig. 2.7 shows the flux lines when only i_a, i_b, or i_c flows. The arrow indicates the direction of flux vector, by which the current vector direction is also set. That is, the direction of the current vector is identified with that of flux. For example, the axis of a-phase current is determined by the direction of outgoing flux from the rotor that would be created by the a-phase current flow. Applying the right hand rule the current axes are easily determined. Fig. 2.7 shows flux lines and the corresponding current vector directions.

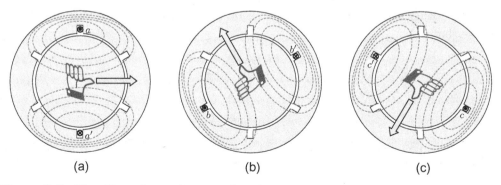

(a)　　　　　　　　　　　(b)　　　　　　　　　　　(c)

Figure 2.7: Flux lines depending on the phase currents.

Using the inverter, rotating fields can be made. Fig. 2.8 shows inverter switch connections and the corresponding current vectors. Note that $i_a + i_b + i_c = 0$ for all time. The inverter consists of three switching arms. The two switches in an arm are cannot be turned at the same time. If the upper switch is on, '1' is assigned to that arm. Otherwise, '0'. Then, nontrivial switching states are (1,0,0), (1,1,0), (0,1,0), (0,1,1), (0,0,1), and (1,0,1). Depending on the switching states, the magnitude and direction of phase currents change. It is assumed that the impedance of phase coil is the same. Thus if $i_a = I$ in the (1,0,0) switching state, the current splits at the neutral point such that $i_b = -I/2$ and $i_c = -I/2$. The current vector sums are shown in Fig. 2.8. It shows that a rotating field can be generated by a sequence of switching states. To make a smooth rotation, pulse width modulation is necessary.

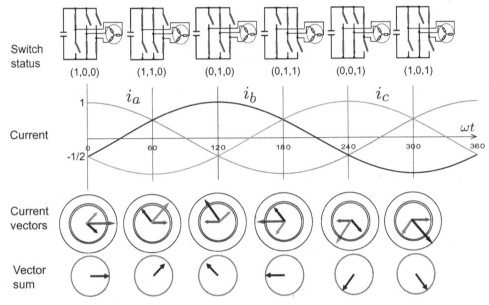

Figure 2.8: Rotation of the current vector corresponding to the balanced three phase currents.

Exercise 2.2   Show that the MMF rotates in the reverse direction (clockwise) if we change the motor terminal connection such that $i_a = I\cos(\omega t)$, $i_b = I\cos(\omega t - \frac{4\pi}{3})$, and $i_c = I\cos(\omega t - \frac{2\pi}{3})$.

Exercise 2.3   Consider a two phase motor as shown in Fig. 2.9. The two stator windings are distributed sinusoidally, and have 90° angle difference; $F_a = \frac{Ni_a}{2}\cos\theta$ and $F_b = \frac{Ni_b}{2}\sin\theta$. Find proper phase currents i_a and i_b as functions of time to generate a rotating MMF.

2.2.2   High Order Space Harmonics

An ideal traveling field is obtained when there are no *space harmonics* and no *time harmonics*. But in practice, harmonics are unavoidable both in time and space. We

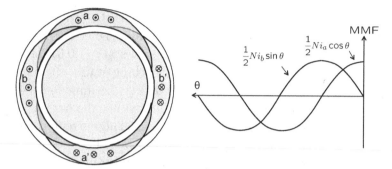

Figure 2.9: Two phase stator windings and their MMF wave forms.

consider here only the space harmonics which are caused by winding cluster in the stator slots. Recall from (2.4) the square MMF is expanded as a Fourier series:

$$F_a(\theta) = \frac{4}{\pi} \frac{Ni_a}{2} \left[\cos\theta - \frac{1}{3}\cos 3\theta + \frac{1}{5}\cos 5\theta - \frac{1}{7}\cos 7\theta \cdots \right]. \qquad (2.30)$$

Since the b-phase MMF is shifted by $\frac{2\pi}{3}$,

$$F_b(\theta) = \frac{4}{\pi}\frac{Ni_b}{2}\left[\cos(\theta - \frac{2}{3}) - \frac{1}{3}\cos 3(\theta - \frac{2\pi}{3}) \right.$$
$$\left. + \frac{1}{5}\cos 5(\theta - \frac{2\pi}{3}) - \frac{1}{7}\cos 7(\theta - \frac{2\pi}{3}) + \cdots \right]. \qquad (2.31)$$

Note that

$$\cos(3\theta - 2\pi) = \cos(3\theta)$$
$$\cos(5\theta - \frac{10}{3}\pi) = \cos(5\theta - \frac{4\pi}{3})$$
$$\cos(7\theta - \frac{14}{3}\pi) = \cos(7\theta - \frac{2\pi}{3}).$$

Thus,

$$F_b(\theta) = \frac{4}{\pi}\frac{Ni_b}{2}\left[\cos(\theta - \frac{\pi}{3}) - \frac{1}{3}\cos(3\theta) + \frac{1}{5}\cos(5\theta - \frac{4\pi}{3}) - \frac{1}{7}\cos(7\theta - \frac{2\pi}{3}) + \cdots \right].$$

Since the c-phase MMF is shifted by $\frac{4\pi}{3}$, it follows that

$$F_c(\theta) = \frac{4}{\pi}\frac{Ni_c}{2}\left[\cos(\theta - \frac{4\pi}{3}) - \frac{1}{3}\cos(3\theta) + \frac{1}{5}\cos(5\theta - \frac{2\pi}{3}) - \frac{1}{7}\cos(7\theta - \frac{4\pi}{3}) + \cdots \right].$$

Suppose that balanced sinusoidal currents are applied, i.e., $i_a = I\cos\omega t$, $i_b = I\cos(\omega t - \frac{2\pi}{3})$, and $i_c = I\cos(\omega t - \frac{4\pi}{3})$. We take the sum of the same harmonic order: $F_{sum(n)} = F_{a(n)} + F_{b(n)} + F_{c(n)}$ for $n = 1, 3, 5, 7, \cdots$, where the subscript '(n)' denotes the nth harmonic component. First, note that the sum of the third order components is equal to zero:

$$F_{sum(3)} = -\frac{4}{3\pi}\frac{NI}{2}\left(\cos(\omega t) + \cos(\omega t - \frac{2\pi}{3}) + \cos(\omega t - \frac{4\pi}{3}) \right)\cos(3\theta) = 0.$$

Likewise, sums of the multiples of three are equal to zero; $F_{sum(6)} = F_{sum(9)} = F_{sum(12)} = \cdots = 0$. Now consider the sum of the 5th order components:

$$F_{sum(5)} = \frac{4}{5\pi}\frac{NI}{2}\left(\cos 5\theta \cos\omega t + \cos(5\theta - \frac{4}{3}\pi)\cos(\omega t - \frac{2\pi}{3}) \right.$$
$$\left. + \cos(5\theta - \frac{2}{3}\pi)\cos(\omega t - \frac{4\pi}{3}) \right)$$
$$= \frac{4}{5\pi}\frac{NI}{2}\frac{3}{2}\cos(\omega t + 5\theta). \tag{2.32}$$

It turns out to be a constant when $\theta = -\frac{1}{5}\omega t$. It means that the 5th-harmonic component rotates clockwise at $\frac{1}{5}$ speed of the fundamental component. Thus, it is referred to as *negative sequence*. Similarly, the sum of the 7th order components is equal to

$$F_{sum(7)} = -\frac{4}{7\pi}\frac{NI}{2}\frac{3}{2}\cos(\omega t - 7\theta). \tag{2.33}$$

It is a positive sequence that rotates at $1/7$ speed of the fundamental component. The total sum is

$$F_{sum} = \frac{4}{\pi}\frac{NI}{2}\frac{3}{2}\left[\cos(\omega t - \theta) + \frac{1}{5}\cos(\omega t + 5\theta) - \frac{1}{7}\cos(\omega t - 7\theta) + \cdots\right].$$

Generalizing the above, it follows that

i) $(\nu - 1 = 6m)$: positive sequence

$$F_{sum(\nu)}' = -\frac{4NI}{\nu\pi 2}\frac{3}{2}\cos(\omega t - \nu\theta), \qquad \nu = 7, 13, 19, \cdots \tag{2.34}$$

ii) $(\nu + 1 = 6m)$: negative sequence

$$F_{sum(\nu)} = \frac{4NI}{\nu\pi 2}\frac{3}{2}\cos(\omega t + \nu\theta), \qquad \nu = 5, 11, 17, \cdots \tag{2.35}$$

iii) $(\nu = 6m - 3)$: zero sequence

$$F_{sum(\nu)} = 0, \qquad \nu = 3, 9, 15, 21\cdots \tag{2.36}$$

Note that the MMF spectra need to be obtained from the discrete-time data in many cases. The discrete-time Fourier series formula for an N-periodic signal, $x[n]$ is given by

$$x[n] = \sum_{k=0}^{N-1} c_k e^{j\frac{2\pi}{N}kn}, \tag{2.37}$$

$$c_k = \frac{1}{N}\sum_{n=0}^{N-1} x[n]e^{-j\frac{2\pi}{N}kn}. \tag{2.38}$$

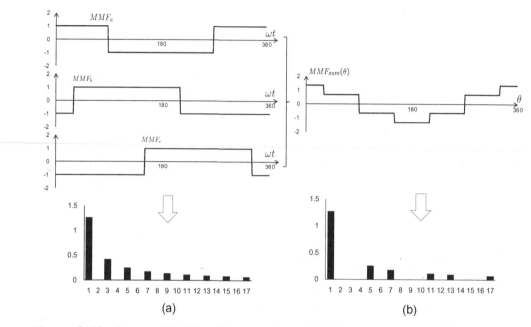

Figure 2.10: Square MMF and harmonics: (a) Phase MMFs and (b) MMF sum.

Fig. 2.10 shows the harmonic spectra for square MMFs. The results were calculated by using Excel based on (2.38). Each phase MMF contains multiples of third order harmonics. But they disappear in the MMF sum. It is an attribute of general three phase systems.

Exercise 2.4   Let $F(\theta) = \cos\theta - \frac{1}{3}\cos 3\theta + \frac{1}{5}\cos 5\theta - \frac{1}{7}\cos 7\theta$. Sample 101 points and make a sequence, $x[n] = F(n\frac{2\pi}{100})$ for $0 \le n \le 100$. Using Excel spreadsheet, find c_k's for $1 \le k \le 20$ and plot them.

Summary

- The coil needs to be distributed sinusoidally. Practically, the MMF contains substantial amount of space harmonics due to the slotted structure of stator core. However, they can be minimized as the slot number increases.

- If balanced sinusoidal three phase currents are applied to a sinusoidally distributed coils, then the rotating field has a single speed component. But with the space harmonics, many positive and negative sequences are generated.

2.3   Change of Coordinates

The current vector is defined in the complex plane so that its direction matches with the outgoing flux from the rotor. Then the flux is visualized by a current vector.

2.3.1 Mapping into Stationary DQ Coordinate

If a periodic three phase current satisfies $i_a(t) + i_b(t) + i_c(t) = 0$ for all t and $\max |i_a(t)| = \max |i_b(t)| = \max |i_c(t)|$, then the current is said to be *balanced*. Since $i_a(t) + i_b(t) + i_c(t) = 0$ acts as a constraint, and it takes away one degree of freedom from the three variables. Thereby, a balanced three phase current has only two degrees of freedom, and it can be mapped into the complex plane uniquely as shown in Fig. 2.11. For general use, definitions are stated in terms of $\mathbf{f} = [f_a, \ f_b, \ f_c]^T$. Note that $\mathbf{f}$ can be current, voltage, or flux vectors. Define a map from $\mathbb{R}^3$ to $\mathbb{C}$ such that [2]

$$
\begin{aligned}
\mathbf{f}^s = f_d^s + j f_q^s &\equiv \frac{2}{3} \left[f_a + e^{j\frac{2\pi}{3}} f_b + e^{j\frac{4\pi}{3}} f_c \right] \\
&= \frac{2}{3}(f_a - \frac{1}{2} f_b - \frac{1}{2} f_c) + j\frac{2}{3}(\frac{\sqrt{3}}{2} f_b - \frac{\sqrt{3}}{2} f_c). \quad (2.39)
\end{aligned}
$$

Note that $\frac{2}{3}$ is multiplied to retain the same peak value after mapping. With the matrix representation, it follows that

$$
\begin{bmatrix} f_d^s \\ f_q^s \end{bmatrix} = \frac{2}{3} \begin{bmatrix} 1 & -\frac{1}{2} & -\frac{1}{2} \\ 0 & \frac{\sqrt{3}}{2} & -\frac{\sqrt{3}}{2} \end{bmatrix} \begin{bmatrix} f_a \\ f_b \\ f_c \end{bmatrix} \quad (2.40)
$$

For example, consider a case, where $[f_a, \ f_b, \ f_c]^T = [\cos \omega t, \ \cos(\omega t - \frac{2\pi}{3}), \ \cos(\omega t - \frac{4\pi}{3})]^T$. Then,

$$
\begin{aligned}
\mathbf{f}^s = f_d^s + j f_q^s &= \frac{2}{3} \left[\cos \omega t + e^{j\frac{2\pi}{3}} \cos(\omega t - \frac{2\pi}{3}) + e^{j\frac{4\pi}{3}} \cos(\omega t - \frac{4\pi}{3}) \right] \\
&= \frac{2}{3} \left[\frac{1}{2}(e^{j\omega t} + e^{-j\omega t}) + \frac{1}{2}e^{j\frac{2\pi}{3}}(e^{j(\omega t - \frac{2\pi}{3})} + e^{-j(\omega t - \frac{2\pi}{3})}) \right. \\
&\qquad \left. + \frac{1}{2}e^{j\frac{4\pi}{3}}(e^{j(\omega t - \frac{4\pi}{3})} + e^{-j(\omega t - \frac{4\pi}{3})}) \right] \\
&= \frac{2}{3} \left[\frac{3}{2}e^{j\omega t} + \underbrace{\frac{1}{2}\left(e^{-j\omega t} + e^{-j(\omega t - \frac{4\pi}{3})} + e^{-j(\omega t - \frac{8\pi}{3})}\right)}_{=0} \right] \\
&= e^{j\omega t} = \cos \omega t + j \sin \omega t. \quad (2.41)
\end{aligned}
$$

Note that negative sequence components are summed to zero. The map sends three phase AC current $[\cos \omega t, \ \cos(\omega t - \frac{2\pi}{3}), \ \cos(\omega t - \frac{4\pi}{3})]$ to a rotating vector $e^{j\omega t}$ in the complex plane. Note again that the peak values do not change under the transformation. To denote the stationary frame, the transformed variable is marked with the superscript 's'.

Exercise 2.5 Using Excel spreadsheet, compute (v_d^s, v_q^s) corresponding to the phase voltages of a six-step inverter: $(v_a, v_b, v_c) = (1, -1/2, -1/2), (1/2, 1/2, -1)$,

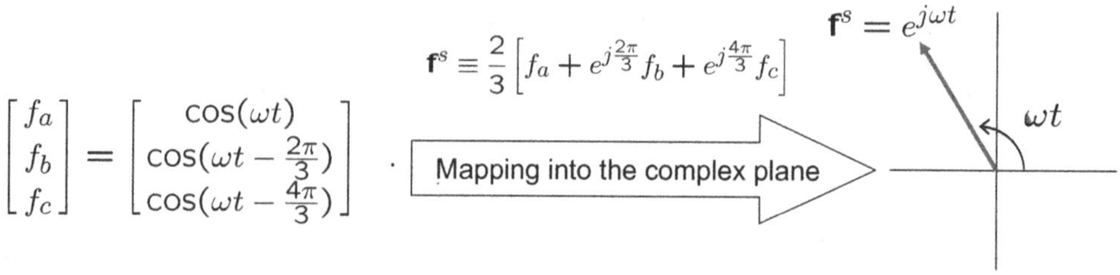

$$\mathbf{f}^s \equiv \frac{2}{3}\left[f_a + e^{j\frac{2\pi}{3}}f_b + e^{j\frac{4\pi}{3}}f_c\right]$$

$$\mathbf{f}^s = e^{j\omega t}$$

$$\begin{bmatrix} f_a \\ f_b \\ f_c \end{bmatrix} = \begin{bmatrix} \cos(\omega t) \\ \cos(\omega t - \frac{2\pi}{3}) \\ \cos(\omega t - \frac{4\pi}{3}) \end{bmatrix} \cdot \boxed{\text{Mapping into the complex plane}} \Longrightarrow$$

$$\omega t$$

Figure 2.11: Mapping of a three phase vector into the stationary $d - q$ frame.

$(-1/2, 1, -1/2)$, $(-1, 1/2, 1/2)$ $(-1/2, -1/2, 1)$, and $(1/2, -1, 1/2)$. Draw v_d^s and v_q^s versus time, and $v_d^s + jv_q^s$ in the complex plane, $\mathbb{C}$.

Solution.

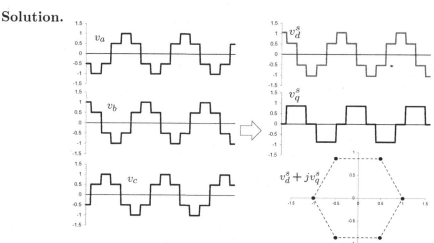

Figure 2.12: Mapping of phase voltages of six-step inverter into the stationary $d-q$ frame (Solution to Exercise 2.5).

Exercise 2.6   Consider a three phase current with 5th order harmonics:

$$\begin{bmatrix} i_a \\ i_b \\ i_c \end{bmatrix} = \begin{bmatrix} \cos(\omega t) \\ \cos(\omega t - \frac{2\pi}{3}) \\ \cos(\omega t - \frac{4\pi}{3}) \end{bmatrix} + \frac{1}{5}\begin{bmatrix} \cos(5\omega t) \\ \cos(5(\omega t - \frac{2\pi}{3})) \\ \cos(5(\omega t - \frac{4\pi}{3})) \end{bmatrix}. \tag{2.42}$$

Compute the corresponding (i_d^s, i_q^s). Draw $i_d^s + ji_q^s$ in $\mathbb{C}$.

Solution. The current vector is obtained such that

$$\mathbf{i}^s = e^{j\omega t} + \frac{2}{3}\frac{1}{5}[\cos(5\omega t) + \cos(5\omega t - \frac{4\pi}{3})e^{j\frac{2\pi}{3}} + \cos(5\omega t - \frac{2\pi}{3})e^{j\frac{4\pi}{3}}] = e^{j\omega t} + \frac{1}{5}e^{-j5\omega t}. \ \blacksquare$$

Note that the 5th order harmonic component has negative sign in the exponent, indicating it constitutes a negative sequence. That is, the 5th order harmonic makes a clockwise rotation along a small circle, while the fundamental component rotates counterclockwise. The trajectory of 5th harmonic component is indicated by dotted line in $\mathbb{C}$ in Fig. 2.13.

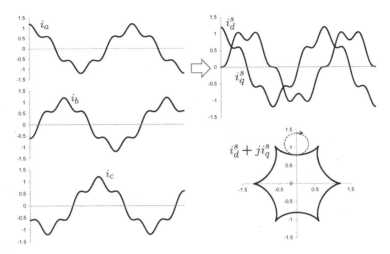

Figure 2.13: Mapping of currents with 5th harmonics into the stationary $d - q$ frame (Solution to Exercise 2.6).

2.3.2 Mapping into Synchronous Frame

Suppose that we want to describe a vector in a frame rotating counterclockwise. Then the angle of a vector appears to be reduced in the new frame as shown in Fig. 2.14. The new representation is obtained by multiplying $e^{-j\theta}$:

$$
\begin{aligned}
\mathbf{f}^e = f_d^e + jf_q^e &\equiv e^{-j\theta}\mathbf{f}^s \\
&= e^{-j\theta}\frac{2}{3}\left[f_a(t) + e^{j\frac{2\pi}{3}}f_b(t) + e^{-j\frac{2\pi}{3}}f_c(t)\right] \\
&= \frac{2}{3}\left[e^{-j\theta}f_a(t) + e^{j(-\theta+\frac{2\pi}{3})}f_b(t) + e^{j(-\theta-\frac{2\pi}{3})}f_c(t)\right]. \quad (2.43)
\end{aligned}
$$

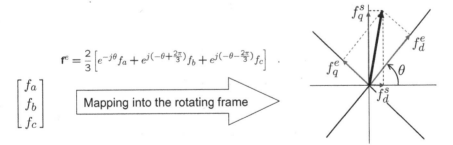

Figure 2.14: Mapping of a three phase vector into a rotating frame.

The transformation is composed of two steps: The first is to transform the *abc*-frame into the stationary *dq*-frame. The second is to rotate it by $e^{-j\theta}$. This transformation is equivalently written as

$$\mathbf{f}^e = f_d^e + j f_q^e = \frac{2}{3}\left[f_a \cos\theta + f_b \cos(\theta - \frac{2\pi}{3}) + f_c \cos(\theta + \frac{2\pi}{3}) \right]$$
$$-j\frac{2}{3}\left[f_a \sin\theta + f_b \sin(\theta - \frac{2\pi}{3}) + f_c \sin(\theta + \frac{2\pi}{3}) \right] . \quad (2.44)$$

Thus in the matrix representation,

$$\begin{bmatrix} f_d^e \\ f_q^e \end{bmatrix} = \frac{2}{3}\begin{bmatrix} \cos\theta & \cos(\theta - \frac{2\pi}{3}) & \cos(\theta + \frac{2\pi}{3}) \\ -\sin\theta & -\sin(\theta - \frac{2\pi}{3}) & -\sin(\theta + \frac{2\pi}{3}) \end{bmatrix}\begin{bmatrix} f_a \\ f_b \\ f_c \end{bmatrix} . \quad (2.45)$$

It is decomposed equivalently as

$$\begin{bmatrix} f_d^e \\ f_q^e \end{bmatrix} = \begin{bmatrix} \cos\theta & \sin\theta \\ -\sin\theta & \cos\theta \end{bmatrix}\frac{2}{3}\begin{bmatrix} 1 & -\frac{1}{2} & -\frac{1}{2} \\ 0 & \frac{\sqrt{3}}{2} & -\frac{\sqrt{3}}{2} \end{bmatrix}\begin{bmatrix} f_a \\ f_b \\ f_c \end{bmatrix} . \quad (2.46)$$

Since the vector appears as a constant, such a transformation simplifies the AC dynamics, gives physical insights, and enables us to establish a high performance controller. It is a transformation into the rotating frame. Since the frame is assumed to rotate at the same speed of a vector, it is called *synchronous frame*. Alternatively, it is also called *reference frame* or *exciting frame*. Vectors in the rotating frame are denoted with superscript 'e'. The inverse map is obtained as

$$f_a = \mathcal{Re}(\mathbf{f}^s)$$
$$f_b = \mathcal{Re}(e^{j\frac{4\pi}{3}}\mathbf{f}^s) \quad (2.47)$$
$$f_c = \mathcal{Re}(e^{j\frac{2\pi}{3}}\mathbf{f}^s).$$

where $\mathcal{Re}(\cdot)$ means taking the real part of a variable in parentheses.

Exercise 2.7   Let $[f_d, f_q]^T = [\cos\omega t, \sin\omega t]^T$. Show that $\mathbf{f}^e = 1$ when $\theta = \omega t$.

Solution. Utilizing (2.41), we obtain $\mathbf{f}^e = e^{-j\theta}\mathbf{f}^s = e^{-j\theta}e^{j\theta} = 1$.   ■

Exercise 2.8   Consider the current (2.42) that has 5th order harmonic component. Draw the current waveform in the synchronous reference frame.

Solution.

Exercise 2.9   Show that $f_a = \mathcal{Re}(\mathbf{f}^s)$.

Solution. $\mathbf{f}^s = \frac{2}{3}\left[f_a - \frac{1}{2}(f_b + f_c) + j\frac{\sqrt{3}}{2}(f_b - f_c) \right] = f_a + j\frac{1}{\sqrt{3}}(f_b - f_c).$ ■

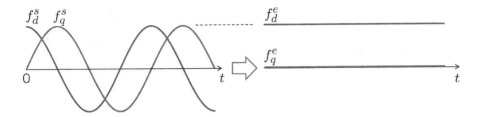

Figure 2.15: Transformation from the stationary dq-frame into a synchronous dq-frame (Solution to Exercise 2.7).

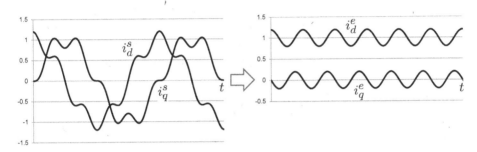

Figure 2.16: Currents with 5th order harmonics: Transformation from the stationary dq-frame into a synchronous reference frame (Solution to Exercise 2.8).

2.3.3 Formulation via Matrices

Transformations into the stationary and the rotating frames can be manipulated by matrix operations. By equating the real and complex parts separately in (2.39), we obtain

$$
\begin{bmatrix} f_d^s \\ f_q^s \\ f_0^s \end{bmatrix} = \frac{2}{3} \begin{bmatrix} 1 & -\frac{1}{2} & -\frac{1}{2} \\ 0 & \frac{\sqrt{3}}{2} & -\frac{\sqrt{3}}{2} \\ \frac{1}{\sqrt{2}} & \frac{1}{\sqrt{2}} & \frac{1}{\sqrt{2}} \end{bmatrix} \begin{bmatrix} f_a \\ f_b \\ f_c \end{bmatrix}.
\tag{2.48}
$$

Note that the last row is introduced to handle the zero sequence, and that $f_0^s = \frac{\sqrt{2}}{3}(f_a + f_b + f_c) = 0$ in balanced systems. Since the matrix is nonsingular, its inverse is well-defined. In some references, $(\frac{1}{2}, \frac{1}{2}, \frac{1}{2})$ is used instead of $(\frac{1}{\sqrt{2}}, \frac{1}{\sqrt{2}}, \frac{1}{\sqrt{2}})$.

Similarly, a transformation matrix into the *synchronous* reference frame follows from (2.44) such that [1]

$$
\begin{bmatrix} f_d^e \\ f_q^e \\ f_0^e \end{bmatrix} = \mathbf{T}(\theta) \begin{bmatrix} f_a \\ f_b \\ f_c \end{bmatrix},
\tag{2.49}
$$

where

$$
\mathbf{T}(\theta) = \frac{2}{3} \begin{bmatrix} \cos\theta & \cos(\theta - \frac{2\pi}{3}) & \cos(\theta + \frac{2\pi}{3}) \\ -\sin\theta & -\sin(\theta - \frac{2\pi}{3}) & -\sin(\theta + \frac{2\pi}{3}) \\ \frac{1}{\sqrt{2}} & \frac{1}{\sqrt{2}} & \frac{1}{\sqrt{2}} \end{bmatrix}.
\tag{2.50}
$$

Note that

$$\mathbf{T}(0) = \frac{2}{3} \begin{bmatrix} 1 & -\frac{1}{2} & -\frac{1}{2} \\ 0 & \frac{\sqrt{3}}{2} & -\frac{\sqrt{3}}{2} \\ \frac{1}{\sqrt{2}} & \frac{1}{\sqrt{2}} & \frac{1}{\sqrt{2}} \end{bmatrix} ;$$

which coincides with the matrix in (2.48). Note further that

$$\mathbf{T}(\theta) = \underbrace{\begin{bmatrix} \cos\theta & \sin\theta & 0 \\ -\sin\theta & \cos\theta & 0 \\ 0 & 0 & 1 \end{bmatrix}}_{=\mathbf{R}(\theta)} \underbrace{\frac{2}{3} \begin{bmatrix} 1 & -\frac{1}{2} & -\frac{1}{2} \\ 0 & \frac{\sqrt{3}}{2} & -\frac{\sqrt{3}}{2} \\ \frac{1}{\sqrt{2}} & \frac{1}{\sqrt{2}} & \frac{1}{\sqrt{2}} \end{bmatrix}}_{=\mathbf{T}(0)} , \tag{2.51}$$

where $\mathbf{R}(\theta)$ is a matrix which represents the rotation of the coordinate frame by θ. That is, map $\mathbf{T}(\theta)$ consists of two parts: One is a map from the abc-frame into the stationary dq-frame. The other is a map from the stationary frame into the rotating dq-frame.

The inverse map is calculated as

$$\mathbf{T}^{-1}(\theta) = \begin{bmatrix} \cos\theta & -\sin\theta & \frac{1}{\sqrt{2}} \\ \cos(\theta - \frac{2\pi}{3}) & -\sin(\theta - \frac{2\pi}{3}) & \frac{1}{\sqrt{2}} \\ \cos(\theta + \frac{2\pi}{3}) & -\sin(\theta + \frac{2\pi}{3}) & \frac{1}{\sqrt{2}} \end{bmatrix} . \tag{2.52}$$

Hence,

$$\mathbf{T}^{-1}(0) = \begin{bmatrix} 1 & 0 & \frac{1}{\sqrt{2}} \\ -\frac{1}{2} & \frac{\sqrt{3}}{2} & \frac{1}{\sqrt{2}} \\ -\frac{1}{2} & -\frac{\sqrt{3}}{2} & \frac{1}{\sqrt{2}} \end{bmatrix} .$$

Similarly to the case of forward transformation, $\mathbf{T}^{-1}$ can be decomposed as

$$\mathbf{T}^{-1}(\theta) = \underbrace{\begin{bmatrix} 1 & 0 & \frac{1}{\sqrt{2}} \\ -\frac{1}{2} & \frac{\sqrt{3}}{2} & \frac{1}{\sqrt{2}} \\ -\frac{1}{2} & -\frac{\sqrt{3}}{2} & \frac{1}{\sqrt{2}} \end{bmatrix}}_{=\mathbf{T}^{-1}(0)} \underbrace{\begin{bmatrix} \cos\theta & -\sin\theta & 0 \\ \sin\theta & \cos\theta & 0 \\ 0 & 0 & 1 \end{bmatrix}}_{=\mathbf{R}(-\theta)} .$$

It should be noted that

$$\mathbf{T}^{-1}(\theta) = \frac{3}{2}\mathbf{T}^T(\theta) \tag{2.53}$$

Formulae for coordinate changes are summarized in Table 2.1.

Exercise 2.10 Show that $\mathbf{R}(\theta)$ is a unitary matrix, i.e., $\mathbf{R}(\theta)^{-1} = \mathbf{R}(\theta)^T$.

Exercise 2.11 Using the matrix formulation, determine $[v_d^s, v_q^s]^T$ and $[v_d^e, v_q^e]^T$ corresponding to $[v_a, v_b, v_c]^T = \left[\cos\theta, \cos(\theta - \frac{2}{3}\pi), \cos(\theta - \frac{4}{3}\pi)\right]^T$.

Solution.

$$\begin{bmatrix} v_d^s \\ v_q^s \end{bmatrix} = \frac{2}{3} \begin{bmatrix} 1 & -\frac{1}{2} & -\frac{1}{2} \\ 0 & \frac{\sqrt{3}}{2} & -\frac{\sqrt{3}}{2} \end{bmatrix} \begin{bmatrix} \cos\theta \\ \cos(\theta - \frac{2\pi}{3}) \\ \cos(\theta - \frac{4\pi}{3}) \end{bmatrix}$$

$$= \frac{2}{3} \begin{bmatrix} \cos\theta - \frac{1}{2}\cos(\theta - \frac{2\pi}{3}) - \frac{1}{2}\cos(\theta - \frac{4\pi}{3}) \\ \frac{\sqrt{3}}{2}\cos(\theta - \frac{2\pi}{3}) - \frac{\sqrt{3}}{2}\cos(\theta - \frac{4\pi}{3}) \end{bmatrix} = \begin{bmatrix} \cos\theta \\ \sin\theta \end{bmatrix}.$$

$$\begin{bmatrix} v_d^e \\ v_q^e \end{bmatrix} = \begin{bmatrix} \cos\theta & \sin\theta \\ -\sin\theta & \cos\theta \end{bmatrix} \begin{bmatrix} \cos\theta \\ \sin\theta \end{bmatrix} = \begin{bmatrix} 1 \\ 0 \end{bmatrix}. \qquad \blacksquare$$

2.3.4 Power Relations

In the three phase system, power is defined as the sum of individual phase powers. The voltage and current vectors are denoted by $\mathbf{v}_{abc} = [v_a, v_b, v_c]^T$ and $\mathbf{i}_{abc} = [i_a, i_b, i_c]^T$. Let $\mathbf{v}_{dq0}^e \equiv \mathbf{T}(\theta)[v_a, v_b, v_c]^T$ and $\mathbf{i}_{dq0}^e \equiv \mathbf{T}(\theta)[i_a, i_b, i_c]^T$. Then, power is equal to

$$\begin{aligned} P &= \mathbf{v}_{abc}^T \mathbf{i}_{abc} = v_a i_a + v_b i_b + v_c i_c \\ &= (\mathbf{T}^{-1}\mathbf{v}_{dq0})^T \mathbf{T}^{-1}\mathbf{i}_{dq0} \\ &= \mathbf{v}_{dq0}^T (\mathbf{T}^{-1})^T \mathbf{T}^{-1}\mathbf{i}_{dq0} \\ &= \mathbf{v}_{dq0}^T (\frac{3}{2}\mathbf{T}^T)^T \mathbf{T}^{-1}\mathbf{i}_{dq0} \\[2mm] &= \frac{3}{2}\mathbf{v}_{dq0}^T \mathbf{i}_{dq0} \\ &= \frac{3}{2}(v_d i_d + v_q i_q + v_0 i_0) \\ &= \frac{3}{2}(v_d i_d + v_q i_q). \end{aligned} \qquad (2.54)$$

In the above calculation, the relation, $\mathbf{T}^{-1} = \frac{3}{2}\mathbf{T}^T$ was utilized. The last equality follows when either $v_0 = 0$ or $i_0 = 0$. Note that the factor, $3/2$ in the transformation map (2.50) also appears in the power equation.

Power Invariant Transformation Matrix

A different power-invariant, right-handed, uniformly-scaled Clarke transformation matrix is defined as

$$\tilde{\mathbf{T}}(\theta) = \sqrt{\frac{2}{3}} \begin{bmatrix} \cos\theta & \cos(\theta - \frac{2\pi}{3}) & \cos(\theta + \frac{2\pi}{3}) \\ -\sin\theta & -\sin(\theta - \frac{2\pi}{3}) & -\sin(\theta + \frac{2\pi}{3}) \\ \frac{1}{\sqrt{2}} & \frac{1}{\sqrt{2}} & \frac{1}{\sqrt{2}} \end{bmatrix}. \qquad (2.55)$$

Table 2.1: Formulae for coordinate changes.

Forward ($abc \Rightarrow dq$)	Inverse ($dq \Rightarrow abc$)
Complex variables for stationary frame	
$\dfrac{2}{3}\left[f_a(t) + e^{j\frac{2\pi}{3}} f_b(t) + e^{-j\frac{2\pi}{3}} f_c(t) \right]$	$\begin{bmatrix} f_a \\ f_b \\ f_c \end{bmatrix} = \mathcal{R}e\left(\begin{bmatrix} \mathbf{f^s} \\ e^{j\frac{4\pi}{3}}\mathbf{f^s} \\ e^{j\frac{2\pi}{3}}\mathbf{f^s} \end{bmatrix} \right)$
Complex variables for rotating frame	
$e^{-j\theta}\dfrac{2}{3}\left[f_a(t) + e^{j\frac{2\pi}{3}} f_b(t) + e^{-j\frac{2\pi}{3}} f_c(t) \right]$	$\begin{bmatrix} f_a \\ f_b \\ f_c \end{bmatrix} = \mathcal{R}e\left(\begin{bmatrix} e^{j\theta}\mathbf{f^s} \\ e^{j(\theta+\frac{4\pi}{3})}\mathbf{f^s} \\ e^{j(\theta+\frac{2\pi}{3})}\mathbf{f^s} \end{bmatrix} \right)$
Matrices for stationary frame	
$\dfrac{2}{3}\begin{bmatrix} 1 & -\frac{1}{2} & -\frac{1}{2} \\ 0 & \frac{\sqrt{3}}{2} & -\frac{\sqrt{3}}{2} \\ \frac{1}{\sqrt{2}} & \frac{1}{\sqrt{2}} & \frac{1}{\sqrt{2}} \end{bmatrix}$	$\begin{bmatrix} 1 & 0 & \frac{1}{\sqrt{2}} \\ -\frac{1}{2} & \frac{\sqrt{3}}{2} & \frac{1}{\sqrt{2}} \\ -\frac{1}{2} & -\frac{\sqrt{3}}{2} & \frac{1}{\sqrt{2}} \end{bmatrix}$
Matrices for rotating frame	
$\dfrac{2}{3}\begin{bmatrix} \cos\theta & \cos(\theta-\frac{2\pi}{3}) & \cos(\theta+\frac{2\pi}{3}) \\ -\sin\theta & -\sin(\theta-\frac{2\pi}{3}) & -\sin(\theta+\frac{2\pi}{3}) \\ \frac{1}{\sqrt{2}} & \frac{1}{\sqrt{2}} & \frac{1}{\sqrt{2}} \end{bmatrix}$	$\begin{bmatrix} \cos\theta & -\sin\theta & \frac{1}{\sqrt{2}} \\ \cos(\theta-\frac{2\pi}{3}) & -\sin(\theta-\frac{2\pi}{3}) & \frac{1}{\sqrt{2}} \\ \cos(\theta+\frac{2\pi}{3}) & -\sin(\theta+\frac{2\pi}{3}) & \frac{1}{\sqrt{2}} \end{bmatrix}$

Then its inverse is given as

$$\tilde{\mathbf{T}}^{-1}(\theta) = \tilde{\mathbf{T}}^{T}(\theta) = \sqrt{\frac{2}{3}}\begin{bmatrix} \cos\theta & -\sin\theta & \frac{1}{\sqrt{2}} \\ \cos(\theta-\frac{2\pi}{3}) & -\sin(\theta-\frac{2\pi}{3}) & \frac{1}{\sqrt{2}} \\ \cos(\theta+\frac{2\pi}{3}) & -\sin(\theta+\frac{2\pi}{3}) & \frac{1}{\sqrt{2}} \end{bmatrix}. \tag{2.56}$$

Note that $\tilde{\mathbf{T}}(\theta)$ is unitary. Now let Let $\tilde{\mathbf{v}}^e_{dq0} \equiv \tilde{\mathbf{T}}(\theta)[v_a, v_b, v_c]^T$ and $\tilde{\mathbf{i}}^e_{dq0} \equiv \tilde{\mathbf{T}}(\theta)$ $[i_a, i_b, i_c]^T$. However, with this definition, we obtain a power invariant relation

$$P = \mathbf{v}^T_{abc}\mathbf{i}_{abc} = \tilde{\mathbf{v}}^T_{dq}\tilde{\mathbf{T}}^T(\theta)\tilde{\mathbf{T}}(\theta)\tilde{\mathbf{i}}_{dq} = \tilde{v}_d\tilde{i}_d + \tilde{v}_q\tilde{i}_q. \tag{2.57}$$

That is, no scaling factor appears in the power relation. However, the magnitude of a vector is not preserved with $\tilde{\mathbf{T}}(\theta)$ for a sinusoidal wave. Specifically,

$\max\{i_a\} \neq \max\{\tilde{i}_d^s\}$ and $\max\{i_a\} \neq \max\{\tilde{i}_q^s\}$ with the transformation, $\tilde{\mathbf{T}}(\theta)$. Instead, $\max\{i_a\} = \sqrt{\frac{2}{3}}\max\{\tilde{i}_d^s\}$. But along with the definition (2.50), we have $\max\{i_a\} = \max\{i_d^s\}$ and $\max\{i_a\} = \max\{i_q^s\}$ which is very useful for checking the correctness of the transformation in experiments.

2.3.5 Transformation of Impedance Matrices

In the transformed coordinates, inductance appears differently. The change is computed via matrix manipulation. Later, the same result is derived using the complex variables.

A. Resistance

Consider a simple three phase resistor load: $\mathbf{v}_{abc} = r\mathbf{i}_{abc}$, where $\mathbf{i}_{abc} = [i_a, i_b, i_c]^T$ and $\mathbf{v}_{abc} = [v_a, v_b, v_c]^T$. Let

$$\begin{aligned} \mathbf{i}_{dq0}^e &= \mathbf{T}(\theta)\mathbf{i}_{abc} \\ \mathbf{v}_{dq0}^e &= \mathbf{T}(\theta)\mathbf{v}_{abc}, \end{aligned}$$

where $\mathbf{i}_{dq0}^e = [i_d, i_q, i_0]^T$ and $\mathbf{v}_{dq0}^e = [v_d, v_q, v_0]^T$. Then,

$$\mathbf{v}_{dq0}^e = \mathbf{T}(\theta)\mathbf{v}_{abc} = r\mathbf{T}(\theta)\mathbf{i}_{abc} = r\mathbf{i}_{dq0}^e .$$

The resistor, being a scalar, is invariant under the coordinate change.

B. Flux Linkage and Inductance

Consider a three phase inductor shown in Fig. 2.17. It has three legs of equal reluctance, and three phase coils are wound around them. Note from the core structure that a major part, $L_{ms}i_a$ of a-phase flux flows through the legs of b and c phases. At this time, some leakage takes place around the a-phase coil, and its amount is denoted by $L_{ls}i_a$. Therefore, a-phase flux produced by a-phase current is given by $\lambda_a = (L_{ms} + L_{ls})i_a$, where L_{ms} and L_{ls} denote the mutual and leakage inductances, respectively. On the other hand, b-phase current also contributes to the a-phase flux generation in such a way that $\lambda_a = -\frac{1}{2}L_{ms}i_b$. Similarly, c-phase contribution is $\lambda_a = -\frac{1}{2}L_{ms}i_c$. Considering all these contributions, we have the same inductance matrix and flux relation $\boldsymbol{\lambda}_{abc} = \mathbf{L}_{abc}\mathbf{i}_{abc}$. The inductance model is the same as (2.16) of the uniform air gap machine. However, the leakage inductance of a transformer is much smaller than that of a rotating machine, since the transformers do not have the air gap.

Let $\boldsymbol{\lambda}_{dq0}^e = [\lambda_d, \lambda_q, \lambda_0]^T$. Then,

$$\boldsymbol{\lambda}_{dq0}^e = \mathbf{T}(\theta)\boldsymbol{\lambda}_{abc} = \mathbf{T}(\theta)\mathbf{L}_{abc}\mathbf{i}_{abc} = \mathbf{T}(\theta)\mathbf{L}_{abc}\mathbf{T}^{-1}(\theta)\mathbf{i}_{dq0}^e = \mathbf{L}_{dq}\mathbf{i}_{dq0}^e. \qquad (2.58)$$

Thus,

$$\mathbf{L}_{dq} = \mathbf{T}(\theta)\mathbf{L}_{abc}\mathbf{T}^{-1}(\theta). \qquad (2.59)$$

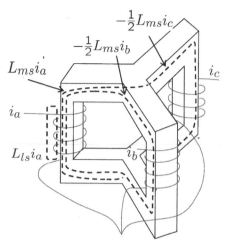

Figure 2.17: Three phase inductor.

When $\mathbf{L}_{abc}$ is a constant matrix, it is sufficient to use $\mathbf{T}(0)$, instead of $\mathbf{T}(\theta)$. It follows from (2.59) that

$$\mathbf{L}_{dq} = \mathbf{T}(0)\mathbf{L}_{abc}\mathbf{T}^{-1}(0)$$

$$= \frac{2}{3}\begin{bmatrix} 1 & -\frac{1}{2} & -\frac{1}{2} \\ 0 & \frac{\sqrt{3}}{2} & -\frac{\sqrt{3}}{2} \\ \frac{1}{\sqrt{2}} & \frac{1}{\sqrt{2}} & \frac{1}{\sqrt{2}} \end{bmatrix}\begin{bmatrix} L_{ls} + L_{ms} & -\frac{1}{2}L_{ms} & -\frac{1}{2}L_{ms} \\ -\frac{1}{2}L_{ms} & L_{ls} + L_{ms} & -\frac{1}{2}L_{ms} \\ -\frac{1}{2}L_{ms} & -\frac{1}{2}L_{ms} & L_{ls} + L_{ms} \end{bmatrix}\begin{bmatrix} 1 & 0 & \frac{1}{\sqrt{2}} \\ -\frac{1}{2} & \frac{\sqrt{3}}{2} & \frac{1}{\sqrt{2}} \\ -\frac{1}{2} & -\frac{\sqrt{3}}{2} & \frac{1}{\sqrt{2}} \end{bmatrix}$$

$$= \begin{bmatrix} L_{ls} + \frac{3}{2}L_{ms} & 0 & 0 \\ 0 & L_{ls} + \frac{3}{2}L_{ms} & 0 \\ 0 & 0 & L_{ls} \end{bmatrix}. \tag{2.60}$$

Note that (2.59) is a similarity transformation and it yields a diagonal matrix [14].

Derivation Using Complex Variables

In this part, the same result is derived using the complex variables. Applying the definition of flux, (2.16) and (2.17) into the mapping rule (2.39), we obtain

$$\begin{aligned}
\lambda_{dq}^s &= \frac{2}{3}\left[\lambda_a(t) + e^{j\frac{2\pi}{3}}\lambda_b(t) + e^{-j\frac{2\pi}{3}}\lambda_c(t)\right] \\
&= \frac{2}{3}\Big[(L_{ms} + L_{ls})i_a - \frac{1}{2}L_{ms}i_b - \frac{1}{2}L_{ms}i_c \\
&\quad + e^{j\frac{2\pi}{3}}\left\{-\frac{1}{2}L_{ms}i_a + (L_{ms} + L_{ls})i_b - \frac{1}{2}L_{ms}i_c\right\} \\
&\quad + e^{-j\frac{2\pi}{3}}\left\{-\frac{1}{2}L_{ms}i_a - \frac{1}{2}L_{ms}i_b + (L_{ms} + L_{ls})i_c\right\}\Big] \\
&= \frac{2}{3}\Big[(\frac{3}{2}L_{ms} + L_{ls} - \frac{1}{2}L_{ms})i_a - \frac{1}{2}L_{ms}i_b - \frac{1}{2}L_{ms}i_c \\
&\quad + e^{j\frac{2\pi}{3}}\left\{-\frac{1}{2}L_{ms}i_a + (\frac{3}{2}L_{ms} + L_{ls} - \frac{1}{2}L_{ms})i_b - \frac{1}{2}L_{ms}i_c\right\}
\end{aligned}$$

$$+e^{-j\frac{2\pi}{3}}\Big\{-\frac{1}{2}L_{ms}i_a-\frac{1}{2}L_{ms}i_b+\Big(\frac{3}{2}L_{ms}+L_{ls}-\frac{1}{2}L_{ms}\Big)i_c\Big\}\Big]$$

$$=\frac{2}{3}\Big[\Big(\frac{3}{2}L_{ms}+L_{ls}\Big)(i_a+e^{j\frac{2\pi}{3}}i_b+e^{-j\frac{2\pi}{3}}i_c)-\frac{1}{2}L_{ms}i_a-\frac{1}{2}L_{ms}i_b-\frac{1}{2}L_{ms}i_c$$

$$+e^{j\frac{2\pi}{3}}\Big\{-\frac{1}{2}L_{ms}i_a-\frac{1}{2}L_{ms}i_b-\frac{1}{2}L_{ms}i_c\Big\}$$

$$+e^{-j\frac{2\pi}{3}}\Big\{-\frac{1}{2}L_{ms}i_a-\frac{1}{2}L_{ms}i_b-\frac{1}{2}L_{ms}i_c\Big\}\Big]$$

$$=\Big(\frac{3}{2}L_{ms}+L_{ls}\Big)\frac{2}{3}(i_a+e^{j\frac{2\pi}{3}}i_b+e^{-j\frac{2\pi}{3}}i_c)=L_s i_{dq}^s, \tag{2.61}$$

where $L_s=\frac{3}{2}L_{ms}+L_{ls}$ and $i_{dq}^s=i_d^s+ji_q^s$. Note that (2.61) is identical with the 2×2 submatrix of (2.60).

C. Voltage Equation

Fig. 2.18 shows a single phase equivalent circuit of a 3-phase LCR circuit. Note that $\mathbf{v}_s=(v_{as},v_{bs},v_{cs})^T$ is a 3-phase voltage source, and $\mathbf{i}_s=(i_{as},i_{bs},i_{cs})^T$, $\mathbf{i}_f=(i_{af},i_{bf},i_{cf})^T$, and $\mathbf{i}_l=(i_{al},i_{bl},i_{cl})^T$ represent 3-phase currents flowing through inductor, capacitor, and resistor, respectively. The voltage equations are given in

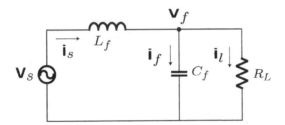

Figure 2.18: LCR circuit with a 3-phase AC source.

the abc-frame such that

$$\frac{d\mathbf{i}_s}{dt}=\frac{1}{L_f}(\mathbf{v}_s-\mathbf{v}_f), \tag{2.62}$$

$$\frac{d\mathbf{v}_f}{dt}=\frac{1}{C_f}(\mathbf{i}_s-\mathbf{i}_l), \tag{2.63}$$

$$\mathbf{i}_l=\frac{\mathbf{v}_f}{R_L}. \tag{2.64}$$

Then, voltage equations (2.62)–(2.64) are transformed into the stationary dq-frame such that

$$\frac{d\mathbf{i}_{dqs}^s}{dt}=\frac{1}{L_f}(\mathbf{v}_{dqs}^s-\mathbf{v}_{dqf}^s), \tag{2.65}$$

$$\frac{d\mathbf{v}_{dqf}^s}{dt}=\frac{1}{C_f}(\mathbf{i}_{dqs}^s-\mathbf{i}_{dql}^s), \tag{2.66}$$

$$\mathbf{i}_{dql}^s=\frac{\mathbf{v}_{dqf}^s}{R_L}. \tag{2.67}$$

Note that

$$e^{-j\omega t}\frac{d\mathbf{i}^s_{dqs}}{dt} = e^{-j\omega t}\frac{d}{dt}\left(e^{j\omega t}e^{-j\omega t}\mathbf{i}^s_{dqs}\right) = e^{-j\omega t}\frac{d}{dt}\left(e^{j\omega t}\mathbf{i}^e_{dqs}\right)$$

$$= e^{-j\omega t}\left(j\omega e^{j\omega t}\mathbf{i}^e_{dqs} + e^{j\omega t}\frac{d\mathbf{i}^e_{dqs}}{dt}\right) = j\omega\mathbf{i}^e_{dqs} + \frac{d\mathbf{i}^e_{dqs}}{dt}.$$

Therefore, the voltage equations are transformed into the synchronous reference frame such that

$$\frac{d\mathbf{i}^e_{dqs}}{dt} + j\omega\mathbf{i}^e_{dqs} = \frac{1}{L_f}(\mathbf{v}^e_{dqs} - \mathbf{v}^e_{dqf}), \tag{2.68}$$

$$\frac{d\mathbf{v}^e_{dqf}}{dt} + j\omega\mathbf{v}^e_{dqf} = \frac{1}{C_f}(\mathbf{i}^e_{dqs} - \mathbf{i}^e_{dql}), \tag{2.69}$$

$$\mathbf{i}^e_{dql} = \frac{\mathbf{v}^e_{dqf}}{R_L}. \tag{2.70}$$

Fig. 2.19 shows a block diagram based on (2.68)–(2.70).

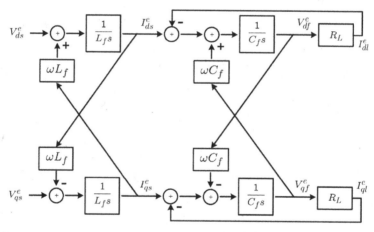

Figure 2.19: Block diagram of an LCR voltage equation in the synchronous reference frame.

2.4   PI Controller in Synchronous Frame

To implement the field oriented control, coordinates should be changed two times: First, the current vector should be changed from the abc to the synchronous dq frame. Then the command voltage vector is computed based on the PI action. Second, the voltage vector should be transformed again into the abc frame. Fig. 2.20 shows the current controller involving coordinate transformations. The question here is what the controller looks like if it is seen from the stationary frame. Specifically, the PI controller in the synchronous frame is interpreted in the stationary frame.

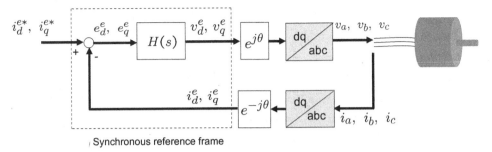

Figure 2.20: Current controller implemented in the synchronous reference.

This part is mainly based on the works of Zmood and Holmes [4],[5]. Consider the transformation of stationary current error e_d^s, e_q^s into the error e_d^e, e_q^e in the synchronous frame.

$$\begin{bmatrix} e_d^e \\ e_q^e \end{bmatrix} = \begin{bmatrix} \cos(\omega t) & \sin(\omega t) \\ -\sin(\omega t) & \cos(\omega t) \end{bmatrix} \begin{bmatrix} e_d^s \\ e_q^s \end{bmatrix} = \begin{bmatrix} \cos(\omega t)e_d^s + \sin(\omega t)e_q^s \\ -\sin(\omega t)e_d^s + \cos(\omega t)e_q^s \end{bmatrix}.$$

The signal passes through a controller whose impulse response is $h(t)$. Then the controller outputs appear as the convolution, $h*$. Further, they are transformed back into the stationary frame:

$$\begin{bmatrix} v_d^s \\ v_q^s \end{bmatrix} = \begin{bmatrix} \cos(\omega t) & -\sin(\omega t) \\ \sin(\omega t) & \cos(\omega t) \end{bmatrix} \begin{bmatrix} h*(\cos(\omega t)e_d^s) + h*(\sin(\omega t)e_q^s) \\ -h*(\sin(\omega t)e_d^s) + h*(\cos(\omega t)e_q^s) \end{bmatrix}$$

$$= \begin{bmatrix} \cos(\omega t)[h*(\cos(\omega t)e_d^s)] + \cos(\omega t)[h*(\sin(\omega t)e_q^s)] \\ +\sin(\omega t)[h*(\sin(\omega t)e_d^s)] - \sin(\omega t)[h*(\cos(\omega t)e_q^s)] \\ \\ \sin(\omega t)[h*(\cos(\omega t)e_d^s)] + \sin(\omega t)[h*(\sin(\omega t)e_q^s)] \\ -\cos(\omega t)[h*(\sin(\omega t)e_d^s)] + \cos(\omega t)[h*(\cos(\omega t)e_q^s)] \end{bmatrix}. \qquad (2.71)$$

For convenience of notation, we let

$$\begin{aligned} f_1(t) &= h*(\cos(\omega t)e_d^s), \\ f_2(t) &= h*(\sin(\omega t)e_d^s). \end{aligned}$$

Utilizing the fact that $\cos(\omega t) = (e^{j\omega t} + e^{-j\omega t})/2$ and $\sin(\omega t) = (e^{j\omega t} - e^{-j\omega t})/2j$, and the frequency shifting property of Laplace transformation, we obtain

$$\begin{aligned} F_1(s) &= \mathcal{L}\{h*(\cos(\omega t)e_d^s)\}\} = H(s) \cdot \frac{1}{2}[E_d^s(s+j\omega) + E_d^s(s-j\omega)] \\ F_2(s) &= \mathcal{L}\{h*(\sin(\omega t)e_d^s)\}\} = H(s) \cdot \frac{1}{2j}[E_d^s(s+j\omega) - E_d^s(s-j\omega)], \end{aligned}$$

where $H(s) = \mathcal{L}(h)$, $E_d^s(s) = \mathcal{L}(e_d^s)$, and $E_q^s(s) = \mathcal{L}(e_q^s)$. Therefore,

$$
\begin{aligned}
\mathcal{L}\left\{\cos(\omega t)[h * (\cos(\omega t)e_d^s)]\right\} &= \frac{1}{2}[F_1(s+j\omega) + F_1(s-j\omega)] \\
&= \frac{1}{4}[H(s+j\omega)E_d^s(s+j2\omega) + H(s+j\omega)E_d^s(s) \\
&\quad + H(s-j\omega)E_d^s(s) + H(s-j\omega)E_d^s(s-j2\omega)] \\
\mathcal{L}\left\{\sin(\omega t)[h * (\sin(\omega t)e_d^s)]\right\} &= \frac{1}{2j}[F_2(s+j\omega) - F_2(s-j\omega)] \\
&= -\frac{1}{4}[H(s+j\omega)E_d^s(s+j2\omega) - H(s+j\omega)E_d^s(s) \\
&\quad - H(s-j\omega)E_d^s(s) + H(s-j\omega)E_d^s(s-j2\omega)].
\end{aligned}
$$

Then the $d-$axis part of the first component of (2.71) is equal to

$$
\begin{aligned}
\mathcal{L}\left\{\cos(\omega t)[h * (\cos(\omega t)e_d^s)]\right\} &+ \mathcal{L}\left\{\sin(\omega t)[h * (\sin(\omega t)e_d^s)]\right\} \\
&= \frac{1}{2}[H(s+j\omega) + H(s-j\omega)]E_d^s(s).
\end{aligned}
$$

Similarly, it follows for e_q that

$$
\begin{aligned}
\mathcal{L}\left\{\cos(\omega t)[h * (\sin(\omega t)e_q^s)]\right\} &- \mathcal{L}\left\{\sin(\omega t)[h * (\cos(\omega t)e_q^s)]\right\} \\
&= -\frac{1}{2j}[H(s+j\omega) - H(s-j\omega)]E_q^s(s).
\end{aligned}
$$

For the I-controller, $H(s) = \frac{K_i}{s}$, therefore, it follows that

$$
\begin{aligned}
V_d^s(s) &= \mathcal{L}\{v_d^s\} \\
&= \frac{K_i}{2}[H(s+j\omega) + H(s-j\omega)]E_d^s(s) - \frac{K_i}{2j}[H(s+j\omega) - H(s-j\omega)]E_q^s(s) \\
&= \frac{K_i}{2}\left[\frac{1}{s+j\omega} + \frac{1}{s-j\omega}\right]E_d^s(s) - \frac{K_i}{2j}\left[\frac{1}{s+j\omega} - \frac{1}{s-j\omega}\right]E_q^s(s) \\
&= \frac{K_i s}{s^2 + \omega^2}E_d^s(s) + \frac{K_i \omega}{s^2 + \omega^2}E_q^s(s).
\end{aligned}
$$

The I-controller in the synchronous frame appears as a resonant controller with the rotating frequency of ω. That is, the I-controller in the synchronous frame has infinite gain at DC, whereas the corresponding controller in the stationary frame has infinite gain at the rotating frequency, ω. For the PI-controller $H(s) = K_p + K_i/s$, it follows that

$$
\begin{bmatrix} V_d^s(s) \\ V_q^s(s) \end{bmatrix} = \begin{bmatrix} K_p + \dfrac{K_i s}{s^2 + \omega^2} & \dfrac{K_i \omega}{s^2 + \omega^2} \\[4mm] -\dfrac{K_i \omega}{s^2 + \omega^2} & K_p + \dfrac{K_i s}{s^2 + \omega^2} \end{bmatrix} \begin{bmatrix} E_d^s(s) \\ E_q^s(s) \end{bmatrix}. \tag{2.72}
$$

The resonant controller discriminates the error signal depending on the frequency: It applies infinite gain to the spectral component of ω. That is, the integral controller

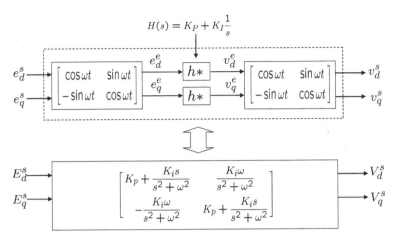

Figure 2.21: Transformation of a PI regulator in the synchronous frame into the stationary frame. Block (a) is equivalent to block (b).

is highly sensitive to the ones that would appear as constants in the synchronous frame. Fig. 2.21 shows an equivalence. The pioneering work for this transformation is shown in [5].

Exercise 2.12
Show that

$$V_q^s(s) = -\frac{K_i\omega}{s^2 + \omega^2}E_d^s(s).$$

References

[1] P. C. Krause, O. Wasynczuk, and S. D. Sudhoff, *Analysis of Electric Machinery*, IEEE Press, 1995.

[2] D. W. Novotny and T. A. Lipo, *Vector Control and Dynamics of AC Drives*, Clarendon Press, 1996.

[3] C.T. Chen, *Linear System Theory and Design*, Oxford Press, 1999.

[4] D. N. Zmood, D. G. Holmes, and G. H. Bode, Frequency-domain analysis of three-phase linear current regulators, *IEEE Trans. Ind. Appl.*, vol. 37, no. 2, pp. 601−609. Mar./Apr. 2001.

[5] D.N. Zmood and D. G. Holmes, Stationary frame current regulation of PWM inverters with zero steady-state error, *IEEE Trans. on Power Elec.*, vol. 18, no. 3, pp. 814−822, May 2003.

Problems

2.1   Consider an inverter with two arms and a three phase motor. The necessary condition for the current balance among the three phase windings is

$$\frac{v_{an} - v_{sn}}{Z} + \frac{v_{bn} - v_{sn}}{Z} - \frac{v_{sn}}{Z} = 0,$$

where Z denotes the impedance of the phase windings. Assume that c phase current is equal to $-\frac{v_{sn}}{Z} = \cos \omega t$. Find v_{an} and v_{bn} such that a rotating field is made.

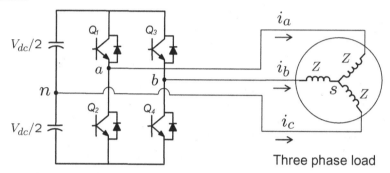

2.2   Consider a one period of a double layered, concentrated winding motor shown below. Assume that both side edges are identical.
a) Sketch the MMF when $i_u = I$ and $i_v = i_w = -I/2$.
b) Express the MMF in Fourier series and plot the magnitude of harmonic components up to $n = 13$.

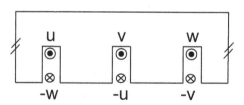

2.3   Consider the following rectangular and step waves. Both of them has the same period, 0.1 sec. Obtain sampled data $x[n]$ based on sampling interval, $\Delta T = 0.001$ sec. Using (2.37) and (2.38), calculate Fourier series coefficients, c_k for $k = 0, \cdots, 20$. Plot the spectra of the two waves up to 20th. Compare the spectra, and discuss the effects of the zero interval provided at each transition.

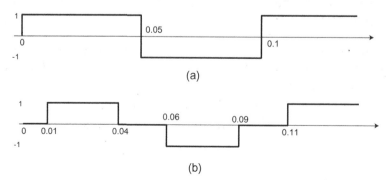

2.4   Let

$$
\begin{bmatrix} v_a(t) \\ v_b(t) \\ v_c(t) \end{bmatrix} = \begin{bmatrix} 100\sin(120\pi t) + 15\sin(360\pi t) \\ 100\sin(120\pi t - 2\pi/3) + 15\sin(360\pi t) \\ 100\sin(120\pi t - 4\pi/3) + 15\sin(360\pi t) \end{bmatrix}.
$$

Using Excel spreadsheet, a) plot $v_a(t)$, $v_b(t)$, and $v_c(t)$ for $0 \le t \le 0.05$;
b) plot $v_d^s(t)$ and $v_q^s(t)$ for $0 \le t \le 0.05$;
c) Compare $\max\{v_a(t)\}$ with $\max\{v_d(t)\}$, and discuss the benefit of adding a third order harmonic component.

2.5   Consider a three phase current:

$$
\begin{bmatrix} i_a \\ i_b \\ i_c \end{bmatrix} = \begin{bmatrix} \cos\omega t - \frac{1}{3}\cos 3\omega t + \frac{1}{5}\cos 5\omega t - \frac{1}{7}\cos 7\omega t \\ \cos(\omega t - \frac{2\pi}{3}) - \frac{1}{3}\cos 3(\omega t - \frac{2\pi}{3}) + \frac{1}{5}\cos 5(\omega t - \frac{2\pi}{3}) - \frac{1}{7}\cos 7(\omega t - \frac{2\pi}{3}) \\ \cos(\omega t - \frac{4\pi}{3}) - \frac{1}{3}\cos 3(\omega t - \frac{4\pi}{3}) + \frac{1}{5}\cos 5(\omega t - \frac{4\pi}{3}) - \frac{1}{7}\cos 7(\omega t - \frac{4\pi}{3}) \end{bmatrix}
$$

a) Transform $[i_a(t),\ i_b(t),\ i_c(t)]^T$ into the stationary dq-frame, and plot the locus of $[i_d^s(t),\ i_q^s(t)]^T$ in the complex plane;
b) Transform $[i_q^s(t),\ i_d^s(t)]^T$ into the synchronous reference frame, and plot $[i_d^e(t),\ i_q^e(t)]^T$ versus t.

2.6   Assume $\mathbf{U} \in \mathbb{R}^{n \times n}$ is a unitary matrix. Show that $det(\mathbf{U}) = \pm 1$, where $det(\mathbf{U})$ denotes the determinant of $\mathbf{U}$.

2.7   Suppose that θ is defined differently as the angle between the q and a axes as shown below: Based on this definition θ, find the transformation map $\mathbf{T}$.

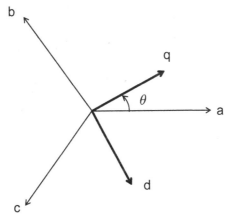

2.8   Consider a five phase system in which five MMFs are

$$
\text{MMF}_a = \frac{Ni_a}{2}\cos\theta, \qquad\qquad \text{MMF}_b = \frac{Ni_b}{2}\cos(\theta - \frac{2\pi}{5}),
$$

$$
\text{MMF}_c = \frac{Ni_c}{2}\cos(\theta - \frac{4\pi}{5}), \qquad \text{MMF}_d = \frac{Ni_d}{2}\cos(\theta + \frac{4\pi}{5}),
$$

$$
\text{MMF}_e = \frac{Ni_e}{2}\cos(\theta + \frac{2\pi}{5}).
$$

Obtain a current set which makes the MMF sum rotate synchronously with the electrical speed, ω, and the MMF sum.

2.9 Consider the following d-axis voltage pulses in the synchronous reference frame. Transform $[i_d^e(t),\ i_d^e(t)]^T$ back into the abc-frame.

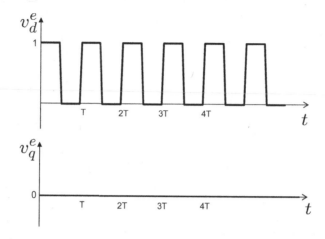

2.10 Derive (2.60).

2.11 Using the complex variable, transform the following rotor PM flux linkage into the stationary $d - q$ frame:

$$\begin{bmatrix} \lambda_a \\ \lambda_b \\ \lambda_c \end{bmatrix} = \psi_m \begin{bmatrix} \sin\theta \\ \sin(\theta - 2\pi/3) \\ \sin(\theta + 2\pi/3) \end{bmatrix}.$$

2.12 (Continued from Exercise 2.12 Draw MATLAB Simulink block diagram based on Fig. 2.19. Apply $v_{ds}^e = 50u_s(t) + 50u_s(t - 0.005)$, and $v_{qs}^e = 0$, where $u_s(t)$ is the unit step function. Obtain current response $\mathbf{i}_l = [i_{al},\ i_{bl},\ i_{cl}]^T$ for $0 \le t \le 0.01$ sec in the stationary frame.

Chapter 3

Induction Motor Basics

An induction motor (IM) can be understood as a polyphase transformer with short-ened and freely rotating secondary winding. It generates torque based on the elec-tromagnetic induction. The amount of induction is proportional to the speed slip between the stator field and the rotor. An IM can therefore work without an elec-trical connection to the rotor. Thus, brushes and commutator are not necessary. Therefore, IMs require minimum maintenance and are robust against mechanical shock and vibration. Furthermore, IMs are made of inexpensive ordinary materi-als like iron, copper, and aluminium. In addition, IMs can be driven by the grid lines. Therefore, they are widely used in industry and domestic appliances; blowers, pumps, cranes, conveyer belts, elevators, refrigerators, traction drives, etc. In fact, 90% of industrial motors are IMs. In this chapter, basic IM operation principle is illustrated using an equivalent circuit model, and the torque speed curve is derived. Various torque profiles are studied with regards to the rotor slot shape. The Hey-land diagram and variable voltage variable frequency (VVVF) control are briefly reviewed.

3.1 IM Construction

The IMs are classified according to the rotor winding types. One type is squirrel cage IM and the other type is wound-rotor IM. The squirrel cage is named after the rotor conductor shape: Looking at rotor conductors only, two circular end rings are connected with multiple straight bars. The rotor circuit is isolated, not requiring any connection to an external circuit. The benefits include relatively simple construction, cost efficiency, and robustness. Almost 95% of the induction motors are squirrel cage. On the other hand, the wound-rotor IM requires a slip ring to connect various external resistors. The external resistors are required to give high starting torque. Nowadays, use of wound-rotor IM is gradually reduced as the inverter applications are widened.

Fig. 3.1 shows states of stator core assembly: slot liner insertion, coil winding, and fitting into a housing. The stator and rotor cores are made by stacking sil-icon steel sheets which are optimized for high permeability and low conductivity.

Through the stator slots, copper coils are wound.

However, the rotor circuit is constructed by aluminium die casting or copper bar brazing with the end-ring. Fig. 3.2 shows a photo of an IM rotor. Cooling fins are often casted on the aluminium end ring.

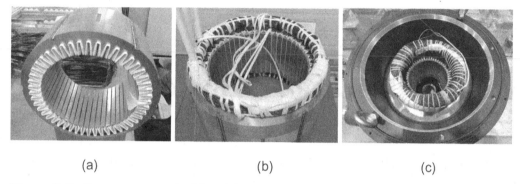

(a)   (b)   (c)

Figure 3.1: Stator core assembly: (a) slot liner insertion, (b) coil winding, and (c) fitting into a housing.

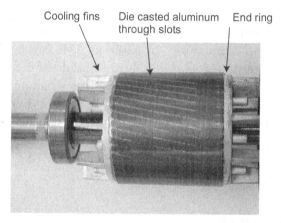

Figure 3.2: Photo of an assembled IM rotor.

Aluminium casting: Aluminium cast is most commonly used in low voltage motors. This kind of construction is relatively simple and cost effective for manufacturing. The rotor slot shape is a critical factor in determining the starting torque, breakdown torque, efficiency, etc. A greater variety of shape design can be achieved with aluminium casting. However, the slot profile must secure a solid casting without cracks or voids which are a main cause of the breakage of rotor bar. Figs. 3.3 show photos of (a) rotor section after casting and (b) die casted end ring.

Copper brazing: Copper bars are welded to copper end rings by high frequency induction brazing or argon welding. In general, copper bars are used for high power or high speed motors. During starting, excessive heat is generated in the short circuit state with a large slip. Then the rotor bars are heated unevenly between upper and lower parts. This causes a bar bending, giving a severe stress to the joint between the bar and the end ring. Compared with aluminium, copper has lower

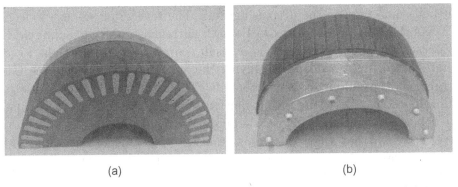

(a) (b)

Figure 3.3: Aluminium die cast: (a) rotor section after casting and (b) die casted end ring.

coefficient of expansion (23% less), higher tensile strength (300% stronger), and higher conductivity (160% higher). Obviously, high conductivity means low heat generation and high efficiency. Lower thermal expansion and high tensile strength are very desirable properties for high rotor durability. Fig. 3.4 shows a photo of a joint section between the copper bar and the end ring after brazing.

Figure 3.4: Joint section between copper bar and end ring after brazing.

3.2 IM Operation Principle

Fig. 3.5 illustrates the torque production principle of IMs [1]. The rotating stator field is modeled as a pair of rotating permanent magnets (PMs). The secondary circuit is shown as a squirrel cage. Suppose that the PMs are turning counterclockwise while the cage conductor remains fixed. We consider Loop 1 and Loop 2 over which the PM field slides. As the PM flux decreases in Loop 1, a clockwise loop current is induced to counteract the decreasing flux. In other words, the induced current adds more flux to the decreasing flux. The reverse phenomenon occurs in Loop 2. As the PM flux increases in Loop 2, current is induced counterclockwise to counteract the increasing flux. As a result, current flows from right to left in the middle bar of the two loops. Applying the left hand rule, we will see Lorentz force acting on the conductor bar to the left. Then, the bar tends to move to the left following the motion of the PM. It illustrates why a non-ferromagnetic material such as copper or aluminium follows the motion of the PMs.

Note, however, that when the cage rotates at the same speed as the PM, no torque is generated. In this synchronism, the flux linkage does not change, thus inducing no current. In summary, the torque is developed proportionally to the slip speed, which is a speed difference between the stator field and the rotor.

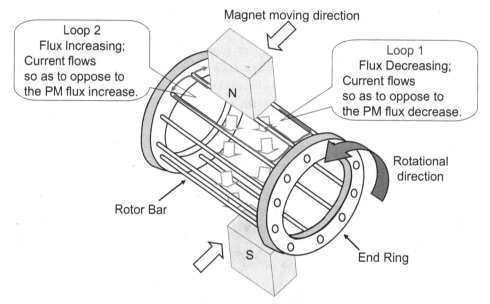

Figure 3.5: Torque production on the squirrel cage circuit.

3.2.1   IM Equivalent Circuit

An IM can be considered as a special three phase transformer. IMs have primary and secondary windings on the cylindrical core structure as shown in Fig. 3.6. But, the IMs have the following differences:

    1) IM cores have cylindrical shapes, while transformer cores E-I types.
    2) IMs have an air gap, while transformers do not.
    3) IM windings are distributed over the air gap,
        whereas transformer windings are concentric.
    4) The IM secondary winding is always shortened.
    5) The IM secondary core and winding are allowed to rotate.

Since the air gap has large reluctance, IMs have larger leakage inductance than transformers. Thus, the IM has a lower power factor than the transformer. The secondary winding is always shorted, but rotating. As a result, the current is limited as long as the slip speed is not excessive.

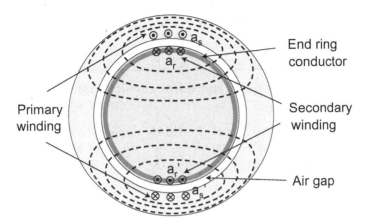

Figure 3.6: Induction motor as a rotational transformer with an air gap.

In the P-pole machine, the synchronism is done if $P/2\omega_r = \omega_e$, where ω_r and ω_e are rotor shaft and electrical speeds, respectively. It is because $P/2$ pair of coils are distributed over the periphery of a geometrical circle. Thus, one revolution of rotor faces $P/2$ electrical periods. The relative speed between the flux and mechanical speeds is referred to as slip speed, $\omega_{sl} \equiv \omega_e - \frac{P}{2}\omega_r$. It is the field speed seen from the rotor frame. The normalized speed difference is defined as the slip:

$$s = \frac{\omega_e - (P/2)\omega_r}{\omega_e}. \tag{3.1}$$

At the synchronous speed, the rotor and the field speeds are the same. As a result, no current is induced in the rotor bars, therefore no torque is developed. Torque is only available if the rotor lags behind the magnetic field. If the rotor speed is higher than $\frac{2}{P}\omega_e$, then the IM goes into regeneration (or braking) mode.

Denote by r_s, r_r, L_{ls}, L_{lr}, and L_m stator coil resistance, rotor coil resistance, stator leakage inductance, rotor leakage inductance, and magnetizing inductance, respectively. Also denote by $\mathbf{I}_s$, $\mathbf{I}_r$ and $\mathbf{V}_s$ stator and rotor phase (rms) current vectors, and stator phase voltage vector, respectively.

At zero slip, $s = 0$ no secondary current is induced. The other extreme is when the slip is equal to one, $s = 1$. It is a case of shaft lock. When locked, a large current flows similarly when the secondary winding is shorted in a normal transformer. Such a current variation can be described by setting the second resistance as r_r/s. A per-phase model of IM is shown in Fig. 3.7. It is a copy of transformer model with the shorted secondary circuit by a slip dependent resistor, $\frac{r_r}{s}$. Note that no induction at the synchronous speed, $s = 0$ is well justified by the open circuit with $\frac{r_r}{s} = \infty$. Rigorous mathematical derivation will be studied in Chapter 4.

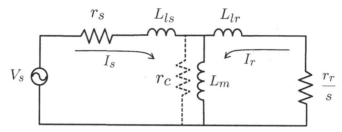

Figure 3.7: Steady state per-phase equivalent circuit of IM.

Based on the T-equivalent circuit, the IM dynamics are described by

$$(r_s + j\omega_e L_{ls})\mathbf{I}_s + j\omega_e L_m(\mathbf{I}_s + \mathbf{I}_r) \quad = \quad \mathbf{V}_s \tag{3.2}$$

$$(\frac{r_r}{s} + j\omega_e L_{lr})\mathbf{I}_r + j\omega_e L_m(\mathbf{I}_s + \mathbf{I}_r) \quad = \quad 0. \tag{3.3}$$

All the power in the secondary side is transferred via the air gap. Thus the air gap power is identified with the secondary power. It is divided into two parts, secondary copper loss and else:

$$P_{ag} = 3|\mathbf{I}'_r|^2 r_r \frac{1}{s} = \underbrace{3|\mathbf{I}'_r|^2 r_r}_{\text{Rotor copper loss}} + \underbrace{3|\mathbf{I}'_r|^2 r_r \frac{1-s}{s}}_{\text{Mechanical power}}. \tag{3.4}$$

It is interpreted that the remaining power is converted into mechanical power.

A Sankey diagram for IM power flow is shown in Fig. 3.8. The losses are divided broadly into copper and iron (core) losses: The copper loss means the conductor losses of the stator and the rotor. The core loss, often called iron loss, consists of hysteresis and eddy current losses. The eddy current loss, resulting from the time-varying magnet field, is often reflected as a parallel resistor, r_c as shown in Fig. 3.7. Stray load loss is mainly the AC loss of the coil caused by skin and proximity effects. Mechanical loss is comprised of windage and friction losses occurring due to air resistance and bearing friction. Typical motor loss components are summarized in Table 3.1 [2].

Table 3.1: Induction Motor Loss Components.

Type of Loss	Motor Horsepower		
	25 HP	50 HP	100 HP
Stator copper loss	42%	38%	28%
Rotor copper loss	21%	22%	18%
Core losses	15%	20%	13%
Friction and windage losses	7%	8%	14%
Stray load losses	15%	12%	27%

Exercise 3.1   A three phase, four pole IM draws 55 kW real power from a 3 phase 60 Hz feeder. The copper and iron losses in the stator amount to 4 kW. Neglect friction and stray losses. If the motor runs at 1740 rpm, calculate

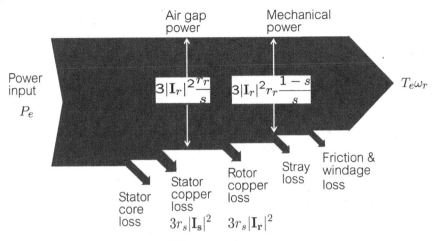

Figure 3.8: Sankey diagram for IM power flow.

    a) the airgap power.
    b) the rotor loss.
    c) mechanical power.
    d) the efficiency.

Solution.

    a) $P_{ag} = 55 - 4 = 51$ (kW).
    b) $s = (1800 - 1740)/1800 = 0.033$. Thus, $3|I_r'|^2 r_r = s P_{ag} = 1.683$ (kW).
    c) $P_m = (1 - s)P_{ag} = 49.3$ (kW).
    d) The efficiency is equal to $49.3/55 = 0.9$.

                                                                 ■

3.2.2 Torque-Speed Curve

The rotor current should be calculated to obtain torque, since the IM torque is equal to

$$T_e = \frac{P_m}{\omega_r} = 3|I_r|^2 \frac{(1-s)r_r}{s\omega_r}. \tag{3.5}$$

To simplify the calculation, the IM equivalent circuit is often modified as shown in Fig. 3.9: The magnetizing inductance L_m is shifted to the source side with the

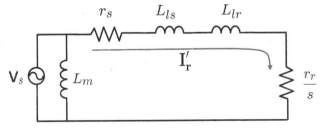

Figure 3.9: Modified equivalent circuit for IM.

assumption that the voltage drop over the stator leakage inductance and resistance is relatively small. Note that the magnetizing inductance is much larger than the leakage inductance. Specifically, the leakage reactance jw_eL_{ls} is about 5% or less of the magnetizing reactance jw_eL_m in IMs.

The rotor current, $\mathbf{I}'_r$ of the modified circuit is calculated as

$$|\mathbf{I}'_r| = \frac{|\mathbf{V}_s|}{\sqrt{\left(r_s + \frac{r_r}{s}\right)^2 + w_e^2(L_{ls} + L_{lr})^2}}. \tag{3.6}$$

From the definition of the slip, it follows that $\frac{P}{2w_e} = \frac{1-s}{w_r}$. Hence, utilizing (3.5) the electro-magnetic torque is calculated such that [3]

$$T_e = \frac{P}{2} \frac{3r_r}{sw_e} \frac{|\mathbf{V}_s|^2}{\left(r_s + \frac{r_r}{s}\right)^2 + w_e^2(L_{ls} + L_{lr})^2}. \tag{3.7}$$

For small slip, i.e., $s \approx 0$, torque equation (3.7) is approximated as

$$T_e \approx \frac{P}{2} \frac{3r_r}{w_e s} \frac{|\mathbf{V}_s|^2}{\left(\frac{r_r}{s}\right)^2} \approx \frac{3P|\mathbf{V}_s|^2}{2w_e r_r}s. \tag{3.8}$$

In a small slip region, the torque is linear in s. Hence, the torque curve appears as a straight line near $s = 0$. On the other hand, when $s \approx 1$, (3.7) is approximated such that

$$T_e = \frac{P}{2} \frac{3r_r}{sw_e} \frac{|\mathbf{V}_s|^2}{(r_s + r_r)^2 + w_e^2(L_{ls} + L_{lr})^2}. \tag{3.9}$$

It is a parabola equation in s. Fig. 3.10 (a) shows a typical torque–speed curve of IM. It can be sketched as a glued curve of (3.8) and (3.9). Fig. 3.10 (b) shows that the stator current as a function of slip. Note that the current rise is steep near $s = 0$. Normally, the rated slip is in the range from $s = 0.01 \sim 0.05$. Note also that the stator current at locked state ($s = 1$) is about several times larger than the rated current.

Exercise 3.2   A three phase, four pole IM for a 440 V (line-to-line) 60 Hz power source has the equivalent circuit shown in Fig. 3.11 (a). A modified circuit is also shown in Fig. 3.11 (b). Utilizing MATLAB, draw the torque-speed curves based on the two circuits. Check the differences near $s = 0$ and 1.

Solution. See Fig. 3.12. The torque is larger with the modified circuit, since $|\mathbf{I}'_r| > |\mathbf{I}_r|$. But the differences are especially small near $s = 0$. Hence, the modified circuit is sufficiently accurate for practical use.   ∎

Exercise 3.3   An IM operates with a load at 1750 rpm when connected to a 60 Hz, 220 V, 3 phase AC source. Assume that the load torque is constant independently of the speed. Calculate the shaft speed if the source voltage is increased to 250 V.

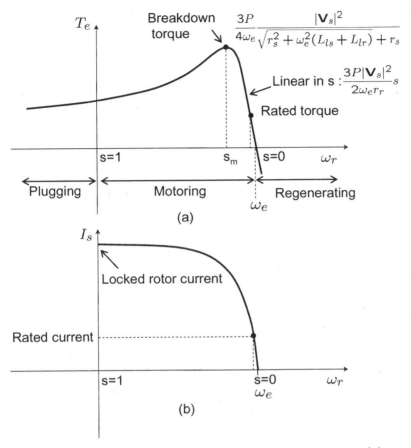

Figure 3.10: Steady state characteristics of an induction machine: (a) torque-speed curve, (b) stator current versus slip.

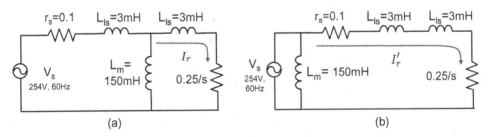

Figure 3.11: Equivalent circuits of an induction machine: (a) original equivalent circuit and (b) modified circuit (Exercise 3.2).

Solution. It follows from (3.8) that the slip is inversely proportional to the square of the voltage. Hence,

$$(\text{slip at } 250) = (\text{slip at } 220)\left(\frac{220}{250}\right)^2 = 50 \times \left(\frac{220}{250}\right)^2 = 38.7 \text{ rpm.}$$

Hence, the operation speed is $1800 - 38.7 = 1761$ rpm.

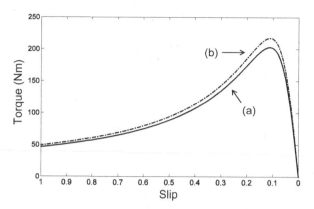

Figure 3.12: Torque–speed curves based on the equivalent circuits shown in Fig. 3.11 (Exercise 3.2).

3.2.3   Breakdown Torque

The peak value of the torque curve is called the *breakdown torque*, since it is the torque at the boundary between the stable and the unstable regions. It will be shown later that the motor cannot be operated steadily in the left region of the torque maximizing slip, s_m, while it works well in the right region. Thus, the high slip range, $s_m < s \leq 1$ is called unstable region, whereas the low slip range, $0 < s < s_m$ is called stable region. Generally, torque reduces in the high slip range, $s_m < s \leq 1$, while current increases. The reason is that the reactive impedance increases rapidly with s.

The breakdown slip and torque are available from $\frac{dT_e}{ds} = 0$:

$$s_m = \frac{r_r}{\sqrt{r_s^2 + \omega_e^2(L_{ls} + L_{lr})^2}}, \tag{3.10}$$

$$T_{max} = \frac{3P}{4\omega_e} \frac{|\mathbf{V}_s|^2}{\sqrt{r_s^2 + \omega_e^2(L_{ls} + L_{lr})^2} + r_s}. \tag{3.11}$$

Normally, the breakdown torque of a machine is three or four times higher than the rated torque.

Fig. 3.13 (a) shows a group of torque-speed curves for different rotor resistances. Note from (3.10) that the breakdown slip, s_m increases with r_r, but from (3.11) that the breakdown torque is independent of r_r. That is, the breakdown torque remains the same for different rotor resistances. On the other hand, note that the motor with a larger rotor resistance has a higher starting torque. For this reason, some IMs are designed to have large r_r intentionally for some applications that require high starting torque. Further, note from Fig. 3.13 (b) that the three different motors require about the same level of currents for the same torque. However, the operation speed is low when the rotor resistance is large. Thereby, the mechanical power, $P_m = T_e \times \omega_r$ is low for the same current, i.e., the IM efficiency drops as the rotor resistance increases. It is caused by the large secondary ohmic loss, $r_r \mathbf{I}_r'^2$.

Exercise 3.4   Consider an IM with parameters listed in Table 3.2. Find the slip, s required to produce $T_e = 80$ Nm for different secondary resistances, $r_r =$

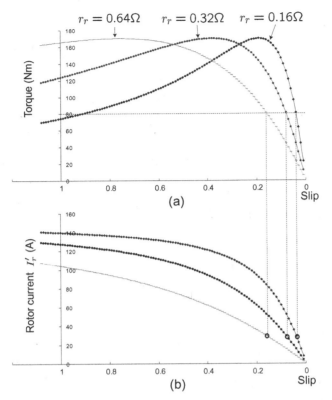

$r_r = 0.64\Omega$ $r_r = 0.32\Omega$ $r_r = 0.16\Omega$

Figure 3.13: (a) Torque and (b) rotor current curves of an IM for different rotor resistances.

0.16, 0.32, 0.64 Ω. Find also efficiencies for the three cases using $\eta = T_e \times \omega_r / (3|\mathbf{V}_s||\mathbf{I}_r'|)$.

Table 3.2: Parameters of an example IM .

Rated power	10 hp (7.46 kW)
Rated stator voltage	220 V (line to line)
Rated frequency	60 Hz
Rated speed	1164 rpm
Number of poles	$P = 6$
Stator resistance, r_s	0.29 Ω
Stator leakage inductance, L_{ls}	1.38 mH
Rotor resistance r_r	0.16 Ω
Rotor leakage inductance, L_{lr}	0.717 mH
Magnetizing inductance, L_m	41 mH

Exercise 3.5   A three phase IM produces 10 hp shaft power at a rated speed, 1750 rpm, when it is connected to a 220 V (line-to-line), 60 Hz line. The IM

parameters are $r_s = 0.2 \, \Omega$, $L_{ls} = L_{lr} = 2.3$ mH, and $L_m=34.2$ mH. Determine the rated torque, rotor resistance, and rotor current ($\mathbf{I}'_r$). Sketch the torque-speed curve.

Solution. $T_e = P_m/\omega_r = 10 \times 746/(1750/60 \times 2\pi) = 40.73$ Nm. Slip $s = 50/1800 = 0.0277$. It follows from (3.8) that

$$r_r = \frac{3P|\mathbf{V}_s|^2}{2\omega_e T_e}s = \frac{3 \times 4 \times 127^2}{2 \times 377 \times 40.73} \times 0.0277 = 0.175 \, \Omega.$$

$$|\mathbf{I}'_r| = \frac{127}{\sqrt{(0.2 + 0.175/0.0277)^2 + 377^2 \times 0.0046^2}} = 18.87 \text{ A.} \qquad \blacksquare$$

Exercise 3.6   Consider a three phase IM listed in Table 3.2. Determine 1) rated slip, 2) rated current, 3) rated power factor, 4) rated torque, and 5) breakdown torque.

Solution.

1) The synchronous speed is equal to 1200 rpm. Thus, $s = \frac{1200-1164}{1200} = 0.03$.

2) Since $r_r/s = 0.16/0.03 = 5.33 \, \Omega$,

$$\frac{(5.33 + j0.28)j15.46}{5.33 + j(15.46 + 0.28)} = 4.61 + j1.84.$$

Thus, the total impedance is equal to

$$(0.29 + j0.52) + (4.61 + j1.84) = 4.9 + j2.36 = 5.44\angle 25.7.$$

The phase voltage is equal to $220/\sqrt{3} = 127$ V. Thus,

$$\mathbf{I}_s = \frac{127}{5.44\angle 25.7°} = 23.3\angle - 25.7° \text{ (A)}.$$

3) $PF = \cos(25.7°) = 0.9$ (lagging).

4)

$$|\mathbf{I}'_r| = \frac{127}{\sqrt{(0.29 + 5.33)^2 + (0.52 + 0.28)^2}} = 22.3 \text{ A}$$

$$T_{e \text{ rated}} = 3 \times \frac{3 \times 0.16}{0.03 \times 377} \times 22.3^2 = 63.3 \text{ Nm}.$$

5)

$$T_{e\max} = \frac{3 \times 6}{4 \times 377} \frac{127^2}{\sqrt{0.29^2 + (0.52 + 0.28)^2} + 0.29} = 189.8 \text{ Nm.} \qquad \blacksquare$$

Exercise 3.7   Using MATLAB Simulink,

1) plot the torque-speed curves for the IM listed in Table 3.2. Repeat it for different rotor resistances: $r_r = 0.32 \; \Omega$ and $0.64 \; \Omega$.
2) determine the breakdown torque for the 3 cases.
3) determine the slip, s and the secondary current, $\mathbf{I}'_r$ for the 3 cases when the shaft torque is 80 Nm,
4) determine the efficiency for the 3 cases when $T_e = 80$ Nm.

3.2.4   Stable and Unstable Regions

The speed is determined at an intersection between motor torque and load curves as shown in Fig. 3.14. Here the stability means the ability to restore the speed when the load torque changes. Otherwise, the operation point is said to be unstable. The stable and unstable regions are separated by the breakdown slip line. IMs have the restore function if $s < s_m$.

Speed restoring capability is illustrated below: When the load increases, the speed drops. It makes the operating point A move to A'. Then, the slip increases with torque. Due to the increased motor torque the speed increases. This restores the original point, A. On the other hand, when the load decreases, the speed increases. Thus, A moves to A'' reducing the slip, as well as the torque. Thus, the operating point returns to the original point, A. Because of this capability, the region including A, A', A'' is called stable region.

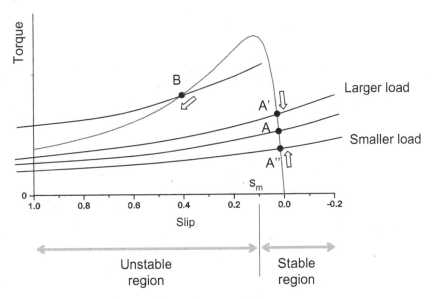

Figure 3.14: Stable and unstable regions.

However, on the left hand side of the breakdown torque, the situation is completely different. If the load increases to point B, then the speed drops. When the speed drops, torque is reduced. As a result, the speed drops more. When this process is repeated, the speed eventually drops to zero. Therefore, a stable operation cannot be realized in $s_m \leq s \leq 1$. This region is called unstable. The stability of an operating point can be determined by the slopes of the motor and the load torque curves, i.e., if $\left(\frac{dT_e}{d\omega_r}\right) \times \left(\frac{dT_L}{d\omega_r}\right) < 0$, then the point is stable, and otherwise unstable, where T_L is the load torque.

3.2.5 Parasitic Torques

It was shown in the previous section that high order MMF harmonics are generated due to the quantized levels. Recall from Chapter 2 that high order fields caused by MMF harmonics travel at reduced speeds. The synchronous speed of harmonics is

$$\omega_{e\nu} = \frac{\omega_e}{\nu}, \tag{3.12}$$

where $\nu = -5, 7, -11, 13, -17, 19 \cdots$. The minus sign in the harmonic number signifies the negative sequence. When the rotor speed is ω_r, the slip for ν^{th} harmonic is

$$s_\nu = \frac{\omega_{e\nu} - \frac{P}{2}\omega_r}{\omega_{e\nu}} = \frac{\omega_e/\nu - (1-s)\omega_e}{\omega_e/\nu} = 1 - \nu(1-s). \tag{3.13}$$

Thus the slip which makes $s_\nu = 0$ is given by

$$s = 1 - \frac{1}{\nu}. \tag{3.14}$$

Therefore, the synchronous slips of 5th and 7th harmonics are $s_5 = 1 - \frac{1}{-5} = 1.2$ and $s_7 = 1 - \frac{1}{7} = 0.86$, respectively. The 5th and 7th parasitic torque and their sum with the fundamental component are depicted in Fig. 3.15. If the 7th harmonic torque is appreciably high, the motor sometimes crawls at a low speed [10]. To reduce such harmonic effects, slot numbers of stator and rotor should be properly selected, and the rotor or stator core needs to be skewed.

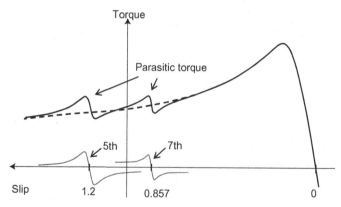

Figure 3.15: Parasitic torque caused by 5th and 7th MMF harmonics.

3.3 Leakage Inductances

The stator flux is expected to cross the air gap and enclose the rotor conductors (bars), building a flux linking. However, not all flux lines create such a complete linking, i.e., a small portion returns to itself without enclosing a rotor conductor. Fig. 3.16 is a finite element method (FEM) simulation result when the stator current flows. Most stator flux passes through the rotor core, making a linking with rotor bar. However, some flux returns to the stator core directly without making a link. Such flux is called *leakage flux*, and the associated inductance *leakage inductance*. It does not contribute to the development of torque or EMF, just adding reactive power. As can be seen from (3.11), the breakdown torque decreases as the leakage inductance increases.

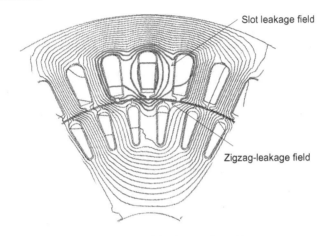

Figure 3.16: FEM simulation result of leakage flux lines.

Obviously, the leakage field increases as the air gap height increases. Due to the air gap, IMs have large leakage inductances compared with transformers that do not have an air gap. The leakage also depends on the slot shape. For example, it is even larger if the rotor slot is closed. Furthermore, it increases as the pole number increases. Therefore, high pole (greater than 6) IMs are rarely made.

Slot Leakage Inductance

Leakage field takes place around the slot, since the slots are normally deep and narrow. Consider a straight slot shown in Fig. 3.17. Assume that the coils are laid evenly, and that the number of conductors in the slot is N_{sl}. Assume further that the permeability of the core is infinite, so that $H = 0$ inside the core. I is the current flowing through each conductor and b_{sl} is the width of the slot. Many loops for line integral are shown in Fig. 3.17. Note that more conductors are enclosed inside the loop as the vertical height, y increases. By the Ampere's law, the horizontal field density, $B_{sl}(y)$ increases along with y.

$$B_{sl}(y) = \mu_0 H_{sl} = \mu_0 \frac{N_{sl} I}{b_{sl}} \frac{y}{h_{sl}}, \qquad 0 \leq y \leq h_{sl}. \qquad (3.15)$$

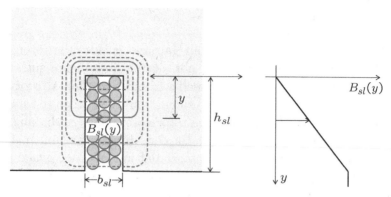

Figure 3.17: Slot leakage inductance.

It follows from Fig. 3.17 that the field energy stored in a slot is equal to

$$W_{sl} = \frac{1}{2}L_{lks}I^2 = b_{sl}l_{st}\int_{B_{sl}(0)}^{B_{sl}(h_{sl})} H_{sl}(y)dB_{sl}(y)$$

$$= \mu_0 b_{sl}l_{st}\int_0^{h_{sl}} \frac{N_{sl}^2 I^2}{b_{sl}^2}\left(\frac{y}{h_{sl}}\right)^2 dy = \mu_0 l_{st}\frac{h_{sl}}{3b_{sl}}N_{sl}^2 I^2,$$

where l_{st} is the stack length. Thus, we have the slot leakage inductance as

$$L_{lks} = 2\mu_0 l_{st}\frac{h_{sl}}{3b_{sl}}N_{sl}^2. \tag{3.16}$$

Exercise 3.8   Consider a slot in which coils are filled partially as shown in Fig. 3.18. Show that the slot leakage inductance is equal to

$$L_{lks} = 2\mu_0 l_{st}\left(\frac{h_{sl1}}{3b_{sl}} + \frac{h_{sl2}}{b_{sl}}\right)N_{sl}^2.$$

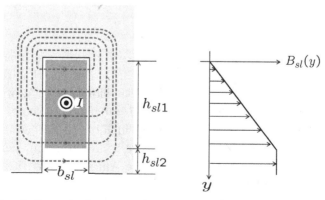

Figure 3.18: Slot leakage inductance (Exercise 3.8).

Solution. According to Ampere's law,

$$H(x) = \begin{cases} I_{bar} \frac{y}{h_{sl1} b_{sl}}, & 0 \le y \le h_{sl1} \\ I_{bar} \frac{1}{b_{sl1}}, & h_{sl1} \le y \le h_{sl1} + h_{sl2}. \end{cases}$$

Thus,

$$
\begin{aligned}
L_{lks} &= \frac{2\mu_0}{I_{bar}^2} \int_0^{h_{sl1}} H(y)^2 b_{sl} l_{st} dy + \frac{2\mu_0}{I_{bar}^2} H(h_{sl1})^2 b_{sl} l_{st} h_{sl2} \\
&= \frac{2\mu_0}{I_{bar}^2} \int_0^{h_{sl1}} \frac{N_{sl}^2}{b_{sl}} \left(\frac{I_{bar} y}{h_{sl1}} \right)^2 l_{st} dy + \frac{2\mu_0}{I_{bar}^2} \frac{N_{sl}^2 I_{bar}^2}{b_{sl}} h_{sl2} l_{st}. \quad \blacksquare
\end{aligned}
$$

End-turn Leakage Inductance

Fig. 3.19 shows a coil end turn and leakage field around it. Various methods to calculate end turn leakage inductance were suggested [7]–[13]. Hanselman [13], for example, simplified the end turn geometry as a half circle to obtain the end turn inductance:

$$L_{end} = \frac{\mu_0 \tau_p N^2}{4} \ln \left(\frac{\tau_p \sqrt{\pi}}{\sqrt{2A_s}} \right), \tag{3.17}$$

where A_s is overhang area, τ_p is the coil pitch, and N is the number of coils.

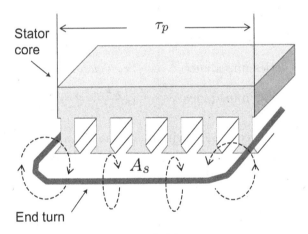

Figure 3.19: Coil end turn and leakage field.

3.3.1 Inverse Gamma Equivalent Circuit

Since the secondary circuit is shorted, it is possible to merge the leakage inductances into one without loss of information [8]. Let the leakage coefficient be defined by

$$\sigma = 1 - \frac{L_m^2}{L_r L_s}. \tag{3.18}$$

Define by $L_s \equiv L_{ls} + L_m$ and $L_r \equiv L_{lr} + L_m$ the stator and rotor inductances, respectively. Then (3.2) and (3.3) are rewritten as

$$\dot{\mathbf{V}}_s = (r_s + j\omega_e L_s)\mathbf{I}_s + j\omega_e L_m \mathbf{I}_r, \tag{3.19}$$

$$0 = (\frac{r_r}{s} + j\omega_e L_r)\mathbf{I}_r + j\omega_e L_m \mathbf{I}_s. \tag{3.20}$$

Replacing $\mathbf{I}_r$ from (3.19) by utilizing (3.20), it follows that

$$Z_s \equiv \frac{\mathbf{V}_s}{\mathbf{I}_s} = r_s + j\omega_e L_s \frac{1 + js\omega_e \sigma L_r/r_r}{1 + js\omega_e L_r/r_r}. \tag{3.21}$$

Further, (3.21) is modified as

$$
\begin{aligned}
Z_s &= r_s + j\omega_e \sigma L_s \frac{1/\sigma + j\omega_e s L_r/r_r}{1 + j\omega_e s L_r/r_r}, \\
&= r_s + j\omega_e \sigma L_s + \frac{(1-\sigma)j\omega_e L_s}{1 + j\omega_e s L_r/r_r}, \\
&= r_s + j\omega_e \sigma L_s + \frac{j\omega_e L_m^2/L_r}{1 + j\omega_e s L_r/r_r}, \\
&= r_s + j\omega_e \sigma L_s + \frac{\frac{r_r}{s}\frac{L_m^2}{L_r} j\omega_e \frac{L_m^2}{L_r^2}}{\frac{r_r}{s}\frac{L_m^2}{L_r^2} + j\omega_e \frac{L_m^2}{L_r}}.
\end{aligned}
\tag{3.22}
$$

We can infer from (3.22) that Z_s is comprised of

series impedance $\qquad r_s + j\omega_e \sigma L_s$

parallel impedance $\qquad \dfrac{r_r}{s}\dfrac{L_m^2}{L_r^2} \parallel j\omega_e \dfrac{L_m^2}{L_r}.$

On this basis we have an equivalent circuit as shown in Fig. 3.20. Note that the total leakage inductance appears in the stator side. It is named *inverse gamma equivalent circuit*, because the inductance elements are arranged similarly.

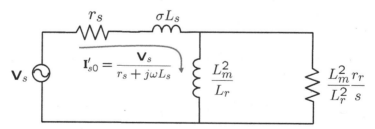

Figure 3.20: Equivalent circuit with σL_s.

Recall that torque is equal to zero at zero slip. The zero slip current is the magnetizing current, and equal to

$$\mathbf{I}_s|_{s=0} = \frac{\mathbf{V}_s}{r_s + j\omega_e L_s}. \tag{3.23}$$

The no-load current is mostly reactive since $r_s \ll w_e L_s$, and amounts to $\frac{1}{4} \sim \frac{1}{3}$ of the rated current. The ratio of no-load to rated current decreases as the motor power rating increases. Note however that the ratio of reactive current increases as the pole number increases since L_{ms} is inversely proportional to p as shown in (2.20). In the case of eight pole IMs, the ratio of no-load to rated current is $\frac{1}{3} \sim \frac{5}{8}$.

When the rotor shaft is locked, large current flows since the slip is equal to one. It is called *starting current, locked current,* or *short circuit current.* It is considered here that $r_r + r_s \ll w\sigma L_s$. Then,

$$\mathbf{I}_s|_{s=1} \approx \frac{\mathbf{V}_s}{j\sigma w_e L_s}. \tag{3.24}$$

Once again, the current is mostly reactive. Therefore, it follows (3.23) and (3.24) that

$$\frac{\mathbf{I}_s(s=1)}{\mathbf{I}_s(s=0)} \approx \frac{1}{\sigma}. \tag{3.25}$$

The locked current is $\frac{1}{\sigma}$ times larger than the no-load current. Since $\sigma = 0.1$ in general case, the locked current would be 10 times larger than the free running current.

3.4 Circle Diagram

The circle diagram is the graphical representation of IM performance based on the machine's voltage and current relation. Since $L_{lr}, L_{ls} \ll L_m$, the leakage coefficient is approximated as

$$\sigma = 1 - \frac{1}{(1+\frac{L_{ls}}{L_m})(1+\frac{L_{lr}}{L_m})} \approx 1 - \frac{1}{1+\frac{L_{ls}+L_{lr}}{L_m}} \approx \frac{L_{lr}+L_{ls}}{L_m}. \tag{3.26}$$

Then the breakdown slip is approximated as

$$s_m = \frac{r_r}{\sqrt{r_s^2 + w_e^2(L_{ls}+L_{lr})^2}} \approx \frac{r_r}{\sqrt{w_e^2\sigma^2 L_m^2}} \approx \frac{r_r}{w_e\sigma L_r} = \frac{1}{\tau_r w_e \sigma}. \tag{3.27}$$

where $\tau_r = L_r/r_r$ is the rotor time constant. Let $r_s = 0$ for convenience of computation. Then, the stator current is obtained from (3.21) and (3.27) such that

$$\mathbf{I}_s \equiv \mathbf{I}_{s0}\frac{1+j\frac{s}{\sigma s_m}}{1+j\frac{s}{s_m}}, \tag{3.28}$$

where $\mathbf{I}_{s0} = \frac{\mathbf{V}_s}{jw_e L_s}$ is the no load current at $s = 0$ [9]. Equation (3.28) is transformed into

$$\begin{aligned}
\mathbf{I}_s &= \mathbf{I}_{s0}\left[\frac{1+\sigma}{2\sigma} - \frac{1-\sigma}{2\sigma}\frac{1-j\frac{s}{s_m}}{1+j\frac{s}{s_m}}\right] \\
&= \mathbf{I}_{s0}\left(\frac{1+\sigma}{2\sigma}\right) - \frac{1-\sigma}{2\sigma}|\mathbf{I}_{s0}|\frac{|1-j\frac{s}{s_m}|}{|1+j\frac{s}{s_m}|}e^{j\angle\frac{1-js/s_m}{1+js/s_m}}. \tag{3.29}
\end{aligned}$$

Note that $\frac{|1-j\frac{s}{s_m}|}{|1+j\frac{s}{s_m}|} = 1$ and

$$\angle\frac{1-j\frac{s}{s_m}}{1+j\frac{s}{s_m}} = \angle(1-j\frac{s}{s_m}) - \angle(1+j\frac{s}{s_m}) = -2\angle(1+j\frac{s}{s_m}) = -2\varphi,$$

where $\varphi = \tan^{-1}\left(\frac{s}{s_m}\right)$. Therefore, (3.29) reduces to

$$\mathbf{I}_s - \mathbf{I}_{s0}\left(\frac{1+\sigma}{2\sigma}\right) = -\frac{1-\sigma}{2\sigma}|\mathbf{I}_{s0}|e^{-j2\varphi} = \frac{1-\sigma}{2\sigma}|\mathbf{I}_{s0}|e^{j(\pi-2\varphi)}. \tag{3.30}$$

Equation (3.30) states that the locus of $\mathbf{I}_s$ for $-\infty < s < \infty$ is a circle centered at $\mathbf{I}_{s0}\frac{1+\sigma}{2\sigma}$ with radius $\frac{1-\sigma}{2\sigma}|\mathbf{I}_{s0}|$.

Let $\mathbf{V}_s$ be a real number. It can be drawn in the vertical axis of complex plane as shown in Fig. 3.21. Note that the vertical axis is the real line and the horizontal axis represents the imaginary numbers. At zero slip, the current is purely imaginary, since $\mathbf{I}_{s0} = -j\frac{\mathbf{V}_s}{w_e L_s}$. Therefore, it is drawn in the horizontal axis.

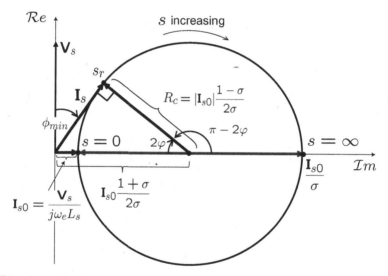

Figure 3.21: A circle diagram for IM with $r_s = 0$.

The second important point is a tangential point between the current contour and a straight line from the origin. It is the point of the rated operation and we denote by $s = s_r$ the rated slip. The power factor (PF) angle is specified by the angle between the voltage (vertical axis) and the current vectors, i.e., $\phi = \angle\mathbf{V}_s - \angle\mathbf{I}_s$. Thus, the minimum PF angle ϕ_{min} is obtained at the rated condition. In that case, it follows that $\phi_{min} = 2\varphi = 2\tan^{-1}(s/s_m)$ from the geometry of the right triangle (solid line). Therefore, the maximum PF is given as

$$PF_{max} = \cos\phi_{min} = \frac{1-\sigma}{1+\sigma}. \tag{3.31}$$

Further, it follows from (3.31) that

$$\frac{s_r}{s_m} = \tan\frac{\phi_{min}}{2} = \frac{\sin\phi_{min}}{1+\cos\phi_{min}} = \frac{\frac{2\sqrt{\sigma}}{1+\sigma}}{1+\frac{1-\sigma}{1+\sigma}} = \sqrt{\sigma}. \tag{3.32}$$

Applying (3.27), we obtain that

$$s_r \approx \frac{1}{T_r \omega_e \sqrt{\sigma}}.$$

That is, the rated slip, s_r decreases as the leakage increases. Further, it also follows from (3.28) and (3.32) that

$$\frac{|\mathbf{I}_s(s_r)|}{|\mathbf{I}_{s0}|} = \frac{\sqrt{1 + \left(\frac{s_r}{\sigma s_m}\right)^2}}{\sqrt{1 + \left(\frac{s_r}{s_m}\right)^2}} \approx \frac{1}{\sqrt{\sigma}}. \tag{3.33}$$

This implies that the ratio of rated current to magnetizing current increases as the leakage decreases [9]. For example, $|\mathbf{I}_s(s_r)|/|\mathbf{I}_{s0}| = 3.16$ when $\sigma = 0.1$, which means that the rated current is three times larger than the magnetizing current.

Exercise 3.9 Derive (3.33) using the property of the right triangle in the circle diagram shown in Fig. 3.21.

Solution.

$$|\mathbf{I}_s(s_r)| = \sqrt{|\mathbf{I}_{s0}|^2 \left(\frac{1+\sigma}{\sigma}\right)^2 - |\mathbf{I}_{s0}|^2 \left(\frac{1-\sigma}{\sigma}\right)^2} = \frac{|\mathbf{I}_{s0}|}{\sqrt{\sigma}}. \qquad ■$$

Exercise 3.10 Consider a three phase IM listed in Table 3.2. Determine the following values or approximations: a) leakage coefficient σ, b) the minimum power factor, c) rotor time constant, d) rated slip, e) no load current, and f) rated current.

Solution. a) $\sigma = \frac{41.717 \times 42.38}{41^2} - 1 = 0.052$;   b) $\cos \phi_{min} = \frac{1-0.052}{1+0.052} = 0.9$;
c) $\frac{41.717}{0.16} = 0.26$;   d) $s_r = \frac{1}{0.26 \times 377 \times \sqrt{0.0524}} = 0.044$;   e) $\frac{127}{15.98} = 7.947$ (A);
f) $\frac{7.947}{\sqrt{0.0524}} = 34.72$ (A).   ■

3.4.1 Torque and Losses

Since mechanical power P_m is equal to $T_e \cdot \omega_r$, it is zero when either torque or speed is zero, i.e., $s = 0$ or $s = 1$. The two operating conditions are marked by 'N' and 'Q' in Fig. 3.22, respectively. The line $\overline{NQ}$ signifies *the zero power line*. The current at infinity slip is marked by 'R'. Note from (3.7) that torque is equal to zero when $s = \infty$. The straight line $\overline{NR}$ connecting two zero torque points is called *the zero torque line*.

Draw a vertical line $\overline{PB}$ from P to the horizontal axis. Geometrically, the real air gap power is equal to $P_{ag} = 3|\mathbf{V}_s||\mathbf{I}_s| \cos \phi = 3|\mathbf{V}_s| \times \overline{PB}$. Recall from (3.4)

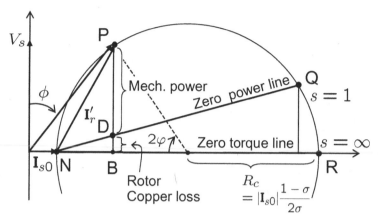

Figure 3.22: Simplified circle (Heyland) diagram with $r_s = 0$ showing zero mechanical power and zero torque lines.

that real air gap power consists of mechanical power and rotor copper loss. Define a point, 'D' on $\overline{PB}$ such that $\overline{PD} = \overline{PB} \times (1 - s)$ and $\overline{DB} = \overline{PB} \times s$. Then

$$P_{ag} = 3V_s\overline{PB} \times (1 - s) + 3V_s\overline{PB} \times s \tag{3.34}$$
$$= \underbrace{3|I'_r|^2\frac{r_r}{s} \times (1 - s)}_{\text{Mechanical power}} + \underbrace{3|I'_r|^2\frac{r_r}{s} \times s}_{\text{Rotor copper loss}} .$$

Thus, the segment $\overline{PD}$ signifies one current component for mechanical power, and $\overline{DB}$ the other component for rotor copper loss. Note that the locus of 'D' forms a straight line $\overline{NQ}$ for $0 \le s \le 1$. In other words, $\overline{NQ}$ is the zero power line, and separates the real current into torque producing and loss components.

Exercise 3.11   Using $\tan \varphi = s/s_m$, show that the slope, $\overline{DB}/\overline{NB}$ is constant independent of the slip, $0 \le s \le 1$.

Solution. Let the radius of the circle be denoted by $R_c \equiv |I_{s0}|\frac{1-\sigma}{2\sigma}$. Then,

$$\overline{PB} = R_c \sin 2\varphi = 2R_c \sin \varphi \cos \varphi \tag{3.35}$$

Note further that
$$\overline{NB} = R_c(1 - \cos 2\varphi) = 2R_c \sin^2 \varphi.$$

Therefore, it follows from (3.35) that

$$\frac{\overline{DB}}{\overline{NB}} = \frac{s\overline{PB}}{\overline{NB}} = \frac{2sR_c \sin \varphi \cos \varphi}{2R_c \sin^2 \varphi} = s\frac{1}{\tan \varphi} = s\frac{1}{s/s_m} = s_m \qquad \blacksquare$$

Fig. 3.23 shows five specific slip points. At the breakdown slip, s_m torque is maximized. The input power, $V_s \cdot I_s$ is the largest at the middle point. However, since the rotor copper loss is also large, the torque is less than the case of the

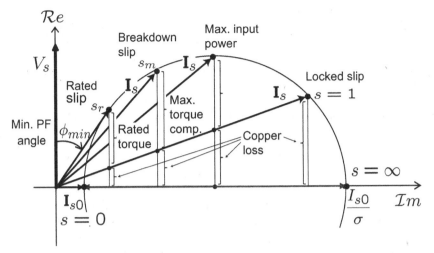

Figure 3.23: Simplified circle diagram for IM with five specific slip points.

breakdown slip. At the locked slip, $s = 1$ all the real component of $\mathbf{I}_s$ is consumed as the rotor bar copper loss. Infinity slip corresponds to a pathological case in which the rotor shaft is rotating infinitely fast by an external force while the electrical speed is limited. Then the secondary circuit is shorted, thereby all the current turns out to be imaginary as in the case of zero slip.

Fig. 3.24 shows a practical circle diagram considering the stator resistance, r_s. The zero slip current $\mathbf{I}_{s0}$ is no more imaginary since $r_s \neq 0$. At infinity slip, the impedance will be $r_r + r_s + jw_e(L_{ls} + L_{lr})$. Thereby the operating point does not lie on the imaginary axis. The horizontal line, being imaginary, is obtained when the slip is negative. A section of real current between the zero torque and the horizontal lines signifies the stator copper loss, $|\mathbf{I}_r|^2 r_s$.

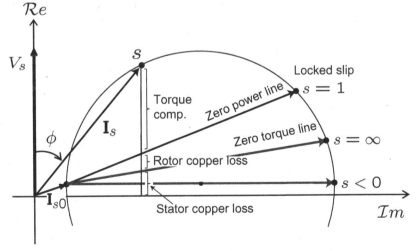

Figure 3.24: Practical circle diagram for IM with $r_s \neq 0$.

Exercise 3.12   Consider a three-phase IM with the following parameters; the rated stator voltage $V_s=220$ V, $r_s = 0\ \Omega$, $r_r = 0.16\ \Omega$, $L_{ls} = 1.38$ mH, $L_{lr} = 0.72$ mH, $L_m = 38$ mH, $P = 6$, $f = 60$ Hz.

   a) Draw the circle diagram.
   b) Find power factors at $s = 0$, $s = 1$, $s = s_m$, $s = s_r$, $s = s_\infty$.
   c) Obtain a torque-speed curve by determining the length $\overline{PD} = \overline{PB} \times (1-s)$ in the circle diagram shown in Fig. 3.22. Obtain the torque-speed curve by using (3.9). Compare the two results.

Solution.

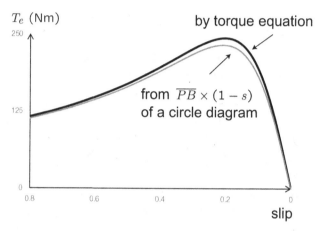

Figure 3.25: Comparison of torque-speed curves when $r_s = 0$ (Solution to Exercise 3.12).

3.5   Current Displacement

In Section 3.3, the slot leakage inductance was calculated assuming that the rotor current flew through the bar with a uniform current density. However, current does

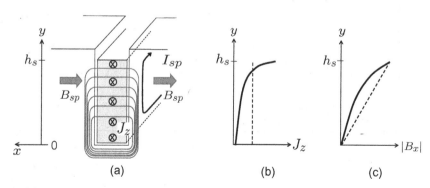

Figure 3.26: Rotor bar current displacement due to slip: (a) rotor slot and conductor with a slip field B_{sp}, (b) rotor bar current density J_z, and (c) field density, $|B_x|$.

not flow evenly through the rotor bar when there is a slip. Instead, the current density depends on the depth of the rotor bar: It is largest at the rotor surface, but decreases as the position deepens.

Fig. 3.26 (a) shows a rotor slot with a conductor bar. It is assumed that rotor current flows through the bar from the front to the back. When it comes to a slip, alternating flux passes through the rotor core: It crosses through the conductor bar from side to side as shown in Fig. 3.26 (a). Since the crossing field B_{sp} alternates at a slip frequency, it causes an eddy current $\mathbf{I}_{sp}$ inside the rotor bar. The current direction is determined so that the resulting field opposes to the change of B_{sp}. Assume that B_{sp} is increasing. Then the induced current $\mathbf{I}_{sp}$ flows clockwise. It adds more to the original bar current in the upper part, while reducing the current density in the bottom part.

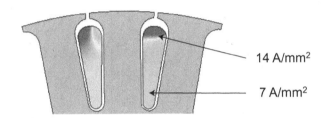

14 A/mm²

7 A/mm²

Figure 3.27: Current density profile in the rotor bar when the slip is 0.015.

Current density and flux density are plotted against y in Fig. 3.26 (a) and (b). The density J_z is no more constant. It has a parabolic profile: Current density is high in the top area. This is commonly known as the *skin effect*, and is often described as rotor bar *current displacement*. The current displacement has the effect of reducing the *effective* conduction area. That is, it increases the rotor resistance. Note from Fig. 3.26 (c) that the flux density B_x also has a parabolic pattern when there is a slip. It is linear without a slip as shown in Fig. 3.17. Fig. 3.27 shows a FEM result of the current distribution on a rotor bar when the slip is equal to 0.015. As stated in the above, the current is localized in the top part.

The resistance increase is described as the ratio of AC and DC resistances of the bar [8]:

$$K_r \equiv \frac{r_{bar}(ac)}{r_{bar}(dc)} = \xi \frac{\sinh 2\xi + \sin 2\xi}{\cosh 2\xi - \cos 2\xi}, \tag{3.36}$$

where $\xi = h_{bar}/\delta$ and δ is the skin depth of the bar. Note that $\delta = \sqrt{\frac{2}{\mu\sigma\omega}}$, where σ is the conductivity. Of course, the resistance increases as the slip frequency increases as shown in Fig. 3.28 (a). At low slip frequency, current displacement is negligible. At higher slip frequencies, however, current displacement is more pronounced. For example, the AC resistance $r_{bar(ac)}$ at 60 Hz is about three times larger than the DC resistance $r_{bar(dc)}$.

Due to the decrease in effective conduction area, the slot leakage inductance decreases by the displacement effect. The inductance correction coefficient is given by [8]

$$K_x \equiv \frac{L_{sl}(ac)}{L_{sl}(dc)} = \frac{3(\sinh 2\xi - \sin 2\xi)}{2\xi(\cosh 2\xi - \cos 2\xi)}. \tag{3.37}$$

Fig. 3.28 (b) shows plot of K_x, showing a decrease as ξ (slip) increases. In summary, the rotor displacement effect increases when

   1) the slip frequency increases;
   2) conductor conductivity increases;
   3) conductor permeability increases;
   4) bar height h_s increases.

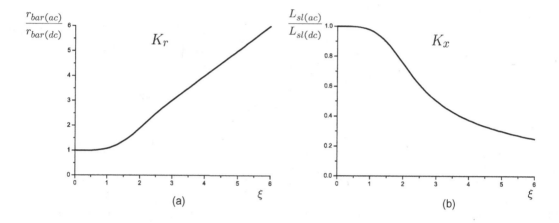

Figure 3.28: Correction coefficients reflecting displacement (skin) effect: (a) rotor bar resistance change (b) slot leakage inductance change.

Exercise 3.13   Determine $r_{bar}(ac)/r_{bar}(dc)$ of a 3 cm high copper bar when the line frequency is 60 Hz and slip is 1. The copper conductivity is $\gamma = 5 \times 10^7$ S/m.

Solution. The skin depth is equal to

$$\delta = \sqrt{2/(4\pi \times 10^{-7} \times 5 \times 10^7 \times 377)} = 9.19 \text{ (mm)}.$$

Thus, $\xi = 30/9.19 = 3.26$. It follows from (3.36) that $r_{bar}(ac)/r_{bar}(dc) = 3.26$. That is, the rotor resistance increases 3.26 times at the locked condition.   ■

3.5.1   Double Cage Rotor

In double cage rotor, each rotor bar is divided into two parts: One is near to the rotor surface, and the other is positioned deep in the shaft side. Fig. 3.29 shows the flux profiles around double cage rotor slots for the two slip conditions: (a) high slip and (b) low slip. When the slip frequency is high, the field cannot penetrate deeply into the rotor core due to the skin effect. Therefore, the flux lines surround only the upper conductors of the rotor slots. Hence, the major current is induced in the

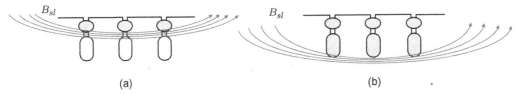

Figure 3.29: Flux lines around double cage rotor slots: (a) high slip ($r_{bar}(ac)$ is large.) and (b) low slip ($r_{bar}(ac)$ is small.).

upper conductors. Since the conduction is restricted to the small upper bars, the rotor resistance r_{bar} is relatively high.

On the other hand, the field penetrates deeply when the slip frequency is low. The flux lines enclose both upper and lower conductors, thereby the induced current conducts both upper and lower conductors. Since large conduction area is provided, the effective rotor resistance becomes low with low slip.

Note from (3.9) that the high rotor resistance is favorable for starting, since the starting torque is high with a high r_r. But, a low r_r is desirable in a normal (low slip) operation for high efficiency and fast torque response. The double cage rotor structure satisfies both requirements because r_r changes depending on the slip conditions. The upper small cage is called *starting cage*, and the lower cage *operation cage*. Fig. 3.30 shows a photo of laminated sheets for stator and rotor.

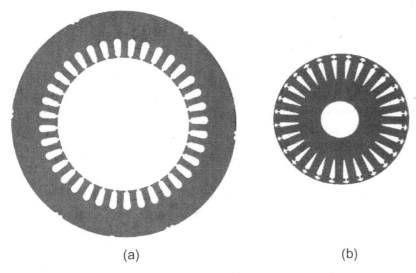

Figure 3.30: Laminated sheets for an IM: (a) stator and (b) rotor.

Slot Types and NEMA Classification

The National Electrical Manufacturers Association (NEMA) of U.S.A. has classified cage-type induction machines into various categories to meet the diversified applications [1]. These categories are characterized by the torque–speed curves, and the rotor slot shape is the most influential factor as shown in Fig. 3.31 (a): The double cage slot motors have a high starting torque. The round or oval slot motors have a low starting torque, but a steep slope near the synchronous speed. Fig. 3.31 (b) shows typical torque-speed curves for NEMA design A, B, C, and D motors.

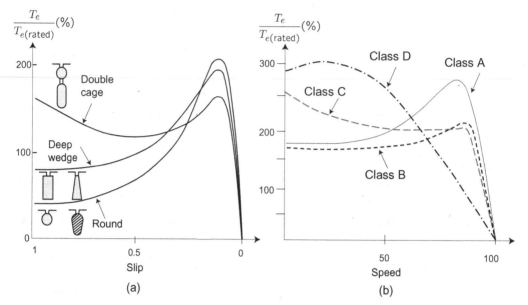

Figure 3.31: (a) Torque–speed curves for various rotor slot shapes and (b) NEMA classification.

NEMA A motors are designed to have low rotor resistance and low rotor leakage inductance with simple deep bar (oval) shapes. They have a high breakdown torque, the steepest slope, and a low operating slip. Therefore, the operating efficiency is high. But large current is required for starting. NEMA A machines are suitable for inverter applications.

NEMA B motors are commonly used for constant-speed applications such as fans, centrifugal pumps, machine tools, and so forth. Slip is 5% or less. The rotor leakage inductance is relatively high, so that the locked-rotor current does not exceed 6.4 times the rated current. The locked torque ranges from 75% to 275% of the full-load torque [1]. Breakdown torque is 175–300% of rated torque

NEMA C motors are designed to have a high starting torque (200% of the full-load torque). On the other hand, the rated torque is obtained with a normal slip which is less than 5%. However, overload capability is low. To meet the requirements, these motors have double-cage rotor slots.

NEMA D motors are special motors designed to have a very high starting torque. To increase the rotor resistance, the rotor conductors are often made with bronze. But these motors exhibit high slip (5 to 13%) at normal operation. They have low efficiency, and are thereby easily overheated. Thus, they are used for intermittent operation [1].

3.5.2 Line Starting

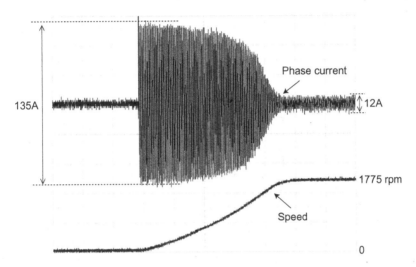

Figure 3.32: IM current and speed responses at the line start: four pole, 1.5kW, 220 V line-to-line.

A three phase power grid is a voltage source that creates a rotating field inside a cylindrical machine. Thus, IM can be run by connecting the motor terminal directly to a power grid. At the starting moment, IM is at the locked state so that a large current flows through the shorted rotor circuit. In the starting period, the IM consumes 5 to 7 times its full load current. This large starting current may cause a large voltage hunting in the grid, which may affect the operation of other devices. It can trip the circuit breaker or blow fuses. Hence, the direct line starting is not recommended for a high power IM (generally above 25 kW). Fig. 3.32 shows a plot of current at the line starting.

To mitigate such a problem, soft starters or inverters are utilized in industry. Inverters are expensive, and over-qualified in some applications. Soft starter is a simple and low cost circuit designed to operate during the starting period [11]. Its main circuit consists of back-to-back thyristors or triacs in each AC supply line. Fig. 3.33 illustrates how the thyristor switches control the rms voltage. Note from Fig. 3.33 (b) that the rms voltage level increases as the gate firing angle, α reduces.

Fig. 3.34 shows the typical soft starter circuit. Initially the turn-on delay, α is large, and it is reduced successively. Then the rms voltage increases progressively until the full line voltage is applied to motor. Once the motor reaches its rated speed, thyristor switches are always turned on.

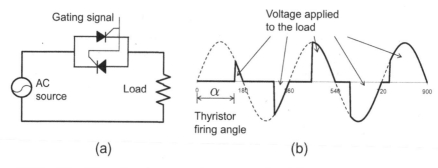

Figure 3.33: Thyristor circuit operation principle: (a) single phase thyristor circuit and (b) the rms voltage increase as the firing angle, α decreases.

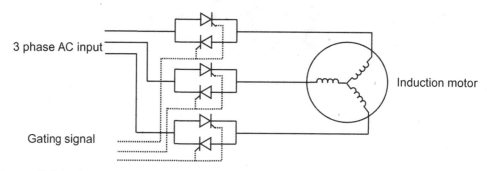

Figure 3.34: Soft starter circuit with thyristors.

3.6   IM Speed Control

With the use of semiconductor switches, it is possible to change the feeding voltage and frequency. A simple speed control method would be to change the voltage only. But this method works only in a small speed range. It is better to change the voltage in accordance with the frequency.

3.6.1   Variable Voltage Control

Fig. 3.35 shows the torque-speed curves of an IM for different voltages while the frequency is fixed at 60 Hz. Recall from (3.11) that the breakdown torque is proportional to the square of the voltage. It also shows a fluidal load curve which is proportional to the square of speed. Considering the intersection (operation) points, one can notice that the speed control range is quite limited.

3.6.2   VVVF Control

With the use of inverters, it is possible to provide a variable voltage-variable frequency (VVVF) source to the motor. Changing the frequency means changing the

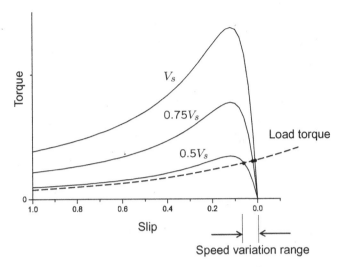

Figure 3.35: Torque-speed curves for variable voltage/fixed frequency operations.

synchronous speed. Note from (3.19) that the stator voltage is approximated as $\mathbf{V}_s = r_s\mathbf{I}_s + j\omega_e(L_{ls} + L_m)\mathbf{I}_s \approx j\omega_e L_s\mathbf{I}_s = j\omega_e\lambda_s$. Therefore, the magnitude of flux satisfies

$$\lambda_s = \frac{\mathbf{V}_s}{\omega_e} = \frac{\mathbf{V}_s}{2\pi f}. \tag{3.38}$$

That is, the voltage to frequency ratio implies the flux. To maintain a constant flux level, the voltage should be increased in accordance with the frequency increase. In contrast, the voltage should be decreased when the frequency decreases. Otherwise, flux saturation occurs.

Torque equation (3.7) is approximated as

$$T_e = \frac{3P}{2}\left(\frac{\mathbf{V}_s}{\omega_e}\right)^2 \frac{\omega_{sl}r_r}{r_r^2 + \omega_{sl}^2 L_{lr}^2} = \frac{3P}{2}\lambda_s^2 \frac{\omega_{sl}r_r}{r_r^2 + \omega_{sl}^2 L_{lr}^2}. \tag{3.39}$$

According to (3.39), torque just depends on the slip independently of the frequency as long as the voltage to frequency ratio (V/F) is kept constant. Hence, as shown in Fig. 3.36, the torque curves in the same shape shift in response to the different frequencies.

The motor control method keeping the V/F constant is called *variable voltage-variable frequency control*. Specifically, consider an operation point marked by 'x'. Note that it is impossible to start the motor by a source of (127 V, 60 Hz), since 'x' belongs to the unstable region. However, 'x' is a stable operation point to the source, (25 V, 12 Hz). Similarly, if both voltage and frequency are doubled at the same time to (50 V, 24 Hz), another stable operation point is obtained at 'xx'. Hence, while keeping V/F constant, it is possible to drive crane like loads that require high start torque.

In the low speed region, the voltage drop over the stator resistance is not negligible. For example, consider the IM in Table 3.2. Note that $r_s/\omega_e L_s = 0.048$ (Ω) at

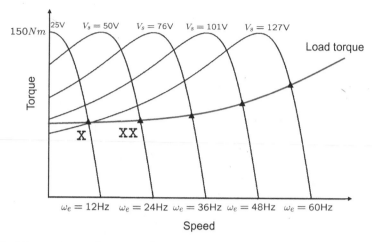

Figure 3.36: Torque-speed curves for a VVVF operation.

60 Hz, whereas $r_s/\omega_e L_s = 0.48$ at 6 Hz, i.e., the ohmic drop takes about a half of the input voltage at 6 Hz. To compensate for this voltage drop, the voltage profile is boosted in a low speed region, as shown in Fig. 3.37.

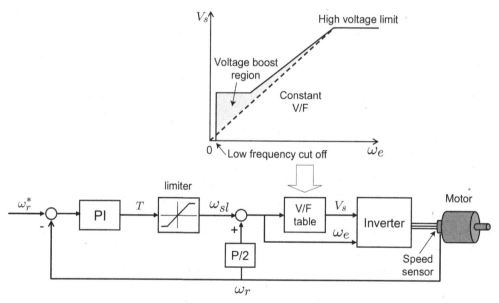

Figure 3.37: VVVF pattern and a slip based VVVF speed controller.

If the speed information is available, it is possible to construct a slip based feedback controller as shown in Fig. 3.37 [11]. The speed controller output is regarded as a torque command, and a required slip is calculated in proportion to the torque command. The calculated slip is added to a measured rotor speed, determining the electrical frequency: $\omega_e = \omega_r + \omega_{sl}$. This electrical speed is used as a command frequency to the VVVF inverter. Based on a V/F table, a voltage level is determined. VVVF inverters are widely used in many applications where precision is not essential. These applications include compressors, blowers, cranes, locomotives, etc.

References

[1] T. Wildi, *Electrical Machines, Drives, and Power Systems*, 2nd Ed., Prentice Hall, 1991.

[2] Advanced Manufacturing Office, *Premium Efficiency Motor Selection and Application Guide*, DOE, 2014.

[3] B. K. Bose, *Modern Power Electronics and AC Drives*, Prentice Hall, 2002.

[4] Hanselman, *Brushless Permanent-Magnet Motor Design*, McGraw-Hill, 1994.

[5] D. Ban, D. Zarko, and I. Mandic, "Turbogenerator end-winding leakage inductance calculation using a 3-D analytical approach based on the solution of Neumann integrals," *IEEE Trans. Energy Convers.*, vol. 20, no. 1, pp. 98−105, Mar. 2005.

[6] Y. Liang, X. Bian, H. Yu, L. Wu, and B. Wang, "Analytical algorithm for strand end leakage reactance of transposition bar in AC machine," *IEEE Trans. Energy Convers.*, vol. 30, no. 2, pp. 533−540, Jun. 2015.

[7] M. Bortolozzi, A. Tessarolo, and C. Bruzzese, "Analytical computation of end-coil leakage inductance of round-rotor synchronous machines field winding," *IEEE Trans. Magn.*, vol. 52, no. 2, Art. no. 8100310, Feb. 2016.

[8] I. Boldea and S. A. Nasar, *The Induction Machine Handbook*, CRC Press, 2002.

[9] W. Leonhard, *Control of Electrical Drives*, Springer, 1996.

[10] P.C. Sen, *Principle of Electric Machines and Power Electronics*, John Wiley & Sons, New York, 1997.

[11] J.M.D. Murphy, *Thyristor Control of AC Motors*, Pergamon Press, 1973.

Problems

3.1 Consider a four pole, three phase IM with the following rated parameters: output power = 15 kW, line voltage (line-to-line) = 440 V, line frequency = 60 Hz, and rated speed = 1748 rpm. The sum of windage and friction loss is 1 kW. Calculate the air gap power and rotor copper loss.

3.2 A three phase IM having a synchronous speed of 900 rpm draws 40 kVA from a three phase feeder. The rated slip speed is 3% of the synchronous speed.

The power factor is 0.9, and the total stator loss is 4 kW. Calculate the torque developed by the motor.

3.3   Consider a four pole, three phase IM with the following parameters: $r_s = 1.6\ \Omega$, $r_r = 0.996\ \Omega$, $L_s = 66$ mH, $L_{ls} = 3.28$ mH, and $L_{lr} = 3.28$ mH. Assume that three phase, 220 V (line-to-line), 60 Hz AC source is applied to the motor.

a) Calculate the slip yielding the breakdown torque.
b) If the rated slip is $s = s_m/2.5$, determine the rotor current, $\mathbf{I}'_r$ and torque using the modified equivalent circuit.
c) Determine the magnetizing current.
d) Calculate the power factor at the rated condition.

3.4   Consider a four pole, three phase IM with the following parameters: $r_s = 0.1\ \Omega$, $r_r = 0.25\ \Omega$, $L_s = 150$ mH, $L_{ls} = 3$ mH, and $L_{lr} = 3$ mH. A 440 V (line-to-line), 60 Hz AC source is applied to the motor. The rated slip is 3%. Determine the rated and starting torques.

3.5   Consider a four pole, three phase IM with the following parameters: $r_s = 0\ \Omega$, $r_r = 0.25\ \Omega$, $L_s = 150$ mH, $L_{ls} = 3$ mH, and $L_{lr} = 3$ mH. Assume that a 440 V (line-to-line), 60 Hz AC source is applied to the motor.

a) Draw a circle (Heyland) diagram.
b) Determine the rated current.
c) Determine the minimum PF angle.

3.6   Consider the thyristor voltage control circuit shown in Fig. 3.34. Assume that the source phase voltage is $V \sin w_e t$. Derive a relationship between the rms output voltage and the firing angle, α.

3.7   A three phase 220 V, four pole IM runs at 1750 rpm. Suppose that the input voltage is increased to 250 V. For the same load, calculate the slip and speed.

3.8   (No load test) A four pole, three phase IM is connected to 220 V(line-to-line), 60 Hz line. It runs at 1750 rpm without a load. The phase current is $\mathbf{I}_s = 2$ A and the reactive power is 50 W.

a) Calculate the power factor angle.
b) Calculate the stator inductance.
c) Calculate the stator resistance.

3.9   (Locked rotor test) While the IM is locked, a 90 V(line-to-line), 60 Hz three phase voltage source is applied. The measured stator current is 7 A, the resistance of the stator is 2.1 Ω and the measured shaft power is 800 W.

a) Calculate the power factor angle.
b) Calculate the impedance.
c) Calculate the rotor resistance, r_r.
d) Assuming $L_{ls} = L_{lr}$, find the leakage impedance.

3.10 Consider a three phase four pole IM with a rated power, 7.5 kW under 60 Hz, 440 V(rms) line-to-line voltage. The rotor resistance and leakage inductance are $r_r = 0.25$ Ω and $L_{lr} = 3.28$ mH, respectively. The rated speed is 1746 rpm. Assume the flux is made to be constant.
a) Calculate the rated torque.
b) Utilizing $T_e = \frac{3P}{2}\lambda_s^2 \frac{\omega_{sl} r_r}{r_r^2 + \omega_{sl}^2 L_{lr}^2}$, calculate the rated flux.
c) Determine the voltages at 45 Hz and 30 Hz such that the flux level is the same as that in the case of 60 Hz. Utilizing MATLAB, plot the torque-speed curves at 60, 45, and 30 Hz.

3.11 Consider a three phase four pole IM with the following parameters: line-to-line voltage = 440 V, $r_s = 0.075$ Ω, L_{ls} : 3 mH, and L_{lr} : 3 mH.
a) Find slip for $r_r = 0.1$ Ω when the load torque is equal to $T_e = 100$ Nm. Repeat the same for $r_r = 0.4$ Ω and $r_r = 0.8$ Ω.
b) Plot the torque-speed curves for $r_r = 0.1$, 0.4, 0.8 Ω.
c) Using modified equivalent circuit, calculate the rotor current I_r' and phase delay ϕ for $r_r = 0.1$, 0.4, 0.8 Ω when the load torque is equal to $T_e = 100$ Nm.
d) Calculate the electrical power supplied from the line for $r_r = 0.1$, 0.4, 0.8 Ω when the load torque is equal to $T_e = 100$ Nm.
e) Calculate mechanical power and efficiencies for $r_r = 0.1$, 0.4, 0.8 Ω when the load torque is equal to $T_e = 100$ Nm.

3.12 Consider a three phase four pole IM with the following parameters: line-to-line voltage = 440 V, rated frequency = 60 Hz, L_s= 12 mH, σ=0.08.
a) Draw a circle diagram. Neglect the ohmic voltage drop.
b) Calculate the rated slip s_r, rated torque, rated output power and the maximum power factor.
c) Calculate the breakdown slip s_m and the breakdown torque.

3.13 Consider a three phase four pole IM with the following parameters: line-to-line voltage = 380 V, rated frequency = 50 Hz, r_s=0.2 Ω, L_s= 12 mH, σ=0.1.
a) Calculate the rated stator flux and no-load current at the rated frequency.
b) Calculate the ohmic voltage drop by the no-load current.
c) Determine the rated current $I_s(s_r)$ using the circle diagram.
d) Assume that the rated current keeps flowing from 0 to 50 Hz with the VVVF control. Determine the voltage/frequency profile from 0 to 50 Hz that keeps the stator flux constant.

Chapter 4

Dynamic Modeling of Induction Motors

Various IM voltage equations are derived in the synchronous reference frame, as well as in the stationary frame. The coordinate transformations are heavily involved in the process of deriving IM dynamic models. The derivation process is somewhat complicated since the rotor speed is different from the synchronous speed. The dynamic model in a synchronous reference frame provides a basis for field oriented control.

4.1 Voltage Equation

Time derivative of flux linkage yields a voltage equation. First, stator and rotor flux linkage equations are described in the stationary frame, and then mapped into the synchronous frame.

4.1.1 Flux Linkage

The stator flux is excited by both stator and rotor currents. In a uniform air gap machine, the stator flux generated by the stator current is equal to

$$
\underbrace{\begin{bmatrix} \lambda_{as} \\ \lambda_{bs} \\ \lambda_{cs} \end{bmatrix}}_{=\lambda_{abcs}^s} = \underbrace{\begin{bmatrix} L_{ls} + L_{ms} & -\frac{1}{2}L_{ms} & -\frac{1}{2}L_{ms} \\ -\frac{1}{2}L_{ms} & L_{ls} + L_{ms} & -\frac{1}{2}L_{ms} \\ -\frac{1}{2}L_{ms} & -\frac{1}{2}L_{ms} & L_{ls} + L_{ms} \end{bmatrix}}_{=\mathbf{L}_{abcs}} \underbrace{\begin{bmatrix} i_{as} \\ i_{bs} \\ i_{cs} \end{bmatrix}}_{=\mathbf{i}_{abcs}^s}, \tag{4.1}
$$

where L_{ls} is the stator leakage inductance. The superscript, 's' of $\mathbf{i}_{abcs}^s$ denotes a variable in the stationary frame. Fig. 4.1 (a) shows a flux linking between the stator coils; linking to the a-phase coil excited by the b phase current. The three phase inductance modeling was illustrated in Section 2.1.2. Fig. 4.1 (b) shows how the rotor current creates flux linking with the stator coil. Some of rotor flux lines (big loops) form a linking, while the others (small loops) do not. Since the rotor coil is fixed on the rotor surface, the linkage changes with the rotor angle, θ_r. Specifically,

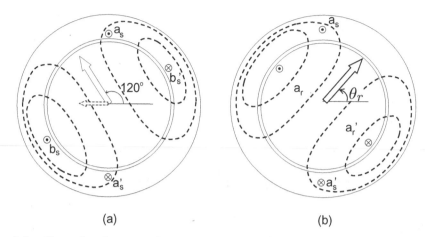

(a)   (b)

Figure 4.1: Contributions to the a-phase flux: (a) by the b-phase stator current and (b) by the a-phase rotor current.

the a-phase flux is described as $\lambda^s_{as} = L_{ms} \cos\theta_r i^r_{ar}$, where the superscript, 'r' signifies the variable in the rotor frame. The flux linking reaches the maximum when $\theta_r = 0$. On the other hand, if $\theta_r = \pi/2$, the flux linking is equal to zero. Since b and c phase windings are $2\pi/3$ and $4\pi/3$ apart in phase, the stator flux contributed by the rotor current is

$$\underbrace{L_{ms}\begin{bmatrix} \cos\theta_r & \cos(\theta_r + \frac{2\pi}{3}) & \cos(\theta_r - \frac{2\pi}{3}) \\ \cos(\theta_r - \frac{2\pi}{3}) & \cos\theta_r & \cos(\theta_r + \frac{2\pi}{3}) \\ \cos(\theta_r + \frac{2\pi}{3}) & \cos(\theta_r - \frac{2\pi}{3}) & \cos\theta_r \end{bmatrix}}_{\equiv \mathbf{M}(\theta_r)} \underbrace{\begin{bmatrix} i^r_{ar} \\ i^r_{br} \\ i^r_{cr} \end{bmatrix}}_{= \mathbf{i}^r_{abcr}} . \tag{4.2}$$

Complete Flux Linkage Equations

Combining $\mathbf{L}_{abcs}\mathbf{i}^s_{abcs}$ with $\mathbf{M}(\theta_r)\mathbf{i}^r_{abcr}$, a complete expression of the stator flux linkage is obtained such that

$$\lambda^s_{abcs} = \mathbf{L}_{abcs}\mathbf{i}^s_{abcs} + \mathbf{M}(\theta_r)\mathbf{i}^r_{abcr}. \tag{4.3}$$

Similarly, the rotor flux linkage seen from the rotor frame is given by

$$\lambda^r_{abcr} = \mathbf{M}(-\theta_r)\mathbf{i}^s_{abcs} + \mathbf{L}_{abcr}\mathbf{i}^r_{abcr}, \tag{4.4}$$

where

$$\lambda^r_{abcr} = \begin{bmatrix} \lambda^r_{ar} \\ \lambda^r_{br} \\ \lambda^r_{cr} \end{bmatrix} \text{ and } \mathbf{L}_{abcr} = \begin{bmatrix} L_{ms} + L_{lr} & -\frac{1}{2}L_{ms} & -\frac{1}{2}L_{ms} \\ -\frac{1}{2}L_{ms} & L_{ms} + L_{lr} & -\frac{1}{2}L_{ms} \\ -\frac{1}{2}L_{ms} & -\frac{1}{2}L_{ms} & L_{ms} + L_{lr} \end{bmatrix}$$

and L_{lr} is the rotor leakage inductance. From the rotor frame, the stator current looks behind by angle θ_r. Thus, $-\theta_r$ appears in $\mathbf{M}$. Note that $\mathbf{L}_{abcr}$ is the same as $\mathbf{L}_{abcs}$ except that the stator leakage inductance L_{ls} is replaced by the rotor leakage inductance L_{lr}.

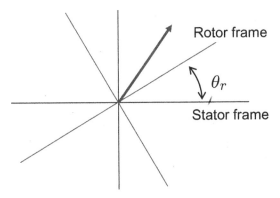

Figure 4.2: Rotor and stator frames.

Transformation into DQ Stationary Frame

As was shown in (2.58) and (2.60), mapping $\boldsymbol{\lambda}^s_{abcs}$ into the stationary $d-q$ frame is executed by multiplying $\mathbf{T}(0)$:

$$\boldsymbol{\lambda}^s_{dq0s} = \mathbf{T}(0)\boldsymbol{\lambda}^s_{abcs} = \mathbf{T}(0)\mathbf{L}_{abcs}\left[\mathbf{T}^{-1}(0)\mathbf{T}(0)\right]\mathbf{i}^s_{abcs} = \underbrace{\mathbf{T}(0)\mathbf{L}_{abc}\mathbf{T}^{-1}(0)}_{\equiv\, \mathbf{L}_{dq0}}\, \mathbf{i}^s_{dqs}, \quad (4.5)$$

where

$$\boldsymbol{\lambda}^s_{dq0s} = \begin{bmatrix} \lambda^s_{ds} \\ \lambda^s_{qs} \\ \lambda_0 \end{bmatrix} \quad \text{and} \quad \mathbf{L}_{dq0} = \begin{bmatrix} L_{ls}+\frac{3}{2}L_m & 0 & 0 \\ 0 & L_{ls}+\frac{3}{2}L_m & 0 \\ 0 & 0 & L_{ls} \end{bmatrix}. \quad (4.6)$$

Recall that the third component of a vector in the dq-frame is equal to zero in the balanced three phase system.

The rotor current contribution, $\mathbf{M}(\theta_r)\mathbf{i}^r_{abcr}$ to the stator flux also needs to be mapped into the same stationary frame. Complex number is more convenient in computing the transformation. When the third component of a three phase vector is ignored, it can be identified with a complex number:

$$\boldsymbol{\lambda}^s_{dq0s} = \begin{bmatrix} \lambda^s_{ds} \\ \lambda^s_{qs} \\ \lambda_0 \end{bmatrix} \quad \Longleftrightarrow \quad \lambda^s_{dqs} = \lambda^s_{ds} + j\lambda^s_{qs} = \frac{2}{3}\left[\lambda^s_{as} + e^{j\frac{2\pi}{3}}\lambda^s_{bs} + e^{j\frac{4\pi}{3}}\lambda^s_{cs}\right]. \quad (4.7)$$

Instead of multiplying by $T(0)$, the mapping rule (4.7) is utilized. The transformation in the stator part was already shown in (2.61); $\boldsymbol{\lambda}^s_{dq} = L_s\mathbf{i}^s_{dq}$, where

$L_s = \frac{3}{2}L_{ms} + L_{ls}$. The rotor coil contribution is transformed such that

$$\mathbf{M}(\theta_r)\mathbf{i}^r_{abcr} \iff \frac{2}{3}L_{ms}\left[\cos\theta_r i_{ar} + \cos(\theta_r + 2\pi/3)i_{br} + \cos(\theta_r - 2\pi/3)i_{cr}\right.$$

$$+e^{j\frac{2\pi}{3}}\left\{\cos(\theta_r - 2\pi/3)i_{ar} + \cos\theta_r i_{br} + \cos(\theta_r + 2\pi/3)i_{cr}\right\}$$

$$\left.+e^{-j\frac{2\pi}{3}}\left\{\cos(\theta_r + 2\pi/3)i_{ar} + \cos(\theta_r - 2\pi/3)i_{br} + \cos\theta_r i_{cr}\right\}\right]$$

$$= \frac{2}{3}\frac{1}{2}L_{ms}\left[(e^{j\theta_r} + e^{-j\theta_r})i_{ar} + (e^{j(\theta_r + \frac{2}{3}\pi)} + e^{-j(\theta_r + \frac{2}{3}\pi)})i_{br}\right.$$

$$+(e^{j(\theta_r - \frac{2}{3}\pi)} + e^{-j(\theta_r - \frac{2}{3}\pi)})i_{cr} + (e^{j\theta_r} + e^{-j(\theta_r - \frac{4}{3}\pi)})i_{ar}$$

$$+(e^{j(\theta_r + \frac{2}{3}\pi)} + e^{-j(\theta_r - \frac{2}{3}\pi)})i_{br} + (e^{j(\theta_r + \frac{2}{3}\pi)} + e^{-j\theta_r})i_{cr}$$

$$+(e^{j\theta_r} + e^{-j(\theta_r + \frac{4}{3}\pi)})i_{ar} + (e^{j(\theta_r - \frac{4}{3}\pi)} + e^{-j\theta_r})i_{br}$$

$$\left.+(e^{j(\theta_r - \frac{2}{3}\pi)} + e^{-j(\theta_r + \frac{2}{3}\pi)})i_{cr}\right]$$

$$= \frac{1}{2}L_{ms}\frac{2}{3}\left[3e^{j\theta_r}i_{ar} + 3e^{j(\theta_r + \frac{2}{3}\pi)}i_{br} + 3e^{j(\theta_r - \frac{2}{3}\pi)}i_{cr}\right]$$

$$= \left(\frac{3}{2}L_{ms}\right)e^{j\theta_r}\underbrace{\frac{2}{3}\left[i_{ar} + e^{j\frac{2}{3}\pi}i_{br} + e^{-j\frac{2}{3}\pi}i_{cr}\right]}_{= \mathbf{i}^r_{dqr}} = L_m e^{j\theta_r}\mathbf{i}^r_{dqr}, \quad (4.8)$$

where $L_m \equiv \frac{3}{2}L_{ms}$. Combining (2.61) with (4.8), we obtain the stator flux in the stationary frame as

$$\boldsymbol{\lambda}^s_{dqs} = L_s\mathbf{i}^s_{dqs} + L_m e^{j\theta_r}\mathbf{i}^r_{dqr}. \quad (4.9)$$

Note from Fig. 4.2 that the rotor reference frame is advanced by θ_r. Thereby, the rotor contribution to $\boldsymbol{\lambda}^s_{dqs}$ appears with $e^{j\theta_r}$.

In a similar manner, the rotor flux equation with reference to the rotor frame is given as

$$\boldsymbol{\lambda}^r_{dqr} = L_m e^{-j\theta_r}\mathbf{i}^s_{dqs} + L_r\mathbf{i}^r_{dqr}, \quad (4.10)$$

where $L_r = \frac{3}{2}L_{ms} + L_{lr}$. Since the rotor frame is referenced on $e^{j\theta_r}$, the stator current is seen behind by θ_r. Therefore, $e^{-j\theta_r}$ is multiplied to $\mathbf{i}^s_{dqs}$. Superscript 'r' is used to denote variables in the rotor frame.

Exercise 4.1   Consider a two pole IM with $L_{ms} = 1$mH and $L_{ls} = 70$ μH. Assume that the rotor flux angle increases according to $\theta_r = 300\,t$, and that

$$\mathbf{i}^s_{abcs} = 100\begin{bmatrix}\cos(312t + \frac{2\pi}{3})\\ \cos(312t)\\ \cos(312t - \frac{2\pi}{3})\end{bmatrix}, \quad \mathbf{i}^r_{abcr} = 30\begin{bmatrix}\cos(12t - \frac{\pi}{2})\\ \cos(12t - \frac{7\pi}{6})\\ \cos(12t - \frac{11\pi}{6})\end{bmatrix}.$$

Note that 12 Hz slip frequency appears in the rotor current. Follow the instructions below using MATLAB Simulink.

1) Build blocks for $\boldsymbol{\lambda}^s_{abcs}$ using (4.1), (4.2), and (4.3).

2) Build blocks for coordinate transformation and obtain $\boldsymbol{\lambda}^s_{dq0s} = T(0)\boldsymbol{\lambda}^s_{abcs}$.

3) Build blocks based on the *dq*-model for $\boldsymbol{\lambda}^s_{dqs}$ according to (4.7), (4.8), and (4.9).

4) Check whether two results obtained in 2) and 3) are the same, i.e., $\boldsymbol{\lambda}^s_{dq0s} = \boldsymbol{\lambda}^s_{dqs}$.

Solution. 1)

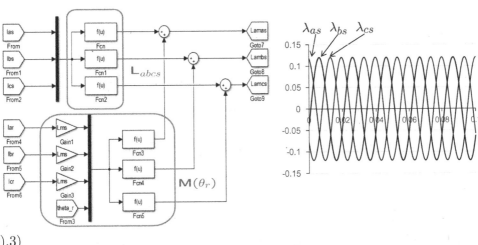

2),3)

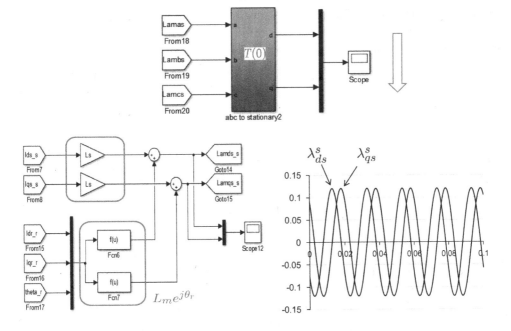

Transformation into Synchronous Reference Frame

The synchronous reference frame is a moving frame that rotates at the same speed as the electrical angular velocity, ω_e. It is the (angular) speed of the stator and

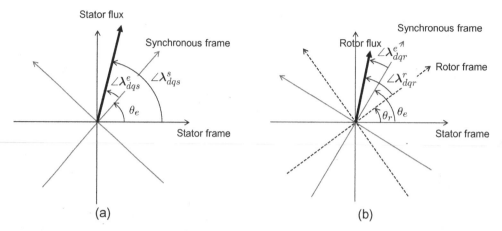

Figure 4.3: Transformation into the synchronous frame: (a) from the stationary (stator) frame and (b) from the rotor frame.

rotor flux. The d-axis angle of the synchronous frame is denoted by 'θ_e' and at time t

$$\theta_e = \int_0^t \omega_e dt = \omega_e t,$$

where ω_e is the electrical speed. In the following, the subscript, 'e' is used to indicate a variable in the synchronous frame. It was originated from the word, 'exciting'. Note that $\omega_e \neq \frac{P}{2}\omega_r$ in IMs, whereas they are the same in synchronous machines.

Since the synchronous frame is θ_e–ahead from the stationary frame, a vector would be represented with a smaller angle in the synchronous frame. Thereby, a vector in the synchronous frame is obtained by multiplying the stationary vector by $e^{-j\theta_e}$. Similarly, it is necessary to multiply the vector in the rotor frame by $e^{-j(\theta_e - \theta_r)}$. Specifically,

$$\boldsymbol{\lambda}_{dqs}^e = e^{-j\theta_e}\boldsymbol{\lambda}_{dqs}^s, \tag{4.11}$$
$$\boldsymbol{\lambda}_{dqr}^e = e^{-j(\theta_e - \theta_r)}\boldsymbol{\lambda}_{dqr}^r, \tag{4.12}$$

Flux angles seen from the different frames are depicted in Fig. 4.3. Note also that

$$\boldsymbol{\lambda}_{dqs}^e = L_{ls}\mathbf{i}_{dqs}^e + L_m(\mathbf{i}_{dqs}^e + \mathbf{i}_{dqr}^e) = L_s\mathbf{i}_{dqs}^e + L_m\mathbf{i}_{dqr}^e \tag{4.13}$$
$$\boldsymbol{\lambda}_{dqr}^e = L_{lr}\mathbf{i}_{dqr}^e + L_m(\mathbf{i}_{dqs}^e + \mathbf{i}_{dqr}^e) = L_m\mathbf{i}_{dqs}^e + L_r\mathbf{i}_{dqr}^e \tag{4.14}$$

since $L_s = L_{ls} + L_m$ and $L_r = L_{lr} + L_m$. The same equations are written in the matrix form such that

$$\begin{bmatrix} \lambda_{ds}^e \\ \lambda_{dr}^e \end{bmatrix} = \begin{bmatrix} L_s & L_m \\ L_m & L_r \end{bmatrix} \begin{bmatrix} i_{ds}^e \\ i_{dr}^e \end{bmatrix}, \tag{4.15}$$

$$\begin{bmatrix} \lambda_{qs}^e \\ \lambda_{qr}^e \end{bmatrix} = \begin{bmatrix} L_s & L_m \\ L_m & L_r \end{bmatrix} \begin{bmatrix} i_{qs}^e \\ i_{qr}^e \end{bmatrix}. \tag{4.16}$$

Flux to current relations are depicted in Fig. 4.4. Using the matrix inversion, we

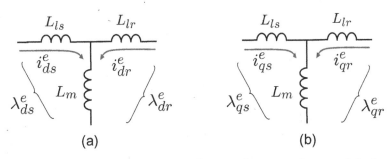

Figure 4.4: Flux to current relation in the synchronous frame: (a) d-axis and (b) q-axis.

obtain

$$\begin{bmatrix} i^e_{ds} \\ i^e_{dr} \end{bmatrix} = \frac{1}{\sigma L_s L_r} \begin{bmatrix} L_r & -L_m \\ -L_m & L_s \end{bmatrix} \begin{bmatrix} \lambda^e_{ds} \\ \lambda^e_{dr} \end{bmatrix}, \tag{4.17}$$

$$\begin{bmatrix} i^e_{qs} \\ i^e_{qr} \end{bmatrix} = \frac{1}{\sigma L_s L_r} \begin{bmatrix} L_r & -L_m \\ -L_m & L_s \end{bmatrix} \begin{bmatrix} \lambda^e_{qs} \\ \lambda^e_{qr} \end{bmatrix}. \tag{4.18}$$

where $\sigma = 1 - \frac{L_m^2}{L_r L_s}$. The flux equations are summarized in Table 4.1.

Table 4.1: Flux vectors described in different reference frames: (a) Stator flux linkage and (b) rotor flux linkage.

	Stator Flux	Rotor Flux
abc	(stationary frame) $\lambda^s_{abcs} = \mathbf{L}_{abcs}\mathbf{i}^s_{abcs} + \mathbf{M}(\theta_r)\mathbf{i}^r_{abcr}$	(rotor frame) $\lambda^r_{abcr} = \mathbf{M}(-\theta_r)\mathbf{i}^s_{abcs} + \mathbf{L}_{abcr}\mathbf{i}^r_{abcr}$
dq	(stationary frame) $\lambda^s_{dqs} = L_s\mathbf{i}^s_{dqs} + L_m e^{j\theta_r}\mathbf{i}^r_{dqr}$	(rotor frame) $\lambda^r_{dqr} = L_m e^{-j\theta_r}\mathbf{i}^s_{dqs} + L_r\mathbf{i}^r_{dqr}$
dq	(synchronous frame) $\lambda^e_{dqs} = L_s e^{-j\theta_e}\mathbf{i}^s_{dqs} + L_m e^{-j(\theta_e - \theta_r)}\mathbf{i}^r_{dqr}$ $= L_s\mathbf{i}^e_{dqs} + L_m\mathbf{i}^e_{dqr}$	(synchronous frame) $\lambda^e_{dqr} = L_m e^{-j\theta_e}\mathbf{i}^s_{dqs} + L_r e^{-j(\theta_e - \theta_r)}\mathbf{i}^r_{dqr}$ $= L_r\mathbf{i}^e_{dqr} + L_m\mathbf{i}^e_{dqr}$

Exercise 4.2 Show that

$$\lambda^e_{dqs} = \sigma L_s\mathbf{i}^e_{dqs} + \frac{L_m}{L_r}\lambda^e_{dqr}, \tag{4.19}$$

$$\lambda^e_{dqr} = \frac{L_m}{L_s}\lambda^e_{dqs} + \sigma L_r\mathbf{i}^e_{dqr}. \tag{4.20}$$

4.1.2 Voltage Equations

Let $\mathbf{p} = \frac{d}{dt}$. Voltage equations for the stator and rotor are given by

$$\mathbf{v}^s_{abcs} = r_s\mathbf{i}^s_{abcs} + \mathbf{p}\lambda^s_{abcs}, \tag{4.21}$$

$$\mathbf{v}^r_{abcr} = r_r\mathbf{i}^r_{abcr} + \mathbf{p}\lambda^r_{abcr}. \tag{4.22}$$

Note again that the rotor equation is written with respect to the rotor frame, whereas the stator equation is given with reference to the stationary frame. By applying $T(0)$ to both sides and deleting the third components, we obtain the voltage equations in the dq stationary and rotor frames such that

$$v_{dqs}^s = r_s i_{dqs}^s + p\lambda_{dqs}^s, \tag{4.23}$$

$$v_{dqr}^r = r_r i_{dqr}^r + p\lambda_{dqr}^r. \tag{4.24}$$

Or, the same equations are obtained via matrix multiplication (2.40). Hereforth, both vectors and complex variables are represented by the same bold face by abusing notation slightly.

In Stationary Frame

Since the stator voltage equation (4.23) is already in the stationary frame, it remains to transform the rotor equation (4.24). It can be done by multiplying $e^{j\theta_r}$ on both sides of (4.24):

$$v_{dqr}^s = e^{j\theta_r} v_{dqr}^r = r_s e^{j\theta_r} i_{dqr}^r + e^{j\theta_r} p\left[e^{-j\theta_r} e^{j\theta_r} \lambda_{dqr}^r\right]$$

$$= r_r i_{dqr}^s + p(-j\omega_r)\lambda_{dqr}^s. \tag{4.25}$$

Note that the differential operator, 'p' and an operand, '$e^{-j\theta_r}$' do not commute, i.e., $pe^{-j\theta_r} \neq e^{-j\theta_r}p$ since θ_r is a function of t. Therefore, a technique of inserting the identity $1 = e^{j\theta_r} e^{-j\theta_r}$ needs to be utilized [1].

When the secondary circuit is shorted as in the squirrel cage IM, $v_{dqr}^r = v_{dqr}^s = 0$. Thus, the rotor equations are given componentwise as

$$0 = r_s i_{dr}^s + p\lambda_{dr}^s + \omega_r \lambda_{qr}^s \tag{4.26}$$

$$0 = r_s i_{qr}^s + p\lambda_{qr}^s - \omega_r \lambda_{dr}^s . \tag{4.27}$$

Together with (4.13) and (4.14), we have an IM model in the stationary frame:

$$\begin{bmatrix} v_{ds}^s \\ v_{qs}^s \\ 0 \\ 0 \end{bmatrix} = \begin{bmatrix} r_s + pL_s & 0 & pL_m & 0 \\ 0 & r_s + pL_s & 0 & pL_m \\ pL_m & \omega_r L_m & r_r + pL_r & \omega_r L_r \\ -\omega_r L_m & pL_m & -\omega_r L_r & r_r + pL_r \end{bmatrix} \begin{bmatrix} i_{ds}^s \\ i_{qs}^s \\ i_{dr}^s \\ i_{qr}^s \end{bmatrix}. \tag{4.28}$$

The block diagram based on (4.28) is shown in Fig. 4.5. The model (4.28) can be rewritten in an ordinary differential form as

$$p\begin{bmatrix} i_{dqs}^s \\ i_{dqr}^s \end{bmatrix} = \frac{1}{\sigma L_s L_r} \begin{bmatrix} L_r I & -L_m I \\ -L_m I & L_s I \end{bmatrix} \left(\begin{bmatrix} r_s I & O \\ -\omega_r L_m J & -\omega_r L_r J \end{bmatrix} \begin{bmatrix} i_{dqs}^s \\ i_{dqr}^s \end{bmatrix} + \begin{bmatrix} v_{dqs}^s \\ 0 \end{bmatrix} \right),$$

$$\tag{4.29}$$

where

$$I = \begin{bmatrix} 1 & 0 \\ 0 & 1 \end{bmatrix} \quad J = \begin{bmatrix} 0 & -1 \\ 1 & 0 \end{bmatrix}, \quad \text{and} \quad O = \begin{bmatrix} 0 & 0 \\ 0 & 0 \end{bmatrix}.$$

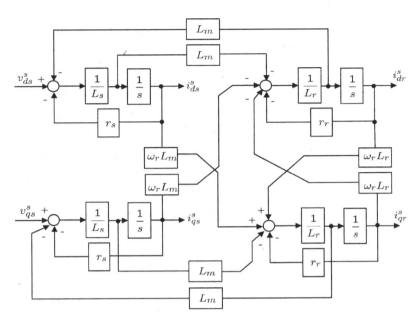

Figure 4.5: An IM model in the stationary frame.

Exercise 4.3 Consider a two pole IM with $r_s = 0.2\,\Omega$, $r_r = 0.16\,\Omega$, $L_{ms} = 1\,\text{mH}$, $L_{ls} = 70\,\mu\text{H}$, and $L_{lr} = 70\,\mu\text{H}$. Construct a MATLAB Simulink block diagram for an IM model in the stationary frame based on Fig. 4.5. Apply $\mathbf{v}^s_{abcs}(t) = 170\,[\cos 312t,\ \cos(312t - 2\pi/3),\ \cos(312t - 4\pi/3)]^T$ when $\omega_r = 300$ rad/sec. Plot $[i^s_{ds}(t),\ i^s_{qs}(t)]$ and $\mathbf{i}^s_{abcs}(t) = T^{-1}(0)[i^s_{ds}(t),\ i^s_{qs}(t),\ 0]^T$ for $0 \le t \le 0.1$ sec.

Solution. See Fig. 4.6 and Fig. 4.7. ∎

In Synchronous Frame

By multiplying $e^{-j\theta_e}$ to $\mathbf{v}^s_{dqs}$ in the stationary frame, we have a new representation $\mathbf{v}^e_{dqs}$ in the synchronous frame, i.e.,

$$
\begin{aligned}
\mathbf{v}^e_{dqs} &= r_s \mathbf{i}^e_{dqs} + e^{-j\theta_e}\mathrm{p}\big[e^{j\theta_e}e^{-j\theta_e}\boldsymbol{\lambda}^s_{dqs}\big] \\
&= r_s \mathbf{i}^e_{dqs} + \mathrm{p}\boldsymbol{\lambda}^e_{dqs} + j\omega_e\boldsymbol{\lambda}^e_{dqs}.
\end{aligned}
\tag{4.30}
$$

Likewise in the previous transformation, the additional term, $j\omega_e\boldsymbol{\lambda}^e_{dqs}$ appears in the synchronous reference frame. It is known as the speed voltage, and called the *coupling voltage*. Let $\mathbf{v}^e_{dqs} = v^e_{ds} + jv^e_{qs}$ and $\boldsymbol{\lambda}^e_{dqs} = \lambda^e_{ds} + j\lambda^e_{qs}$. Then, (4.30) is written componentwise as

$$
v^e_{ds} = r_s i^e_{ds} + \mathrm{p}\lambda^e_{ds} - \omega_e\lambda^e_{qs}
\tag{4.31}
$$

$$
v^e_{qs} = r_s i^e_{qs} + \mathrm{p}\lambda^e_{qs} + \omega_e\lambda^e_{ds}.
\tag{4.32}
$$

The rotor voltage equation can be transformed in the same way. But, the transforming angle is $\theta_e - \theta_r$. Multiplying both sides of (4.24) by $e^{-j(\theta_e-\theta_r)}$, we have

$$
\mathbf{v}^e_{dqr} = r_r \mathbf{i}^e_{dqr} + \mathrm{p}\boldsymbol{\lambda}^e_{dqr} + j(\omega_e - \omega_r)\boldsymbol{\lambda}^e_{dqr}.
\tag{4.33}
$$

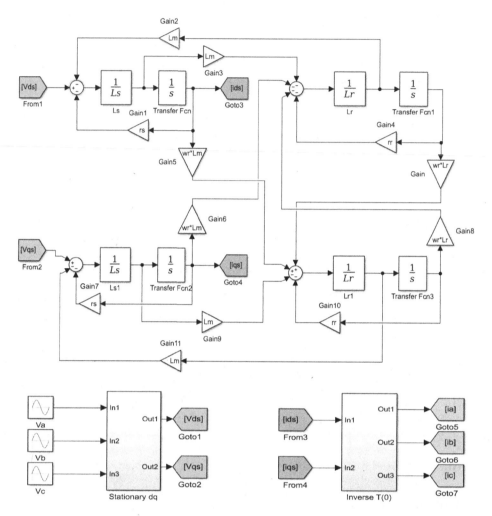

Figure 4.6: Simulink IM model in the stationary frame (Solution to Exercise 4.3).

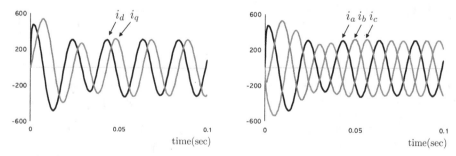

Figure 4.7: Simulation results (Solution to Exercise 4.3).

Componentwise, it is equal to

$$0 = r_r i_{dr}^e + \mathrm{p}\lambda_{dr}^e - (\omega_e - \omega_r)\lambda_{qr}^e \tag{4.34}$$

$$0 = r_r i_{qr}^e + \mathrm{p}\lambda_{qr}^e + (\omega_e - \omega_r)\lambda_{dr}^e . \tag{4.35}$$

Comparing the stator and rotor voltage equations, one can see the identical structure. However, the slip speed, $\omega_e - \omega_r$ appears in (4.34) and (4.35), instead of ω_r.

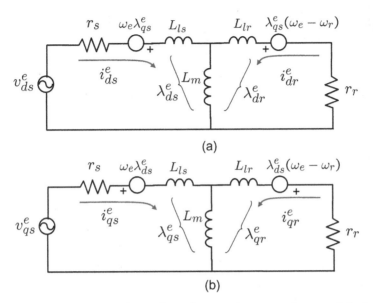

(a)

(b)

Figure 4.8: Equivalent circuit for IM in the synchronous frame: (a) d-axis and (b) q-axis.

Based on (4.13), (4.14), and (4.31)–(4.35), the equivalent circuits in the synchronous frame can be drawn as Fig. 4.8. The above voltage equations are mixed since both flux and current variables appear in (4.31)–(4.35). Fig. 4.9 shows a simulation block diagram for IM in the synchronous frame that includes the flux to current relation, (4.17) and (4.18). Utilizing (4.13) and (4.14), we can express the IM model only with the current variables:

$$\begin{bmatrix} v_{ds}^e \\ v_{qs}^e \\ 0 \\ 0 \end{bmatrix} = \begin{bmatrix} r_s + \mathrm{p}L_s & -\omega_e L_s & \mathrm{p}L_m & -\omega_e L_m \\ \omega_e L_s & r_s + \mathrm{p}L_s & \omega_e L_m & \mathrm{p}L_m \\ \mathrm{p}L_m & -\omega_{sl} L_m & r_r + \mathrm{p}L_r & -\omega_{sl} L_r \\ \omega_{sl} L_m & \mathrm{p}L_m & \omega_{sl} L_r & r_r + \mathrm{p}L_r \end{bmatrix} \begin{bmatrix} i_{ds}^e \\ i_{qs}^e \\ i_{dr}^e \\ i_{qr}^e \end{bmatrix}. \qquad (4.36)$$

Exercise 4.4 (Continued from Exercise 4.3) Construct a MATLAB Simulink block diagram for the IM model in the synchronous frame using (4.31)–(4.35). Apply the same voltage in Exercise 4.3. Use the inverse transformation, obtain $\mathbf{i}_{abcs}^s(t) = T^{-1}(\omega_e t)[i_{ds}^e(t), i_{qs}^e(t), 0]^T$. Check that the resulting currents are the same.

4.1.3 Transformation via Matrix Multiplications

In this section we will repeat the same coordinate transformation via matrix multiplications. Some mathematical preliminaries are given in the following exercises:

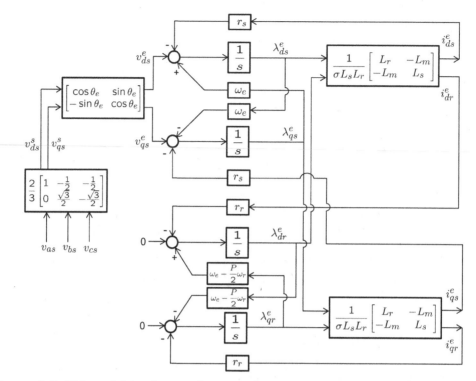

Figure 4.9: IM model in the synchronous frame including the flux to current relation, (4.17) and (4.18).

Exercise 4.5   Show that

$$\begin{bmatrix} \cos\theta & -\sin\theta \\ \sin\theta & \cos\theta \end{bmatrix} = e^{\mathbf{J}\theta}, \qquad \text{where} \qquad \mathbf{J} = \begin{bmatrix} 0 & -1 \\ 1 & 0 \end{bmatrix}. \tag{4.37}$$

Solution

$$e^{\mathbf{J}\theta} = \mathcal{L}^{-1}\left\{(s\mathbf{I}-\mathbf{J})^{-1}\right\} = \mathcal{L}^{-1}\left\{\frac{1}{s^2+1}\begin{bmatrix} s & -1 \\ 1 & s \end{bmatrix}\right\} = \begin{bmatrix} \cos\theta & -\sin\theta \\ \sin\theta & \cos\theta \end{bmatrix}.$$

Note that $\mathbf{J}$ is skew symmetric, i.e., $\mathbf{J}^{-1} = \mathbf{J}^T$ and that $\mathbf{J}$ is an operator that rotates a vector 90 degrees counterclockwise [2]. Hence, multiplying $\mathbf{J}$ to a vector is equivalent to multiplying j to its corresponding complex variable.   ∎

Exercise 4.6   Let $\theta = \omega t$. Show that

$$\mathbf{T}(\theta)\mathbf{p}\mathbf{T}^{-1}(\theta) = \omega\begin{bmatrix} 0 & -1 & 0 \\ 1 & 0 & 0 \\ 0 & 0 & 1 \end{bmatrix}.$$

Exercise 4.7   Show that $\mathbf{A}e^{\mathbf{A}} = e^{\mathbf{A}}\mathbf{A}$ for $\mathbf{A} \in \mathbb{R}^{n\times n}$.

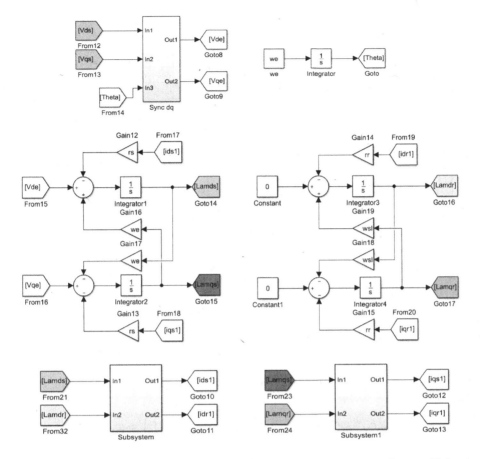

Figure 4.10: Simulink block diagram for IM in the synchronous frame (Solution to Exercise 4.4).

Exercise 4.8   Show that

$$e^{\mathbf{J}\theta_e}\mathrm{p}e^{-\mathbf{J}\theta_e} = -\omega_e\mathbf{J}. \tag{4.38}$$

Solution

$$e^{\mathbf{J}\theta_e}\mathrm{p}e^{-\mathbf{J}\theta_e} = e^{\mathbf{J}\theta_e}(-\mathbf{J})e^{-\mathbf{J}\theta_e}\frac{d\theta}{dt} = -\omega_e\mathbf{J}e^{\mathbf{J}\theta_e}e^{-\mathbf{J}\theta_e} = -\omega_e\mathbf{J}$$

Alternatively,

$$e^{\mathbf{J}\theta_e}\mathrm{p}e^{-\mathbf{J}\theta_e} = \omega_e\begin{bmatrix}\cos\theta_e & -\sin\theta_e\\ \sin\theta_e & \cos\theta_e\end{bmatrix}\begin{bmatrix}-\sin\theta_e & \cos\theta_e\\ -\cos\theta_e & -\sin\theta_e\end{bmatrix} = \omega_e\begin{bmatrix}0 & 1\\ -1 & 0\end{bmatrix} = -\omega_e\mathbf{J}. \blacksquare$$

From the above example, we can see that $e^{\mathbf{J}\theta_e}e^{-\mathbf{J}\theta_e} = e^{-\mathbf{J}\theta_e}e^{\mathbf{J}\theta_e} = \mathbf{I}$ and $e^{\mathbf{J}\theta_e}\mathbf{J} = \mathbf{J}e^{\mathbf{J}\theta_e}$. Note from (4.9) and (4.23) that

$$\begin{bmatrix}v_{ds}^s\\ v_{qs}^s\end{bmatrix} = r_s\begin{bmatrix}i_{ds}^s\\ i_{qs}^s\end{bmatrix} + L_s\mathrm{p}\begin{bmatrix}i_{ds}^s\\ i_{qs}^s\end{bmatrix} + L_m e^{\mathbf{J}\theta_r}\mathrm{p}\begin{bmatrix}i_{dr}^r\\ i_{qr}^r\end{bmatrix} + \omega_r L_m\mathbf{J}e^{\mathbf{J}\theta_r}\begin{bmatrix}i_{dr}^r\\ i_{qr}^r\end{bmatrix}. \tag{4.39}$$

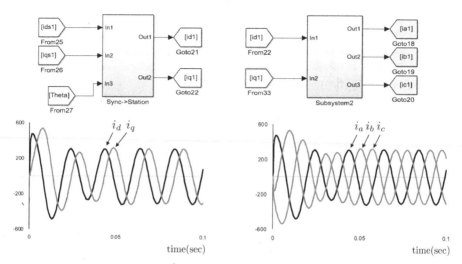

Figure 4.11: Coordinate transformation and resulting currents (Solution to Exercise 4.4).

Multiplying both sides of (4.39) by $e^{-\mathbf{J}\theta_e}$, we obtain

$$
e^{-\mathbf{J}\theta_e}\begin{bmatrix} v_{ds}^s \\ v_{qs}^s \end{bmatrix} = r_s e^{-\mathbf{J}\theta_e}\begin{bmatrix} i_{ds}^s \\ i_{qs}^s \end{bmatrix} + L_s e^{-\mathbf{J}\theta_e}\mathrm{p}\left(e^{\mathbf{J}\theta_e} e^{-\mathbf{J}\theta_e}\begin{bmatrix} i_{ds}^s \\ i_{qs}^s \end{bmatrix}\right)
$$

$$
+ L_m e^{-\mathbf{J}\theta_e} e^{\mathbf{J}\theta_r}\mathrm{p}\left(e^{-\mathbf{J}\theta_r} e^{\mathbf{J}\theta_e} e^{-\mathbf{J}\theta_e} e^{\mathbf{J}\theta_r}\begin{bmatrix} i_{dr}^r \\ i_{qr}^r \end{bmatrix}\right) + \omega_r L_m \mathbf{J} e^{-\mathbf{J}\theta_e} e^{\mathbf{J}\theta_r}\begin{bmatrix} i_{dr}^r \\ i_{qr}^r \end{bmatrix}
$$

$$
= r_s\begin{bmatrix} i_{ds}^e \\ i_{qs}^e \end{bmatrix} + L_s e^{-\mathbf{J}\theta_e}\mathrm{p}\left(e^{\mathbf{J}\theta_e}\right)\begin{bmatrix} i_{ds}^e \\ i_{qs}^e \end{bmatrix} + L_s\mathrm{p}\begin{bmatrix} i_{ds}^e \\ i_{qs}^e \end{bmatrix}
$$

$$
+ L_m e^{-\mathbf{J}(\theta_e-\theta_r)}\mathrm{p}\left(e^{\mathbf{J}(\theta_e-\theta_r)}\right) e^{-\mathbf{J}(\theta_e-\theta_r)}\begin{bmatrix} i_{dr}^r \\ i_{qr}^r \end{bmatrix}
$$

$$
+ L_m\mathrm{p}\left(e^{-\mathbf{J}(\theta_e-\theta_r)}\begin{bmatrix} i_{dr}^r \\ i_{qr}^r \end{bmatrix}\right) + \omega_r L_m \mathbf{J}\begin{bmatrix} i_{dr}^e \\ i_{qr}^e \end{bmatrix}.
$$

$$
\begin{bmatrix} v_{ds}^e \\ v_{qs}^e \end{bmatrix} = r_s\begin{bmatrix} i_{ds}^e \\ i_{qs}^e \end{bmatrix} + \omega_e L_s\mathbf{J}\begin{bmatrix} i_{ds}^e \\ i_{qs}^e \end{bmatrix} + L_s\mathrm{p}\begin{bmatrix} i_{ds}^e \\ i_{qs}^e \end{bmatrix} + (\omega_e-\omega_r) L_m\mathbf{J}\begin{bmatrix} i_{dr}^e \\ i_{qr}^e \end{bmatrix}
$$

$$
+ L_m\mathrm{p}\begin{bmatrix} i_{dr}^e \\ i_{qr}^e \end{bmatrix} + \omega_r L_m\mathbf{J}\begin{bmatrix} i_{dr}^e \\ i_{qr}^e \end{bmatrix}
$$

$$
= r_s\begin{bmatrix} i_{ds}^e \\ i_{qs}^e \end{bmatrix} + L_s(\mathrm{p}\mathbf{I}+\omega_e\mathbf{J})\begin{bmatrix} i_{ds}^e \\ i_{qs}^e \end{bmatrix} + L_m(\mathrm{p}\mathbf{I}+\omega_e\mathbf{J})\begin{bmatrix} i_{dr}^e \\ i_{qr}^e \end{bmatrix}
$$

$$
= r_s\begin{bmatrix} i_{ds}^e \\ i_{qs}^e \end{bmatrix} + (\mathrm{p}\mathbf{I}+\omega_e\mathbf{J})\begin{bmatrix} \lambda_{ds}^e \\ \lambda_{qs}^e \end{bmatrix}. \tag{4.40}
$$

One can see that the matrix manipulation yields the same result as (4.30).

Using the unit vectors, $\mathbf{i}_x$, $\mathbf{i}_y$, and $\mathbf{i}_z$, it follows that

$$\boldsymbol{\omega} \times \boldsymbol{\lambda}_{dqs}^e = \begin{vmatrix} \mathbf{i}_x & \mathbf{i}_y & \mathbf{i}_z \\ 0 & 0 & \omega_e \\ \lambda_{ds}^e & \lambda_{qs}^e & 0 \end{vmatrix} = -\omega_e \lambda_{qs}^e \mathbf{i}_x + \omega_e \lambda_{ds}^e \mathbf{i}_y \Leftrightarrow \begin{bmatrix} -\omega_e \lambda_{qs}^e \\ \omega_e \lambda_{ds}^e \\ 0 \end{bmatrix}$$

Therefore, utilizing the cross product, the voltage equation can be expressed alternatively as

$$\mathbf{v}_{dqs}^e = r_s \mathbf{i}_{dqs}^e + \frac{d}{dt} \boldsymbol{\lambda}_{dqs}^e + \boldsymbol{\omega} \times \boldsymbol{\lambda}_{dqs}^e. \tag{4.41}$$

The three expressions are summarized in Table 4.2.

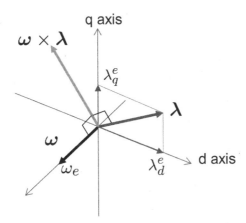

Figure 4.12: Coupling voltage associated with the coordinate rotation.

Table 4.2: Speed voltage representation

Complex number	Skew matrix	Cross product
$j\omega_e \boldsymbol{\lambda}_{dqs}^e$	$\omega_e \mathbf{J} \boldsymbol{\lambda}_{dqs}^e$	$\boldsymbol{\omega} \times \boldsymbol{\lambda}_{dqs}^e$

Summary

The voltage equation changes according to the reference frame where it is described. Another voltage term, called the *speed voltage*, takes place when the voltage equations are expressed in a rotating frame. With the complex variables, it appears as $j\omega_e \boldsymbol{\lambda}_{dqs}^e$. With the vector notation, it is shown to be $\omega_e \mathbf{J} \boldsymbol{\lambda}_{dqs}^e$. Finally, it can be described by the vector *cross product*, '$\times$'. Note from Fig. 4.12 that the speed vector $\boldsymbol{\omega} = [0, \, 0, \, \omega_e]^T$ is defined as an orthogonal vector to the plane where the flux vectors are defined.

4.2   IM Dynamic Models

As we change the state variables, various dynamic models are obtained. The starting equation is the following simple model. But the variables are redundant since the model contains both flux and current variables:

$$v_{dqs}^e = r_s i_{dqs}^e + \frac{d\lambda_{dqs}^e}{dt} + jw_e \lambda_{dqs}^e \tag{4.42}$$

$$0 = r_r i_{dqr}^e + \frac{d\lambda_{dqr}^e}{dt} + j(w_e - w_r)\lambda_{dqr}^e. \tag{4.43}$$

To numerically solve IM dynamics or to construct a state observer, the equations must be changed in the form of an ordinary differential equation (ODE).

4.2.1   ODE Model with Current Variables

In this subsection, we derive an IM model in the ODE form with the current variables. Note from the third line of (4.36) that

$$p i_{dr}^e = -\frac{1}{L_r} \left(L_m p i_{ds}^e - w_{sl} L_m i_{qs}^e + r_r i_{dr}^e - w_{sl} L_r i_{qr} \right). \tag{4.44}$$

Substituting (4.44) into the first line of (4.36), we have

$$\sigma L_s p i_{ds}^e = -r_s i_{ds}^e + (w_e L_s - w_{sl} \frac{L_m^2}{L_r}) i_{qs}^e + \frac{L_m}{L_r} r_r i_{dr}^e + L_m (w_e - w_{sl}) i_{qr}.$$

Note that

$$\frac{w_e}{\sigma} - \frac{L_m^2}{\sigma L_s L_r} w_{sl} = w_e + w_r \frac{L_m^2}{\sigma L_s L_r} = w_e + w_r \frac{1-\sigma}{\sigma}. \tag{4.45}$$

Therefore, it follows that

$$p i_{ds}^e = -\frac{r_s}{\sigma L_s} i_{ds}^e + (w_e + w_r \frac{1-\sigma}{\sigma}) i_{qs}^e + \frac{L_m}{\sigma L_s} \frac{r_r}{L_r} i_{dr}^e + \frac{L_m}{\sigma L_s} w_r i_{qr}$$

In the similar fashion, expressions for $p i_{qs}^e$, $p i_{dr}^e$, and $p i_{qr}^e$ can be derived. Summarizing the result, we obtain an IM dynamic model in the ODE form:

$$\frac{d}{dt} \begin{bmatrix} i_{ds}^e \\ i_{qs}^e \\ i_{dr}^e \\ i_{qr}^e \end{bmatrix} = \begin{bmatrix} -\frac{r_s}{\sigma L_s} & w_e + w_r \frac{1-\sigma}{\sigma} & \frac{L_m r_r}{\sigma L_s L_r} & w_r \frac{L_m}{\sigma L_s} \\ -w_e - w_r \frac{1-\sigma}{\sigma} & -\frac{r_s}{\sigma L_s} & -w_r \frac{L_m}{\sigma L_s} & \frac{L_m r_r}{\sigma L_s L_r} \\ \frac{L_m r_s}{\sigma L_r L_s} & -w_r \frac{L_m}{\sigma L_r} & -\frac{r_r}{\sigma L_r} & w_e - \frac{w_r}{\sigma} \\ w_r \frac{L_m}{\sigma L_r} & \frac{L_m r_s}{\sigma L_r L_s} & -w_e + \frac{w_r}{\sigma} & -\frac{r_r}{\sigma L_r} \end{bmatrix} \begin{bmatrix} i_{ds}^e \\ i_{qs}^e \\ i_{dr}^e \\ i_{qr}^e \end{bmatrix}$$

$$+ \begin{bmatrix} \frac{1}{\sigma L_s} & 0 \\ 0 & \frac{1}{\sigma L_s} \\ -\frac{L_m}{\sigma L_r L_s} & 0 \\ 0 & -\frac{L_m}{\sigma L_r L_s} \end{bmatrix} \begin{bmatrix} v_{ds}^e \\ v_{qs}^e \end{bmatrix}. \tag{4.46}$$

Equivalently, (4.46) can be expressed as

$$\frac{d}{dt}\begin{bmatrix} \mathbf{i}^e_{dqs} \\ \mathbf{i}^e_{dqr} \end{bmatrix} = \begin{bmatrix} -\frac{r_s}{\sigma L_s}\mathbf{I} - \left(\omega_e + \omega_r\frac{1-\sigma}{\sigma}\right)\mathbf{J} & \frac{L_m}{\sigma L_s}\left(\frac{1}{\tau_r}\mathbf{I} - \omega_r\mathbf{J}\right) \\ \frac{L_m}{\sigma L_r}\left(\frac{r_s}{L_s}\mathbf{I} + \omega_r\mathbf{J}\right) & -\frac{1}{\sigma\tau_r}\mathbf{I} - \left(\omega_e - \frac{\omega_r}{\sigma}\right)\mathbf{J} \end{bmatrix}\begin{bmatrix} \mathbf{i}^e_{dqs} \\ \mathbf{i}^e_{dqr} \end{bmatrix} + \begin{bmatrix} \frac{1}{\sigma L_s}\mathbf{I} \\ -\frac{L_m}{\sigma L_r L_s}\mathbf{I} \end{bmatrix}\mathbf{v}^e_{dqs}$$

$$(4.47)$$

where $\tau_r = L_r/r_r$ is rotor time constant.

The above ODE model can be derived more systematically: The IM dynamics (4.36) is written as

$$p\begin{bmatrix} L_s\mathbf{I} & L_m\mathbf{I} \\ L_m\mathbf{I} & L_r\mathbf{I} \end{bmatrix}\begin{bmatrix} \mathbf{i}^e_{dqs} \\ \mathbf{i}^e_{dqr} \end{bmatrix} = \begin{bmatrix} -r_s\mathbf{I} - \omega_e\mathbf{J} & -L_m\mathbf{J} \\ -L_m\mathbf{J} & -r_r\mathbf{I} - \omega_{sl}\mathbf{J} \end{bmatrix}\begin{bmatrix} \mathbf{i}^e_{dqs} \\ \mathbf{i}^e_{dqr} \end{bmatrix} + \begin{bmatrix} \mathbf{v}^e_{dqs} \\ 0 \end{bmatrix}. \qquad (4.48)$$

Note that

$$\begin{bmatrix} L_s\mathbf{I} & \omega_e L_m\mathbf{I} \\ \omega_e L_m\mathbf{I} & L_r\mathbf{I} \end{bmatrix}^{-1} = \frac{1}{L_s L_r \sigma}\begin{bmatrix} L_r\mathbf{I} & -L_m\mathbf{I} \\ -L_m\mathbf{I} & L_s\mathbf{I} \end{bmatrix}. \qquad (4.49)$$

Multiplying (4.49) on both sides of (4.48), the model (4.47) is obtained.

4.2.2 IM ODE Model with Current-Flux Variables

For the field oriented control that appears later, it is better to select flux as the rotor variable. It is desired here to use $[\mathbf{i}^{e\,T}_{dqs}, \boldsymbol{\lambda}^{e\,T}_{dqr}]^T$ as the stator variable. It follows from (4.14) that

$$\begin{bmatrix} \mathbf{i}^e_{dqs} \\ \mathbf{i}^e_{dqr} \end{bmatrix} = \begin{bmatrix} \mathbf{I} & \mathbf{O} \\ -\frac{L_m}{L_r}\mathbf{I} & \frac{1}{L_r}\mathbf{I} \end{bmatrix}\begin{bmatrix} \mathbf{i}^e_{dqs} \\ \boldsymbol{\lambda}^e_{dqr} \end{bmatrix}. \qquad (4.50)$$

Substituting (4.50) into (4.36), we obtain

$$\begin{bmatrix} \mathbf{v}^e_{dqs} \\ 0 \end{bmatrix} = \begin{bmatrix} (r_s + pL_s)\mathbf{I} + \omega_e L_s\mathbf{J} & pL_m\mathbf{I} + \omega_e L_m\mathbf{J} \\ pL_m\mathbf{I} + \omega_{sl}L_m\mathbf{J} & (r_r + pL_r)\mathbf{I} + \omega_e L_r\mathbf{J} \end{bmatrix}\begin{bmatrix} \mathbf{I} & \mathbf{O} \\ -\frac{L_m}{L_r}\mathbf{I} & \frac{1}{L_r}\mathbf{I} \end{bmatrix}\begin{bmatrix} \mathbf{i}^e_{dqs} \\ \boldsymbol{\lambda}^e_{dqr} \end{bmatrix}$$

$$= \begin{bmatrix} (r_s + p\sigma L_s)\mathbf{I} + \omega_e\sigma L_s\mathbf{J} & p\frac{L_m}{L_r}\mathbf{I} + \omega_e\frac{L_m}{L_r}\mathbf{J} \\ -\frac{r_r L_m}{L_r}\mathbf{I} & \left(\frac{r_r}{L_r} + p\right)\mathbf{I} + \omega_{sl}\mathbf{J} \end{bmatrix}\begin{bmatrix} \mathbf{i}^e_{dqs} \\ \boldsymbol{\lambda}^e_{dqr} \end{bmatrix}. \qquad (4.51)$$

Collecting the terms with differential operator on the left hand side, we obtain

$$\begin{bmatrix} \sigma L_s\mathbf{I} & \frac{L_m}{L_r}\mathbf{I} \\ \mathbf{O} & \mathbf{I} \end{bmatrix}p\begin{bmatrix} \mathbf{i}^e_{dqs} \\ \boldsymbol{\lambda}^e_{dqr} \end{bmatrix} = \begin{bmatrix} \mathbf{v}^e_{dqs} \\ 0 \end{bmatrix} - \begin{bmatrix} r_s\mathbf{I} + \omega_e\sigma L_s\mathbf{J} & +\omega_e\frac{L_m}{L_r}\mathbf{J} \\ -\frac{r_r L_m}{L_r}\mathbf{I} & \frac{r_r}{L_r}\mathbf{I} + \omega_{sl}\mathbf{J} \end{bmatrix}\begin{bmatrix} \mathbf{i}^e_{dqs} \\ \boldsymbol{\lambda}^e_{dqr} \end{bmatrix}. \qquad (4.52)$$

In taking the inverse of the matrix in the left hand side of (4.52), we utilize the following identity

$$\begin{bmatrix} \mathbf{A} & \mathbf{B} \\ \mathbf{O} & \mathbf{C} \end{bmatrix}^{-1} = \begin{bmatrix} \mathbf{A}^{-1} & -\mathbf{A}^{-1}\mathbf{B}\mathbf{C}^{-1} \\ \mathbf{O} & \mathbf{C}^{-1} \end{bmatrix},$$

where $\mathbf{A}, \mathbf{C} \in \mathbb{R}^{n\times n}$ are invertible, and $\mathbf{O} \in \mathbb{R}^{n\times n}$ is a zero matrix. Then,

$$\begin{bmatrix} \sigma L_s\mathbf{I} & \frac{L_m}{L_r}\mathbf{I} \\ \mathbf{O} & \mathbf{I} \end{bmatrix}^{-1} = \begin{bmatrix} \frac{1}{\sigma L_s}\mathbf{I} & -\frac{L_m}{\sigma L_r L_s}\mathbf{I} \\ \mathbf{O} & \mathbf{I} \end{bmatrix}. \qquad (4.53)$$

Applying (4.53) to (4.52), we obtain

$$\frac{d}{dt}\begin{bmatrix} \mathbf{i}^e_{dqs} \\ \boldsymbol{\lambda}^e_{dqr} \end{bmatrix} = \begin{bmatrix} -\left(\frac{1}{\sigma\tau_s}+\frac{1-\sigma}{\sigma\tau_r}\right)\mathbf{I}-\omega_e\mathbf{J} & \frac{L_m}{\sigma L_r L_s}\left(\frac{1}{\tau_r}\mathbf{I}-\omega_r\mathbf{J}\right) \\ \frac{L_m}{\tau_r}\mathbf{I} & -\frac{1}{\tau_r}\mathbf{I}-\omega_{sl}\mathbf{J} \end{bmatrix}\begin{bmatrix} \mathbf{i}^e_{dqs} \\ \boldsymbol{\lambda}^e_{dqr} \end{bmatrix}+\frac{1}{\sigma L_s}\begin{bmatrix} \mathbf{v}^e_{dqs} \\ 0 \end{bmatrix},$$

(4.54)

or

$$\frac{d}{dt}\begin{bmatrix} i^e_{ds} \\ i^e_{qs} \\ \lambda^e_{dr} \\ \lambda^e_{qr} \end{bmatrix} = \begin{bmatrix} -\frac{1}{\sigma\tau_s}-\frac{(1-\sigma)}{\sigma\tau_r} & \omega_e & \frac{L_m}{\sigma\tau_r L_r L_s} & \frac{\omega_r L_m}{\sigma L_s L_r} \\ -\omega_e & -\frac{1}{\sigma\tau_s}-\frac{(1-\sigma)}{\sigma\tau_r} & -\frac{\omega_r L_m}{\sigma L_s L_r} & \frac{L_m}{\sigma\tau_r L_r L_s} \\ \frac{L_m}{\tau_r} & 0 & -\frac{1}{\tau_r} & \omega_{sl} \\ 0 & \frac{L_m}{\tau_r} & -\omega_{sl} & -\frac{1}{\tau_r} \end{bmatrix}\begin{bmatrix} i^e_{ds} \\ i^e_{qs} \\ \lambda^e_{dr} \\ \lambda^e_{qr} \end{bmatrix}+\frac{1}{\sigma L_s}\begin{bmatrix} v^e_{ds} \\ v^e_{qs} \\ 0 \\ 0 \end{bmatrix},$$

(4.55)

where $\tau_s = L_s/r_s$.

The stationary frame ODE model in current-flux variables is obtained from (4.54) by substituting $\omega_e = 0$. Then, ω_{sl} should be replaced by $\omega_e - \omega_r = -\omega_r$. Thereby, we have a stationary model as

$$\frac{d}{dt}\begin{bmatrix} \mathbf{i}^s_{dqs} \\ \boldsymbol{\lambda}^s_{dqr} \end{bmatrix} = \begin{bmatrix} -\left(\frac{r_s}{\sigma L_s}+\frac{1-\sigma}{\sigma\tau_r}\right)\mathbf{I} & \frac{L_m}{\sigma L_r L_s}\left(\frac{1}{\tau_r}\mathbf{I}-\omega_r\mathbf{J}\right) \\ \frac{L_m}{\tau_r}\mathbf{I} & -\frac{1}{\tau_r}\mathbf{I}+\omega_r\mathbf{J} \end{bmatrix}\begin{bmatrix} \mathbf{i}^s_{dqs} \\ \boldsymbol{\lambda}^s_{dqr} \end{bmatrix}+\frac{1}{\sigma L_s}\begin{bmatrix} \mathbf{v}^s_{dqs} \\ 0 \end{bmatrix}. \quad (4.56)$$

Exercise 4.9   Write a program using C-code for Runge-Kutta fourth method to solve the IM dynamics based on (4.54). Apply the same IM parameters and input voltage $\mathbf{v}^e_{dq0s} = T(\theta_e)\mathbf{v}^s_{abcs}$ as in Exercise 4.3, and compare the results of $\mathbf{i}^s_{dqs}$.

IM Model in General Reference Frame

Comparing (4.54) with (4.56), the IM dynamics can be generalized independently of the frame such that

$$\frac{d}{dt}\begin{bmatrix} \mathbf{i}^g_{dqs} \\ \boldsymbol{\lambda}^g_{dqr} \end{bmatrix} = \begin{bmatrix} -\left(\frac{r_s}{\sigma L_s}+\frac{1-\sigma}{\sigma\tau_r}\right)\mathbf{I} & \frac{L_m}{\sigma L_r L_s}\frac{1}{\tau_r}\mathbf{I} \\ \frac{L_m}{\tau_r}\mathbf{I} & -\frac{1}{\tau_r}\mathbf{I} \end{bmatrix}\begin{bmatrix} \mathbf{i}^g_{dqs} \\ \boldsymbol{\lambda}^g_{dqr} \end{bmatrix} - \omega_A\begin{bmatrix} \mathbf{J} & \frac{L_m}{\sigma L_r L_s}\mathbf{J} \\ 0 & 0 \end{bmatrix}\begin{bmatrix} \mathbf{i}^g_{dqs} \\ \boldsymbol{\lambda}^g_{dqr} \end{bmatrix}$$
$$+\omega_B\begin{bmatrix} 0 & -\frac{L_m}{\sigma L_r L_s}\mathbf{J} \\ 0 & \mathbf{J} \end{bmatrix}\begin{bmatrix} \mathbf{i}^g_{dqs} \\ \boldsymbol{\lambda}^g_{dqr} \end{bmatrix}+\frac{1}{\sigma L_s}\begin{bmatrix} \mathbf{v}^g_{dqs} \\ 0 \end{bmatrix}, \quad (4.57)$$

where the superscript 'g' denotes the generalized frame. Note that ω_A and ω_B can be arbitrary values as long as they satisfy the constraint, $\omega_A - \omega_B = \omega_r$. The equation becomes the synchronous frame IM model with $\omega_A = \omega_e$ and $\omega_B = \omega_{sl}$. On the other hand, a stationary IM model results with $\omega_A = 0$ and $\omega_B = -\omega_r$. Previously, all the dynamic equations were derived based on two pole machine. In general, ω_r should be replaced by $\frac{P}{2}\omega_r$, and the slip must be defined as $\omega_{sl} = \omega_e - \frac{P}{2}\omega_r$. General IM dynamic equations are summarized in Table 4.3.

Table 4.3: P-pole IM dynamic equations.

Stationary frame
$$\begin{bmatrix} \mathbf{v}^s_{dqs} \\ \mathbf{0} \end{bmatrix} = \begin{bmatrix} (r_s + L_s\mathbf{p})\mathbf{I} & \mathbf{p}L_m\mathbf{I} \\ \mathbf{p}\mathbf{I} + \frac{P}{2}\omega_r L_m\mathbf{J} & (r_s + L_s\mathbf{p})\mathbf{I} + \frac{P}{2}\omega_r L_r\mathbf{J} \end{bmatrix} \begin{bmatrix} \mathbf{i}^s_{dqs} \\ \mathbf{i}^s_{dqr} \end{bmatrix}$$
Synchronous frame
$$\begin{bmatrix} \mathbf{v}^e_{dqs} \\ \mathbf{0} \end{bmatrix} = \begin{bmatrix} (r_s + L_s\mathbf{p})\mathbf{I} + \omega_e L_s\mathbf{J} & \mathbf{p}L_m\mathbf{I} + \omega_e L_m\mathbf{J} \\ \mathbf{p}L_m\mathbf{I} + \omega_{sl} L_m\mathbf{J} & (r_s + L_r\mathbf{p})\mathbf{I} + \omega_{sl} L_r\mathbf{J} \end{bmatrix} \begin{bmatrix} \mathbf{i}^e_{dqs} \\ \mathbf{i}^e_{dqr} \end{bmatrix}$$
ODE with current variables (Stationary)
$$\frac{d}{dt} \begin{bmatrix} \mathbf{i}^s_{dqs} \\ \mathbf{i}^s_{dqr} \end{bmatrix} = \begin{bmatrix} -\frac{r_s}{\sigma L_s}\mathbf{I} - \frac{P}{2}\omega_r \frac{1-\sigma}{\sigma}\mathbf{J} & \frac{L_m}{\sigma L_s}\left(\frac{1}{\tau_r}\mathbf{I} - \frac{P}{2}\omega_r\mathbf{J}\right) \\ \frac{L_m}{\sigma L_r}\left(\frac{r_s}{L_s}\mathbf{I} + \frac{P}{2}\omega_r\mathbf{J}\right) & -\frac{1}{\sigma}\left(\frac{1}{\tau_r}\mathbf{I} - \frac{P}{2}\omega_r\mathbf{J}\right) \end{bmatrix} \begin{bmatrix} \mathbf{i}^s_{dqs} \\ \mathbf{i}^s_{dqr} \end{bmatrix} + \begin{bmatrix} \frac{1}{\sigma L_s}\mathbf{I} \\ -\frac{L_m}{\sigma L_r L_s}\mathbf{I} \end{bmatrix} \mathbf{v}^s_{dqs}$$
ODE with current variables (Synchronous)
$$\frac{d}{dt} \begin{bmatrix} \mathbf{i}^e_{dqs} \\ \mathbf{i}^e_{dqr} \end{bmatrix} = \begin{bmatrix} -\frac{r_s}{\sigma L_s}\mathbf{I} - \left(\omega_e + \frac{P}{2}\omega_r\frac{1-\sigma}{\sigma}\right)\mathbf{J} & \frac{L_m}{\sigma L_s}\left(\frac{1}{\tau_r}\mathbf{I} - \frac{P}{2}\omega_r\mathbf{J}\right) \\ \frac{L_m}{\sigma L_r}\left(\frac{r_s}{L_s}\mathbf{I} + \frac{P}{2}\omega_r\mathbf{J}\right) & -\frac{1}{\sigma\tau_r}\mathbf{I} - \left(\omega_e - \frac{P\omega_r}{2\sigma}\right)\mathbf{J} \end{bmatrix} \begin{bmatrix} \mathbf{i}^e_{dqs} \\ \mathbf{i}^e_{dqr} \end{bmatrix} + \begin{bmatrix} \frac{1}{\sigma L_s}\mathbf{I} \\ -\frac{L_m}{\sigma L_r L_s}\mathbf{I} \end{bmatrix} \mathbf{v}^e_{dqs}$$
ODE with current-flux variables (Stationary)
$$\frac{d}{dt} \begin{bmatrix} \mathbf{i}^s_{dqs} \\ \boldsymbol{\lambda}^s_{dqr} \end{bmatrix} = \begin{bmatrix} -\left(\frac{r_s}{\sigma L_s} + \frac{1-\sigma}{\sigma\tau_r}\right)\mathbf{I} & \frac{L_m}{\sigma L_r L_s}\left(\frac{1}{\tau_r}\mathbf{I} - \frac{P}{2}\omega_r\mathbf{J}\right) \\ \frac{L_m}{\tau_r}\mathbf{I} & -\frac{1}{\tau_r}\mathbf{I} + \omega_r\mathbf{J} \end{bmatrix} \begin{bmatrix} \mathbf{i}^s_{dqs} \\ \boldsymbol{\lambda}^s_{dqr} \end{bmatrix} + \frac{1}{\sigma L_s}\begin{bmatrix} \mathbf{v}^s_{dqs} \\ \mathbf{0} \end{bmatrix}$$
ODE with current-flux variables (Synchronous)
$$\frac{d}{dt} \begin{bmatrix} \mathbf{i}^e_{dqs} \\ \boldsymbol{\lambda}^e_{dqr} \end{bmatrix} = \begin{bmatrix} -\left(\frac{r_s}{\sigma L_s} + \frac{1-\sigma}{\sigma\tau_r}\right)\mathbf{I} - \omega_e\mathbf{J} & \frac{L_m}{\sigma L_r L_s}\left(\frac{1}{\tau_r}\mathbf{I} - \frac{P}{2}\omega_r\mathbf{J}\right) \\ \frac{L_m}{\tau_r}\mathbf{I} & -\frac{1}{\tau_r}\mathbf{I} - \omega_{sl}\mathbf{J} \end{bmatrix} \begin{bmatrix} \mathbf{i}^e_{dqs} \\ \boldsymbol{\lambda}^e_{dqr} \end{bmatrix} + \frac{1}{\sigma L_s}\begin{bmatrix} \mathbf{v}^e_{dqs} \\ \mathbf{0} \end{bmatrix}$$

4.2.3 Alternative Derivations

More IM models can be developed by selecting the state variables differently. Furthermore, gamma and inverse gamma models are obtained by exploiting the property that the rotor circuit is shorted. These variants are used mainly in the speed sensorless control.

A. State Variables: $(\boldsymbol{\lambda}^e_{dqs}, \mathbf{i}^e_{dqs})$

The state variables are comprised of the state flux and state current. None of the rotor variables are used. But, their influence is reflected in the state flux. It follows from (4.20) and (4.43) that

$$0 = (r_r + j\omega_{sl}\sigma L_r)\mathbf{i}^e_{dqr} + \frac{L_m}{L_s}\frac{d\lambda^e_{dqs}}{dt} + \sigma L_r\frac{d\mathbf{i}^e_{dqr}}{dt} + j\omega_{sl}\frac{L_m}{L_s}\lambda^e_{dqs}. \quad (4.58)$$

Since $\mathbf{i}^e_{dqr} = \frac{1}{L_m}(\lambda^e_{dqs} - L_s\mathbf{i}_{dqs})$, we have

$$\begin{aligned}
0 = &\left(\frac{r_r}{L_m} + j\omega_{sl}(\frac{\sigma L_r}{L_m} + \frac{L_m}{L_s})\right)\lambda^e_{dqs} - \left(\frac{r_r}{L_m} + j\omega_{sl}\frac{\sigma L_r}{L_m}\right)L_s\mathbf{i}^e_{dqs} \\
&+ \left(\frac{\sigma L_r}{L_m} + \frac{L_m}{L_s}\right)\frac{d\lambda^e_{dqs}}{dt} - \frac{\sigma L_r L_s}{L_m}\frac{d\mathbf{i}^e_{dqs}}{dt}. \quad (4.59)
\end{aligned}$$

That is,

$$\frac{\sigma L_r L_s}{L_m}\frac{d\mathbf{i}^e_{dqs}}{dt} = \frac{L_r}{L_m}\frac{d\lambda^e_{dqs}}{dt} + \left(\frac{r_r}{L_m} + j\omega_{sl}\frac{L_r}{L_m}\right)\lambda^e_{dqs} - j\omega_{sl}\frac{\sigma L_r L_s}{L_m}\mathbf{i}^e_{dqs} - \frac{r_r L_s}{L_m}\mathbf{i}^e_{dqs}.$$

Utilizing (4.42), we have the following IM model:

$$\frac{d\mathbf{i}^e_{dqs}}{dt} = -\left(\frac{1}{\sigma\tau_s} + \frac{1}{\sigma\tau_r} + j\omega_{sl}\right)\mathbf{i}^e_{dqs} + \frac{1}{\sigma L_s}\left(\frac{1}{\tau_r} - j\omega_r\right)\lambda^e_{dqs} + \frac{1}{\sigma L_s}\mathbf{v}^e_{dqs}, \quad (4.60)$$

$$\frac{d\lambda^e_{dqs}}{dt} = -r_s\mathbf{i}^e_{dqs} - j\omega_e\lambda^e_{dqs} + \mathbf{v}^e_{dqs}. \quad (4.61)$$

This model was used in the construction of sliding mode adaptive observer [3].

B. State Variables: $(\lambda^e_{dqs}, \lambda^e_{dqr})$

The stator and rotor flux variables can be a set of state variables. Replacing $(\mathbf{i}_{dqs}, \mathbf{i}_{dqr})$ by $(\lambda_{dqs}, \lambda_{dqr})$ using (4.17) and (4.18) we have from (4.42) and (4.43)

$$\frac{d\lambda^e_{dqs}}{dt} = -\left(\frac{1}{\sigma\tau_s} + j\omega_e\right)\lambda^e_{dqs} + \frac{L_m}{\sigma\tau_s L_r}\lambda^e_{dqr} + \mathbf{v}^e_{dqs}, \quad (4.62)$$

$$\frac{d\lambda^e_{dqr}}{dt} = \frac{L_m}{\sigma\tau_r L_s}\lambda^e_{dqs} - \left(\frac{1}{\sigma\tau_r} + j\omega_{sl}\right)\lambda^e_{dqr}. \quad (4.63)$$

C. Gamma Model

Two leakage inductances, L_{ls} and L_{lr} can be merged into one, since the rotor circuit is shorted [4]. The IM model is rewritten as

$$\mathbf{v}^e_{dqs} = r_s\mathbf{i}^e_{dqs} + L_s\mathbf{p}\mathbf{i}^e_{dqs} + L_m\mathbf{p}\mathbf{i}^e_{dqr} + j\omega_e\lambda^e_{dqs} \quad (4.64)$$

$$0 = r_r\mathbf{i}^e_{dqr} + L_r\mathbf{p}\mathbf{i}^e_{dqr} + L_m\mathbf{p}\mathbf{i}^e_{dqs} + j\omega_{sl}\lambda^e_{dqr}. \quad (4.65)$$

Multiplying (4.65) by a nonzero constant a, we obtain

$$
\begin{aligned}
0 &= a^2 r_r \frac{\mathbf{i}_{dqr}^e}{a} + a^2 L_r \mathbf{p}\frac{\mathbf{i}_{dqr}^e}{a} + aL_m \mathbf{pi}_{dqs}^e + ja^2 \omega_{sl} L_r \frac{\mathbf{i}_{dqr}^e}{a} + ja\omega_{sl} L_m \mathbf{i}_{dqs}^e \\
&= a^2 r_r \frac{\mathbf{i}_{dqr}^e}{a} + a^2 L_r \mathbf{p}\frac{\mathbf{i}_{dqr}^e}{a} - \left(\mathbf{p}\frac{\mathbf{i}_{dqr}^e}{a} aL_m - \mathbf{p}\frac{\mathbf{i}_{dqr}^e}{a} aL_m\right) + aL_m \mathbf{pi}_{dqs}^e \\
&\quad + \left(-j\omega_{sl} L_m \mathbf{i}_{dqr}^e + j\omega_{sl} L_m \mathbf{i}_{dqr}^e\right) + ja^2 \omega_{sl} L_r \frac{\mathbf{i}_{dqr}^e}{a} + ja\omega_{sl} L_m \mathbf{i}_{dqs}^e \\
&= a^2 r_r \frac{\mathbf{i}_{dqr}^e}{a} + a\mathbf{p}\left[(aL_r - L_m)\frac{\mathbf{i}_{dqr}^e}{a} + L_m(\mathbf{i}_{dqs}^e + \frac{\mathbf{i}_{dqr}^e}{a})\right] \\
&\quad + j\omega_{sl} a \left[(aL_r - L_m)\frac{\mathbf{i}_{dqr}^e}{a} + L_m(\mathbf{i}_{dqs}^e + \frac{\mathbf{i}_{dqr}^e}{a})\right]
\end{aligned}
\tag{4.66}
$$

The stator equation, (4.64) is written equivalently as

$$
\mathbf{v}_{dqs}^e = r_s \mathbf{i}_{dqs}^e + (L_s - aL_m)\mathbf{pi}_{dqs}^e + aL_m \mathbf{p}\left(\frac{\mathbf{i}_{dqr}^e}{a} + \mathbf{i}_{dqs}^e\right) + j\omega_e \boldsymbol{\lambda}_{dqs}^e.
\tag{4.67}
$$

The corresponding equivalent circuit is shown in Fig. 4.13 (a).

If we let $a = L_s/L_m$, then the stator leakage inductance disappears and all the leakage inductance is located in the rotor side.

$$
\mathbf{v}_{dqs}^e = r_s \mathbf{i}_{dqs}^e + (\mathbf{p} + j\omega_e)\boldsymbol{\lambda}_{dqs}^e,
\tag{4.68}
$$

$$
0 = \frac{L_s^2}{L_m^2} r_r \frac{L_m}{L_s} \mathbf{i}_{dqr}^e + (\mathbf{p} + j\omega_{sl})\left[L_s \frac{\sigma}{1-\sigma} \frac{L_m}{L_s} \mathbf{i}_{dqr}^e + \boldsymbol{\lambda}_{dqs}^e\right].
\tag{4.69}
$$

The Γ-equivalent circuit is shown in Fig. 4.13 (b).

D. Inverse Gamma Model

If we let $a = L_m/L_r$, then (4.66) and (4.67) turn out to be

$$
\mathbf{v}_{dqs}^e = r_s \mathbf{i}_{dqs}^e + \sigma L_s \mathbf{p}\,\mathbf{i}_{dqs}^e + \frac{L_m^2}{L_r}\mathbf{p}\,(\mathbf{i}_{dqs}^e + \frac{L_r}{L_m}\mathbf{i}_{dqr}^e) + j\omega_e \boldsymbol{\lambda}_{dqs}^e,
\tag{4.70}
$$

$$
0 = \frac{L_m^2}{L_r^2} r_r \frac{L_r}{L_m}\mathbf{i}_{dqr}^e + \frac{L_m^2}{L_r}\mathbf{p}\,(\mathbf{i}_{dqs}^e + \frac{L_r}{L_m}\mathbf{i}_{dqr}^e) + j\omega_{sl}\frac{L_m}{L_r}\boldsymbol{\lambda}_{dqr}^e.
\tag{4.71}
$$

Based on (4.70) and (4.71), the inverse Γ circuit is drawn as Fig. 4.13 (c). Note that the rotor leakage inductance disappeared and all the leakage inductance is located in the stator side. This equivalent circuit provides the rotor flux oriented dynamic model, which is widely used in the field oriented control.

E. Inverse Gamma Model with Back EMF

Note from (4.19) that

$$
j\omega_e \boldsymbol{\lambda}_{dqs}^e = j\omega_e \sigma L_s \mathbf{i}_{dqs}^e + j\omega_e \frac{L_m}{L_r}\boldsymbol{\lambda}_{dqr}^e
\tag{4.72}
$$

Therefore, it follows from (4.70) and (4.71) that

$$\mathbf{v}_{dqs}^e = r_s\mathbf{i}_{dqs}^e + \sigma L_s\mathbf{p}\mathbf{i}_{dqs}^e + \frac{L_m}{L_r}\mathbf{p}\lambda_{dqr}^e + j\omega_e\sigma L_s\mathbf{i}_{dqs}^e + j\omega_e\frac{L_m}{L_r}\lambda_{dqr}^e \quad (4.73)$$

$$0 = \frac{L_m}{L_r}r_r\mathbf{i}_{dqr}^e + \frac{L_m}{L_r}\mathbf{p}\lambda_{dqr}^e + j\omega_{sl}\frac{L_m}{L_r}\lambda_{dqr}^e. \quad (4.74)$$

Subtracting (4.74) from (4.73), it follows that

$$\sigma L_s\mathbf{p}\mathbf{i}_{dqs}^e + j\omega_e\sigma L_s\mathbf{i}_{dqs}^e = \mathbf{v}_{dqs}^e - r_s\mathbf{i}_{dqs}^e - \mathbf{e}, \quad (4.75)$$

where

$$\mathbf{e} = -\frac{L_m}{L_r}r_r\mathbf{i}_{dqr}^e + j\omega_r\frac{L_m}{L_r}\lambda_{dqr}^e \quad (4.76)$$

and it is called IM back EMF. The equivalent circuit is shown in Fig. 4.14. It is basically the same as the inverse Γ-equivalent circuit model. This model was utilized for constructing an IM sensorless drive by Harnefors [5]. The above alternative derivations are summarized in Table 4.4.

Exercise 4.10   Derive (4.70) and (4.69) from (4.66) and (4.67).

Exercise 4.11   Show that

$$\mathbf{e} = \frac{L_m^2}{L_r^2}\left[r_r\mathbf{i}_{dqs}^e - \left(\frac{r_r}{L_m} - j\omega_r\frac{L_r}{L_m}\right)\lambda_{dqr}^e\right].$$

4.2.4   Steady State Models

A steady state model follows straightforwardly from (4.36). The steady state variables are denoted by capital letters, V_{dqs}, I_{dqs}, and I_{dqr}. Letting $\mathbf{p} = 0$ and substituting $s\omega_e = \omega_e - \omega_r$, it follows that

$$V_{dqs} = r_sI_{dqs} + j\omega_e(L_sI_{dqs} + L_mI_{dqr}) \quad (4.77)$$

$$0 = \frac{r_s}{s}I_{dqr} + j\omega_e(L_rI_{dqr} + L_mI_{dqs}). \quad (4.78)$$

The T-equivalent and its modified circuits are shown in Fig. 4.15. Note that the steady state model shown in Fig. 4.15 (a) is the same as the one treated in Section 3.2.1.

4.3   Power and Torque Equations

Since $v_{ar}^e = v_{br}^e = v_{cr}^e = 0$, the rotor side source power is equal to zero. Recall that the power equation is written as (2.54) in terms of dq variables. Electrical power

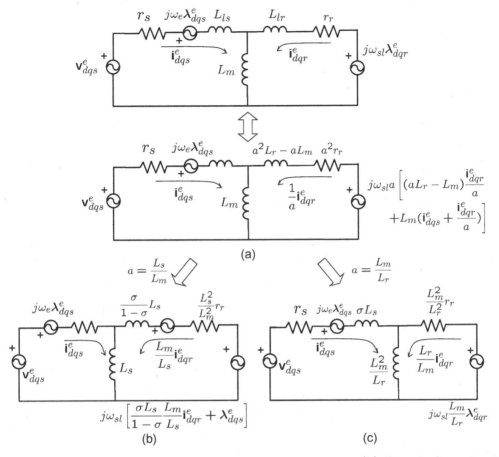

(a)

(b)

(c)

Figure 4.13: (a) Transformation of IM equivalent circuits, (b) Γ-equivalent circuit, and (c) inverse Γ-equivalent circuit.

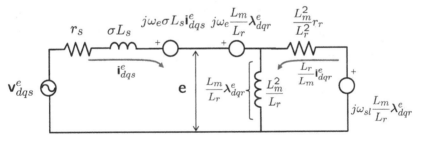

Figure 4.14: Inverse-Γ equivalent circuit with an EMF.

applied to IM is obtained from (4.36) such that

$$
\begin{aligned}
P_e &= \mathbf{v}_{abcs}^T \mathbf{i}_{abcs} = \frac{3}{2}\left(v_{ds}^e i_{ds}^e + v_{qs}^e i_{qs}^e\right) \\
&= \frac{3}{2} i_{ds}^e \left((r_s + \mathrm{p}L_s)i_{ds}^e + \mathrm{p}L_s i_{dr}^e - \omega_e L_m i_{qs}^e - \omega_e L_m i_{qr}^e\right) \\
&\quad + \frac{3}{2} i_{qs}^e \left((r_s + \mathrm{p}L_s)i_{qs}^e + \omega_e L_m i_{ds}^e + \omega_e L_m i_{dr}^e + \mathrm{p}L_m i_{qr}^e\right). \quad (4.79)
\end{aligned}
$$

Table 4.4: Alternative IM dynamic equations.

State Variables: $(\boldsymbol{\lambda}^e_{dqs},\, \mathbf{i}^e_{dqs})$
$\dfrac{d}{dt}\begin{bmatrix}\mathbf{i}^e_{dqs}\\[2pt]\boldsymbol{\lambda}^e_{dqs}\end{bmatrix}=\begin{bmatrix}-\frac{1}{\sigma}\left(\frac{1}{\tau_s}+\frac{1}{\tau_r}\right)\mathbf{I}-\omega_{sl}\mathbf{J} & \frac{1}{\tau_r\sigma L_s}\mathbf{I}-\frac{P}{2}\frac{\omega_r}{\sigma L_s}\mathbf{J}\\[6pt] -r_s\mathbf{I} & -\omega_e\mathbf{J}\end{bmatrix}\begin{bmatrix}\mathbf{i}^e_{dqs}\\[2pt]\boldsymbol{\lambda}^e_{dqs}\end{bmatrix}+\begin{bmatrix}\frac{1}{\sigma L_s}\mathbf{I}\\[2pt]\mathbf{I}\end{bmatrix}\mathbf{v}^e_{dqs}$
State Variables: $(\boldsymbol{\lambda}^e_{dqs},\, \boldsymbol{\lambda}^e_{dqr})$
$\dfrac{d}{dt}\begin{bmatrix}\boldsymbol{\lambda}^e_{dqs}\\[2pt]\boldsymbol{\lambda}^e_{dqr}\end{bmatrix}=\begin{bmatrix}-\frac{1}{\sigma\tau_s}\mathbf{I}-\omega_e\mathbf{J} & \frac{L_m}{\sigma\tau_s L_r}\mathbf{I}\\[6pt] \frac{L_m}{\sigma\tau_r L_s}\mathbf{I} & -\frac{1}{\sigma\tau_r}\mathbf{I}-\omega_{sl}\mathbf{J}\end{bmatrix}\begin{bmatrix}\boldsymbol{\lambda}^e_{dqs}\\[2pt]\boldsymbol{\lambda}^e_{dqr}\end{bmatrix}+\begin{bmatrix}\mathbf{I}\\[2pt]\mathbf{0}\end{bmatrix}\mathbf{v}^e_{dqs}$
Inverse Gamma Model with Back EMF
$\dfrac{d}{dt}\begin{bmatrix}\mathbf{i}^e_{dqs}\\[2pt]\boldsymbol{\lambda}^e_{dqr}\end{bmatrix}=\begin{bmatrix}-\frac{r_s}{\sigma L_s}\mathbf{I}-\omega_e\mathbf{J} & 0\\[6pt] 0 & -\omega_e\mathbf{J}\end{bmatrix}\begin{bmatrix}\mathbf{i}^e_{dqs}\\[2pt]\boldsymbol{\lambda}^e_{dqr}\end{bmatrix}+\begin{bmatrix}\frac{1}{\sigma L_s}\mathbf{I} & -\frac{1}{\sigma L_s}\mathbf{I}\\[6pt] 0 & \frac{L_r}{L_m}\mathbf{I}\end{bmatrix}\begin{bmatrix}\mathbf{v}^e_{dqs}\\[2pt]\mathbf{e}\end{bmatrix}$
where $\quad \mathbf{e}=\frac{L_m^2}{L_r\tau_r}\mathbf{i}^e_{dqs}-\left(\frac{L_m}{L_r\tau_r}\mathbf{I}-\frac{P}{2}\omega_r\frac{L_m}{L_r}\mathbf{J}\right)\boldsymbol{\lambda}^e_{dqr}.$

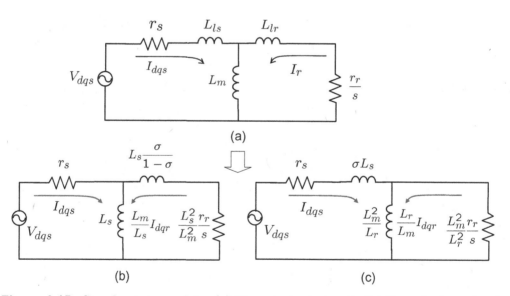

Figure 4.15: Steady state models: (a) T-equivalent circuit, (b) Γ-equivalent circuit, and (c) inverse Γ-equivalent circuit.

Note that the shaft torque of a rotating machine is obtained as a gradient of the electromagnetic power with respect to the shaft speed, i.e., $T_e = \frac{\partial P_e}{\partial \omega_r}$. Assume that the slip is constant and that the machine is in the steady state. Then, the shaft torque follows from (4.79) such that

$$T_e = \frac{\partial P_e}{\partial \omega_r} = \frac{P}{2}\frac{\partial P_e}{\partial \omega_e} = \frac{P}{2}\frac{3}{2}L_m\left(i_{qs}^e i_{dr}^e - i_{ds}^e i_{qr}^e\right). \tag{4.80}$$

A different way of expressing (4.80) is

$$T_e = \frac{3}{2}\frac{P}{2}L_m Im\{\mathbf{i}_{dqs}^e \mathbf{i}_{dqr}^{e*}\}, \tag{4.81}$$

where $Im\{z\}$ implies the complex part of $z \in \mathbb{C}$. Note that $\boldsymbol{\lambda}_{dqr}^e = L_r\mathbf{i}_{dqr}^e + L_m\mathbf{i}_{dqs}^e$ and $Im\{\mathbf{i}_{dqs}^e \mathbf{i}_{dqs}^{e*}\} = 0$. Therefore, $Im\{\mathbf{i}_{dqs}^e \mathbf{i}_{dqr}^{e*}\} = \frac{1}{L_r}Im\{\mathbf{i}_{dqs}^e \boldsymbol{\lambda}_{dqr}^{e*}\}$. Hence, (4.81) is expressed equivalently as

$$T_e = \frac{3}{2}\frac{P}{2}\frac{L_m}{L_r}Im\{\mathbf{i}_{dqs}^e \boldsymbol{\lambda}_{dqr}^{e*}\}. \tag{4.82}$$

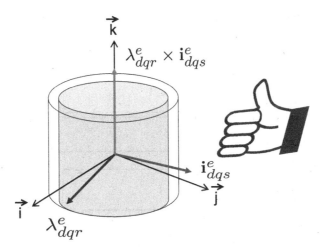

Figure 4.16: Cross product of rotor flux and stator current yields torque.

The other method of expressing $Im\{\mathbf{i}_{dqs}^e \boldsymbol{\lambda}_{dqr}^{e*}\}$ is to use the cross product of vectors. Note that

$$T_e = \frac{3}{2}\frac{P}{2}\frac{L_m}{L_r}\left(\boldsymbol{\lambda}_{dqr}^e \times \mathbf{i}_{dqs}^e\right)_k \tag{4.83}$$

$$= \frac{3}{2}\frac{P}{2}\frac{L_m}{L_r}\left(\begin{bmatrix}\lambda_{dr}^e\\\lambda_{qr}^e\\0\end{bmatrix} \times \begin{bmatrix}i_{ds}^e\\i_{qs}^e\\0\end{bmatrix}\right)_k$$

$$= \frac{3}{2}\frac{P}{2}\frac{L_m}{L_r}\begin{vmatrix}\overrightarrow{i} & \overrightarrow{j} & \overrightarrow{k}\\\lambda_{dr}^e & \lambda_{qr}^e & 0\\i_{ds}^e & i_{qs}^e & 0\end{vmatrix}_k$$

$$= \frac{3}{2}\frac{P}{2}\frac{L_m}{L_r}(\lambda_{dr}^e i_{qs}^e - \lambda_{qr}^e i_{ds}^e), \tag{4.84}$$

where the subscript 'k' implies the k-th component of a vector. Fig. 4.16 is a graphical description of IM torque as a cross product of $\boldsymbol{\lambda}^e_{dqr}$ and $\mathbf{i}^e_{dqs}$.

The mechanical system is modeled typically as

$$J\frac{d}{dt}\omega_r + B\omega_r = T_e - T_L, \tag{4.85}$$

where J is the inertia of a rotating object, B is the friction coefficient, and T_L is an external load. Fig. 4.17 shows a complete IM block diagram based on (4.31)−(4.35) along with the torque equation, (4.84) and a mechanical model, (4.85).

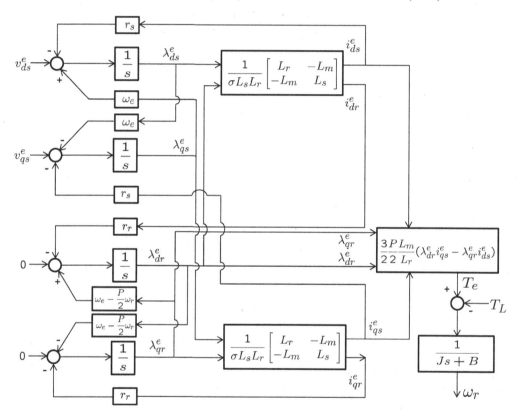

Figure 4.17: IM block diagram with a mechanical load model.

Exercise 4.12   Consider an IM with the parameters: the number of poles $P = 6$, $r_s = 0.282\,\Omega$, $r_r = 0.151\,\Omega$, $L_{ls} = 1.358\,\mathrm{mH}$, $L_{lr} = 0.711\,\mathrm{mH}$, $L_m = 39.43\,\mathrm{mH}$, $J = 0.4\,\mathrm{Kg \cdot m^2}$, $B = 0.124\,\mathrm{Nm \cdot s}$. Construct a MATLAB Simulink model for the IM based on the ODE-model with mixed variables, $\mathbf{i}^e_{dqs}$ and $\boldsymbol{\lambda}^e_{dqr}$ in the synchronous frame. Also, construct blocks for torque and mechanical model. Assume that the line to line voltage is $V_{ll} = 270$ V(rms), and the frequency is $f = 60$ Hz. When the line voltage is applied directly to the IM, obtain waveforms of ω_r, T_e, $\mathbf{i}_{abc}$, and $\mathbf{i}^e_{dqs}$.

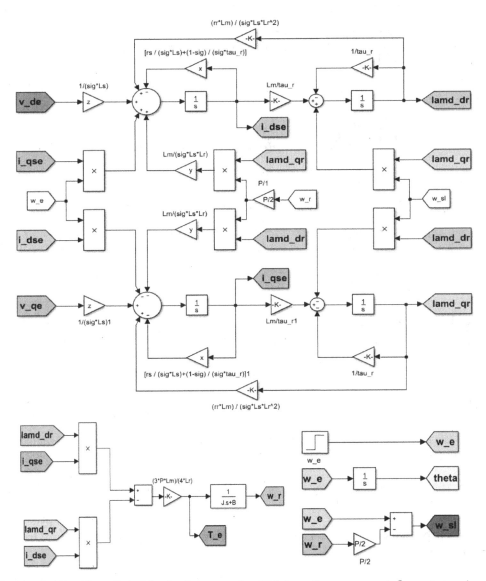

Figure 4.18: Simulink block diagram for IM based on current-flux states in the synchronous frame (Solution to Exercise 4.12).

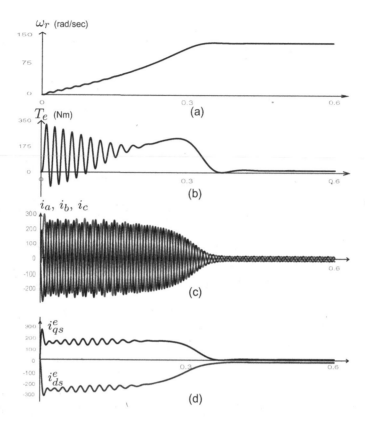

Figure 4.19: At the line start of IM: (a) speed, (b) torque, (c) phase currents, and (d) dq-currents (Solution to Exercise 4.12).

References

[1] P. C. Krause, O. Wasynczuk, and S. D. Sudhoff, *Analysis of Electric Machinery*, IEEE Press, 1995.

[2] C. T. Chen, *Linear System Theory and Design*, Oxford University Press, 1999.

[3] C. Lascu, I. Boldea, and F. Blaabjerg, "Comparative Study of Adaptive and Inherently Sensorless Observers for Variable-Speed Induction-Motor Drives," *IEEE Trans. Ind. Electron.*, vol. 53, no. 1, pp.57−65, Feb. 2006.

[4] G. R. Slemon, "Modelling of induction machines for electric drives," *IEEE Trans. Ind. Appl.*, vol. 25, no. 6, pp. 1126−1131, Nov./Dec. 1989.

[5] M. Hinkkanen, L. Harnefors, and J. Luomi, "Reduced-order flux observers with stator-resistance adaptation for speed-sensorless induction motor drives," *IEEE Trans. Power Electron.*, vol. 25, no. 5, pp. 1173−1183, May 2010.

Problems

4.1 Show that for $\mathbf{A} \in \mathbb{R}^{n \times n}$

$$e^{\mathbf{A}t} = \mathcal{L}^{-1}\left\{(s\mathbf{I} - \mathbf{A})^{-1}\right\}, \qquad (4.86)$$

where $\mathbf{I} \in \mathbb{R}^{n \times n}$ is the identity matrix, $\mathcal{L}$ denotes Laplace transformation, and s is the Laplace variable.

4.2 Show that

$$i_{ds}^e = \frac{1}{\sigma L_s}\left(\lambda_{ds}^e - \frac{L_m}{L_r}\lambda_{dr}^e\right)$$

$$i_{qs}^e = \frac{1}{\sigma L_s}\left(\lambda_{qs}^e - \frac{L_m}{L_r}\lambda_{qr}^e\right)$$

$$i_{dr}^e = \frac{1}{\sigma L_r}\left(\lambda_{dr}^e - \frac{L_m}{L_s}\lambda_{ds}^e\right)$$

$$i_{qr}^e = \frac{1}{\sigma L_r}\left(\lambda_{qr}^e - \frac{L_m}{L_s}\lambda_{qs}^e\right)$$

4.3 Utilizing the results of **4.2**, derive the following from (4.30) and (4.33):

$$\frac{d}{dt}\begin{bmatrix}\lambda_{ds}^e \\ \lambda_{qs}^e \\ \lambda_{dr}^e \\ \lambda_{qr}^e\end{bmatrix} = \begin{bmatrix} -\frac{r_s}{\sigma L_s} & \omega_e & \frac{r_r L_m}{\sigma L_r L_s} & 0 \\ -\omega_e & -\frac{r_s}{\sigma L_s} & 0 & \frac{r_s L_m}{\sigma L_r L_s} \\ \frac{r_s L_m}{\sigma L_r L_s} & 0 & -\frac{r_r}{\sigma L_r} & \omega_{sl} \\ 0 & \frac{r_s L_m}{\sigma L_r L_s} & -\omega_{sl} & -\frac{r_r}{\sigma L_r}\end{bmatrix}\begin{bmatrix}\lambda_{ds}^e \\ \lambda_{qs}^e \\ \lambda_{dr}^e \\ \lambda_{qr}^e\end{bmatrix} + \begin{bmatrix}v_{ds}^e \\ v_{qs}^e \\ 0 \\ 0\end{bmatrix}.$$

4.4 Consider a six phase IM shown below. Note that there are two sets of stator coils, abc and xyz which are displaced 30°. The flux is modeled as

$$\lambda_{abcs}^s = \mathbf{L}i_{abcs}^s + \mathbf{M}i_{xyzs}^s + \mathbf{A}(\theta)i_{abcr}^r + \mathbf{A}\left(\theta - \frac{\pi}{12}\right)i_{abcr}^r.$$

Find $\mathbf{L}$, $\mathbf{M}$, and $\mathbf{A}(\theta)$. Also, derive λ_{dq}^s.

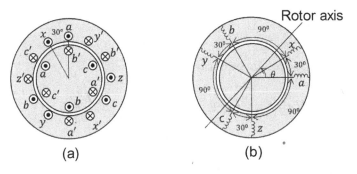

(a)   (b)

Six phase IM: (a) coil configuration and (b) phase current axes.

4.5   Derive

$$\lambda_{dqr}^r = L_m e^{-j\theta_r} i_{dqs}^s + L_r i_{dqr}^{ir}$$

using $\lambda_{abcr}^r = \mathbf{M}(-\theta_r)i_{abcs}^s + \mathbf{L}_{abcr}i_{abcr}^r$, where $L_r = \frac{3}{2}L_{ms} + L_{lr}$.

4.6   In the stationary frame, the IM model is described as

$$
\begin{bmatrix} v_{ds}^s \\ v_{qs}^s \\ 0 \\ 0 \end{bmatrix} =
\begin{bmatrix}
r_s + pL_s & 0 & pL_m & 0 \\
0 & r_s + pL_s & 0 & pL_m \\
pL_m & \omega_r L_m & r_r + pL_r & \omega_r L_r \\
-\omega_r L_m & pL_m & -\omega_r L_r & r_r + pL_r
\end{bmatrix}
\begin{bmatrix} i_{ds}^s \\ i_{qs}^s \\ i_{dr}^s \\ i_{qr}^s \end{bmatrix}.
$$

Using the above equation, derive the following dynamic model of IM in the stationary frame:

$$
\frac{d}{dt}
\begin{bmatrix} i_{ds}^s \\ i_{qs}^s \\ i_{dr}^s \\ i_{qr}^s \end{bmatrix} =
\begin{bmatrix}
-\frac{r_s}{\sigma L_s} & \omega_r\frac{1-\sigma}{\sigma} & \frac{r_r L_m}{\sigma L_r L_s} & \omega_r\frac{L_m}{\sigma L_s} \\
-\omega_r\frac{1-\sigma}{\sigma} & -\frac{r_s}{\sigma L_s} & -\omega_r\frac{L_m}{\sigma L_s} & \frac{r_r L_m}{\sigma L_r L_s} \\
\frac{r_s L_m}{\sigma L_s L_r} & -\omega_r\frac{L_m}{\sigma L_r} & -\frac{r_r}{\sigma L_r} & -\frac{\omega_r}{\sigma} \\
\omega_r\frac{L_m}{\sigma L_r} & \frac{r_s L_m}{\sigma L_s L_r} & \frac{\omega_r}{\sigma} & -\frac{r_r}{\sigma L_r}
\end{bmatrix}
\begin{bmatrix} i_{ds}^s \\ i_{qs}^s \\ i_{dr}^s \\ i_{qr}^s \end{bmatrix}
$$

$$
+
\begin{bmatrix}
\frac{1}{\sigma L_s} & 0 \\
0 & \frac{1}{\sigma L_s} \\
-\frac{L_m}{\sigma L_s L_r} & 0 \\
0 & -\frac{L_m}{\sigma L_s L_r}
\end{bmatrix}
\begin{bmatrix} v_{ds}^s \\ v_{qs}^s \end{bmatrix}.
$$

4.7   Consider the steady state voltage equations, (4.77) and (4.78):

$$\mathbf{V}_{dqs} = r_s \mathbf{I}_{dqs} + j\omega_e \mathbf{\Lambda}_{dqs}$$

$$0 = \frac{r_r}{s}\mathbf{I}_{dqr} + j\omega_e \mathbf{\Lambda}_{dqr}.$$

a) Assume that the d−axis coincides with the rotor flux.
    Justify $i_{dr}^e = 0$ and $L_m i_{qs}^e + L_r i_{qr}^e = 0$.
b) Show that

$$\lambda_{qs}^e = \sigma L_s i_{qs}^e,$$

$$\lambda_{ds}^e = \sigma L_s i_{ds}^e + \frac{L_m^2}{L_r}i_{ds}^e.$$

c) Show that the voltage equations become

$$\mathbf{V}_{dqs} = (r_s + j\omega_e\sigma L_s)\mathbf{I}_{dqs} + j\omega_e\frac{L_m^2}{L_r}i_{ds}^e$$

$$0 = \frac{r_r}{s}i_{qr}^e + j\omega_e L_m i_{ds}^e.$$

d) Draw an equivalent circuit based on the equations in c).

4.8 Construct a MATLAB Simulink block diagram for the IM model in the synchronous frame using (4.55) as shown below. Apply the same voltage as the one in Exercise 4.3. Using the inverse transformation, obtain $\mathbf{i}^s_{abcs}(t) = T^{-1}(\omega_e t)\mathbf{i}^e_{dq0s} = T^{-1}(\omega_e t)[i^s_{ds}(t),\, i^s_{qs}(t),\, 0]^T$. Check that the current is the same.

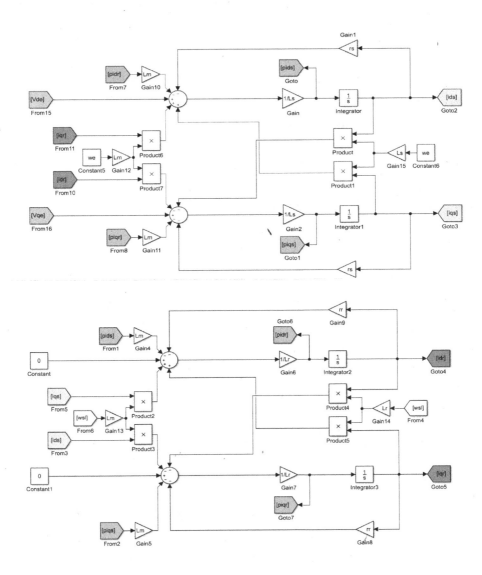

4.9 Show that

$$\text{i)} \quad T_e = \frac{3}{2}\frac{P}{2}(\lambda^e_{ds}i^e_{qs} - \lambda^e_{qs}i^e_{ds}),$$

$$\text{ii)} \quad T_e = \frac{3}{2}\frac{P}{2}\frac{L_m}{\sigma L_s L_r}(\lambda^e_{dr}\lambda^e_{qs} - \lambda^e_{ds}\lambda^e_{qr}).$$

4.10 Consider a three phase, four pole IM with following rated parameters: line to line voltage = 440 V, line frequency = 60 Hz, rated speed = 1746 rpm,

$r_s = 1\ \Omega$, $r_r = 2.256\ \Omega$, $L_m = 572$ mH, $L_{ls} = 32$ mH, and $L_{lr} = 32$ mH.

a) Construct a MATLAB Simulink block diagram for an IM model in the stationary frame.

b) Plot $[i_{ds}^s(t),\ i_{qs}^s(t)]$ and $\mathbf{i}_{abcs}^s(t) = T^{-1}(0)[i_{ds}^s(t),\ i_{qs}^s(t),\ 0]^T$.

4.11 Consider a three phase four pole IM with the following rated parameters: line to line voltage $= 440$ V(rms), line frequency $= 60$ Hz, $r_s = 1\ \Omega$, $r_r = 2.256\ \Omega$, $L_m = 572$ mH, $L_{ls} = 32$ mH, $L_{lr} = 32$ mH, $J = 0.2$ Kg $\cdot$ m^2, and $B = 0.1$ Nm $\cdot$ s.

Write a program using C-code for Runge-Kutta fourth method to obtain the IM currents $(i_{ds}^s,\ i_{qs}^s)$ based on (4.56) for $0 \leq t \leq 0.8$ sec.

4.12 Show the following identity:

$$\frac{3}{2}(v_{ds}^e i_{ds}^e + v_{qs}^e i_{qs}^e) - \frac{3}{2}r_s(i_{ds}^{e\,2} + i_{qs}^{e\,2}) - \frac{3}{2}r_r(i_{dr}^{e\,2} + i_{qr}^{e\,2})$$
$$= \frac{3P}{4}\frac{L_m}{L_r}(\lambda_{dr}^e i_{qs}^e - \lambda_{qr}^e i_{ds}^e)\,\omega_r$$

Chapter 5

Induction Motor Control

The electromagnetic torque is proportional to the product of gap field and armature current. In separately excited DC motor, the orthogonality is maintained between the armature and field currents with the aid of the brush and commutator, and the field current and the armature current are controlled independently.

The field oriented control, often called *the vector control* refers to all AC machine control methods that split the current vector into flux and torque components and control each of them independently. The basic control principle is the same as that of DC machine except the rotating reference frame. The phase currents are transformed into a synchronous reference frame whose d-axis is aligned with a flux vector. The d-axis current determines the flux magnitude, whereas the q-axis current, being orthogonal to the d-axis, controls the torque. In the case of IM, the control methods are sorted as rotor, stator, and air gap flux oriented schemes depending on the flux on which the frame is referenced. Among them *the rotor flux oriented scheme* is most popular for simplicity and good decoupling nature.

A technical hurdle is the difficulty in measuring or estimating the flux angle. Based on the flux angle access methods, field oriented controls are categorized as direct and indirect methods. The direct field oriented control relies on the flux estimates or measurements. The flux can be measured by using flux sensing coils or Hall devices. But, the flux measurement is rarely used in practice since it is not easy to install a sensing device in the air gap or slot, and the signal processing circuitry adds an extra cost. Usually, flux is estimated by utilizing flux observers. On the other hand, the flux angle is estimated by integrating the sum of calculated slip and measured rotor speed in the indirect field oriented control.

Sensorless control is a collective term of field oriented control methods that do not utilize speed or position sensors. The vector control without a speed sensor is attractive in the practical application, since it eliminates many problems in sensor installation, cabling, processing circuit, and low signal to noise ratio. Enhancing the torque and speed performance in the low speed region is an on-going problem in the sensorless control.

5.1   Rotor Field Oriented Scheme

The rotor field oriented scheme is favored, because the algorithm yields a natural decomposition in the roles of d and q axis currents. As the name says, the d-axis of the reference frame is aligned with the rotor flux. Then, it results in $\lambda_{qr}^e = 0$ automatically. The axis aligning effort is realized by regulating λ_{qr}^e to zero by a PI controller in the reference frame. It should be remembered that the rotor flux rotates at the synchronous speed, ω_e, not in accordance with the shaft (mechanical) speed, $\frac{P}{2}\omega_r$.

Utilizing (4.19), the voltage equation is derived as (4.51):

$$v_{ds}^e = (r_s + pL_s\sigma)i_{ds}^e + \frac{L_m}{L_r}p\lambda_{dr}^e - \omega_e\left(L_s\sigma i_{qs}^e + \frac{L_m}{L_r}\lambda_{qr}^e\right), \tag{5.1}$$

$$v_{qs}^e = (r_s + pL_s\sigma)i_{qs}^e + \frac{L_m}{L_r}p\lambda_{qr}^e + \omega_e\left(L_s\sigma i_{ds}^e + \frac{L_m}{L_r}\lambda_{dr}^e\right). \tag{5.2}$$

Fig. 5.1 shows an alignment of the d-axis to the rotor flux. Then it follows that $\lambda_{qr}^e = 0$, as well as $\dot{\lambda}_{qr}^e = 0$.

Stator Equation

By letting $\lambda_{qr}^e = 0$, we obtain from (5.1) and (5.2)

$$v_{ds}^e = (r_s + p\sigma L_s)i_{ds}^e - \omega_e\sigma L_s i_{qs}^e + \frac{L_m}{L_r}p\lambda_{dr}^e, \tag{5.3}$$

$$v_{qs}^e = (r_s + p\sigma L_s)i_{qs}^e + \omega_e\sigma L_s i_{ds}^e + \omega_e\frac{L_m}{L_r}\lambda_{dr}^e. \tag{5.4}$$

Note that $-\omega_e\sigma L_s i_{qs}^e$ and $\omega_e\sigma L_s i_{ds}^e$ are the coupling terms between d and q axes, and that $\omega_e\frac{L_m}{L_r}\lambda_{dr}^e$ is the back EMF.

Rotor Equation

Applying $\lambda_{qr}^e = 0$ and $\dot{\lambda}_{qr}^e = 0$ to (4.34) and (4.34), it follows that

$$0 = r_r i_{dr}^e + p\lambda_{dr}^e = (r_r + L_r p)i_{dr}^e + pL_m i_{ds}^e, \tag{5.5}$$
$$0 = r_r i_{qr}^e + (\omega_e - \omega_r)\lambda_{dr}^e. \tag{5.6}$$

Therefore, we have from (5.5) that

$$i_{dr}^e = -\frac{L_m p i_{ds}^e}{r_r + pL_r} \tag{5.7}$$

Utilizing (5.7), the d-axis rotor flux turns out to be

$$\lambda_{dr}^e = L_m i_{ds}^e + L_r i_{dr}^e = L_m i_{ds}^e - \frac{L_r L_m p i_{ds}^e}{r_r + pL_r} = \frac{L_m}{1 + p\tau_r}i_{ds}^e, \tag{5.8}$$

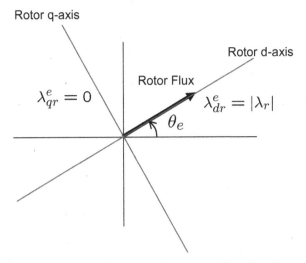

Figure 5.1: Aligning the d-axis frame to the rotor flux λ_r. It results in $\lambda_{dr} = |\lambda_{\mathbf{r}}|$ and $\lambda_{qr} = 0$.

where $\tau_r = \frac{L_r}{r_r}$ is the rotor time constant. In the steady state, (5.8) reduces to

$$\lambda_{dr}^e = L_m i_{ds}^e . \tag{5.9}$$

Since $\lambda_{qr}^e = 0$ in the rotor field oriented scheme, it follows that

$$0 = L_m i_{qs}^e + L_r i_{qr}^e . \tag{5.10}$$

Then, the slip equation follows from (5.6) and (5.10):

$$\omega_{sl} \equiv \omega_e - \frac{P}{2}\omega_r = s\omega_e = -r_r \frac{i_{qr}^e}{\lambda_{dr}} = \frac{1}{\tau_r}\frac{L_m}{\lambda_{dr}}i_{qs}^e . \tag{5.11}$$

The slip is often written as $\omega_{sl} = \frac{1}{\tau_r}\frac{i_{qs}^e}{i_{ds}^e}$ in the steady state.

Roles of i_{ds} and i_{qs}

Comparing $\lambda_{dr}^e = L_m i_{ds}^e$ with $\lambda_{dr}^e = L_m i_{ds}^e + L_r i_{dr}^e$, it is observed that $i_{dr}^e = 0$. With $\lambda_{qr}^e = 0$, the torque equation (4.84) reduces to

$$T = \frac{3P}{4}\frac{L_m}{L_r}\lambda_{dr}^e i_{qs}^e .$$

This equation is comparable to the torque equation of the DC motor. The similarities with the DC motor are:

λ_{dr}^e corresponds to the field.
i_{ds}^e corresponds to the field current.
i_{qs}^e corresponds to the armature current.

Now, the roles of d and q axes stator currents become clear:

Note that no d-axis current flows in the rotor. Only the stator current flows in the d-direction, and it is responsible for the gap field generation. In other words, the stator d-axis current, i_{ds} is used solely to generate λ_{dr}^e.

On the other hand, the q-axis stator current, i_{qs} flows to generate torque. Note however that no flux is generated along the q-axis, although a huge q-axis stator current, i_{qs}^e flows. It is because about the same amount of rotor current, i_{qr}^e flows in the opposite direction, i.e., $i_{qr}^e = -\frac{L_m}{L_r}i_{qs}^e$. As a result, a possible generation of λ_{qr}^e is suppressed. It can be viewed as a correction of the *armature reaction*, which refers to the gap field distortion by the armature current.

Fig. 5.2 shows the current and field distribution and illustrates the torque production based on the Lorentz force. The rotor field is applied to the q-axis stator current, i_{qs}^e. Then Lorentz force is developed on the q-axis coil. However, the stator coils, being fixed in the stator slots, are not allowed to move. Instead, the rotor rotates in the reverse direction due to the principle of force couple. Fig. 5.2 (b) shows current vector cancelation in the q-axis.

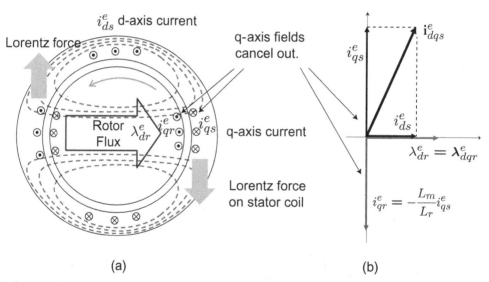

(a)   (b)

Figure 5.2: (a) Torque production and (b) current and flux vectors with the rotor field oriented scheme.

The roles of currents are summarized in the following:

1. i_{ds}^e is used solely to generate the rotor flux $\Leftarrow$ (5.8). It acts like the field current of a DC motor.

2. $i_{dr}^e = 0$ in the steady state. Only the q-axis current, i_{qr}^e flows in the rotor in order to nullify a possible generation of the q-axis rotor field caused by i_{qs}^e $\Leftarrow$ (5.10).

3. i_{qs}^e is used to generate the torque $\Leftarrow$ (5.11). The q-axis current is the armature current in the AC motor.

Vector Diagram in Steady State

In the steady state, $p\lambda_{dr} = 0$ and $pi_{ds}^e = pi_{qs}^e = 0$. With the complex variables, (5.3) and (5.4) are rewritten as

$$v_{dqs}^e = r_s i_{dqs}^e + j\omega_e \sigma L_s i_{dqs}^e + j\omega_e \frac{L_m}{L_r} \lambda_{dr}^e. \tag{5.12}$$

Based on (5.12), a vector diagram for the rotor field oriented control can be drawn as Fig. 5.3. Note that the voltage vector leads the current vector by ϕ.

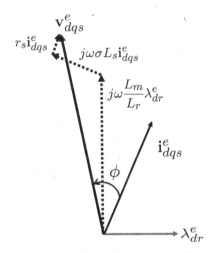

Figure 5.3: Voltage vector diagram for the rotor field oriented control.

Exercise 5.1   Consider a 220 V (line-to-line) 60 Hz, three phase, four pole IM. The parameters of the IM are $r_s = 0.2 \ \Omega$, $L_{ls} = L_{lr} = 2.3$ mH, and L_m=34.2 mH. Assume that the IM produces T_e =40 Nm at 60 Hz electrical speed when the rotor field oriented control is applied to the IM with $i_{ds}^e = 7$ A. Sketch the voltage and current vector diagram.

Block Diagram for Rotor Field Oriented Scheme

Substituting Laplace operator s for p, we obtain from (5.3) and (5.4) that

$$i_{ds}^e = \frac{\frac{1}{\sigma L_s}}{s + \frac{r_s}{\sigma L_s}} v_{ds}^e + \frac{1}{s + \frac{r_s}{\sigma L_s}} \omega_e i_{qs}^e, \tag{5.13}$$

$$i_{qs}^e = \frac{\frac{1}{\sigma L_s}}{s + \frac{r_s}{\sigma L_s}} \left(v_{qs}^e - \omega_e \frac{L_m}{L_r} \lambda_{dr}^e \right) - \frac{1}{s + \frac{r_s}{\sigma L_s}} \omega_e i_{ds}^e. \tag{5.14}$$

A block diagram based on the reduced model (5.13) and (5.14) is shown in Fig. 5.4. The IM model contains just coupling terms and the back EMF.

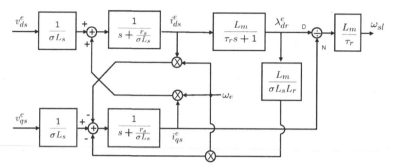

Figure 5.4: IM model under the rotor field oriented scheme.

Field Oriented Control Implementation

The IM dynamic model mimics the DC motor dynamic model in the rotor flux reference frame in which the roles of the dq-axis currents are separated. Specifically, the d-axis current, functioning as the field current, should be regulated to maintain a desired rotor field level. The q-axis current, as the armature current, needs to be set in accordance with the torque command. Thus, the torque control can be realized by setting both d and q axis currents constants. For this purpose, PI controllers are employed to keep the desired DC level. In brief, the torque control turns out to be a DC regulation problem in the synchronous reference frame, and the DC leveling by the PI controller is a well established control method. It gives an answer to why the control in a synchronous reference frame is preferred. Otherwise, AC currents must be controlled in the stationary frame. However, it is hard to be realized. Fig. 5.5 shows a general control block diagram for the field oriented control which involves coordinate transformations in the current control loop.

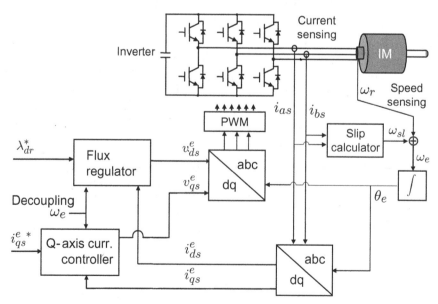

Figure 5.5: Field oriented control block diagram involving coordinate change.

Current Controller in Synchronous Frame

The controller is comprised of submodules: PI controllers, current transformation from abc to dq, voltage (inverse) transformation from dq to abc, and decoupling controller. Two PI controllers are employed for d and q axes. Normally, the same gains are applied. The coupling terms, $-\omega_e \sigma L_s i_{qs}^e$ and $\omega_e \sigma L_s i_{ds}^e$ and the back EMF, $\omega_e(L_m/L_r)\lambda_{dr}$ are compensated. Summarizing the above, the current controller is constructed as

$$v_{ds}^e = K_p(i_{ds}^{e*} - i_{ds}^e) + K_i \int_0^t (i_{ds}^{e*} - i_{ds}^e)dt - \omega_e \sigma L_s i_{qs}^e, \qquad (5.15)$$

$$v_{qs}^e = K_p(i_{qs}^{e*} - i_{qs}^e) + K_i \int_0^t (i_{qs}^{e*} - i_{qs}^e)dt + \omega_e \sigma L_s i_{ds}^e + \omega_e \frac{L_m}{L_r}\lambda_{dr}^e, \qquad (5.16)$$

where K_p and K_i are proportional and integral gains, respectively. The d-axis current command, i_d^{e*}, is selected to provide sufficient flux in the air gap. Normally, the d-axis current command, i_{ds}^{e*} is about 30% of the rated current level. The q-axis current command, i_q^{e*} determines torque, so that it comes out from a higher level control loop, e.g. torque or speed controller.

Angle Estimation

The rotor flux angle, θ_e is an essential parameter for coordinate transformation. In the indirect vector (or field oriented) control, the electrical angular velocity is obtained by adding the slip speed estimate to the motor shaft speed, i.e., $\omega_e = \omega_r + \omega_{sl}$. Encoders or resolvers are most commonly used to detect the rotor speed, ω_r. The angular position θ_e of the rotor flux is obtained by integrating ω_e:

$$\theta_e = \int_0^t \omega_e dt = \int_0^t (\omega_{sl} + \frac{P}{2}\omega_r)dt = \int_0^t (\frac{L_m i_{qs}^e}{\tau_r \lambda_{dr}^e} + \frac{P}{2}\omega_r)\, dt. \qquad (5.17)$$

Control Block Diagram

A detailed block diagram for the rotor field oriented control is shown in Fig. 5.6. The control sequence is illustrated as follows:

1) Measure phase currents.
2) Estimate the rotor flux angle, θ_e according to (5.17).
3) Transform (i_{as}, i_{bs}) into (i_{ds}^e, i_{qs}^e) using the coordinate transformation map, $\mathbf{T}(\theta_e)$.
4) Apply *PI* controllers with the decoupling feedback.
5) Transform the voltage vector, (v_{ds}^e, v_{qs}^e) into on-duties (T_a, T_b, T_c) of pulse width modulation (PWM).

The phase currents are measured by utilizing Hall current sensor or shunt resistor. Since the current sum is equal to zero, it is normal to measure only two phase currents, for example, (i_{as}, i_{bs}). Then, we let $i_{cs} = -(i_{as} + i_{bs})$. Step 5) and 6) are practically merged into a single processing block. That is, the inverse transformation is fused into the *space vector modulation* in normal practice, which is treated in Chapter 10. The current controller is constructed based on (5.15) and (5.16), and the slip is calculated according to (5.11). Within a rated speed range, the flux level is kept constant, but it is reduced as the speed increases (field weakening). The field reference command, λ_{dr}^{e*} is depicted as a Mexican hat shape.

The microcontroller or digital signal processor (DSP) takes care of all the computing burden needed for coordinate transformation, executing current controller, and the space vector modulation. Since the above process is repeated every PWM cycle, high performance DSP is required. However, with the advances in DSP design and manufacturing technology, some motor control DSPs are now available in less than 10 US dollars.

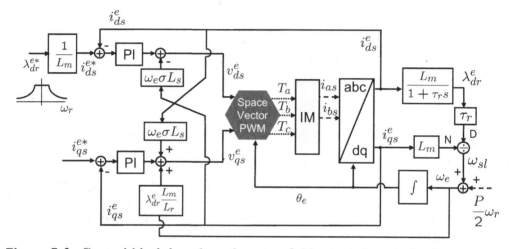

Figure 5.6: Control block based on the rotor field oriented control scheme.

Table 5.1: IM parameters for simulation. (Exercise 5.2)

Parameter	Value	Parameter	Value
Rated power	7.5 kW	Stator resistance, r_s	0.28 Ω
Voltage, V_{ll}	220 V$_{rms}$	Rotor resistance, r_r	0.15 Ω
Rated speed	1160 rpm	Leakage inductance, L_{ls}	1.4 mH
Frequency, f	60 Hz	Leakage inductance, L_{lr}	0.7 mH
Number of poles, P	6	Magnetizing inductance,L_{ms}	39.4 mH
Moment of inertia, J	0.4 kgm^2	Friction coefficient, B	10^{-4} kg·s

Exercise 5.2   Consider an IM with the parameters shown in the Table 5.1.

a) Calculate rated slip, s_r and rated current, $I_{s\,rated}$.

b) Build an IM control block diagram using MATLAB Simulink based on the rotor flux oriented scheme.

c) Set $i_{ds}^{e*} = 0.3 \times I_{s\,rated}$ and $i_{qs}^{e*} = 20 \times u_s(t - 0.01)$. Determine the current I controller, K_{ic} such that current control bandwidth is [0, 200] Hz.

Find K_{pc} such that q-axis current overshoot is about 10% or less.

d) Construct a speed PI controller and determine K_{ps} and K_{is} so that the speed bandwidth is $[0, 20]$ Hz and that no overshoot takes place. Plot i_{ds}^e, i_{qs}^e, and ω_r for $0 \le t \le 0.03$.

e) Plot the speed response to the following speed reference:

$$\begin{cases} 125t, & \text{if } t < 0.4 \\ 50, & \text{if } 0.4 \le t < 0.8 \\ -125t + 150, & \text{if } 0.8 \le t < 1.6 \\ -50, & \text{if } 1.6 \le t < 1.8 \end{cases}$$

Plot T_e, ω_r, and i_a, i_b, i_c for $0 \le t \le 1.8$.

Solution.

a) $s_r = (1200 - 1160)/1200 = 1/30$. $r_r/s_r = 4.53$, $X_{ls} = 377 \times 1.4 \times 10^{-3} = 0.53$, $X_{lr} = 377 \times 0.7 \times 10^{-3} = 0.264$, and $X_m = 377 \times 39.4 \times 10^{-3} = 14.85$.

$$Z = r_s + jX_{ls} + \frac{jX_m(jX_{lr} + r_r/s_r)}{jX_m + jX_{lr} + r_r/s_r} = 4.3 + j2.$$

Thus, $\mathbf{I}_{s\,rated} = 220/\sqrt{3}/(4.3 + j2)$ and $I_{s\,rated} = 26.6$ A.

b) See Figs. 5.7, 5.8, 5.9.

c) Note that $L_m = \frac{3}{2}L_{ms} = 59.1$ mH and $L_s = L_{ls} + \frac{3}{2}L_{ms} = 60.5$ mH. On the

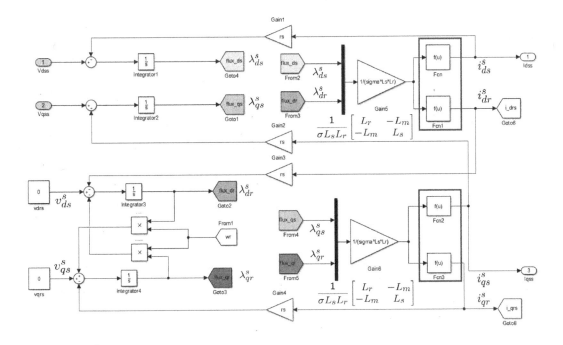

Figure 5.7: Simulink sub-block for IM (Solution to Exercise 5.2).

other hand, $\sigma = 1 - \frac{L_m^2}{L_s L_r} = 1 - 59.1^2/(60.5 \times 59.8) = 0.0346$. Thus, $K_{ic} = \sigma L_s \omega_n^2 = 0.0346 \times 0.0605 \times (2\pi \times 200)^2 = 3302$. Due to the effect of a zero in the transfer function, it is proper to select $\zeta = 1.2$ to limit the overshoot less than 10%. Therefore, $K_{pc} = 2\zeta\omega_n\sigma L_s - r_s = 2 \times 1.2 \times (2\pi \times 200) \times 0.0346 \times 0.0605 - 0.282 = 6.03$.

d) The nominal d-axis current is $i_{ds}^{e*} = 0.3 \times I_{s\,rated} = 7.965$ A. At this time, $\lambda_{dr} = L_m i_{dref} = 0.0591 \times 7.965 = 0.471$ Wb. Therefore, $K_t = \frac{3P}{4}\frac{L_m}{L_r}\lambda_{dr} = 4.5 \times 0.0591/0.0598 \times 0.471 = 2.09$. Thereby, $K_{is} = \frac{\omega_{ns}^2 J}{K_t} = \frac{(2\pi \times 20)^2 \times 0.4}{2.09} = 3019$. For a sufficient damping, a large damping coefficient, $\zeta_s = 8$ is selected. Thus, $K_{ps} = \frac{2\zeta_s\omega_{ns}J - B}{K_t} = \frac{2 \times 8 \times (2\pi \times 20) \times 0.4 - 0.0001}{2.09} = 385$. For current and speed responses, see Fig. 5.10 (a).

e) See Fig. 5.10 (b).

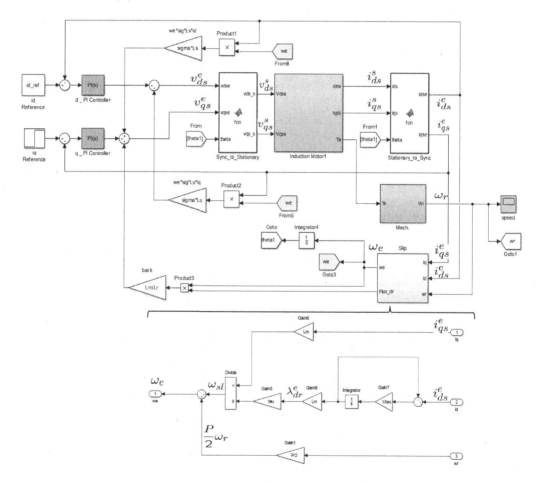

Figure 5.8: Simulink blocks for the rotor field oriented control (Solution to Exercise 5.2).

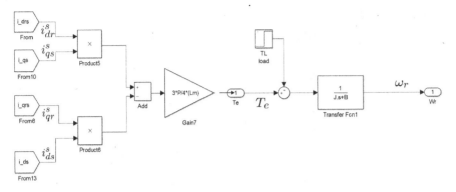

Figure 5.9: Simulink blocks for the mechanical part (Solution to Exercise 5.2).

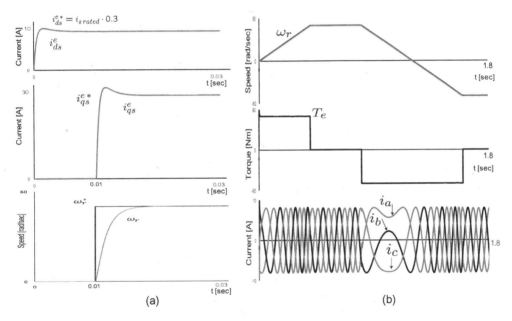

Figure 5.10: Current and speed step responses: (a) to a step speed command and (b) to the command of a speed profile (Solution to Exercise 5.2).

5.2 Stator Field Oriented Scheme

The stator field oriented scheme is the control method of aligning the d-axis with the stator flux. It is known to be less sensitive to machine parameters and produces more torque than the rotor field oriented scheme [1],[2]. But, the torque-flux decoupling is more difficult.

With the stator field oriented scheme, the q-axis stator flux is zero, i.e., $\lambda_{qs}^e = 0$. Therefore, it follows from (4.13), (4.14), (4.34), and (4.35) that

$$\lambda_{dr}^e = \frac{L_r}{L_m}(\lambda_{ds}^e - \sigma L_s i_{ds}^e), \tag{5.18}$$

$$\lambda_{qr}^e = -\frac{L_r}{L_m}\sigma L_s i_{qs}^e, \tag{5.19}$$

$$i_{dr}^e = \frac{1}{L_m}\lambda_{ds}^e - \frac{L_s}{L_m}i_{ds}^e, \tag{5.20}$$

$$i_{qr}^e = -\frac{L_s}{L_m}i_{qs}^e. \tag{5.21}$$

The stator flux and slip equations can be derived by substituting (5.18)−(5.21) into (4.34) and (4.35): From the rotor d-axis equation,

$$
\begin{aligned}
0 &= r_r i_{dr}^e + \mathrm{p}\lambda_{dr}^e - \omega_{sl}\lambda_{qr}^e \\
&= \frac{r_r}{L_m}(\lambda_{ds}^e - L_s i_{ds}^e) + \mathrm{p}\frac{L_r}{L_m}(\lambda_{ds}^e - \sigma L_s i_{ds}^e) + \frac{L_r}{L_m}\omega_{sl}\sigma L_s i_{qs}^e \\
&= \frac{1}{\tau_r}(\lambda_{ds}^e - L_s i_{ds}^e) + \mathrm{p}(\lambda_{ds}^e - \sigma L_s i_{ds}^e) + \omega_{sl}\sigma L_s i_{qs}^e \\
&= (\frac{1}{\tau_r} + \mathrm{p})\lambda_{ds}^e - L_s(\frac{1}{\tau_r} + \mathrm{p}\sigma)i_{ds}^e + \omega_{sl}\sigma L_s i_{qs}^e
\end{aligned}
$$

we obtain

$$\lambda_{ds}^e = \frac{(\sigma\mathrm{p} + \frac{1}{\tau_r})L_s}{\mathrm{p} + \frac{1}{\tau_r}}i_{ds}^e - \frac{\sigma L_s i_{qs}^e}{\mathrm{p} + \frac{1}{\tau_r}}\omega_{sl}. \tag{5.22}$$

From the rotor q-axis equation,

$$
\begin{aligned}
0 &= r_r i_{qr}^e + \mathrm{p}\lambda_{qr}^e + \omega_{sl}\lambda_{dr}^e \\
&= -\frac{r_r L_s}{L_m}i_{qs}^e - \mathrm{p}\frac{L_r}{L_m}\sigma L_s i_{qs}^e + \frac{L_r}{L_m}\omega_{sl}(\lambda_{ds}^e - \sigma L_s i_{ds}^e) \\
&= -L_s i_{qs}^e - \mathrm{p}\tau_r\sigma L_s i_{qs}^e + \tau_r\omega_{sl}(\lambda_{ds}^e - \sigma L_s i_{ds}^e)
\end{aligned}
$$

we obtain [3]

$$\omega_{sl} = \frac{(\frac{1}{\tau_r} + \mathrm{p}\sigma)L_s i_{qs}^e}{\lambda_{ds}^e - \sigma L_s i_{ds}^e} = \frac{(\mathrm{p} + \frac{1}{\sigma\tau_r})i_{qs}^e}{\frac{\lambda_{ds}^e}{\sigma L_s} - i_{ds}^e}. \tag{5.23}$$

Note that flux and slip equations are not decoupled in the stator field oriented scheme. Specifically, ω_{sl} is involved in the flux calculation, (5.22). In turn, λ_{ds}^e is used for slip calculation, (5.23). Further, the slip equation (5.23) contains differential operator, p. Thus, it requires differentiation of i_{qs}^e, which is disadvantageous when it is compared with (5.11). Hence, the rotor field oriented scheme is commonly used in practical applications. The torque equation corresponding to the stator reference frame is $T_e = \frac{3}{2}\frac{P}{2}\lambda_{ds}^e i_{qs}^e$.

5.3 Field Weakening Control

When the speed increases, the back EMF grows. At a rated condition, the source (inverter) voltage reaches its limit. To operate the machine above the rated speed, both torque and field have to be reduced.

5.3.1 Current and Voltage Limits

The current rating is determined mostly by the thermal capacity of the motor. Let the peak value of the maximum phase current be denoted by I_m. Then,

$$i_{ds}^{e2} + i_{qs}^{e2} \leq I_m^2. \tag{5.24}$$

Thus, the current limitation is described by a circle in the current plane.

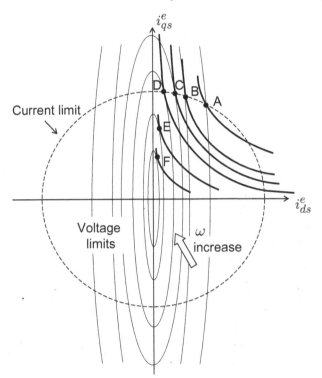

Figure 5.11: Current and speed limits with constant torque curves.

Let V_m be the peak value of the maximum phase voltage. Note that $V_m = V_{DC}/\sqrt{3}$ in the space vector modulation, where V_{DC} is the DC link voltage. The voltage limit is

$$v_{ds}^{e2} + v_{qs}^{e2} \leq V_m^2. \tag{5.25}$$

Suppose that the machine is controlled according to the rotor field oriented scheme, and assume that the machine is in the steady state. The steady state model (5.12) can be rewritten as

$$v_{ds} = \cdot r_s i_{ds}^e - \omega_e \sigma L_s i_{qs}^e, \tag{5.26}$$

$$v_{qs} = r_s i_{qs}^e + \omega_e L_s i_{ds}^e. \tag{5.27}$$

By substituting (5.26) and (5.27) into (5.25) we obtain an ellipse:

$$\left(\frac{r_s i_{qs}^e}{\omega_e} + L_s i_{ds}^e\right)^2 + \left(\frac{r_s i_{ds}^e}{\omega_e} - \sigma L_s i_{qs}^e\right)^2 \leq \frac{V_m^2}{\omega_e^2}. \tag{5.28}$$

A further simplification is made by neglecting the stator resistance. The voltage drop over the stator resistance is relatively small especially in the high speed region. Then, (5.28) reduces to an elliptical equation:

$$\frac{i_{ds}^{e\,2}}{\left(\frac{V_m}{\omega_e L_s}\right)^2} + \frac{i_{qs}^{e\,2}}{\left(\frac{V_m}{\omega_e \sigma L_s}\right)^2} \leq 1. \tag{5.29}$$

Note that $\frac{V_m}{\omega_e L_s} \ll \frac{V_m}{\omega_e \sigma L_s}$, since $\sigma \approx 0.1$. Thus, the major axis is aligned with the q–axis, whereas the minor axis lies on the d–axis. Hence, the ellipse of (5.29) has an upright shape, as shown in Fig. 5.11. The center is at the origin. As the frequency increases, the voltage constraint will shrink towards the origin.

With the rotor field oriented scheme, the torque equation is expressed as

$$T_e = \frac{3}{2}\frac{P}{2}\frac{L_m^2}{L_r} i_{ds}^e i_{qs}^e. \tag{5.30}$$

Hence, constant torque curves appear as parabolic curves of the type $y = \frac{c}{x}$. Fig. 5.11 shows the current and voltage limits with several constant torque curves.

5.3.2 Torque–Speed Curve

The machine can work only in the intersected area between the current limit circle and the voltage limit ellipse, i.e, in the overlapped area between the current circle and the voltage ellipse. The maximum torque and the maximum power are feasible if the voltage and current are used to the maximum. Therefore, the torque–speed curve needs to be interpreted with the voltage and current limit curves. The operation range is separated into three parts depending on the operation modes. The typical profiles of voltage, current, torque, and flux are shown in Fig. 5.12.

Constant Torque Limit Region

The constant torque region indicates the speed range from zero to a rated speed. In this region, torque can be increased to the maximum by utilizing the maximum current. The power increases in proportion to the speed. At the rated speed, the motor terminal voltage reaches the limit.

Constant Power Limit Region

To take advantage of the maximum efficiency of the machine, the maximum voltage and current should be used. For maximum power operation the operating points must be determined at the intersections of the voltage and current limits. Note again that the current limit circle has nothing to do with the speed. Meanwhile,

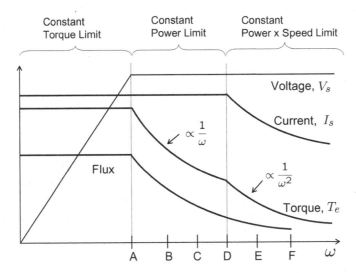

Figure 5.12: IM characteristics in constant torque and field-weakening region.

the voltage limit ellipse shrinks as the speed increases. Therefore, the operating points migrate along the circle to the left (Points 'B', 'C', and 'D' of Fig. 5.11). Along this migration, the torque reduces. In this region, the torque is inversely proportional to $1/\omega$.

Along the contour, the d-axis current reduces along with the flux. Since the maximum voltage and maximum current are utilized, power remains nearly constant if the power factor does not vary significantly. Thus, this region is often called *the constant power limit region* or *field weakening region I*.

Second Field Weakening Region

In the very high speed region above the point 'D', the voltage ellipses shrink inside the circle. In this region, the highest torque is found at the tangential intersection between torque and voltage limits. Note also that current also reduces as the speed increases. Therefore, torque reduction is accelerated. Representing torque as $T_e = k i_d^e i_q^e$, it follows that

$$i_{ds}^{e\,2} + \left(\frac{\sigma T_e}{k i_{ds}^e}\right)^2 = \left(\frac{V_m}{\omega_e L_s}\right)^2 .$$

Applying inequality $A^2 + B^2 \geq 2AB$, we obtain

$$i_{ds}^{e\,2} + \left(\frac{\sigma T_e}{k i_{ds}^e}\right)^2 \geq \frac{2\sigma T_e}{k_e} .$$

Therefore, we have

$$\left(\frac{V_m}{\omega_e L_s}\right)^2 \geq \frac{2}{k}\sigma T_e. \tag{5.31}$$

This implies that torque reduces in proportion to $1/\omega_e^2$ in this region. This region is often called *second field weakening region* or *field weakening region II*.

Exercise 5.3   Consider a 220 V (line-to-line) 60 Hz, three phase, four pole IM. The parameters of the IM are $r_s = 0.2$ Ω, $L_{ls} = L_{lr} = 2$ mH, and $L_m = 34$ mH. Draw the voltage limit ellipse for different speeds with the current limit, $I_m = 40$ A. Draw the maximum torque versus speeds and determine the speed, ω_e where the field weakening regions I and II are separated.

5.3.3   Torque and Power Maximizing Solutions

Stator and leakage inductances are the major factors determining the maximum torque and power under a given DC-link voltage. The constant power speed range is also an important decision factor for high speed IMs. High inductance is good for reducing the current, whereas harmful in securing a wide speed range.

Maximum Torque per Ampere

The maximum torque per ampere (MTPA) is a current-minimizing scheme for a given torque. That is, it is a control strategy that minimizes the current magnitude, $I = \sqrt{i_{ds}^{e\,2} + i_{qs}^{e\,2}}$ for a given torque level $T_e = \frac{3P}{4}\frac{L_m^2}{L_r}i_{ds}^e i_{qs}^e$. Note that the constant torque curve is symmetric about the 45° line and that the set of a constant current magnitude is a circle. Therefore, the solution is always found at the 45° line:

$$i_{ds}^e = i_{qs}^e = \sqrt{\frac{4T_e}{3P}\frac{L_r}{L_m^2}} = \frac{I_s}{\sqrt{2}}, \quad (I_s \le I_m) \tag{5.32}$$

In Fig. 5.13 the MTPA trajectory is marked by $\square$ and point A is a boundary point between the MTPA and the constant d-axis current control contour.

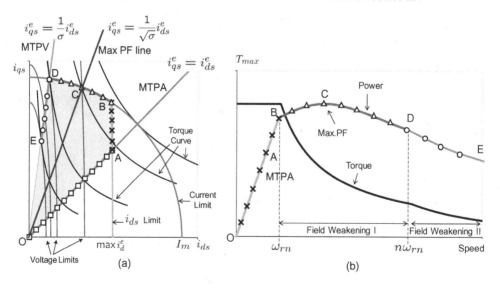

Figure 5.13: Maximum torque trajectory under current and voltage limits: (a) operation contour and limits in the current plane and (b) torque-speed curve.

D-Axis Current Limit Contour

The flux level is determined by the d-axis current, so that it should be limited to avoid core saturation. Note that a rated current is selected at the border point, B between the constant torque and the field weakening regions. Thus, the maximum voltage is utilized at the rated condition. Since σ is small, we have the following approximation at B

$$i_{ds}^{e\,2} + \frac{i_{qs}^{e\,2}}{\sigma^2} \approx i_{ds}^{e\,2} \leq \left(\frac{V_m}{\omega_{rn} L_s}\right)^2. \tag{5.33}$$

where ω_{rn} is the rated speed. Thus, the maximum d-axis current is set such that

$$\max\{i_{ds}^e\} \leq \frac{V_m}{\omega_{rn} L_s}. \tag{5.34}$$

The d-axis current limit contour appears as a vertical line $\overline{AB}$.

Maximum Power Contour

The maximum power is feasible when the maximum voltage and current are utilized. Therefore, the maximum power is found at the intersection between the current and voltage limits. Solving (5.24) and (5.28) simultaneously it follows that

$$(i_{ds}^e,\ i_{qs}^e) = \left(\sqrt{\frac{1}{1-\sigma^2}\left(\frac{V_m^2}{\omega_e^2 L_s^2} - \sigma^2 I_m^2\right)},\ \sqrt{\frac{1}{1-\sigma^2}\left(I_m^2 - \frac{V_m^2}{\omega_e^2 L_s^2}\right)}\right) \tag{5.35}$$

The solution contour is an arc, $\overline{BD}$ of the current limit circle which is marked by $\triangle$ in Fig. 5.13(a). Note from the torque-speed curve shown in Fig. 5.13(b) that the power is not constant during the journey from B to D. Fig. 5.14(a) shows two sets of current and voltage vectors corresponding to B and D. It shows different power factor angles, ϕ_B and ϕ_D. Fig. 5.14(b) shows a locus of power factor when the speed increases from B to D. Expressing power factor in terms of currents, we obtain

$$PF = \frac{v_{ds}^e i_{ds}^e + v_{qs}^e i_{qs}^e}{\sqrt{v_{ds}^{e\,2} + v_{qs}^{e\,2}}\sqrt{i_{ds}^{e\,2} + i_{qs}^{e\,2}}} = \frac{1-\sigma}{\sqrt{(i_{ds}^e/i_{qs}^e)^2 + (\sigma i_{qs}^e/i_{ds}^e)^2 + 1 + \sigma^2}}. \tag{5.36}$$

It is a function of i_{ds}^e/i_{qs}^e. Thus, the PF is constant on $\overline{OA}$ and $\overline{DE}$ where the current ratio is constant.

Power factor is maximized by minimizing the denominator of (5.36). To minimize $(i_{ds}^e/i_{qs}^e)^2 + (\sigma i_{qs}^e/i_{ds}^e)^2$ under the constraint, $I_m^2 = i_{ds}^{e\,2} + i_{qs}^{e\,2}$ we define the following Lagrangian:

$$\mathcal{L} = \left(\frac{i_{ds}^e}{i_{qs}^e}\right)^2 + \left(\frac{\sigma i_{qs}^e}{i_{ds}^e}\right)^2 + \mu\left\{I_m^2 - i_{ds}^{e\,2} - i_{qs}^{e\,2}\right\}, \tag{5.37}$$

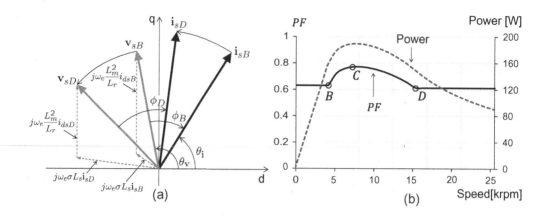

Figure 5.14: Power factor change depending on speed: (a) voltage and current vectors at points B and D and (b) power factor angle versus speed.

where μ is a Lagrange coefficient. Then, the necessary conditions are

$$\frac{\partial \mathcal{L}}{\partial i_{ds}^e} = \frac{2i_{ds}^e}{i_{qs}^{e\,2}} - \frac{2\sigma^2 i_{qs}^{e\,2}}{i_{ds}^{e\,3}} - 2\mu i_{ds}^e = 0 \tag{5.38}$$

$$\frac{\partial \mathcal{L}}{\partial i_{qs}^e} = -\frac{2i_{ds}^{e\,2}}{i_{qs}^{e\,3}} + \frac{2\sigma^2 i_{qs}^e}{i_{ds}^2} - 2\mu i_{qs}^e = 0 \tag{5.39}$$

$$\frac{\partial \mathcal{L}}{\partial \mu} = I_m^2 - i_{ds}^{e\,2} - i_{qs}^{e\,2} = 0. \tag{5.40}$$

Solving (5.38)−(5.40) simultaneously, the maximum PF condition is obtained as

$$i_{qs}^e = \frac{1}{\sqrt{\sigma}} i_{ds}^e. \tag{5.41}$$

Substituting (5.41) into (5.36), the maximum power factor is calculated as

$$PF = \frac{1 - \sigma}{1 + \sigma}. \tag{5.42}$$

Note that (5.42) is identical to (3.31) which was obtained from a geometrical relation. Another approach is to use inequality, $A^2 + B^2 \geq 2AB$. It directly follows that $\left(i_{ds}^e / i_{qs}^e\right)^2 + \left(\sigma i_{qs}^e / i_{ds}^e\right)^2 \geq 2\sigma$. Substituting it into (5.36), the same value is obtained as (5.42). Note from Fig. 5.13 that the power factor reaches the apex at C when (5.41) is satisfied, and decreases as the point approaches to D.

Maximum Torque per Voltage

There are two possible choices for operation: a) intersection between current and voltage limits and b) intersection between torque curve and voltage limit. Above a certain speed in the field weakening range, case b) gives a larger torque. Specifically, the maximum torque for a given voltage is obtained at the tangential intersection between the voltage and torque curves. Such a solution is called *maximum torque per voltage* (MTPV). In the high speed region after point D, the MTPV solutions yield the maximum torque, and it corresponds to the second field weakening region. The MTPV locus can be found by solving (5.29) and (5.30) simultaneously.

$$i_{ds}^{e\,4} - \left(\frac{V_m}{\omega_e L_s}\right)^2 i_{ds}^{e\,2} + \left(\frac{4T\sigma}{3P(1-\sigma)L_s}\right)^2 = 0. \tag{5.43}$$

Substituting $\chi = i_{ds}^{e\,2}$ into (5.43), it follows that

$$\chi^2 - \left(\frac{V_m}{\omega_e L_s}\right)^2 \chi + \left(\frac{4T\sigma}{3PL_s}\right)^2 = 0. \tag{5.44}$$

Note that the tangential condition turns out to be the double root condition, i.e., the discriminant of (5.44) must be zero. It yields

$$\chi = \frac{V_m{}^2}{2(\omega_e L_s)^2}. \tag{5.45}$$

Since negative d-axis current is inappropriate, we have $i_{ds}^e = \frac{V_m}{\sqrt{2}\omega_e L_s}$. Then, it follows from (5.29) that $i_{qs}^e = \frac{V_m}{\sqrt{2}\omega_e \sigma L_s}$. Thus the MTPV locus is a straight line

$$i_{qs} = \frac{1}{\sigma}\, i_{ds}. \tag{5.46}$$

Alternatively, L_s can be eliminated from the MTPV solution such that

$$(i_{ds},\, i_{qs}) = \left(\frac{\sigma}{1-\sigma}\frac{4\sqrt{2}T_e\omega_e}{3PV_m},\ \frac{1}{1-\sigma}\frac{4\sqrt{2}T_e\omega_e}{3PV_m}\right). \tag{5.47}$$

Exercise 5.4 Using (5.31) and (5.46), derive $i_{ds}^e = \frac{V_m}{\sqrt{2}\omega_e L_s}$.

5.4 IM Sensorless Control

Encoder and resolver are usually used for speed measurements. Differently from the current sensors, the speed sensor must be mounted on the rotor shaft. However, mounting a sensor on a moving object is more difficult, it needs a mechanical integrity such as alignment. In addition, sensors are often under harsh environments like vibration, dust, water, ice, etc. Sensor cabling is another nuisance as the distance between motor and inverter increases. It forms a noise conduction loop with the inverter. In some home appliances, cost pressure prohibits the use of encoders. The capability of flux angle estimation should be provided to establish a field oriented control without a sensor. Usually, observers are devised to estimate the flux angle.

5.4.1   Voltage Model Estimator

Most sensorless schemes rely on the stator voltage and current measurements, which are not hard to obtain. It follows from (4.23) that the stator flux is obtained by integrating $\mathbf{v}^s_{dqs} - r_s \mathbf{i}^s_{dqs}$, i.e.,

$$\hat{\boldsymbol{\lambda}}^s_{dqs} = \int \left(\mathbf{v}^s_{dqs} - r_s \mathbf{i}^s_{dqs} \right) dt, \tag{5.48}$$

where $\hat{\boldsymbol{\lambda}}^s_{dqs}$ is the stator flux in the stationary frame. Then the rotor flux is obtained via

$$\hat{\boldsymbol{\lambda}}^s_{dqr} = \frac{L_r}{L_m} \left(\hat{\boldsymbol{\lambda}}^s_{dqs} - \sigma L_s \mathbf{i}^s_{dqs} \right) . \tag{5.49}$$

Finally, the rotor flux angle is estimated such that

$$\hat{\theta}_r = \tan^{-1} \left(\frac{\hat{\lambda}^s_{qr}}{\hat{\lambda}^s_{dr}} \right) . \tag{5.50}$$

The flux estimator based on the open loop voltage model, is called *voltage model* (VM) *based estimator* or simply VM estimator. This VM based estimator is simple and inherently sensorless but is marginally stable due to the open-loop integration: The integrator output tends to grow indefinitely when there is a small DC offset error. The VM based estimator shows a better performance in a high speed range where the back EMF is dominant over the ohmic drop or the noise level. In the low frequency ($0 \sim 10$ Hz) region, $r_s \mathbf{i}^s_{dqs}$ is relatively large among many voltage constituents. Furthermore, the exact value of r_s is hardly known due to the temperature dependence and the AC effects.

The classic approach to maintaining stability is to replace the integrator with a low-pass filter. Utilizing a first order filter, $\frac{1}{s+1/\tau_0}$, we obtain

$$\tau_0 \frac{d\hat{\boldsymbol{\lambda}}^s_{dqs}}{dt} + \hat{\boldsymbol{\lambda}}^s_{dqs} = \tau_0 (\mathbf{v}^s_{dqs} - r_s \mathbf{i}^s_{dqs}). \tag{5.51}$$

It is obvious the filter behaves like an integrator for small τ_0. In particular, the pseudo integration is effective in the high frequency range greater than $1/\tau_0$. Combining (5.48) and (5.49), the rotor flux can be written directly as

$$\hat{\boldsymbol{\lambda}}^s_{dqr} = \frac{L_r}{L_m} \left(\frac{1}{s} (\mathbf{v}^s_{dqs} - r_s \mathbf{i}^s_{dqs}) - \sigma L_s \mathbf{i}^s_{dqs} \right) \tag{5.52}$$

$$\approx \frac{L_r}{L_m} \left(\frac{\tau_0}{s + \tau_0} (\mathbf{v}^s_{dqs} - r_s \mathbf{i}^s_{dqs}) - \sigma L_s \mathbf{i}^s_{dqs} \right) . \tag{5.53}$$

The whole VM estimator is depicted graphically in Fig. 5.15. But the solution fails at frequencies lower than several times of the filter cutoff frequency, $1/\tau_0$ [11]. The VM performance can be significantly improved if explicit compensation of the measurement offset, noise, and inverter nonlinearity is added. In general, it cannot be used solely in the low frequency region (under 3 Hz) [4].

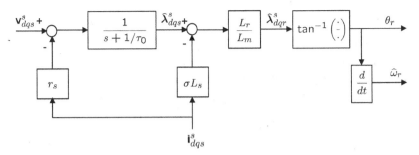

Figure 5.15: Open loop flux estimator based on the VM.

5.4.2   Current Model Estimator

The current model may also be used for the realization of sensorless observers although it requires the use of angle or speed information. In the rotor frame, the rotor voltage equation is given by

$$v_{dqr}^r = r_r i_{dqr}^r + \frac{d\lambda_{dqr}^r}{dt}.$$

To transform to the stationary frame, we multiply $e^{j\theta_r}$. Then, it follows that

$$
\begin{aligned}
e^{j\theta_r} v_{dqr}^r &= r_r e^{j\theta_r} i_{dqr}^r + e^{j\theta_r} \frac{d}{dt}\left(e^{-j\theta_r} e^{j\theta_r} \lambda_{dqr}^r\right) \\
0 &= r_r i_{dqr}^s - j\omega_r \lambda_{dqr}^s + \frac{d\lambda_{dqr}^s}{dt} \\
&= \frac{r_r}{L_r}(\lambda_{dqr}^s - L_m i_{dqs}^s) - j\omega_r \lambda_{dqr}^s + \frac{d\lambda_{dqr}^s}{dt}.
\end{aligned}
$$

Therefore, the rotor flux in the stationary frame satisfies

$$\tau_r \frac{d\hat{\lambda}_{dqr}^s}{dt} + \hat{\lambda}_{dqr}^s = j\omega_r \tau_r \hat{\lambda}_{dqr}^s + L_m \hat{i}_{dqs}^s. \tag{5.54}$$

Note that in the stationary frame the mechanical speed ω_r appears instead of ω_{sl}
Therefore, the rotor flux estimator is constructed such that

$$\tau_r \frac{d\hat{\lambda}_{dqr}^s}{dt} + \hat{\lambda}_{dqr}^s = j\hat{\omega}_r \tau_r \hat{\lambda}_{dqr}^s + L_m i_{dqs}^s. \tag{5.55}$$

Since it is driven by the current, it is referred to as *current model estimator*. Note that λ_{dqr}^s, being viewed from the stationary frame, includes the rotor speed, ω_r. It is unknown, i.e., the estimator includes an unknown parameter. Therefore, the current model (CM) based estimator cannot be used alone. Instead, it requires other means to estimate $\hat{\omega}_r$.

5.4.3   Closed-Loop MRAS Observer

We denote by $\hat{\lambda}_{dqr}^{sC}$, the rotor flux estimate obtained from the CM (5.55). Similarly, we denote by $\hat{\lambda}_{dqr}^{sV}$ the one obtained from the VM (5.53). The model reference adaptive system (MRAS) observer utilizes both estimates in parallel [4]. Note that the two estimates must be the same, i.e., $\hat{\lambda}_{dqr}^{sV} = \hat{\lambda}_{dqr}^{sC}$ if both estimations are correct. However if they do not match, the CM estimate is regarded incorrect due to error in the speed estimate, $\hat{\omega}_r$ in (5.55). Based on this reasoning, the speed estimate, $\hat{\omega}_r$ is adjusted based on the difference between $\hat{\lambda}_{dqr}^{sV}$ and $\hat{\lambda}_{dqr}^{sC}$. In particular, the angle difference of the two vectors is utilized for speed updating [4]:

$$\hat{\omega}_r = \left(K_1 + K_2 \frac{1}{s} \right) e_\lambda = \left(K_1 + K_2 \frac{1}{s} \right) (\hat{\lambda}_{dr}^{sC} \hat{\lambda}_{qr}^{sV} - \hat{\lambda}_{qr}^{sC} \hat{\lambda}_{dr}^{sV}), \qquad (5.56)$$

where

$$e_\lambda = \hat{\lambda}_{dqr}^{sC} \times \hat{\lambda}_{dqr}^{sV} \Big|_z = |\hat{\lambda}_{dqr}^{sC}| |\hat{\lambda}_{dqr}^{sV}| \sin \delta \approx |\hat{\lambda}_{dqr}^{sC}| |\hat{\lambda}_{dqr}^{sV}| \delta, \qquad (5.57)$$

and δ is the angle between the two flux estimates. It is equivalently written as

$$e_\lambda = Im \left(\hat{\lambda}_{dqr}^{sV} \cdot \hat{\lambda}_{dqr}^{sC *} \right), \qquad (5.58)$$

where $Im(\cdot)$ denotes the imaginary part and superscript $*$ on a complex variable denotes the complex conjugate. Two flux estimates and δ are depicted in Fig. 5.16.

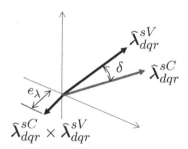

Figure 5.16: Cross product of two estimates to derive an angle error.

Fig. 5.17 shows the closed loop rotor flux model: Two models yield two rotor flux estimates in the stationary frame, and the outer product gives the angle difference. Passing it through a PI regulator, a rotor speed estimate is obtained. The speed estimate is used to update $j\hat{\omega}_r \tau_r \lambda_{dqr}^{sC}$ in the rotor model, completing a closed loop. Similarly to the above open loop scheme, this closed loop scheme has a fundamental limitation in the low speed region, since the VM estimator utilizes the same low pass filter.

5.4.4 Dual Reference Frame Observer

Many feasible methods of linking two types of observers were studied in [5], [6]. The dual reference frame observer is considered here that employs two reference frames, so that the speed-dependent terms within the motor model do not appear in its implementation. Fig. 5.18 shows a Gopinath style observer that utilizes two models described in different frames [6], [7]. The CM is modeled in the rotor frame,

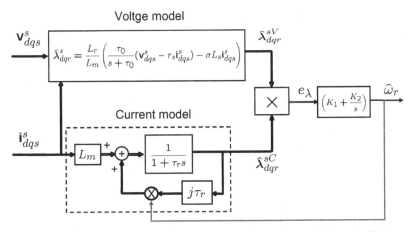

Figure 5.17: MRAS for rotor speed estimation based on VM and CM.

whereas the VM in the stationary frame. Therefore, the CM is accompanied by the coordinate transformation maps, $e^{j\hat{\theta}_r}$ and $e^{-j\hat{\theta}_r}$ in the upstream and downstream, where $\hat{\theta}_r$ is an estimate of θ_r.

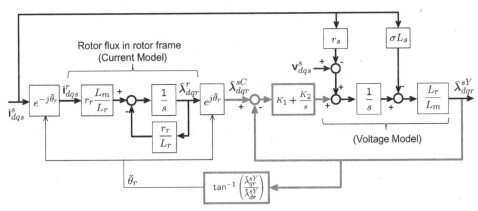

Figure 5.18: Gopinath flux observer incorporating the CM and VM via a PI regulator.

In the rotor frame the voltage equation is equal to

$$0 = r_r \mathbf{i}^r_{dqr} + \frac{d\boldsymbol{\lambda}^r_{dqr}}{dt} = \frac{r_r}{L_r}(\boldsymbol{\lambda}^r_{dqr} - L_m \mathbf{i}^r_{dqs}) + \frac{d\boldsymbol{\lambda}^r_{dqr}}{dt}.$$

Therefore, the CM estimator is given as

$$\frac{d\hat{\boldsymbol{\lambda}}^r_{dqr}}{dt} = -\frac{r_r}{L_r}\hat{\boldsymbol{\lambda}}^r_{dqr} + r_r \frac{L_m}{L_r}e^{-j\hat{\theta}_r}\mathbf{i}^s_{dqs} \tag{5.59}$$

$$\text{or} \quad \frac{d\hat{\boldsymbol{\lambda}}^r_{dqr}}{dt} = \frac{1}{\tau_r}(L_m \mathbf{i}^r_{dqs} - \hat{\boldsymbol{\lambda}}^r_{dqr}) = -r_r \mathbf{i}^r_{dqr}. \tag{5.60}$$

The estimate in the stationary frame is obtained via the transformation:

$$\hat{\boldsymbol{\lambda}}^{sC}_{dqr} = e^{j\hat{\theta}_r}\hat{\boldsymbol{\lambda}}^r_{dqr}. \tag{5.61}$$

The VM is based on (5.48), and the difference between the two estimates is fed to a PI-type regulator.

$$\left(K_1 + \frac{K_2}{s}\right)\left(\hat{\lambda}_{dqr}^{sC} - \hat{\lambda}_{dqr}^{sV}\right) \tag{5.62}$$

The output of (5.62) is used to correct the signal to the voltage model.

To visualize the convergence between the two flux estimates, we let $E(s) = \frac{L_r}{L_m}\left(K_1 + \frac{K_2}{s}\right)\frac{1}{s}$. Then, it follows that

$$\hat{\lambda}_{dqr}^{sY} = \underbrace{\frac{E(s)}{1 + E(s)}}_{= 1 - F(s)}\hat{\lambda}_{dqr}^{sC} + \underbrace{\frac{1}{1 + E(s)}}_{= F(s)}\hat{\lambda}_{dqr}^{sV}. \tag{5.63}$$

Based on (5.63), the dynamic linking between the two estimates is depicted as shown in Fig. 5.19. Note that $E(s)$ is *system Type 2*, since it has two open-loop poles at

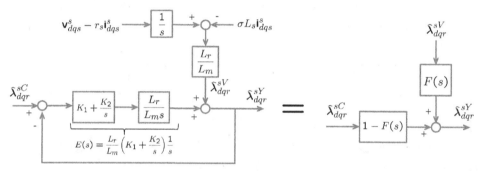

Figure 5.19: Dynamic linking between the current and voltage models.

the origin [8]. Thereby its unity feedback system,

$$F(s) = \frac{1}{1 + E(s)} = \frac{s^2}{s^2 + \frac{L_r}{L_m}K_1 s + \frac{L_r}{L_m}K_2} \tag{5.64}$$

suppresses the steady-state error even to a ramp input. It acts as a weighting function depending on frequency. Fig. 5.20 (a) shows Bode plots of $F(s)$ and $1-F(s)$ when $F(s) = s^2/(s^2 + 20s + 100)$. Note that $F(s)$ is a high pass filter, whereas $1 - F(s)$ is a low pass filter. However, their sum is equal to zero. Thus, the resulting closed-loop flux observer provides an automatic transition between the two open-loop flux observers. Since $|1 - F(j\omega)| \approx 1$ and $|F(j\omega)| \approx 0$ for small ω, the CM is weighted more than the VM in the low frequency region. It looks reasonable, since the VM is inaccurate in the low speed region. But as the frequency increases the VM is weighted more gradually. Finally, $|1 - F(j\omega)| \approx 0$ and $|F(j\omega)| \approx 1$, i.e., only the VM estimate is respected in the high frequency region. The crossover frequency can be determined by PI gains, K_1 and K_2.

However, the weight, $F(j\omega)$ is not real number, so that $|1-F(j\omega)|+|F(j\omega)| \neq 1$ for intermediate frequencies. For example, see $|1 - F(j\omega)|$ and $|F(j\omega)|$ at 10 Hz shown in Fig. 5.20 (b). Kim et al. [7] proposed a method to make the transition

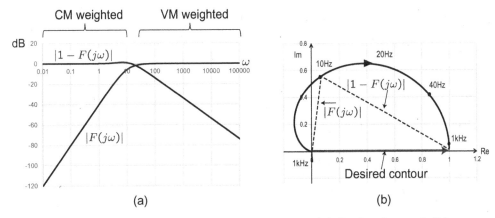

Figure 5.20: Frequency based weighting function: (a) Bode plot and (b) contour.

contour straight. They suggested a phase compensation $e^{-j\alpha}$ to make $F(j\omega)e^{-j\alpha}$ real number for all ω, i.e., $\angle F(j\omega)e^{-j\alpha} = 0$. The desired α can be easily calculated as

$$\alpha = \pi - \tan^{-1}\left(\frac{\frac{L_r}{L_m}K_1\omega}{\frac{L_r}{L_m}K_2 - \omega^2}\right). \tag{5.65}$$

To make $1 - F(s)$ real, a feedforward path of $1 - e^{-j\alpha}$ is added. Fig. 5.21 shows the angle compensation blocks. Then, the trajectory contour becomes a straight line as shown in Fig. 5.20 (b). For other types of observer integration, see [9] and [10].

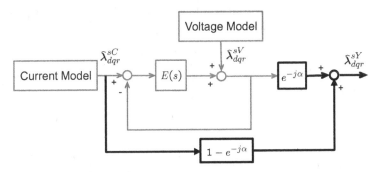

Figure 5.21: Combination of CM and VM with a phase compensation.

Exercise 5.5 Consider block diagram shown in Fig. 5.21. The input to output relation is given by $\hat{\boldsymbol{\lambda}}_{dqr}^{sY} = A(s)\hat{\boldsymbol{\lambda}}_{dqr}^{sC} + B(s)\hat{\boldsymbol{\lambda}}_{dqr}^{sV}$.
a) Find $A(s)$ and $B(s)$.
b) Show that both $A(s)$ and $B(s)$ are real and $A(s) + B(s) = 1$ for all s.

5.4.5   Full Order Observer

An adaptive observer is different from a classical state (Luenberger) observer [12] in that it contains a parameter adaptive scheme. The full order observer is based on a complete IM model in the stationary frame. Since the system matrix contains a rotor speed estimate, it requires a speed adaptation algorithm.

Using the stator current and rotor flux in the stationary frame, a stator vector is defined such that

$$\mathbf{x} = \begin{bmatrix} \mathbf{i}^s_{dqs} \\ \boldsymbol{\lambda}^s_{dqr} \end{bmatrix} \in \mathbb{R}^4 .$$

Recall from (4.56) that the IM dynamics are given in the stationary frame as

$$\frac{d}{dt}\mathbf{x} = \mathbf{A}(\omega_r)\mathbf{x} + \mathbf{B}\mathbf{v}^s_{dqs}, \tag{5.66}$$

$$\mathbf{i}^s_{dqs} = \mathbf{C}\mathbf{x} \tag{5.67}$$

where

$$\mathbf{A}(\omega_r) \equiv \begin{bmatrix} \mathbf{A}_{11} & \mathbf{A}_{12} \\ \mathbf{A}_{21} & \mathbf{A}_{22} \end{bmatrix} = \begin{bmatrix} -\left(\frac{r_s}{\sigma L_s} + \frac{(1-\sigma)}{\sigma \tau_r}\right)\mathbf{I} & \frac{L_m}{\sigma L_r L_s}\left(\frac{1}{\tau_r}\mathbf{I} - \omega_r \mathbf{J}\right) \\ \frac{L_m}{\tau_r}\mathbf{I} & -\frac{1}{\tau_r}\mathbf{I} + \omega_r\mathbf{J} \end{bmatrix}, \tag{5.68}$$

$$\mathbf{B} \equiv \begin{bmatrix} \mathbf{B}_1 \\ \mathbf{0} \end{bmatrix} = \begin{bmatrix} \frac{1}{\sigma L_s}\mathbf{I} \\ \mathbf{0} \end{bmatrix}, \quad \text{and} \quad \mathbf{C} = [\mathbf{I}\ \ \mathbf{0}] . \tag{5.69}$$

Note that the stationary frame IM model contains ω_r. Now, ω_r is regarded as the only unknown parameter in the system matrix, and an adaptive observer is constructed to estimate $\hat{\omega}_r$. A full order observer for (5.66) and (5.67) can be constructed such that

$$\frac{d}{dt}\hat{\mathbf{x}} = \mathbf{A}(\hat{\omega}_r)\hat{\mathbf{x}} + \mathbf{B}\mathbf{v}^s_{dqs} + \mathbf{G}(\mathbf{i}^s_{dqs} - \mathbf{C}\hat{\mathbf{x}}), \tag{5.70}$$

where $\mathbf{G} \in \mathbb{R}^{4\times2}$ is an observer gain matrix. The state observer copies the target system dynamics, and also contains output error injection term, $\mathbf{G}(\mathbf{i}^s_{dqs} - \mathbf{C}\hat{\mathbf{x}})$. The current estimation error, $\mathbf{i}^s_{dqs} - \hat{\mathbf{i}}^s_{dqs}$ is used to force $[\hat{\mathbf{i}}^{sT}_{dqs}, \hat{\boldsymbol{\lambda}}^{sT}_{dqr}]^T$ to converge to $[\mathbf{i}^{sT}_{dqs}, \boldsymbol{\lambda}^{sT}_{dqr}]^T$. Let the observer error be defined by $\Delta\mathbf{x} \equiv \mathbf{x} - \hat{\mathbf{x}}$ [13]. Then, the error equation turns out to be

$$\frac{d}{dt}\Delta\hat{\mathbf{x}} = (\mathbf{A}(\omega_r) - \mathbf{GC})\Delta\mathbf{x} + (\mathbf{A}(\omega_r) - \mathbf{A}(\hat{\omega}_r))\hat{\mathbf{x}},$$

$$= (\mathbf{A}(\omega_r) - \mathbf{GC})\Delta\mathbf{x} + \Delta\omega_r\begin{bmatrix} -\frac{L_m}{\sigma L_r L_s}\mathbf{J}\hat{\boldsymbol{\lambda}}^s_{dqr} \\ \mathbf{J}\hat{\boldsymbol{\lambda}}^s_{dqr} \end{bmatrix}, \tag{5.71}$$

where $\Delta\omega_r = \omega_r - \hat{\omega}_r$. By using $\mathbf{G}$, the poles of error dynamics can be assigned to any desired location in $\mathbb{C}$ if $(\mathbf{A}, \mathbf{C})$ is an observable pair [14]. That is, a gain matrix $\mathbf{G}$ can be selected such that all the eigenvalues of $\mathbf{A} - \mathbf{GC}$ have negative real parts. Then, the convergence of the autonomous part of (5.71) is established.

Lemma. An equilibrium, $\mathbf{0}$ of $\Delta\dot{\mathbf{x}} = (\mathbf{A} - \mathbf{GC})\Delta\mathbf{x}$, is uniformly asymptotically stable, if and only if there exists a positive definite matrix, $\mathbf{P} \in \mathbb{R}^{n\times n}$ satisfying

$$(\mathbf{A} - \mathbf{GC})^T\mathbf{P} + \mathbf{P}(\mathbf{A} - \mathbf{GC}) = -\mathbf{Q} \tag{5.72}$$

for a positive definite matrix, $\mathbf{Q} \in \mathbb{R}^{n\times n}$. Here, a symmetric matrix $\mathbf{Q}$ is said to be *positive definite* if $\mathbf{x}^T\mathbf{Q}\mathbf{x} > \alpha_0\mathbf{x}^T\mathbf{x}$ for all $\mathbf{x} \neq \mathbf{0} \in \mathbb{R}^n$ and some $\alpha_0 > 0$. If $\mathbf{x}^T\mathbf{Q}\mathbf{x} \geq 0$, then $\mathbf{Q}$ is called *positive semi-definite* [14].

Speed Update Law in Full Order Observer

The following theory is developed with the assumption that ω_r is constant. Assume further that there exists a positive definite matrix, $\mathbf{P} \in \mathbb{R}^{4 \times 4}$ such that (5.72) holds for a positive definite matrix $\mathbf{Q} \in \mathbb{R}^{4 \times 4}$. Choose a Lyapunov function candidate such that

$$\mathcal{V} = \Delta \mathbf{x}^T \mathbf{P} \Delta \mathbf{x} + \frac{1}{2\gamma} \Delta \omega_r^2,$$

where $\gamma > 0$ is a constant. Then,

$$\dot{\mathcal{V}} = -\Delta \mathbf{x}^T \mathbf{Q} \Delta \mathbf{x} + 2 \Delta \omega_r \Delta \mathbf{x}^T \mathbf{P} \begin{bmatrix} -\frac{L_m}{\sigma L_r L_s} \mathbf{J} \\ \mathbf{J} \end{bmatrix} \hat{\lambda}_{dqr}^s - \frac{1}{\gamma} \Delta \omega_r \dot{\hat{\omega}}_r \qquad (5.73)$$

To make the right hand side less than or equal to zero, it should follow that

$$2 \Delta \mathbf{x}^T \mathbf{P} \begin{bmatrix} -\frac{L_m}{\sigma L_r L_s} \mathbf{J} \\ \mathbf{J} \end{bmatrix} \hat{\lambda}_{dqr}^s - \frac{1}{\gamma} \dot{\hat{\omega}}_r = 0.$$

That is,

$$\dot{\hat{\omega}}_r = 2\gamma \Delta \mathbf{x}^T \mathbf{P} \begin{bmatrix} -\frac{L_m}{\sigma L_r L_s} \mathbf{J} \\ \mathbf{J} \end{bmatrix} \hat{\lambda}_{dqr}^s. \qquad (5.74)$$

Further, assume that there is no error in the rotor flux estimate, i.e., $\hat{\lambda}_{dqr}^s = \lambda_{dqr}^s$. Then, the third and fourth components of $\Delta \mathbf{x}$ are equal to zero. If $\mathbf{P}$ is an identity matrix, (5.74) is simplified as [13]

$$\dot{\hat{\omega}}_r = \gamma' \left[(i_{ds}^s - \hat{i}_{ds}^s)\hat{\lambda}_{qr}^s - (i_{qs}^s - \hat{i}_{qs}^s)\hat{\lambda}_{dr}^s \right], \qquad (5.75)$$

where $\gamma' = 2\gamma L_m/(\sigma L_r L_s)$. Further, (5.75) can be extended to

$$\hat{\omega}_r = \gamma_0' \left[(i_{ds}^s - \hat{i}_{ds}^s)\hat{\lambda}_{qr}^s - (i_{qs}^s - \hat{i}_{qs}^s)\hat{\lambda}_{dr}^s \right] + \gamma' \int_0^t \left[(i_{ds}^s - \hat{i}_{ds}^s)\hat{\lambda}_{qr}^s - (i_{qs}^s - \hat{i}_{qs}^s)\hat{\lambda}_{dr}^s \right] dt, \qquad (5.76)$$

where $\gamma_0' > 0$.

Remark. The speed update law, (5.74), yields

$$\dot{\mathcal{V}} = -\Delta \mathbf{x}^T \mathbf{Q} \Delta \mathbf{x} \leq 0.$$

Note that $\dot{\mathcal{V}}$ is not negative definite in the augmented space $[\Delta \mathbf{x}^T, \Delta \omega_r]^T \in \mathbb{R}^5$. Therefore, (5.74) does not guarantee asymptotic convergence. For asymptotic convergence, we may need LaSalle's theorem [15].

5.4.6   Reduced Order Observer

Reduced-order observers estimate only the stator or rotor flux. Reduced-order observers can be made inherently sensorless i.e., the speed does not appear in the

observer equations. Such an observer is decoupled from the speed estimator, and it
is expected to provide an increased accuracy in a wide-speed-range operation [10].

The reduced order observer is constructed by using a half of the state variable
such that

$$\dot{\hat{\boldsymbol{\lambda}}}_{dqr}^s \;=\; \mathbf{A}_{22}\hat{\boldsymbol{\lambda}}_{dqr}^s + \mathbf{A}_{21}\mathbf{i}_{dqs}^s + \mathbf{G}_r(\dot{\mathbf{i}}_{dqs}^s - \dot{\hat{\mathbf{i}}}_{dqs}^s). \tag{5.77}$$

Note that $\dot{\mathbf{i}}_{dqs}^s - \dot{\hat{\mathbf{i}}}_{dqs}^s$ is utilized, instead of $\mathbf{i}_{dqs}^s - \hat{\mathbf{i}}_{dqs}^s$, as the driving term for the
observer. By incorporating the voltage model

$$\frac{d}{dt}\hat{\mathbf{i}}_{dqs}^s = \mathbf{A}_{12}\hat{\boldsymbol{\lambda}}_{dqr}^s + \mathbf{A}_{11}\hat{\mathbf{i}}_{dqs}^s + \mathbf{B}_1\mathbf{v}_{dqs}^s \tag{5.78}$$

the observer is written equivalently as [16]

$$\frac{d}{dt}\left(\hat{\boldsymbol{\lambda}}_{dqr}^s - \mathbf{G}_r\mathbf{i}_{dqs}^s\right) \;=\; \mathbf{A}_{22}\hat{\boldsymbol{\lambda}}_{dqr}^s + \mathbf{A}_{21}\mathbf{i}_{dqs}^s - \mathbf{G}_r(\mathbf{A}_{12}\hat{\boldsymbol{\lambda}}_{dqr}^s + \mathbf{A}_{11}\hat{\mathbf{i}}_{dqs}^s + \mathbf{B}_1\mathbf{v}_{dqs}^s), \tag{5.79}$$

or

$$\hat{\boldsymbol{\lambda}}_{dqr}^s = \mathbf{G}_r\mathbf{i}_{dqs}^s + \int_0^t \left[\mathbf{A}_{22}\hat{\boldsymbol{\lambda}}_{dqr}^s + \mathbf{A}_{21}\mathbf{i}_{dqs}^s - \mathbf{G}_r(\mathbf{A}_{12}\hat{\boldsymbol{\lambda}}_{dqr}^s + \mathbf{A}_{11}\hat{\mathbf{i}}_{dqs}^s + \mathbf{B}_1\mathbf{v}_{dqs}^s)\right]dt, \tag{5.80}$$

Note that the flux observer (5.77) can be implemented in the form of (5.79) or
(5.80) without any differential action. Also, the reduced observer does not involve
any speed estimation algorithm. Thus, it is a kind of inherently sensorless observer.
The block diagram is shown in Fig. 5.22.

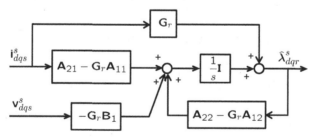

Figure 5.22: Reduced order observer.

5.4.7 Sliding Mode Observer

A sliding mode observer is a high gain observer that utilizes only two extreme
positive and negative values depending on the observation error polarity. In the
sliding mode observer, an appropriate switching function is defined, and a switching
rule is set that enforces the observation error to reach the sliding surface [17]. It is
inherently robust to the parameter mismatch and to disturbances. Other advantages
of sliding mode observer are its fast dynamic response and simplicity in design and
implementation [18], [19]. Unlike the sliding mode controller, it is easily realized

without incurring a high cost because a high gain setting is done only by software. But, it has a chattering problem.

The sliding mode observer is applied to detect the rotor flux in the IM sensorless control. From the perspective of dynamics, the back EMF is a disturbance for the stator equation, since it has nothing to do with the electricity. Thereby, the back EMF is usually ignored in the sliding mode observer. As a result, it does not require a speed estimation or update algorithm. Therefore, it is a kind of inherently sensorless observer.

Consider the IM dynamics in the stationary frame that utilizes the stator current and the rotor flux as the state variables.

$$\frac{d}{dt}\mathbf{i}^s_{dqs} = -\left(\frac{r_s}{\sigma L_s} + \frac{1-\sigma}{\sigma \tau_r}\right)\mathbf{i}^s_{dqs} + \frac{L_m}{\sigma L_s L_r}\left(\frac{1}{\tau_r}\mathbf{I} - \omega_r\mathbf{J}\right)\boldsymbol{\lambda}^s_{dqr} + \frac{1}{\sigma L_s}\mathbf{v}^s_{dqs} \quad (5.81)$$

$$\frac{d}{dt}\boldsymbol{\lambda}^s_{dqr} = \left(\omega_r\mathbf{J} - \frac{1}{\tau_r}\mathbf{I}\right)\boldsymbol{\lambda}^s_{dqr} + \frac{L_m}{\tau_r}\mathbf{i}^s_{dqs} \quad (5.82)$$

Both equations contain the back-EMF term $\omega_r\mathbf{J}\boldsymbol{\lambda}^s_{dqr}$. This term, being considered as a disturbance, is not considered in the observer:

$$\frac{d}{dt}\hat{\mathbf{i}}^s_{dqs} = -\left(\frac{r_s}{\sigma L_s} + \frac{1-\sigma}{\sigma \tau_r}\right)\hat{\mathbf{i}}^s_{dqs} + \frac{L_m}{\sigma L_s L_r}\frac{1}{\tau_r}\hat{\boldsymbol{\lambda}}^s_{dqr} + \frac{1}{\sigma L_s}\mathbf{v}^s_{dqs} + \frac{L_m}{\sigma L_s L_r}K\boldsymbol{\nu} \quad (5.83)$$

$$\frac{d}{dt}\hat{\boldsymbol{\lambda}}^s_{dqr} = -\frac{1}{\tau_r}\hat{\boldsymbol{\lambda}}^s_{dqr} + \frac{L_m}{\tau_r}\hat{\mathbf{i}}^s_{dqs} + K\boldsymbol{\nu}, \quad (5.84)$$

where K is the observer gain and the correction vector $\boldsymbol{\nu}$ is the output of a sliding-mode controller. Let $\Delta\mathbf{i}^s_{dqs} = \mathbf{i}^s_{dqs} - \hat{\mathbf{i}}^s_{dqs}$ and $\Delta\boldsymbol{\lambda}^s_{dqr} = \boldsymbol{\lambda}^s_{dqr} - \hat{\boldsymbol{\lambda}}^s_{dqr}$. Denote by $\mathbf{S} = [S_d, S_q]^T$ the sliding surfaces of d and q axis currents. Further, let

$$\boldsymbol{\nu} = \text{sgn}(\mathbf{S}) \equiv \begin{bmatrix} \text{sgn}(S_d) \\ \text{sgn}(S_q) \end{bmatrix} \quad (5.85)$$

$$\mathbf{S} = \Delta\mathbf{i}^s_{dqs}. \quad (5.86)$$

It follows from (5.81)–(5.85) that

$$\frac{d}{dt}\Delta\mathbf{i}^s_{dqs} = -\left(\frac{r_s}{\sigma L_s} + \frac{1-\sigma}{\sigma \tau_r}\right)\Delta\mathbf{i}^s_{dqs} + \frac{L_m}{\sigma L_s L_r}\frac{1}{\tau_r}\Delta\boldsymbol{\lambda}^s_{dqr} - \omega_r\frac{L_m}{\sigma L_s L_r}\mathbf{J}\boldsymbol{\lambda}^s_{dqr}$$

$$- \frac{L_m}{\sigma L_s L_r}K\text{sgn}(\mathbf{S}), \quad (5.87)$$

$$\frac{d}{dt}\Delta\boldsymbol{\lambda}^s_{dqr} = -\frac{1}{\tau_r}\Delta\boldsymbol{\lambda}^s_{dqr} + \frac{L_m}{\tau_r}\Delta\mathbf{i}^s_{dqs} + \omega_r\mathbf{J}\boldsymbol{\lambda}^s_{dqr} - K\text{sgn}(\mathbf{S}). \quad (5.88)$$

Choose a Lyapunov function candidate $\mathcal{V} = \frac{1}{2}\mathbf{S}^T\mathbf{S}$. Then

$$\dot{\mathcal{V}} = \mathbf{S}^T\dot{\mathbf{S}} = -\left(\frac{r_s}{\sigma L_s} + \frac{1-\sigma}{\sigma \tau_r}\right)\mathbf{S}^T\mathbf{S} + \frac{L_m}{\sigma L_s L_r}\mathbf{S}^T\left(\frac{1}{\tau_r}\Delta\boldsymbol{\lambda}^s_{dqr} - \omega_r\mathbf{J}\boldsymbol{\lambda}^s_{dqr} - K\text{sgn}(\mathbf{S})\right)$$

$$(5.89)$$

Note that $\mathbf{S}^T \mathrm{sgn}(\mathbf{S}) = \mathrm{sgn}(\Delta i^s_{ds})\Delta i^s_{ds} + \mathrm{sgn}(\Delta i^s_{qs})\Delta i^s_{qs} = |\Delta i^s_{ds}| + |\Delta i^s_{qs}| = \|\mathbf{S}\|_1$, where $\| \cdot \|_1$ is a one-norm defined by $\|\mathbf{x}\|_1 = \sum_{i=1}^{n} |x_i|$. Therefore, it follows that

$$-K\mathbf{S}^T\mathrm{sgn}(\mathbf{S}) + \mathbf{S}^T \left(\frac{1}{\tau_r}\Delta\lambda^s_{dqr} - \omega_r\mathbf{J}\lambda^s_{dqr} \right) \le \|\mathbf{S}\|_1 \left(-K + \|\frac{1}{\tau_r}\Delta\lambda^s_{dqr} - \omega_r\mathbf{J}\lambda^s_{dqr}\|_1 \right).$$

Assume that K is large enough such that for some $\epsilon > 0$ and all $t \ge 0$

$$K - \|\frac{1}{\tau_r}\Delta\lambda^s_{dqr} - \omega_r\mathbf{J}\lambda^s_{dqr}\|_1 \ge \epsilon. \tag{5.90}$$

Then,

$$\dot{\mathcal{V}} \le - \left(\frac{r_s}{\sigma L_s} + \frac{1-\sigma}{\sigma\tau_r} \right) \mathbf{S}^T\mathbf{S} - \epsilon\frac{L_m}{\sigma L_s L_r}\|\mathbf{S}\|_1 \le 0. \tag{5.91}$$

The equality holds whenever $\mathbf{S} = \mathbf{0}$. It means that the error state Δi^s_{dqs} converges to the sliding surface, $S_d = 0$ and $S_q = 0$. In case of $\mathbf{S} = \mathbf{0}$, K approximates the back-EMF, i.e., $K \approx \|\omega_r\mathbf{J}\lambda^s_{dqr}\|_1$. However, the controller has a limited bandwidth due to the sampling time. Therefore, chattering takes place along the sliding surfaces, which is an undesired oscillation. By increasing the sampling frequency, the magnitude of chattering can be reduced. As a chattering reduction method, the PI type is widely used:

$$\mathbf{S} = \left(K_{ps} + \frac{K_{is}}{s} \right) (\mathbf{i}^s_{dqs} - \hat{\mathbf{i}}^s_{dqs}). \tag{5.92}$$

The integrating action cumulates the past error history, thereby reduces the number of switchings significantly.

It was reported in [11] that the above sliding observer was sensitive to stator resistance detuning and to measurement offset and was insensitive to rotor resistance and magnetizing inductance errors. The instability in convergence can be understood by considering (5.88). The convergence rate of the rotor-flux error is, at best,

$$\frac{d}{dt}\Delta\lambda^s_{dqr} = -\frac{1}{\tau_r}\Delta\lambda^s_{dqr}. \tag{5.93}$$

Note however that the right hand side is small since τ_r is relatively large. Thus, the dynamic behavior is almost like an integrator, that causes instability in the presence of measurement offset [11].

For a more complete analysis, let $\mathcal{V} = \frac{1}{2}\Delta\mathbf{i}^{s\,T}_{dqs}\Delta\mathbf{i}^s_{dqs} + \frac{1}{2\sigma L_s L_r}\Delta\lambda^{s\,T}_{dqr}\Delta\lambda^s_{dqr}$. Then,

$$\begin{aligned}
\dot{\mathcal{V}} = &- \left(\frac{r_s}{\sigma L_s} + \frac{1-\sigma}{\sigma\tau_r} \right) \Delta\mathbf{i}^{s\,T}_{dqs}\Delta\mathbf{i}^s_{dqs} - \frac{1}{\sigma L_s L_r\tau_r}\Delta\lambda^{s\,T}_{dqr}\Delta\lambda^s_{dqr} \\
&+ \frac{L_m}{\sigma L_s L_r}\Delta\mathbf{i}^{s\,T}_{dqs} \left(\frac{2}{\tau_r}\Delta\lambda^s_{dqr} - \omega_r\mathbf{J}\lambda^s_{dqr} - K\mathrm{sgn}(\Delta\mathbf{i}^s_{dqs}) \right) \\
&+ \frac{1}{\sigma L_s L_r} \left(\omega_r\Delta\lambda^{s\,T}_{dqr}\mathbf{J}\hat{\lambda}^s_{dqr} - K\Delta\lambda^{s\,T}_{dqr}\mathrm{sgn}(\Delta\mathbf{i}^s_{dqs}) \right)
\end{aligned} \tag{5.94}$$

To make $\dot{V} < 0$, it is required to have $\Delta\lambda_{dqr}^{sT}\text{sgn}(\Delta\mathbf{i}_{dqs}^s) > 0$. It means $\Delta\lambda_{dqr}^s$ needs to be parallel with $\Delta\mathbf{i}_{dqs}^s$. However, there is no such a guarantee. Thereby, $\Delta\lambda_{dqr}^s$ has a weak convergence property. Nonetheless, some successful experimental results were reported in [10] and [11].

5.4.8 Reduced Order Observer by Harnefors

Harnefors and Hinkkanen [20],[21] constructed a reduced observer utilizing a neat IM model. Their IM model has a kind of inverse Γ model containing a generalized back EMF, $\mathbf{e}$, as shown in Section 4.2.3. Their observer is simple, while encompassing both voltage and current models.

The stator-current equation is derived from the upper half of (4.54) as

$$
\begin{aligned}
\sigma L_s \frac{d\mathbf{i}_{dqs}^e}{dt} + \omega_e \sigma L_s \mathbf{J}\mathbf{i}_{dqs}^e &= \mathbf{v}_{dqs}^e - r_s\mathbf{i}_{dqs}^e - \frac{L_s(1-\sigma)}{\tau_r}\mathbf{i}_{dqs}^e - \frac{L_m}{L_r}\left(\frac{1}{\tau_r}\mathbf{I} - \omega_r\mathbf{J}\right)\lambda_{dqr}^e \\
&= \mathbf{v}_{dqs}^e - r_s\mathbf{i}_{dqs}^e - \underbrace{\frac{L_m}{L_r}\left(\frac{L_m}{L_r}r_r\mathbf{i}_{dqs}^e - \frac{1}{\tau_r}\lambda_{dqr}^e + \omega_r\mathbf{J}\lambda_{dqr}^e\right)}_{\equiv\, \mathbf{e}}.
\end{aligned}
$$

$$(5.95)$$

Note that $\mathbf{e}$ here is identical to (4.76). Fig. 5.23 shows an equivalent circuit based on (5.95). Now, we have two estimates of $\mathbf{e}$ based on voltage and current information:

$$
\text{VM}: \quad \mathbf{e}' = \mathbf{v}_{dqs}^e - r_s\mathbf{i}_{dqs}^e - \sigma L_s\frac{d\mathbf{i}_{dqs}^e}{dt} - \hat{\omega}_e\sigma L_s\mathbf{J}\mathbf{i}_{dqs}^e, \tag{5.96}
$$

$$
\text{CM}: \quad \hat{\mathbf{e}} = \frac{L_m}{L_r}\left[\frac{L_m}{L_r}r_r\mathbf{i}_{dqs}^e - \left(\frac{1}{\tau_r}\mathbf{I} - \hat{\omega}_r\mathbf{J}\right)\hat{\lambda}_{dqr}^e\right], \tag{5.97}
$$

where $\hat{\omega}$ is an estimate of ω.

Since $\omega_e = \omega_{sl} + \omega_r$, the back EMF is rewritten as

$$
\frac{L_r}{L_m}\mathbf{e} = \frac{L_m}{\tau_r}\mathbf{i}_{dqs}^e - \frac{1}{\tau_r}\lambda_{dqr}^e - \omega_{sl}\mathbf{J}\lambda_{dqr}^e + \omega_e\mathbf{J}\lambda_{dqr}^e. \tag{5.98}
$$

From the second row of (4.54), the rotor flux equation follows such that

$$
\frac{d\lambda_{dqr}^e}{dt} = \frac{L_m}{\tau_r}\mathbf{i}_{dqs}^e - \frac{1}{\tau_r}\lambda_{dqr}^e - \omega_{sl}\mathbf{J}\lambda_{dqr}^e. \tag{5.99}
$$

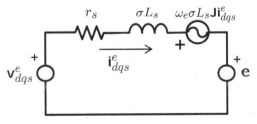

Figure 5.23: IM voltage model with back EMF, $\mathbf{e}$.

Subtracting (5.98) from (5.99), it follows that

$$\frac{d\boldsymbol{\lambda}_{dqr}^e}{dt} + \omega_e \mathbf{J} \boldsymbol{\lambda}_{dqr}^e = \frac{L_r}{L_m} \mathbf{e}. \tag{5.100}$$

It is the rotor flux equation driven by the back EMF. Based on (5.96), (5.97), (5.100) a reduced observer for rotor flux is derived such that

$$\frac{d\hat{\boldsymbol{\lambda}}_{dqr}^e}{dt} + \hat{\omega}_e \mathbf{J} \hat{\boldsymbol{\lambda}}_{dqr}^e = \frac{L_r}{L_m} \mathbf{e}' + \frac{L_r}{L_m} \mathbf{K}(\hat{\mathbf{e}} - \mathbf{e}'), \tag{5.101}$$

where $\mathbf{K}$ is an observer gain. Note that all terms on the right hand side of (5.101) are available for measurements except $\hat{\boldsymbol{\lambda}}_{dqr}^e$.

It looks different from the previous reduced observer that employs $\mathbf{i}_{dqs}^s - \hat{\mathbf{i}}_{dqs}^s$. Selecting $\mathbf{K} = \mathbf{I}$, the CM estimate follows from (5.101) and (5.97) such that

$$\frac{d\hat{\boldsymbol{\lambda}}_{dqr}^e}{dt} + \hat{\omega}_{sl} \mathbf{J} \hat{\boldsymbol{\lambda}}_{dqr}^e = -r_r \mathbf{i}_{dqr}^e \qquad \text{in the synchronous frame,} \tag{5.102}$$

$$\frac{d\hat{\boldsymbol{\lambda}}_{dqr}^r}{dt} = -r_r \mathbf{i}_{dqr}^r \qquad \text{in the rotor frame.} \tag{5.103}$$

Note that (5.103) is identical to (5.60). The VM in the stationary frame is obtained by selecting the null matrix, i.e., $\mathbf{K} = \mathbf{O}$:

$$\frac{d\hat{\boldsymbol{\lambda}}_{dqr}^s}{dt} = \frac{L_r}{L_m} \left(\mathbf{v}_{dqs}^s - r_s \mathbf{i}_{dqs}^s - \sigma L_s \frac{d\mathbf{i}_{dqs}^s}{dt} \right), \tag{5.104}$$

which is the same as (5.52).

In order to utilize (5.102) it is necessary to have a slip estimate. It is obtained using

$$\omega_{sl} = \frac{L_m}{\tau_r} \frac{\mathbf{i}_{dqs}^{eT} \mathbf{J} \hat{\boldsymbol{\lambda}}_{dqr}^e}{\|\hat{\boldsymbol{\lambda}}_{dqr}^e\|^2}. \tag{5.105}$$

Note that $\lambda_{qr}^e = 0$ and $\lambda_{dr}^e = \|\boldsymbol{\lambda}_{dqr}^e\|$ in the rotor field oriented control. Then,

$$\frac{L_m}{\tau_r} \frac{\mathbf{i}_{dqs}^{eT} \mathbf{J} \hat{\boldsymbol{\lambda}}_{dqr}^e}{\|\hat{\boldsymbol{\lambda}}_{dqr}^e\|^2} = \frac{L_m i_{qs}^e}{\tau_r \lambda_{dr}^e},$$

which is the same as (5.11). The gain can be structured as [20]

$$\mathbf{K} = (g_1 \mathbf{I} + g_2 \mathbf{J}) \frac{\hat{\boldsymbol{\lambda}}_{dqr}^e \hat{\boldsymbol{\lambda}}_{dqr}^{eT}}{\|\hat{\boldsymbol{\lambda}}_{dqr}^e\|^2}. \tag{5.106}$$

Since $\hat{\boldsymbol{\lambda}}_{dqr}^{eT} \mathbf{J} \hat{\boldsymbol{\lambda}}_{dqr}^e = 0$, the dependence on $\hat{\omega}_r$ disappears from $\mathbf{K}\hat{\mathbf{e}}$. Specifically, it is reduced as $\hat{\boldsymbol{\lambda}}_{dqr}^{eT} \hat{\mathbf{e}} = \frac{L_m}{\tau_r} \hat{\boldsymbol{\lambda}}_{dqr}^{eT} \mathbf{i}_{dqr}^e$. Then the observer (5.101) turns out to be inherently sensorless in the stationary frame. If the rotor flux oriented frame is used, then $\hat{\boldsymbol{\lambda}}_{dqr}^e = [\hat{\lambda}_{dr}^e, 0]^T$, and (5.106) is further simplified to

$$\mathbf{K} = \begin{bmatrix} g_1 & 0 \\ g_2 & 0 \end{bmatrix}. \tag{5.107}$$

Further, Harnefors and Hinkkanen [20], [21] proposed a reduced-order observer augmented with a stator-resistance adaptation algorithm, and derived stability conditions.

Exercise 5.6   Show that

$$\hat{\omega}_e = \frac{(\mathbf{v}_{dqs}^s - r_s \mathbf{i}_{dqs}^s)^T \mathbf{J} \hat{\boldsymbol{\lambda}}_{dqs}^s}{\|\hat{\boldsymbol{\lambda}}_{dqs}^s\|^2}. \tag{5.108}$$

Solution.

$$\hat{\omega}_e = \frac{d\hat{\theta}_e}{dt} = \frac{d}{dt}\tan^{-1}\left(\frac{\hat{\lambda}_{qr}^s}{\hat{\lambda}_{dr}^s}\right) = \frac{\dot{\hat{\lambda}}_{qs}^s \hat{\lambda}_{ds}^s - \hat{\lambda}_{qs}^s \dot{\hat{\lambda}}_{ds}^s}{\hat{\lambda}_{qs}^{s\,2} + \hat{\lambda}_{ds}^{s\,2}} = \frac{\dot{\hat{\boldsymbol{\lambda}}}_{dqs}^{sT} \mathbf{J} \hat{\boldsymbol{\lambda}}_{dqs}^s}{\|\hat{\boldsymbol{\lambda}}_{dqs}^s\|^2} \qquad \blacksquare$$

5.4.9   Robust Sensorless Algorithm

Various sensorless approaches are introduced in the above. But all the algorithms are marginally stable in the low speed or frequency range due to the open loop integration or its similar nature. The disturbing elements in practical applications are the measurement offset, noise, coil resistance variation, and inverter nonlinearity. Typical inverter nonlinearities are the IGBT on-drop, which changes during the current transition and the dead time interval provided to prevent arm-short. They are compensated, although calculation based, to increase the accuracy of a VM based estimator [23].

The accuracy of coil resistance, r_s affects the performance of sensorless control algorithm especially in the low speed region. Note that the exact value of r_s is hardly known due to the temperature dependence and the AC effects. The coil resistance increases by 50% when the temperature increases from 25 to 150°C. The AC effect refers to the resistance increase due to the current localization caused by the skin and the proximity effects [22]. Due to the proximity effects, the current displacement occurs in the slot conductors: More current tends to flow near the forefront of air gap, and the reason was studied in Section 3.5.

Many sensorless algorithms were compared experimentally by the author, and the one appeared in [24] looked most robust in the low speed and heavy load condition. It is developed based on a stator flux estimator, in which the flux magnitude and angle are compensated depending on the polarities of speed and the q-axis current. It is being used in commercial drives.

In the stator field oriented frame, the reference axis is aligned with stator flux $\boldsymbol{\lambda}_{dqs}^s$. Thereby,

$$v_{ds}^e = r_s i_{ds}^e + \mathrm{p}\lambda_{ds}^e \tag{5.109}$$

$$v_{qs}^e = r_s i_{qs}^e + \omega_e \lambda_{ds}^e. \tag{5.110}$$

In [24], a robust sensorless algorithm was proposed considering the change of r_s. The flux is estimated via

$$\hat{\lambda}_{dqs}^s = \hat{\lambda}_{ds}^s + j\hat{\lambda}_{qs}^s = \int \left(\mathbf{v}_{dqs}^s - \hat{r}_s \mathbf{i}_{dqs}^s \right) dt = \hat{\lambda}_{ds}^s e^{j\hat{\theta}_e}, \tag{5.111}$$

where $\theta_e = \tan^{-1}(\hat{\lambda}_{qs}^s/\hat{\lambda}_{ds}^s)$ and $\hat{r}_s$ is an estimate of stator coil. Also, recall flux and slip equations from (5.22) and (5.23):

$$\lambda_{ds}^e = \frac{1 + \sigma\tau_r \mathrm{p}}{1 + \tau_r \mathrm{p}}(L_s i_{ds}^e) - \frac{\sigma\tau_r}{1 + \tau_r \mathrm{p}}(L_s \omega_{sl} i_{qs}^e) \tag{5.112}$$

$$\omega_{sl} = \frac{\sigma L_s \mathrm{p} + \frac{L_s}{\tau_r}}{\lambda_{ds}^e - \sigma L_s i_{ds}^e} i_{qs}^e \tag{5.113}$$

The torque equation is given by

$$T_e = \frac{3}{2}\frac{P}{2}\lambda_{ds}^e i_{qs}^e \tag{5.114}$$

D-Axis Current Controller

Equation (5.112) can be rearranged such that

$$\mathrm{p}\lambda_{ds}^e = -\frac{1}{\tau_r}\lambda_{ds}^e + \left(\frac{L_s}{\tau_r} + \sigma L_s \mathrm{p} \right) i_{ds}^e - \sigma L_s \omega_{sl} i_{qs}^e \tag{5.115}$$

Substituting (5.115) into (5.109), it follows that

$$\mathrm{p}i_{ds}^e = -\alpha i_{ds}^e + \frac{1}{\sigma L_s}\left(\frac{\lambda_{ds}^e}{\tau_r} + \sigma L_s \omega_{sl} i_{qs}^e + v_{ds}^e \right) \tag{5.116}$$

where $\alpha = (r_s + L_s/\tau_r)/\sigma L_s$. The d-axis controller can be designed such that

$$v_{ds}^{e*} = \left(k_p + k_i\frac{1}{\mathrm{p}} \right)(i_{ds}^{e*} - \hat{i}_{ds}^e) - \frac{\hat{\lambda}_{ds}^e}{\tau_r} - \sigma L_s \hat{\omega}_{sl} \hat{i}_{qs}^e \tag{5.117}$$

where $\hat{\omega}_{sl} = \left(\sigma L_s \mathrm{p} + \frac{L_s}{\tau_r} \right) / \left(\hat{\lambda}_{ds}^e - \sigma L_s \hat{i}_{ds}^e \right) \hat{i}_{qs}^e$. The value with superscript * denotes a command value. Here, $\frac{\hat{\lambda}_{ds}^e}{\tau_r}$ and $\sigma L_s \hat{\omega}_{sl} \hat{i}_{qs}^e$ are considered as disturbances and canceled out by their estimates. The PI gains are chosen to be sufficiently high, so that the transfer function i_{ds}^e/i_{ds}^{e*} is approximated as unity.

Flux Regulator

When i_{ds}^{e*} is a constant, (5.112) turns out to be

$$i_{ds}^{e*} = \frac{\lambda_{ds}^e}{L_s} + \frac{\sigma\tau_r}{1 + \sigma\tau_r \mathrm{p}}(\omega_{sl} i_{qs}^e). \tag{5.118}$$

To let $\hat{\lambda}_{ds}^e$ track λ_{ds}^{e*} while keeping the relation (5.118), the d-axis current command is chosen such that

$$i_{ds}^{e*} \equiv \left(\bar{k}_p + \bar{k}_i\frac{1}{\mathrm{p}} \right)(\lambda_{ds}^{e*} - \hat{\lambda}_{ds}^e) + \frac{\lambda_{ds}^{e*}}{L_s} + \frac{\sigma\tau_r}{1 + \sigma\tau_r \mathrm{p}}(\hat{\omega}_{sl} \hat{i}_{qs}^e), \tag{5.119}$$

where $\bar{k}_p$ and $\bar{k}_i$ denote proportional and integral gains, respectively.

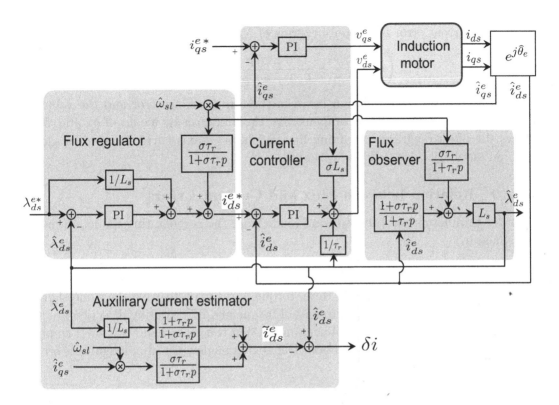

Figure 5.24: Flux regulator and current controller along with an auxiliary d-axis current estimator.

Auxiliary Current Estimator from Stator Flux

Rearranging (5.112) for i_{ds}^e, an estimate can be defined such that

$$\tilde{i}_{ds}^e \equiv \frac{1+\tau_r \mathrm{p}}{1+\sigma\tau_r \mathrm{p}}\left(\frac{\hat{\lambda}_{ds}^e}{L_s}\right) + \frac{\sigma\tau_r}{1+\sigma\tau_r \mathrm{p}}(\hat{\omega}_{sl}\hat{i}_{qs}^e) \tag{5.120}$$

in which the flux estimate is utilized. In addition, we define d-axis current error by subtracting it from the measured value, $\hat{i}_{ds}^e$:

$$\delta i \equiv \hat{i}_{ds}^e - \tilde{i}_{ds}^e. \tag{5.121}$$

Note that the measured value is denoted with $\hat{\ }$, since it is a value based on the estimated synchronous frame.

Due to the high gain property of the current controller, we treat $i_{ds}^{e*} = \hat{i}_{ds}^e$. Then, subtracting (5.120) from (5.119), we obtain

$$\delta i = \left(\bar{k}_p + \bar{k}_i \frac{1}{\mathrm{p}}\right)(\lambda_{ds}^{e*} - \hat{\lambda}_{ds}^e) + \frac{1}{L_s}\left(\lambda_{ds}^{e*} - \frac{1+\tau_r \mathrm{p}}{1+\sigma\tau_r \mathrm{p}}\hat{\lambda}_{ds}^e\right) \tag{5.122}$$

Further, since λ_{ds}^{e*} is constant, we presume

$$\frac{\lambda_{ds}^{e*}}{L_s} \approx \frac{1+\tau_r \mathrm{p}}{1+\sigma\tau_r \mathrm{p}}\left(\frac{\lambda_{ds}^{e*}}{L_s}\right) \tag{5.123}$$

Next, the current error is written in terms of $\lambda_{ds}^{e*} - \hat{\lambda}_{ds}^{e}$ such that

$$\delta i = \left(\bar{k}_p + \bar{k}_i \frac{1}{p} + \frac{1}{L_s} \frac{1 + \tau_r p}{1 + \sigma \tau_r p} \right) (\lambda_{ds}^{e*} - \hat{\lambda}_{ds}^{e}) \tag{5.124}$$

This equation states the relation between the flux tracking error and the value of δi. It shows the possibility of compensating the flux error by means of δi. Fig. 5.24 shows the block diagram containing flux regulator, d-axis current controller along with an auxiliary d-axis current estimator.

5.4.10   Relation between Flux and Current Errors

In order to differentiate the motoring mode from the regenerating mode, an index is defined by

$$P_f \equiv \operatorname{sgn}(\omega_e i_{qs}^e), \tag{5.125}$$

where 'sgn' is the sign function. Here, torque is assumed to have the same sign as i_{qs}^e. Then, $P_f = 1$ implies that the signs of torque and flux rotating direction are the same, i.e, the motor is in the motoring mode. On the other hand, $P_f = -1$ implies the motor is in the regenerating mode. Define flux angle error, flux magnitude error, and current error by

$$\delta\theta(k) \equiv \theta_e(k) - \hat{\theta}_e(k), \tag{5.126}$$
$$\delta\lambda(k) \equiv \|\lambda_s(k)\| - \|\hat{\lambda}_s(k)\|, \tag{5.127}$$
$$\delta i(k) \equiv \hat{i}_{ds}^e(k) - \tilde{i}_{ds}^e(k). \tag{5.128}$$

Note that the auxiliary current estimate is utilized in δi. There are four different operating modes depending on the directions of rotation and power flow. We let $\omega_e > 0$ when the field rotates counterclockwise, whereas $\omega_e < 0$ when clockwise.

Kim et al. [24] observed that Δr_s affected the flux and current errors differently depending on the operation modes; versus motoring/regeneration and rotating directions. For example, if $\Delta r_s > 0$, $\delta i > 0$ and $\delta\lambda > 0$ when $P_f = 1$, $\omega_e > 0$, However, if $\Delta r_s < 0$, $\delta i < 0$ and $\delta\lambda < 0$ in the same environments. The extended results are summarized in Table 5.2. It is desired to access unknown flux error via the

Table 5.2: Relation between flux errors ($\delta\theta$, $\delta\lambda$) and current error (δi).

Quadrants	$\Delta r_s > 0$	$\Delta r_s < 0$
Quad. I ($i_{qs} > 0$, $\omega_e > 0$)	$\delta\theta > 0$, $\delta\lambda > 0$, $\delta i > 0$	$\delta\theta < 0$, $\delta\lambda < 0$, $\delta i < 0$
Quad. II ($i_{qs} > 0$, $\omega_e < 0$)	$\delta\theta > 0$, $\delta\lambda > 0$, $\delta i < 0$	$\delta\theta > 0$, $\delta\lambda < 0$, $\delta i > 0$
Quad. III ($i_{qs} < 0$, $\omega_e < 0$)	$\delta\theta < 0$, $\delta\lambda > 0$, $\delta i > 0$	$\delta\theta > 0$, $\delta\lambda < 0$, $\delta i < 0$
Quad. IV ($i_{qs} < 0$, $\omega_e > 0$)	$\delta\theta < 0$, $\delta\lambda > 0$, $\delta i < 0$	$\delta\theta < 0$, $\delta\lambda < 0$, $\delta i > 0$

current error δi which is obtainable by (5.120). The various polarity combinations in Table 5.2 are simplified as

$$\delta\lambda \cdot [P_f \cdot \delta i] > 0 \tag{5.129}$$
$$\delta\theta \cdot [\omega_e \cdot \delta i] > 0. \tag{5.130}$$

The two inequalities are sorted in the (i_{qs}^e, ω_e) plane as shown in Table 5.3. Note that it is similar to the four quadrants of torque-speed plane.

Table 5.3: Separation of cases by the polarities of (i_{qs}^e, ω_e).

Quad. I $\quad(i_{qs} > 0, \omega_e > 0)$	$\delta\lambda\,\delta i > 0, \quad \delta\theta\,\delta i > 0$
Quad. II $\ (i_{qs} > 0, \omega_e < 0)$	$\delta\lambda\,\delta i < 0, \quad \delta\theta\,\delta i < 0$
Quad. III $(i_{qs} < 0, \omega_e < 0)$	$\delta\lambda\,\delta i > 0, \quad \delta\theta\,\delta i < 0$
Quad. IV $(i_{qs} < 0, \omega_e > 0)$	$\delta\lambda\,\delta i < 0, \quad \delta\theta\,\delta i < 0$

Error Compensation Rule

The stator flux is estimated via $\hat{\boldsymbol{\lambda}}_{dqs}^s = \int (\mathbf{v}_{dqs}^s - r_s \mathbf{i}_{dqs}^s)\, dt$ in the stationary frame, and from which the flux angle is estimated by $\hat{\theta}_e = \tan^{-1}(\hat{\lambda}_{qs}^s / \hat{\lambda}_{ds}^s)$. Based on this angle estimate, currents are transformed into the reference frame, and a controller is implemented. However, precise value of r_s is hardly obtainable, which governs the accuracy of the flux estimate. Ultimately, resistance error, $\Delta r_s \equiv \hat{r}_s - r_s$ dominates the performance of a whole sensorless algorithm. Furthermore, there are unmodeled errors.

To improve the performance, the effects of Δr_s have to be compensated in the flux estimation. For this purpose, γ_λ and γ_θ are defined as the flux correction terms such that

$$\hat{\boldsymbol{\lambda}}_{dqs}^{s+} = (\|\hat{\boldsymbol{\lambda}}_{dqs}^s\| + \gamma_\lambda)\, e^{j(\hat{\theta}_e + \gamma_\theta)}, \tag{5.131}$$

where $\hat{\boldsymbol{\lambda}}_{dqs}^{s+}$ is an updated flux estimate. Note that the magnitude is compensated by γ_λ, and the angle by γ_θ.

Table 5.4: Flux magnitude and angle compensation rule depending on torque-speed quadrants.

	Magnitude comp. (γ_λ)	Angle comp. (γ_θ)
Quad. I $\quad(i_{qs} > 0, \omega_e > 0)$	$\kappa_1\,\delta i$	$\kappa_2\,\delta i$
Quad. II $\ (i_{qs} > 0, \omega_e < 0)$	$-\kappa_1\,\delta i$	$-\kappa_2\,\delta i$
Quad. III $(i_{qs} < 0, \omega_e < 0)$	$\kappa_1\,\delta i$	$-\kappa_2\,\delta i$
Quad. IV $(i_{qs} < 0, \omega_e > 0)$	$-\kappa_1\,\delta i$	$\kappa_2\,\delta i$

Table 5.2 gives us a guide for the flux error correction, since magnitude and angle errors of $\hat{\boldsymbol{\lambda}}_{dqs}^s$ are assessed indirectly by δi. Specifically, if $\delta i > 0$ in Quadrant I, then $\delta\lambda > 0$ and $\delta\theta > 0$ meaning that the flux estimate's magnitude and angle need to be increased to get closer to the real flux, $\boldsymbol{\lambda}_{dqs}^s$. Therefore, it is useful to set the compensation in proportion to δi, i.e., $\gamma_\lambda = \kappa_1 \delta i$ and $\gamma_\theta = \kappa_2 \delta i$, where κ_1 and κ_2 are positive constants. According to (5.129) and (5.130), the signs of compensating terms are determined depending on the operation modes as shown in Table 5.4. Fig. 5.25 shows a stator flux estimation algorithm with the quadrant compensation rule.

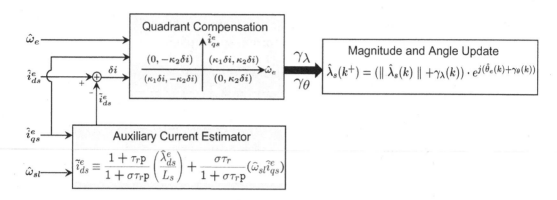

Figure 5.25: Stator flux estimation algorithm with the quadrant compensation rule.

Table 5.5: Specifications of IM under test.

V_{rms}	I_{rms}	Power	Poles	Frequency	Rated speed
415 V	35 A	18.5 kW	4	60 Hz	1465 rpm

Figure 5.26: Speed control response under ±100% (120 Nm) step load change at 50 rpm.

Experimental Results

Specifications of an IM under test are listed in Table 5.5. Fig. 5.26 shows the speed control performance when a 100% step load was applied at 50 rpm. The maximum speed error was about 100 rpm. Note that the synchronous speed of a four pole motor is 1800 rpm. Therefore, 50 rpm is 0.028 p.u. Previously, it was stressed that the low speed performance was important. It showed a fairly good performance since it undertook 1 p.u. load at 0.03 p.u. speed.

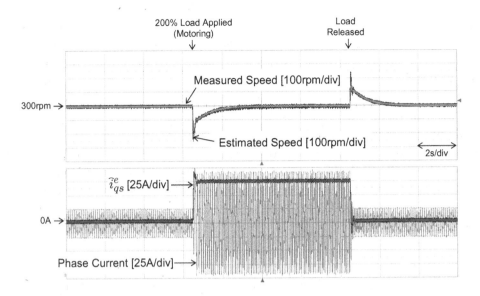

Figure 5.27: Speed control response under 200% (240 Nm) step load change at 300 rpm.

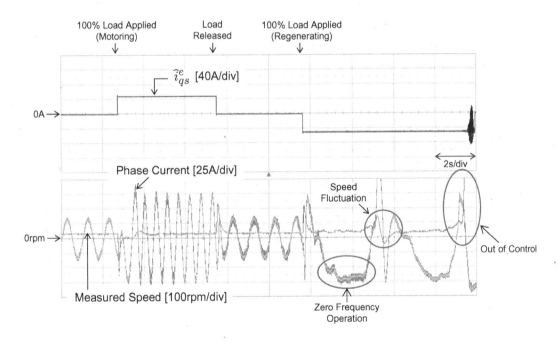

Figure 5.28: Speed control responses under ±200% (240 Nm) step load change at 300 rpm.

Fig. 5.27 shows the speed control performance under a 200% step load change at 300 rpm. The maximum speed error was about 250 rpm, however it did not show any instability.

If a regenerating load is applied at the low speed, the slip frequency is almost nullified. In that case, DC voltage and DC current are supplied to a motor. Since the signals are DC, the flux estimator is easily saturated by a small DC offset. For example, a small error in Δr_s would deteriorate greatly the stability of whole system along with the inverter nonlinearity caused by dead time and switch on-drop. Fig. 5.28 shows such an empirical evidence. It displays speed and current responses when a 100% load changed its polarity at 30 rpm. Correspondingly, the motor changed its operation mode from motoring to regeneration. During the motoring load, the proposed sensorless algorithm worked well. However, in the generation mode, the electric frequency became nearly zero and control was lost after speed fluctuation.

References

[1] J. Holtz and A. Khambadkone, "A vector controlled induction motor drive with self commissioning scheme," *IEEE Trans. Ind. Elect.*, vol. 38, no. 5, pp. 322−327, 1991.

[2] X. Xu and D. W. Novotny, "Selection of the flux reference for induction machine drive in the field weakening region", *IEEE Trans. Ind. Appl.*, vol. 28, no. 6, pp. 1353−1358, 1992.

[3] A. M. Trzynadlowski, *The Field Orientaion Principle in Control of Induction Motors*, Kluwer Academic Publishers, 1994.

[4] J. Holtz, "Sensorless control of induction machines−with or without signal injection?" *IEEE Trans. Ind. Electron.*, vol. 53, No. 1, pp. 7−30, Feb. 2006.

[5] P. L. Jansen and R. D. Lorenz, "A physically insightful approach to the design and accuracy assessment of flux observers for field oriented induction machine drives," *IEEE Trans. Ind. Appl.*, vol. 30, pp. 101−110, Jan./Feb. 1994.

[6] P. L. Jansen, R. D. Lorenz, and D. W. Novotny, "Observer-based direct field orientation: Analysis and comparison of alternative methods," *IEEE Trans. Ind. Appl.*, vol. 30, pp. 945−953, Jul./Aug. 1994.

[7] J. Kim, J. Choi, and S. Sul, "Novel Rotor-Flux Observer Using Observer Characteristic Function in Complex Vector Space for Field-Oriented Induction Motor Drives," *IEEE Trans. Ind. Appl.*, vol. 38, no. 5, pp. 1334−1343, Sep./Oct. 2002.

[8] B. C. Kuo, *Automatic Control System*, 9th Ed., Wiley, 2016.

[9] L. Harnefors, "Design and analysis of general rotor-flux-oriented vector control systems," *IEEE Trans. Ind. Electron.*, vol. 48, no. 2, pp. 383−390, Apr. 2001.

[10] C. Lascu, I. Boldea, and F. Blaabjerg, "Comparative study of adaptive and inherently sensorless observers for variable-speed induction-motor drives," *IEEE Trans. Ind. Electron.*, vol. 53, no. 1, pp. 57−65, Feb. 2006.

[11] C. Lascu, I. Boldea, and F. Blaabjerg, "A Class of Speed-Sensorless Sliding-Mode Observers for High-Performance Induction Motor Drives," *IEEE Trans. Ind. Electron.*, vol. 59, no. 9, pp. 3394–3402, Feb. 2009.

[12] D. G. Luenberger, "An introduction to observers," *IEEE Trans. Autom. Control*, vol. AC-16, no. 6, pp. 596–602, Dec. 1971.

[13] H. Kubota, K. Matsuse, and T. Nakano, "DSP based speed adaptive flux observer of induction motor," *IEEE Trans. Ind. Appl.*, Vol. 29, pp. 344–348, Mar./Apr. 1993.

[14] C.T. Chen, *Linear System Theory and Design*, Oxford University Press, 3rd Ed., 1999.

[15] M. Vidyasagar, *Nonlinear Systems Analysis*, 2nd Ed., Prentice Hall, 1993.

[16] H. Tajima and Y. Hori, "Speed sensorless field-orientation control of the induction machines," *IEEE Trans. Ind. Appl.*, vol. 29, pp. 175–180, Jan./Feb. 1993.

[17] V. Utkin, J. Guldner, and J. Shi, *Sliding Mode Control in Electromechanical Systems.* Taylor & Francis, 1999.

[18] Z. Yan, C. Jin, and V. I. Utkin, "Sensorless Sliding-Mode Control of Induction Motors," *IEEE Trans. Ind. Electron.*, vol. 47, no. 6, pp. 1286–1296, Dec. 2000.

[19] H. Kim, J. Son, and J. Lee, "A high-speed sliding-mode observer for the sensorless speed control of a PMSM," *IEEE Trans. Ind. Electron.*, vol. 58, no. 9, pp. 4069–4077, Sep. 2011.

[20] M. Hinkkanen, L. Harnefors, and J. Luomi, "Reduced-order flux observers with stator-resistance adaptation for speed-sensorless induction motor drives," *IEEE Trans. Power Electron.*, vol. 25, no. 5, pp. 1173–1183, May 2010.

[21] L. Harnefors and M. Hinkkanen, "Stabilization of sensorless induction motor drives: a Survey," *IEEE Trans. Power Electron.*, vol. 25, no. 5, pp. 183–192, Jan. 2013.

[22] M. Popescu and D. G. Dorrell, "Proximity Losses in the Windings of High Speed Brushless Permanent Magnet AC Motors with Single Tooth Windings and Parallel Paths," *IEEE Int. Magn. Conf., Intermag*, 2013.

[23] J. Holtz and J. Quan, "Sensorless vector control of induction motors at very low speed using a nonlinear inverter model and parameter identification," *IEEE Trans. Ind. Appl.*, vol. 38, no. 4, pp. 1087–1095, Jul./Aug. 2002.

[24] J. Kim, K. Nam, J. Chung, and S. Hong, "Sensorless vector control scheme for induction motors based on a stator flux estimator with quadrant error compensation rule," *IEEE Trans. Ind. Appl.*, vol. 39, no. 2, pp. 492–503, Mar./Apr. 2003.

Problems

5.1   Consider an IM with the parameters

Rated power	10 hp (7.46 kW)
Rated stator voltage	220 V
Rated frequency	60 Hz
Rated speed	1160 rpm
Number of poles	$P = 6$
Stator resistance, r_s	0.33 Ω
Stator leakage inductance, L_{ls}	1.38 mH
Rotor resistance r_r	0.16 Ω
Rotor leakage inductance, L_{lr}	0.717 mH
Magnetizing inductance, L_m	38 mH

a)   Determine the rated current if the power factor is 0.86 and efficiency is 0.84 at the rated condition.

b)   Assume that the motor is in the steady state with the above conditions. Suppose that the d-axis current is regulated to be $I_{ds}^e = 10$ A. Calculate I_{qs} at the rated condition.

c)   Calculate I_{qr} and λ_{dr}, and draw a current vector diagram.

d)   Using the dq currents obtained in b), calculate the slip, s at $w_e = 377$ rad/sec when the motor is controlled with the rotor field oriented scheme.

e)   Calculate the rated torque.

5.2   An IM is controlled with the rotor field oriented scheme. Let τ_{rn} be the nominal rotor time constant. Suppose that the rotor resistance increased two times from the nominal value. Then, the rotor time constant is halved. However, the slip is calculated with the nominal value, τ_{rn}. Then, an error will take place in the rotor flux angle estimation. This angle error affects all controlled variables.

a)   Let $\Delta\theta_e = \bar{\theta}_e - \theta_e$, where θ_e is the real rotor flux angle and $\vec{\theta}_e$ is the calculated angle based on τ_{rn}. Determine whether $\Delta\theta_e > 0$ or $\Delta\theta_e < 0$.

b)   The rotor flux magnitude does not change immediately with a misalignment, since the rotor time constant, τ_r, is relatively large. Assuming that λ_{dr}^e is constant, show that the IM model is approximated in the misaligned coordinate system $(\hat{\theta}_e)$ as

$$\bar{i}_{ds}^e = \frac{1}{s + \frac{r_s}{\sigma L_s}} \left(\frac{1}{\sigma L_s} \bar{v}_{ds}^e + \omega_e \frac{L_m}{\sigma L_s L_r} \lambda_{dr}^e \sin \Delta\theta_e \right) + \frac{1}{s + \frac{r_s}{\sigma L_s}} \omega_e \bar{i}_{qs}^e,$$

$$(5.132)$$

$$\bar{i}_{qs}^e = \frac{1}{s + \frac{r_s}{\sigma L_s}} \left(\frac{1}{\sigma L_s} \bar{v}_{qs}^e - \omega_e \frac{L_m}{\sigma L_s L_r} \lambda_{dr}^e \cos \Delta\theta_e \right) - \frac{1}{s + \frac{r_s}{\sigma L_s}} \omega_e \bar{i}_{ds}^e.$$

$$(5.133)$$

c)   Let $\Delta i_{ds}^e = \bar{i}_{ds}^e - i_{ds}^e$ and $\Delta i_{qs}^e = \bar{i}_{qs}^e - i_{qs}^e$. By subtracting (5.13) and (5.14) from (5.132) and (5.133), show that in the steady state

$$\Delta i_{ds}^e = \omega_e \frac{L_m}{r_s L_r} \lambda_{dr}^e \sin \Delta\theta_e + \frac{\sigma L_s}{r_s} \omega_e \Delta i_{qs}^e$$

$$\Delta i_{qs}^e = \omega_e \frac{L_m}{r_s L_r} \lambda_{dr}^e (1 - \cos \Delta\theta_e) - \frac{\sigma L_s}{r_s} \omega_e \Delta i_{ds}^e.$$

It is assumed that the applied voltages are the same in both cases, i.e., $\overline{\mathbf{v}}_{dq}^e = \mathbf{v}_{dq}^e$.

d)   Neglect the coupling term effects for small ω_e. Based on the results obtained in c), explain why the field oriented control does not lose the synchronism even though there is an angle error.

5.3   Show that

$$\lambda_r^e = \frac{L_r}{L_m} (\lambda_{ds}^e - \sigma L_s i_{ds}^e).$$

5.4   Consider equations (5.22) and (5.23) for the stator field oriented scheme. .

a)   By setting $\mathbf{p} = 0$ and eliminating i_{ds}^e, derive a quadratic equation in ω_{sl}.

b)   Obtain a relation so that the discriminant is not negative.

c)   Using $T_e = \frac{3P}{4} \lambda_{ds}^e i_{qs}^e$, calculate the upper bound of $|T_e|/|\lambda_s|^2$.

5.5   Consider the IM in Problem **5.1**. Assume that $I_m = 33$ A. Calculate the maximum torque at $\omega_e = 377$ rad/sec by utilizing (5.24), (5.29), and (5.30).

5.6   Consider a 220V (line-to-line), three phase, four pole IM. The parameters of the IM are $r_s = 0.2$ Ω, $L_{ls} = L_{lr} = 2.3$ mH, and $L_m = 34.2$ mH. Assume that the rotor field oriented control is achieved with $i_{ds}^e = 14$ A and $i_{qs}^e = 20$ A at 60 Hz.

a) Draw a voltage-current vector diagram, and obtain power factor angle, ϕ.

b) Obtain torque T_e when $i_{ds}^e = 14$ A and $i_{qs}^e = 20$ A. Suppose that the motor shaft speed is 1740 rpm. Calculate the mechanical power, $P_m = T_e \times \omega_r$.

c) Calculate the motor efficiency using $\eta = P_m/(V_{dqs}I_{dqs}\cos\phi)$.

5.7 Suppose that rotor flux is not aligned with an estimated reference frame. The estimated d-axis is behind the actual rotor flux by $\Delta\theta_e$. See angle error, $\Delta\theta_e = \hat\theta_e - \theta_e$ shown below and assume that it is maintained to be a constant. Let $\hat\lambda_{dr}^e = L_m i_{ds}^e$ and $\hat\omega_{sl} = \frac{1}{T_r}\frac{L_m}{\hat\lambda_{dr}^e}i_{qs}^e$. When $\Delta\theta_e$ is small, develop expressions for $\Delta\lambda_{dr}^e \equiv \hat\lambda_{dr}^e - \lambda_{dr}^e$ and $\Delta\omega_{sl} \equiv \hat\omega_{sl} - \omega_{sl}$ in terms of $\Delta\theta_e$.

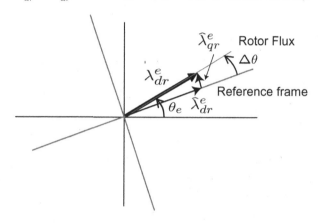

5.8 (Flying start) Construct a Simulink IM model and a current controller as shown in Fig. 5.7 and Fig. 5.8, respectively. Be sure that the IM model is built in the stationary frame. Note that the controller is constructed based on $\hat\theta_e = \int_0^t (\frac{P}{2}\omega_r + \hat\omega_{sl})dt$, where $\hat\omega_{sl} = \frac{L_m}{T_r \hat\lambda_{dr}^e}i_{qs}^e$ and $\hat\lambda_{dr}^e + j\hat\lambda_{qr}^e = e^{j\hat\theta_e}(\lambda_{dr}^s + j\lambda_{qr}^s)$. Use the IM parameters shown in Table 5.1.

Assume that the IM shaft is rotating at 1000 rpm by another machine while its own inverter is not activated. Apply current commands, $i_{ds}^{e*} = 8u_s(t)$ and $i_{qs}^{e*} = 20u_s(t - 0.01)$.

a) Plot current responses, i_{ds}^e and i_{qs}^e and torque, T_e.

b) Plot $\omega_{sl} = \omega_e - \frac{P}{2}\omega_r$.

c) Plot λ_{dr}^e and $\hat\lambda_{dr}^e$.

5.9 Consider a 220V (line-to-line), three phase, four pole IM. The parameters of the IM are $r_s = 0.2$ Ω, $L_{ls} = L_{lr} = 2.3$ mH, and $L_m = 34.2$ mH. Assume that the rated torque $T_e = 40$ Nm is produced when $i_{ds}^e = 14$ A.

a) Determine i_{qs}^e at the rated condition.

b) Assume that the current limit is determined by the rated current, i.e.,

$I_m = \sqrt{i_{ds}^{e\,2} + i_{qs}^{e\,2}}$. Let the phase voltage maximum be $V_m = 220\sqrt{2}/\sqrt{3}$. Using Excel spreadsheet, find the boundary speed w_e between field weakening region I and II.

c) Find the maximum power, the maximum power factor, and the corresponding speed.

5.10 Consider a 220 V (line-to-line), three phase, four pole IM. The IM parameters are $r_s = 0.2\ \Omega$, $L_{ls} = L_{lr} = 2.3$ mH, and $L_m = 34.2$ mH. Build a stator field oriented controller using MATLAB Simulink. Construct a torque controller. Plot current and torque responses to a step torque command, $T_e^* = 40u_s(t - 0.01)$.

5.11 Consider an IM with the following parameters: $P_{rated} = 7.5$ kW, $P=4$, $r_s = 0.28\ \Omega$, $L_{ls} = 1.4$ mH, $r_r = 0.15\ \Omega$, $L_{lr} = 0.7$ mH, $L_{ms} = 39.4$ mH.
a) Show that the MTPA condition is satisfied on the line, $i_{ds}^e = i_{qs}^e$.
b) Let $I_m = 32$ A be the current limit. When a voltage limit curve intersects the MTPA point on the current limit circle, find w_e.
c) Determine w_e at which an MTPV point intersects the current limit.
d) Find the maximum power factor and its operation point (i_d^e, i_q^e) on the current limit curve.

5.12 Consider an IM with the following specifications: $P = 4$, $r_s = 0.55\ \Omega$, $r_r = 0.75\ \Omega$, $L_{ls} = 5$ mH, $L_{lr} = 5$ mH, $L_m = 63$ mH, $J = 0.0179$ kg $\cdot$ m^2.
a) Build an IM dynamic model in the stationary frame using MATLAB/ Simulink similarly to Fig. 5.7.
b) Construct a current controller similarly to Fig. 5.8 based on the rotor flux oriented scheme. Determine PI gains, K_{ic} and K_{pc} such that the current control bandwidth is equal to [0, 200] Hz and current overshoot is about 15%.
c) Construct a speed controller and determine PI gains, K_{ps} and K_{is} so that the speed bandwidth is [0, 20] Hz and that no overshoot takes place.

5.13 (Continued from Problem **5.12**)
a) If $T_L = 50$ Nm, plot the speed response to the following speed reference:

$$w_r^*(t) = \begin{cases} -50, & \text{if } \ t < 0.4 \\ 125t - 100, & \text{if } \ 0.4 < t < 1.2 \\ 50, & \text{if } \ 1.2 < t \end{cases}$$

b) Plot the current responses i_a, i_b, and i_c.

5.14 Consider the dual reference frame observer discussed in Section 5.4.4. It is desired to make $F(jw)e^{-j\alpha}$ real for all w, where $F(s) = 1/(1 + E(s))$ and $E(s) = \frac{L_r}{L_m}\left(K_1 + \frac{K_2}{s}\right)\frac{1}{s}$. Show that

$$\alpha = \pi - \tan^{-1}\left(\frac{\frac{L_r}{L_m}K_1 w_e}{\frac{L_r}{L_m}K_2 - w_e^2}\right).$$

5.15 The observability is determined by

$$\text{rank} \begin{bmatrix} \mathbf{C} \\ \mathbf{CA}(\omega_r) \end{bmatrix} = 4 \tag{5.134}$$

Check the observability for (5.68) and (5.69) when $\omega_r \neq 0$ and $\omega_r = 0$.

5.16 (Continued from Problem **5.12**) Construct a full order observer using MAT-LAB Simulink based on (5.68)–(5.70) and (5.75). Note that $2\omega_r$ should be used in (5.68) instead of ω_r, since the number of poles are $P = 4$. Plot the real speed, $\hat{\omega}_r$ and speed estimate, $\hat{\omega}_r$.

5.17 (MTPA and MTPV)
a) Show that the tangential intersection between a torque curve and a voltage limit ellipse is laid on the line, $\frac{1}{\sigma} i_{ds}^e = i_{qs}^e$.
b) By using (5.36), derive the power factors when $i_{ds}^e = i_{qs}^e$ and $\frac{1}{\sigma} i_{ds}^e = i_{qs}^e$. Check whether they are the same.

Chapter 6

Permanent Magnet AC Motors

Neodymium and cobalt-samarium magnets are widely used in the permanent magnet synchronous motors (PMSMs). They have very suitable properties for motor use; high remnant (residual) flux density and high resilience against demagnetization threat. Since the field winding is replaced by PMs, the PMSMs have low rotor inertia and high power density. Further, since there is no secondary copper loss, the PMSMs have a higher efficiency than IMs. In addition, PMSMs are more favorable in incorporating the reluctance torque especially in the field weakening region. As a result, PMSMs have a wide constant power speed range (CPSR). Because of such advantages, PMSMs are popularly used in industrial drives, home appliances, and electrical vehicles (EV). Fig. 6.1 shows a taxonomy of AC motors. They are classified by the synchronism and the structure.

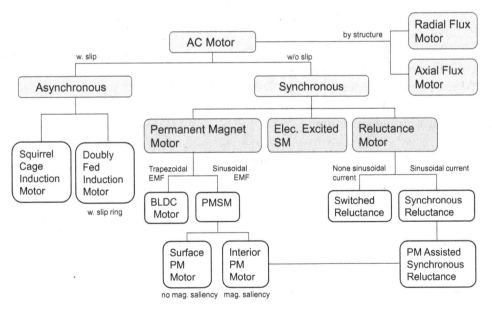

Figure 6.1: Taxonomy of AC motors.

6.1   PMSM and BLDCM

A three phase AC motor can be described simply as a three phase circuit consisting of inductors and EMF's as shown in Fig. 6.2. In the motor equivalent circuit, the back EMFs are denoted by e_a, e_b, and e_c. Further, it is assumed that the motor is connected to a three phase voltage source, (v_a, v_b, v_c). The neutral points of the voltage source and motor are denoted by 'n' and 's', respectively. As discussed in Section 2, the two neutral points are virtually connected, as far as the source voltages and the load impedances are balanced. Hence, the three phase circuit can be seen as a cluster of triple single phase systems.

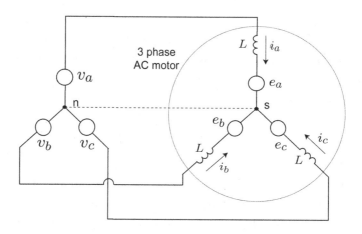

Figure 6.2: Simplified equivalent circuit of three phase motors.

PM motors can be roughly divided into two categories according to the patterns of back EMFs. One category is characterized by sinusoidal back EMF, and the other category by trapezoidal or square back EMF. The former is called *PMAC motor* or PMSM, and the latter *brushless DC motor* (BLDCM), They are compared in Fig. 6.3. The back EMF patterns are shaped depending on the magnet arrangement and the coil winding structure.

6.1.1   PMSM Torque Generation

Both PMSM and BLDCM have the same design goal, which is to establish a linear relationship between torque and the phase current magnitude, independently of the rotor angle. Fig. 6.4 shows how a constant torque is made from the PMSM. It is assumed that a two pole PMSM rotates at a constant speed ω, and that the back EMF is sinusoidal: $[e_a, e_b, e_c] = [E\cos(\omega t), \ E\cos(\omega t - \frac{2\pi}{3}), \ E\cos(\omega t - \frac{4\pi}{3})]$. Assume further that the power source (inverter) provides balanced three phase sinusoidal currents that are in-phase with the back EMFs: $[i_a, i_b, i_c] = [I\cos(\omega t),$

	PMSM or PMAC	BLDC
Control method	Vector control to the field weakening range.	Current level with phase angle adjustment
Applications	High power/precise motion control	Small power/low cost drives
Back EMF		
Current		

Figure 6.3: Classification of PM motors according to the back EMF.

$I\cos(\omega t - \frac{2\pi}{3})$, $I\cos(\omega t - \frac{4\pi}{3})]$. Then, the total electrical power is equal to

$$
\begin{aligned}
P_{tot} &= e_a i_a + e_b i_b + e_c i_c \\
&= EI[\cos^2(\omega t) + \cos^2(\omega t - \frac{2\pi}{3}) + \cos^2(\omega t - \frac{4\pi}{3})] \\
&= \frac{EI}{2}[1 + \cos 2(\omega t) + 1 + \cos 2(\omega t - \frac{2\pi}{3}) + 1 + \cos 2(\omega t - \frac{4\pi}{3})] = \frac{3EI}{2}.
\end{aligned}
$$

Note that each phase power is not constant, but the sum turns out to be constant.

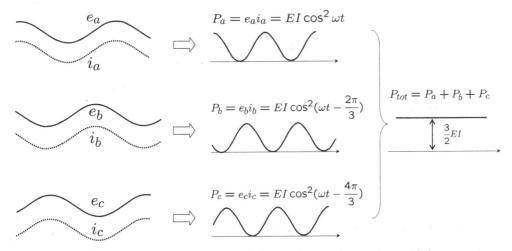

Figure 6.4: Constant power generated from three phase sinusoidal currents and sinusoidal back EMFs.

All the electrical power, $P_{tot} = e_a i_a + e_b i_b + e_c i_c$ is assumed to be converted into the mechanical power in the motor. Then, the shaft torque is given by dividing the motor power by the mechanical speed:

$$
T_e = \frac{P_{tot}}{\omega} = \frac{3EI}{2\omega}. \tag{6.1}
$$

Further, since the back EMF is proportional to the rotor speed, we may let $E = k_b\omega$ for some constant, $k_b > 0$. Therefore, it follows from (6.1) that

$$T_e = \frac{3k_b}{2}I.$$ (6.2)

That is, the torque is proportional only to the current magnitude independently of the angle, like in a DC motor.

Exercise 6.1   Consider a PMSM which has the following back EMFs: $[e_a, e_b, e_c]$ $= [E\cos(\omega t),\ E\cos(\omega t - \frac{2\pi}{3}),\ E\cos(\omega t - \frac{4\pi}{3})]$. Assume that an external power source supplies balanced three phase sinusoidal currents: $[i_a, i_b, i_c] = [I\cos(\omega t - \phi),$ $I\cos(\omega t - \frac{2\pi}{3} - \phi),\ I\cos(\omega t - \frac{4\pi}{3} - \phi)]$. Determine the total electrical power and torque when the rotor speed is ω.

Solution. Electrical power is

$$P_{tot} = e_a i_a + e_b i_b + e_c i_c = \frac{3EI}{2}\cos\phi.$$

Thus torque is equal to $T_e = \frac{3EI}{2\omega}\cos\phi = \frac{3k_b}{2}I\cos\phi.$ ∎

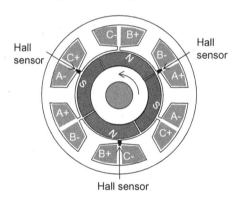

Figure 6.5: Four pole 6 slot BLDCM with 120° coil span in electrical angle.

6.1.2   BLDCM Torque Generation

Fig. 6.5 shows a schematic diagram of a typical four pole BLDCM with 120° coil span in electrical angle. If the PMs cover the whole rotor surface, the back EMF would exhibit a trapezoidal waveform with 120° flat top and bottom. Suppose that an external power source provides a square wave current with a constant amplitude in accordance with the back EMF. Specifically, the phase current keeps a nonzero constant amplitude in each interval where the back EMF is flat top or bottom. Further, assume that the current polarity is the same as the back EMF, and that current is zero during the EMF transition. Then, each phase generates periodic square power. The nonzero power lasts during 120° a half period. Since each phase is shifted 120°, the sum of phase powers turns out to be constant. In every moment, a phase power is zero, while the other two are equal to EI. Fig. 6.6 shows the back EMF along with the phase current. It also shows how a constant power sum is made. Therefore, the torque is proportional to the current:

$$T_e = \frac{P_{tot}}{\omega_r} = \frac{2EI}{\omega_r} = 2k_b I.$$

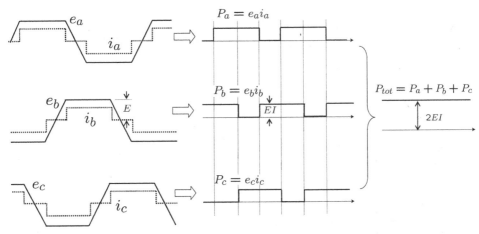

Figure 6.6: Three phase trapezoidal back EMFs and square wave currents making a constant power.

Position Sensing and BLDCM Control

The BLDCM has a simple winding structure: Each stator tooth has a concentrated winding. In the BLDCM, three Hall sensors are positioned at the slot openings, to monitor the rotor position. The back EMF states are identified by the Hall sensor signals, (H_1, H_2, H_3). Fig. 6.7 shows the location of three Hall sensors in a four pole, 6 slot BLDCM, which are displaced by 120° in mechanical angle. The Hall sensors are discrete type and each Hall sensor detects the radial field of PMs. As the rotor rotates 180°, three Hall sensors provide six sets of signals: $(H_1, H_2, H_3)=(1,0,1)$, $(1,0,0)$, $(1,1,0)$, $(0,1,0)$, $(0,1,1)$, and $(0,0,1)$. Fig. 6.7 shows instances when the sensor signal transition occurs. Note that the signal combination changes each moment of back EMF transition. Specifically, consider a state shown in Fig. 6.7 (a). The stator tooth of a–phase winding sees only a PM south pole during first 30° rotation. Then PM flux linkage to the a-phase coil will be constant, thereby $e_a = 0$ for 60° in electrical angle.

Fig. 6.7 (b) shows a rotor position at 60°. In the next 30° rotation, the south pole is replaced by the north pole steadily over the stator tooth of a-phase winding. Then PM flux linkage to the a-phase coil increases, thereby $e_a < 0$ by the Lenz law. Similarly, Fig. 6.7 (c) shows a period during which $e_a = 0$.

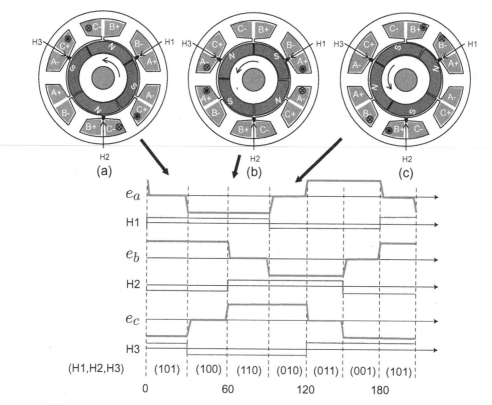

Figure 6.7: Three Hall sensor signals with the EMFs and applied currents.

Each set of the Hall signal indicates one of six sector positions, based on which the gating signals are generated. An inverter circuit and PWM signals for a BLDCM are shown in Fig. 6.8. Consider the interval of (101), where $e_b > 0$ and $e_c < 0$. Since gating signals to PWM3 and PWM2 are active, $i_b > 0$ and $i_c < 0$. Thus, positive torque is generated. The current level is determined by the on-duty interval of the PWM.

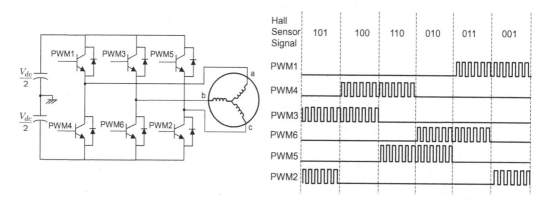

Figure 6.8: Inverter and PWM gating signals for the BLDCM.

Exercise 6.2   Consider a BLDCM with a fractional pole coverage shown in Fig. 6.9. The pole coverage ratio is defined as $\alpha = \tau_M/\tau_p$, where τ_M is the PM arc length and τ_p is the pole pitch.
 a) Sketch the a-phase back EMF;

Figure 6.9: Fractional PM coverage of BLDCM rotor (Exercise 6.2).

 b) Assume that each phase current has a typical rectangular shape with 120° conduction. Sketch the a-phase power.

Solution. See Fig. 6.10.

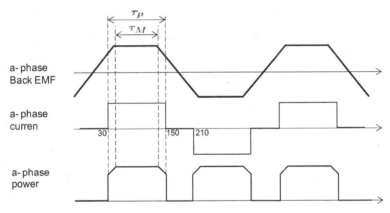

Figure 6.10: A-phase back EMF and power of a BLDCM with a fractional PM coverage (Solution to Exercise 6.2).

Torque Ripple of BLDCM

In case of BLDCM, the torque ripple gets bigger as the speed increases. It is because the current cannot change sharply as shown in Fig. 6.6. The presence of inductance limits the rate of current rise. It takes time to reach a set value I, since the current increases with a certain slope and the rate is determined by the inductance and the available voltage. Here, the coil phase voltage minus back EMF is can be used for current change. Based on the motor model shown in Fig. 6.2, the rate of current change (for example, a-phase) is determined by

$$\left|\frac{di_a}{dt}\right| = \frac{\left|\frac{V_{dc}}{2} - k_e\omega_r\right|}{L} \tag{6.3}$$

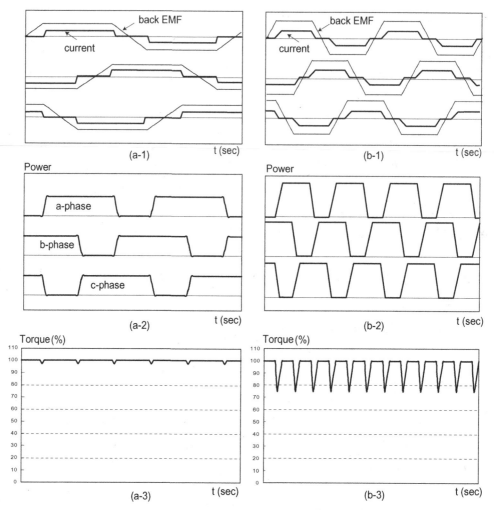

Figure 6.11: Torque ripples of a BLDCM depending on speeds: (a-1), (b-1) back EMFs and phase currents; (a-2), (b-2) power of each phase; (a-3), (b-3) shaft torque.

where k_e is the back EMF constant and V_{dc} is the inverter DC link voltage. In the low speed region, $V_{dc}/2 \gg k_e\omega_r$ so that $|\frac{di_a}{dt}|$ can be made large. But in the high speed region, $V_{dc}/2 \approx k_e\omega_r$ so that $|\frac{di_a}{dt}|$ is low. In such a case, current transition time becomes long. It can be observed from Fig. 6.6 that constant torque is available only if the phase current has a sharp edge at each transition. If current changes with a slope, the torque ripple is maintained.

Fig. 6.11 shows schematic figures that illustrates how the torque ripple is generated: In column (b), the motor speed is about two times higher than that in column (a). Therefore, the marginal voltage for current increase is reduced. As a result, the current slope is low in the high speed operation. Therefore, the power sum, $P_{tot} = e_a i_a + e_b i_b + e_c i_c$, is no more constant. It has dents which deepen as the speed increases. Fig. 6.11 (a-3) and (b-3) show a torque ripple increase of a BLDCM with speed.

6.1.3 Comparison between PMSM and BLDCM

In general, PMSMs are better speed and positional accuracy than BLDCMs. In addition, PMSMs do not produce torque ripple, as in the case with BLDCM. But benefits of BLDCM are in the simplicity and cost competitiveness. BLDCMs are normally used for low cost, low power (less than 5 kW) applications such as blowers, material handling equipments, home appliances, etc. Comparisons between BLDCMs and PMSMs are listed in Table 6.1.

Table 6.1: Characteristics of BLDCM and PMSMs

	BLDCM	PMSM
Torque ripple	high	low
Position sensor	Hall sensors (inexpensive)	resolver (expensive)
Stator winding	concentrated (less copper)	distributed (more copper)
PM usage	large	relatively small
Eddy loss in PMs	large	relatively small
Control complexity	simple	complicated
Speed range	narrow	wide
Inverter price	low	high

6.2 PMSM Dynamic Modeling

PMSMs are classified broadly into two categories depending on the PM location in the rotor. If the PMs are mounted on the surface of the rotor, the motor is called surface mounted PMSM, or in short 'SPMSM.' On the other hand, if PMs are embedded inside the rotor core, the motor is called interior PMSM, or 'IPMSM.' This classification has more meaning than structural differences. PM imbedding causes a magnetic anisotropy, which results in the difference in the voltage and torque equations.

Note that the relative recoil permeabilities are ferrite: $1.05 \sim 1.15$, Nd-Fe-B : $1.04 \sim 1.11$, and Sm-Co : $1.02 \sim 1.07$ [1]. That is, the permeabilities are close to that of air although the PMs have high residual flux density. Therefore, PMs in the motor look like the air in the view of the stator coil.

6.2.1 Types of PMSMs

Fig. 6.12 shows cross-sectional views of four pole SPMSMs and IPMSMs in which the PMs are marked by dark areas. Fig. 6.12(a) and (b) show SPMSMs, while Fig. 6.12(c) and (d) IPMSMs. In Fig. 6.12 (b), the PMs are inserted on the groove of the rotor surface. The inset magnet motor, though PMs are on the surface, has a different magnetic saliency. Specifically, the inset motor has a larger q-axis inductance, like IPMSM. The flux concentrating arrangement is shown in Fig. 6.12(d), in which the air gap flux density can be increased higher than the remnant flux

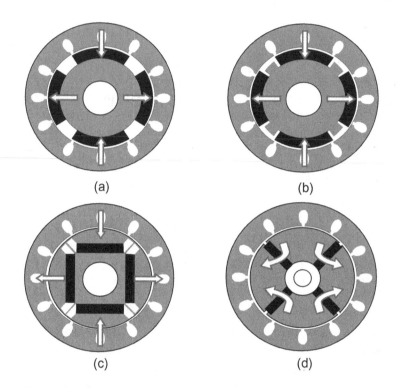

Figure 6.12: Typical PMSM structures: (a) surface magnet, (b) inset magnet, (c) interior magnet, (d) interior magnet (flux concentration).

density of PM. For example, it is possible to achieve 0.8 T air gap field density with 0.4 T ferrite magnets. It is important to note that none ferromagnetic material (e.g. stainless steel, beryllium-copper) needs to be used at the shaft area to penalize flux flow through the center part.

Usually the SPMSM has a problem fixing the PM on the rotor surface. Glues are widely used, but they have aging effects under the stress of thermal cycles and large centrifugal force. If stainless band is used for fixing the PMs, eddy current loss will take place on its surface due to slot and PWM harmonics. Further, glass (or carbon) fiber band is often used for high speed machine, but it requires a larger air gap. In IPMSMs, no fixation device is required since the PMs are inserted in the cavities. Further, the steel core protects PMs from the demagnetization threat caused by stator MMF harmonics and slot harmonics. Differences between SPMSM and IPMSM are listed in Table 6.2. Fig. 6.13 shows photos of SPMSM and IPMSM rotors.

SPMSM Inductance

Consider the SPMSM schematics shown in Fig. 6.14. The two diagrams show two different flux paths corresponding to the different coil groups. Similarly to the case of IM, the d-axis is determined by the rotor field axis. The lines shown in Fig. 6.14 (a) denote d-axis stator flux corresponding to the d-axis current. Note on

(a)

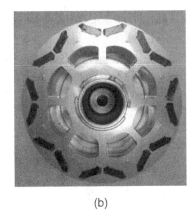
(b)

Figure 6.13: Rotors of (a) SPMSM and (b) IPMSM.

Table 6.2: Comparisons between SPMSM and IPMSM

	SPMSM	IPMSM
PM location	surface	cavities
PM fixation	glue or band	insertion
Field harmonics on PM	large	small
PM usage	large	relatively small
Saliency ratio	1	> 1
Reluctance torque usage	no	yes
Power density	low	high
Speed range (Field weakening)	small	large

the other hand that the d-axis winding is positioned at the q-axis. Applying the Ampere's law and integrating along the designated loop in Fig. 6.14 (a), it follows that

$$\oint \frac{\mathbf{B}}{\mu} \cdot d\boldsymbol{\ell} = \frac{B}{\mu_{PM}} 2h_m + \frac{B}{\mu_0} 2g + \frac{B}{\mu_{Fe}} \ell_{core} = N i_d, \qquad (6.4)$$

where N is the number of (effective) coil turns of the d-axis winding, and μ_{PM} and μ_{Fe} are permeabilities of PM and core material, respectively. In addition, h_m is the PM height and ℓ_{core} is the total length of flux paths in the steel core. However, the core permeability $\mu_{Fe} \approx 4000 \sim 5000 \, \mu_0$ is so high that $\frac{B}{\mu_{Fe}} \ell_{core}$ is relatively small, thereby neglected. On the other hand since $\mu_{PM} \approx \mu_0$, the PM height is also counted as a part of the air gap. That is,

$$B = \frac{\mu_0 N}{2(g + h_m)} i_d. \qquad (6.5)$$

Further, let A be the air gap area through which the flux passes. In this case, $A \approx \frac{1}{2}\pi D_r l_{st}$, where D_r and l_{st} are the diameter and stacked length of the rotor, respectively. Note that $N\Phi = NB \times A = L_d i_d$. Thus, the d-axis inductance is

$$L_d = \frac{\mu_0 N^2 A}{2(g + h_m)}. \qquad (6.6)$$

The loops shown in Fig. 6.14 (b) describe the q-axis flux. Note that the flux does not pass through PMs. Applying Ampere's law, we obtain $L_q = \frac{\mu_0 N^2 A}{2(g+h_m)}$ which is the same as L_d, i.e., $L_d = L_q$. The presence of PMs does not affect the reluctance of the stator coils. In other words, the PM is also counted the same as the air, so that the effective air gap is uniform. Thus, the inductance is the same independently of the axis direction.

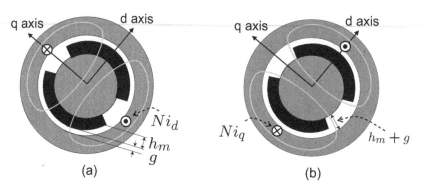

Figure 6.14: Flux paths of SPMSM ($L_d = L_q$): (a) d-axis and (b) q-axis.

IPMSM Inductance

The reluctance is different depending on the flux paths in IPMSMs. According to Fig. 6.15 (a) and (b), PMs are encountered along the d-axis flux, whereas no PM is found along the q-axis flux. Hence, the effective air gap is $2(g + h_m)$ along the d-axis, whereas $2g$ along the q-axis. Therefore, the d-axis inductance is smaller than that of the q-axis, i.e., $L_d < L_q$ where

$$L_d = \frac{\mu_0 N^2 A}{2(g + h_m)}, \tag{6.7}$$

$$L_q = \frac{\mu_0 N^2 A}{2g}. \tag{6.8}$$

This magnetic anisotropy is quantified by the saliency ratio defined by $\xi = L_q/L_d$. Fig. 6.16 shows the FEM simulation results of an eight pole IPMSM with the PM . Note that the number of flux lines are less with the d-axis current.

Summary

1. The inductance is determined as the rate of flux increase per current, i.e., $L = \mu \times \text{MMF}/\mathcal{R}$, where $\mathcal{R}$ is the reluctance. The PM materials like Nd-Fe-B, Sm-Co, or ferrites have as low permeability as air, so that the PMs are counted as the empty space in the stator inductance calculation.

2. The SPMSM is characterized by a uniform gap height, and the reluctance is independent of the PM existence or PM magnetization. Therefore, $L_d = L_q$. On the other hand if the PMs are set inside the

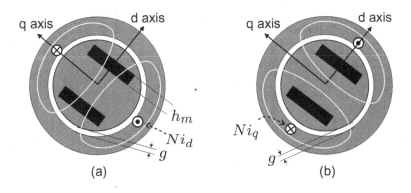

Figure 6.15: Flux paths of IPMSM ($L_d < L_q$): (a) d-axis and (b) q-axis.

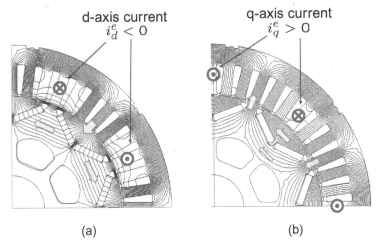

Figure 6.16: Simulation results of flux lines: (a) d-axis current with PM unmagnetized and (b) q-axis current with PM unmagnetized.

rotor cavities, the reluctance differs depending on the direction. The cavities for PM are seen as an additional air gap to the stator winding. Specifically, $L_d < L_q$ in IPMSMs.

6.2.2 SPMSM Voltage Equations

The SPMSM voltage equation is described in the *abc*-frame firstly, and then description in the *dq* frame follows. In the synchronous motor, $\theta_e = \frac{P}{2}\theta_r$ and $\omega_e = \frac{P}{2}\omega_r$. The theory will be developed for two pole motors ($P = 2$). Thus, θ and ω are used instead of θ_e and ω_e. The rotor PM and the stator currents are sources of the flux linkage. Note, however, that the PM flux linkage to a phase winding varies as the rotor rotates, i.e., it is described a function of θ. Fig. 6.17 shows how the PM flux is linked to *a*-phase winding: The linking is maximum at $\theta = 0$ and zero at $\theta = \pi/2$ as shown in (b) and (c). Therefore, the fundamental component is described as $\psi_m \cos\theta$, where ψ_m is a constant. Considering 120° difference among the phase coils, the PM flux linkage is described as $\psi_m\left[\cos\theta,\ \cos(\theta-2\pi/3),\ \cos(\theta+2\pi/3)\right]^T$. It is assumed here that the PM field has no harmonics. The self-inductance of

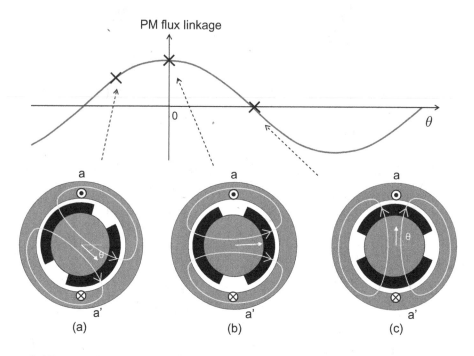

Figure 6.17: PM flux linkage variation to the a-phase winding as the rotor rotates.

a uniform air gap machine is described in Section 2.1.2. Taking into account all contributions, the flux linkage of SPMSM is described as

$$
\begin{bmatrix}\lambda_a\\\lambda_b\\\lambda_c\end{bmatrix}=\underbrace{\begin{bmatrix}L_{ms}+L_{ls}&-\frac{1}{2}L_{ms}&-\frac{1}{2}L_{ms}\\-\frac{1}{2}L_{ms}&L_{ms}+L_{ls}&-\frac{1}{2}L_{ms}\\-\frac{1}{2}L_{ms}&-\frac{1}{2}L_{ms}&L_{ms}+L_{ls}\end{bmatrix}}_{=\ \mathbf{L}_{abcs}}\begin{bmatrix}i_a\\i_b\\i_c\end{bmatrix}+\psi_m\underbrace{\begin{bmatrix}\cos\theta\\\cos(\theta-2\pi/3)\\\cos(\theta+2\pi/3)\end{bmatrix}}_{=\ \boldsymbol{\psi}(\theta)}.\quad(6.9)
$$

Since the SPMSM belongs to the class of uniform machine, $\mathbf{L}_{abcs}$ is constant independently of the rotor angle, θ. The inductance element, L_{ms} is estimated via (2.20).

Exercise 6.3　　Consider a two pole six phase SPMSM shown in Fig. 6.18. Show that the flux linkage equation is equal to

$$
\begin{bmatrix}\boldsymbol{\lambda}_{abc}\\\boldsymbol{\lambda}_{xyz}\end{bmatrix}=\begin{bmatrix}\mathbf{L}&\mathbf{M}\\\mathbf{M}^T&\mathbf{L}\end{bmatrix}\begin{bmatrix}\mathbf{i}_{abc}\\\mathbf{i}_{xyz}\end{bmatrix}+\begin{bmatrix}\boldsymbol{\psi}(\theta)\\\boldsymbol{\psi}(\theta-\frac{\pi}{6})\end{bmatrix},\qquad(6.10)
$$

where

$$
\mathbf{M}=\frac{\sqrt{3}L_m}{2}\begin{bmatrix}1&-1&0\\0&1&-1\\-1&0&1\end{bmatrix}.\qquad(6.11)
$$

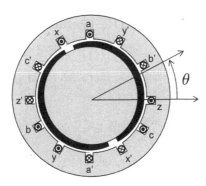

Figure 6.18: Two pole six phase SPMSM.

SPMSM Dynamics in ABC Frame

The voltage equation is given by

$$\mathbf{v}_{abc} = r_s \mathbf{i}_{abc} + \frac{d}{dt}\boldsymbol{\lambda}_{abc} = r_s \mathbf{i}_{abc} + \mathbf{L}_{abcs}\frac{d}{dt}\mathbf{i}_{abc} - \omega\psi_m \begin{bmatrix} \sin\theta \\ \sin(\theta - 2\pi/3) \\ \sin(\theta + 2\pi/3) \end{bmatrix}. \qquad (6.12)$$

In the form of ordinary differential equation,

$$\frac{d}{dt}\mathbf{i}_{abc} = -r_s\mathbf{L}_{abcs}^{-1}\mathbf{i}_{abc} + \omega\psi_m\mathbf{L}_{abcs}^{-1}\begin{bmatrix} \sin\theta \\ \sin(\theta - 2\pi/3) \\ \sin(\theta + 2\pi/3) \end{bmatrix} + \mathbf{L}_{abcs}^{-1}\mathbf{v}_{abc}. \qquad (6.13)$$

Note that

$$\mathbf{L}_{abcs}^{-1} = \frac{2}{L_{ms}((2+\gamma)^3 - 3(2+\gamma) - 2)} \begin{bmatrix} (2+\gamma)^2 - 1 & 3+\gamma & 3+\gamma \\ 3+\gamma & (2+\gamma)^2 - 1 & 3+\gamma \\ 3+\gamma & 3+\gamma & (2+\gamma)^2 - 1 \end{bmatrix},$$
$$(6.14)$$

where $\gamma = 2L_{ls}/L_{ms}$, and it is used in the simulation of SPMSM in the *abc* frame Fig. 6.19.

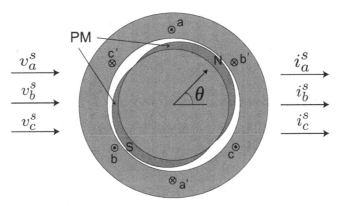

Figure 6.19: A schematic model for SPMSM.

SPMSM Dynamics in Stationary DQ Frame

Using the transformation map $f_{dq}^s = \frac{2}{3}\left[f_a(t) + e^{j\frac{2\pi}{3}}f_b(t) + e^{-j\frac{2\pi}{3}}f_c(t)\right]$, $\mathbf{L}_{abcs}\mathbf{i}_{abc}$ of (6.9) yields $L_s\mathbf{i}_{dq}^s$, as was shown in (4.5). Here, the PM part of (6.9) is transformed as follows:

$$\frac{2}{3}\psi_m\left[\cos\theta + e^{j\frac{2\pi}{3}}\cos(\theta - \frac{2\pi}{3}) + e^{-j\frac{2\pi}{3}}\cos(\theta + \frac{2\pi}{3})\right]$$

$$= \frac{2}{3}\frac{1}{2}\psi_m\left[e^{j\theta} + e^{-j\theta} + (e^{j(\theta-\frac{2}{3}\pi)} + e^{-j(\theta-\frac{2}{3}\pi)})e^{j\frac{2\pi}{3}}\right.$$

$$\left. +(e^{j(\theta+\frac{2}{3}\pi)} + e^{-j(\theta+\frac{2}{3}\pi)})e^{-j\frac{2\pi}{3}}\right]$$

$$= \frac{2}{3}\frac{1}{2}\psi_m\left[e^{j\theta} + e^{-j\theta} + e^{j\theta} + e^{j(\theta-\frac{4}{3}\pi)} + e^{j\theta} + e^{j(\theta+\frac{4}{3}\pi)}\right]$$

$$= \frac{2}{3}\frac{1}{2}\psi_m 3e^{j\theta} = \psi_m e^{j\theta}.$$

That is, the rotor flux vector appears as a rotating vector with a constant magnitude, ψ_m. Hence, the flux linkage in the stationary dq frame is given by

$$\boldsymbol{\lambda}_{dq}^s = L_s\mathbf{i}_{dq}^s + \psi_m e^{j\theta}, \tag{6.15}$$

$$\text{or} \qquad \begin{bmatrix}\lambda_d^s \\ \lambda_q^s\end{bmatrix} = L_s\begin{bmatrix}i_d^s \\ i_q^s\end{bmatrix} + \psi_m\begin{bmatrix}\cos\theta \\ \sin\theta\end{bmatrix}, \tag{6.16}$$

where $L_s = \frac{3}{2}L_{ms} + L_{ls}$.

Then the SPMSM dynamics in the stationary dq frame are given as follows:

$$\mathbf{v}_{dq}^s = r_s\mathbf{i}_{dq}^s + L_s\frac{d}{dt}\mathbf{i}_{dq}^s + j\omega\psi_m e^{j\theta}, \tag{6.17}$$

Equivalently,

$$v_d^s = r_s i_d^s + L_s\frac{d}{dt}i_d^s - \omega\psi_m\sin\theta$$

$$v_q^s = r_s i_q^s + L_s\frac{d}{dt}i_q^s + \omega\psi_m\cos\theta$$

Or, we have in the matrix form

$$\frac{d}{dt}\begin{bmatrix}i_d^s \\ i_q^s\end{bmatrix} = -\frac{r_s}{L_s}\begin{bmatrix}i_d^s \\ i_q^s\end{bmatrix} - \frac{\psi_m\omega}{L_s}\begin{bmatrix}-\sin\theta \\ \cos\theta\end{bmatrix} + \frac{1}{L_s}\begin{bmatrix}v_d^s \\ v_q^s\end{bmatrix}. \tag{6.18}$$

SPMSM Dynamics in Synchronous Reference Frame

Consider transforming $\boldsymbol{\lambda}_{dq}^s$ into the one in a synchronous frame via multiplication by $e^{-j\theta}$, i.e., $\boldsymbol{\lambda}_{dq}^e = e^{-j\theta}\boldsymbol{\lambda}_{dq}^s$. The dynamics will be described in the rotating coordinates that are aligned with the physical rotor. It follows from (6.15) that

$$\boldsymbol{\lambda}_{dq}^e = L_s\mathbf{i}_{dq}^e + \psi_m \tag{6.19}$$

$$\text{or} \qquad \begin{bmatrix}\lambda_d^e \\ \lambda_q^e\end{bmatrix} = L_s\begin{bmatrix}i_d^e \\ i_q^e\end{bmatrix} + \psi_m\begin{bmatrix}1 \\ 0\end{bmatrix}.$$

Now, we embark on the voltage equation in the synchronous frame:

$$
\begin{aligned}
e^{-j\theta}\mathbf{v}^s_{dq} &= r_s e^{-j\theta}\mathbf{i}^s_{dq} + e^{-j\theta}\mathbf{p}e^{j\theta}e^{-j\theta}\boldsymbol{\lambda}^s_{dq}\\
\mathbf{v}^e_{dq} &= r_s\mathbf{i}^e_{dq} + e^{-j\theta}\mathbf{p}(e^{j\theta}\boldsymbol{\lambda}^e_{dq})\\
&= r_s\mathbf{i}^e_{dq} + e^{-j\theta}j\omega e^{j\theta}\boldsymbol{\lambda}^e_{dq} + e^{-j\theta}e^{j\theta}\mathbf{p}\boldsymbol{\lambda}^e_{dq}\\
&= r_s\mathbf{i}^e_{dq} + j\omega\boldsymbol{\lambda}^e_{dq} + \mathbf{p}\boldsymbol{\lambda}^e_{dq}\\
&= r_s\mathbf{i}^e_{dq} + L_s\frac{d}{dt}\mathbf{i}^e_{dq} + j\omega L_s\mathbf{i}^e_{dq} + j\omega\psi_m. \quad (6.20)
\end{aligned}
$$

Then, the voltage equation for SPMSM turns out to be

$$
v^e_d = r_s i^e_d + L_s\frac{d}{dt}i^e_d - \omega L_s i^e_q \quad (6.21)
$$

$$
v^e_q = r_s i^e_q + L_s\frac{d}{dt}i^e_q + \omega L_s i^e_d + \omega\psi_m. \quad (6.22)
$$

Note again that $\omega\psi_m$ is the back EMF which depends only on speed, and $\omega L_s i^e_q$ and $-\omega L_s i^e_q$ are the coupling terms which are induced while transforming into the rotating frame. In the ODE form, (6.21) and (6.22) are written equivalently as [2]

$$
\frac{d}{dt}\begin{bmatrix} i^e_d \\ i^e_q \end{bmatrix} = \begin{bmatrix} \frac{-r_s}{L_s} & \omega \\ -\omega & -\frac{r_s}{L_s} \end{bmatrix}\begin{bmatrix} i^e_d \\ i^e_q \end{bmatrix} - \frac{\psi_m\omega}{L_s}\begin{bmatrix} 0 \\ 1 \end{bmatrix} + \frac{1}{L_s}\begin{bmatrix} v^e_d \\ v^e_q \end{bmatrix}. \quad (6.23)
$$

Exercise 6.4 Consider an eight pole SPMSM is mounted on a dynamo system. Open circuit voltage is measured when the motor rotates at 3600 rpm. The fundamental component of line to line voltage is read 280 V(peak). Find the back EMF coefficient, ψ_m.

Solution. The electrical speed is equal to $\omega_e = 4 \times 3600/60 \times 2\pi = 1508$ rad/sec. On the other hand, the phase voltage is equal to $280/\sqrt{3} = 162$ V. Therefore, $\psi_m = 162/1508 = 0.107$ Wb. ∎

Validity of the model is checked through computer simulation by comparing the current responses. In Fig. 6.20, two SPMSM dynamic models are compared: one is constructed in the *abc* frame, and the other is realized in the synchronous *dq* frame. The *abc* frame model is based on (6.13) and (6.14), whereas the *dq* model (6.23). They should yield the same results via coordinate transformations, $T(\theta)$ and $T^{-1}(\theta)$ if the input is same. The comparison process is detailed in the following exercise.

Exercise 6.5 Consider an eight pole SPMSM with $L_m = 1$ mH, $L_m = 0.05$ mH, $\psi_m = 0.08$ Wb, and $r = 0.1\ \Omega$. Assume that $\mathbf{v}^s_{abc} = 150\big[\cos 377t,\ \cos(377t - 2\pi/3),$ $\cos(377t - 4\pi/3)\big]^T$ is applied to the terminal, and the motor runs in synchronism.
a) Using MATLAB Simulink, compute $[i_a,\ i_b,\ i_c]$ for $t \in [0,\ 0.1]$ sec using (6.13) in the *abc* frame.
b) Compute $[i^e_d,\ i^e_q]$ for $t \in [0,\ 0.1]$ sec using (6.23) in the synchronous reference frame.

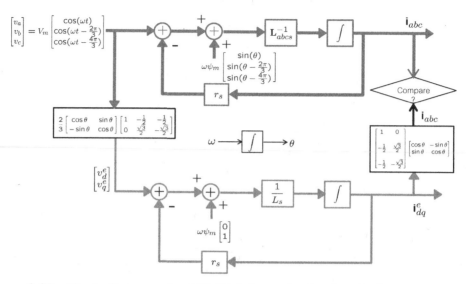

Figure 6.20: Block diagrams for SPMSM dynamics in the abc frame and dq synchronous frame.

c) Obtain $[i_a, i_b, i_c]$ using the following inverse transformation:

$$\begin{bmatrix} i_a \\ i_b \\ i_c \end{bmatrix} = \begin{bmatrix} 1 & 0 \\ -\frac{1}{2} & \frac{\sqrt{3}}{2} \\ -\frac{1}{2} & -\frac{\sqrt{3}}{2} \end{bmatrix} \begin{bmatrix} \cos\theta & -\sin\theta \\ \sin\theta & \cos\theta \end{bmatrix} \begin{bmatrix} i_d^e \\ i_q^e \end{bmatrix}.$$

Check whether the two current responses are the same.

Solution. See Fig. 6.21. Two current responses match perfectly. ∎

Summary

1. The fundamental component of back EMF appears as $\omega\psi_m[-\sin\theta, \cos\theta]^T$ in the stationary dq frame. On the other hand, it appears as $\omega\psi_m[0, 1]^T$ in the synchronous frame.

2. While transforming into the rotating frame, the coupling voltage, $-\omega L_s i_q^e$ and $\omega L_s i_d^e$ take place additionally.

6.2.3   IPMSM Dynamic Model

In IPMSM, the PMs are buried inside the rotor core. The cavities for PM insertion rotate while the stator coils are fixed. Correspondingly, the feasible flux routes differently depending on the cavity position and it appears as the reluctance variation.

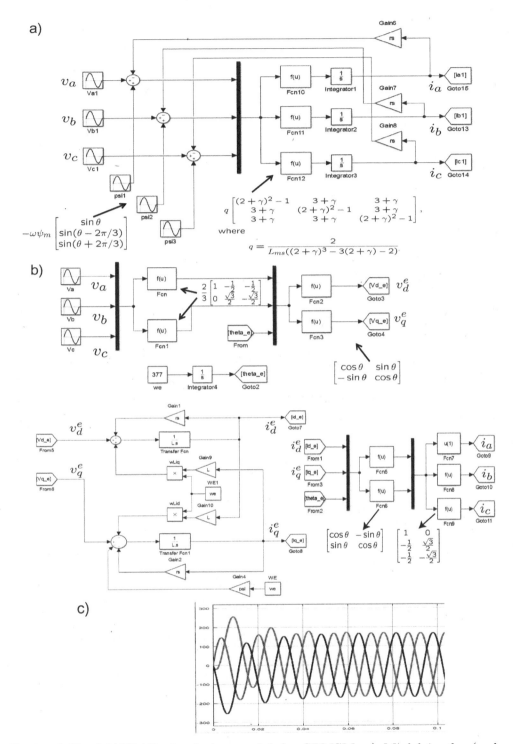

Figure 6.21: MATLAB simulation model for SPMSM: a) Model in the (a, b, c) frame, b) model in the (d, q) frame and coordinate changes from (a, b, c) to (d, q) and from (d, q) to (a, b, c). c) plots of i_a, i_b, and i_c based on the *abc* and *dq* models.

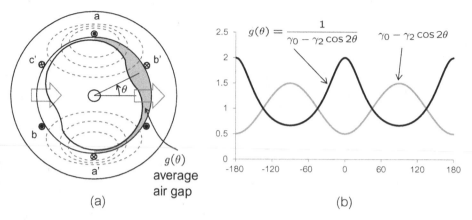

(a)

(b)

Figure 6.22: (a) Non-uniform air gap machine and (b) average gap function.

Inductance of Non-uniform Gap Machine

As a leading example, consider a non-uniform air gap machine with a peanut shaped rotor core shown in Fig. 6.22 (a). The flux lines of the a-phase coil should cross the air gap in the range of $-\frac{\pi}{2} < \theta < \frac{\pi}{2}$. The gap to the a-phase coil is shaded as shown in Fig. 6.22 (a). Here, the average air gap length is the mean gap distance over $[-\frac{\pi}{2}, \frac{\pi}{2}]$. It varies as the rotor angle changes. Note that the average gap for a-phase coil is the largest at $\theta = 0$, and the smallest at $\theta = \frac{\pi}{2}$. Based on this observation, the average air gap length is modeled such that

$$g(\theta) = \frac{1}{\gamma_0 - \gamma_2 \cos 2\theta} \tag{6.24}$$

using two constants γ_0 and γ_2, where $\gamma_0 > \gamma_2 > 0$. The air gap function and its inverse are shown in Fig. 6.22 (b). Note that $g(\theta)$ is a function of 2θ, since the same position is repeated two times each revolution. Assume that the coil has N-turn. Then, the inductance is equal to

$$L(\theta) = \frac{\mu_0}{2}(\gamma_0 - \gamma_2 \cos 2\theta)AN^2, \tag{6.25}$$

where A is the area of air gap.

In the three phase cylindrical machine, the a-phase inductance with respect to the a-phase coil is

$$L_a(\theta) = L_{ms} + L_{ls} - L_\delta \cos 2\theta, \tag{6.26}$$

where

$$L_{ms} = \mu_0 \frac{4}{\pi^2} \frac{\tau_p l_{st}}{p} k_w^2 N_{ph}^2 \gamma_0 \tag{6.27}$$

$$L_\delta = \mu_0 \frac{4}{\pi^2} \frac{\tau_p l_{st}}{p} k_w^2 N_{ph}^2 \gamma_2. \tag{6.28}$$

In (6.27), $1/g$ is replaced by γ_0. Since b and c phase coils are shifted by $\frac{2\pi}{3}$ and $\frac{4\pi}{3}$, it follows that

$$
\begin{aligned}
\lambda_a &= (L_{ms} + L_{ls} - L_\delta \cos 2\theta)i_a + \left(-\frac{1}{2}L_{ms} - L_\delta \cos(2\theta - \frac{2\pi}{3})\right)i_b \\
&+ \left(-\frac{1}{2}L_{ms} - L_\delta \cos(2\theta - \frac{4\pi}{3})\right)i_c
\end{aligned}
\tag{6.29}
$$

Repeating the same to other phases, we have the following inductance:

$$
\overline{\mathbf{L}}_{abcs} = \mathbf{L}_{abcs} - L_\delta \mathbf{L}_{rlc}(\theta),
\tag{6.30}
$$

where

$$
\mathbf{L}_{rlc}(\theta) =
\begin{bmatrix}
\cos 2\theta & \cos(2\theta - 2\pi/3) & \cos(2\theta + 2\pi/3) \\
\cos(2\theta - 2\pi/3) & \cos(2\theta + 2\pi/3) & \cos 2\theta \\
\cos(2\theta + 2\pi/3) & \cos 2\theta & \cos(2\theta - 2\pi/3)
\end{bmatrix}.
\tag{6.31}
$$

Note that $\mathbf{L}_{abcs}$ is the inductance corresponding to an average uniform air gap while $-L_\delta \mathbf{L}_{rlc}(\theta)$ represents the reluctance caused by the rotor saliency. This approach was in the modeling of classical salient pole synchronous motors [2].

Flux Linkage of IPMSM

Consider flux linkage of $a-$phase winding for different rotor positions shown in Fig. 6.23. The effective air gap becomes the largest, when the flux lines cross the cavities at the right angle, $\theta = 0$. However, it reduces to a minimum at $\theta = \frac{\pi}{2}$, when the lines do not cross the cavities. The effective air gap fluctuates depending whether the PM cavities are added to the physical air gap or not.

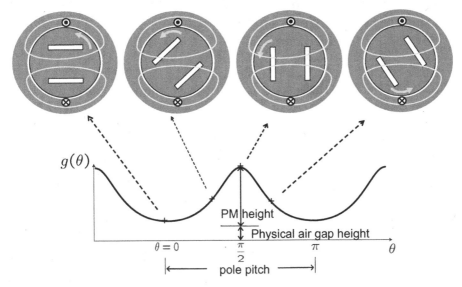

Figure 6.23: Effective air gap as a function of θ.

Similarly to the above peanut type rotor, the average air gap of IPMSM can be approximated as (6.24). The fundamental component of IPMSM inductance is also described by (6.30) and (6.31). With the PM flux linkage, the IPMSM flux linkage is equal to

$$\lambda_{abc} = (\mathbf{L}_{abc} - L_\delta \mathbf{L}_{rlc}(\theta))\mathbf{i}_{abc} + \psi(\theta). \tag{6.32}$$

Transformation of Reluctance Matrix

Let $\Delta\boldsymbol{\lambda} = [\Delta\lambda_a, \Delta\lambda_b, \Delta\lambda_c]^T \equiv \mathbf{L}_{rlc}(\theta)\mathbf{i}_{abc}$. Using $f_{dq}^s = \frac{2}{3}[f_a(t) + e^{j\frac{2\pi}{3}} f_b(t) + e^{-j\frac{2\pi}{3}} f_c(t)]$, we consider mapping $\Delta\boldsymbol{\lambda}$ into a vector in the complex plane [3]:

$$\frac{2}{3}\left[\Delta\lambda_a(t) + e^{j\frac{2\pi}{3}}\Delta\lambda_b(t) + e^{-j\frac{2\pi}{3}}\Delta\lambda_c(t)\right]$$

$$= \frac{2}{3}\frac{1}{2}L_\delta\left[(e^{j2\theta} + e^{-j2\theta})i_a + (e^{j(2\theta-\frac{2}{3}\pi)} + e^{-j(2\theta-\frac{2}{3}\pi)})i_b + (e^{j(2\theta+\frac{2}{3}\pi)}\right.$$
$$+ e^{-j(2\theta+\frac{2}{3}\pi)})i_c + e^{j\frac{2}{3}\pi}(e^{j(2\theta-\frac{2}{3}\pi)} + e^{-j(2\theta-\frac{2}{3}\pi)})i_a + e^{j\frac{2}{3}\pi}(e^{j(2\theta+\frac{2}{3}\pi)}$$
$$+ e^{-j(2\theta+\frac{2}{3}\pi)})i_b + e^{j\frac{2}{3}\pi}(e^{j2\theta} + e^{-j2\theta})i_c + e^{-j\frac{2}{3}\pi}(e^{j(2\theta+\frac{2}{3}\pi)} + e^{-j(2\theta+\frac{2}{3}\pi)})i_a$$
$$\left.+ e^{-j\frac{2}{3}\pi}(e^{j2\theta} + e^{-j2\theta})i_b + e^{-j\frac{2}{3}\pi}(e^{j(2\theta-\frac{2}{3}\pi)} + e^{-j(2\theta+\frac{2}{3}\pi)})i_c\right]$$

$$= \frac{1}{2}L_\delta\frac{2}{3}\left[3e^{j2\theta}i_a + 3e^{j(2\theta-\frac{2}{3}\pi)}i_b + 3e^{j(2\theta+\frac{2}{3}\pi)}i_c\right]$$

$$= \frac{3}{2}L_\delta e^{j2\theta}\frac{2}{3}\left[i_a + e^{-j\frac{2}{3}\pi}i_b + e^{j\frac{2}{3}\pi}i_c\right]$$

$$= \frac{3}{2}L_\delta e^{j2\theta}(\mathbf{i}_{dq}^s)^*. \tag{6.33}$$

where $(\mathbf{i}_{dq}^s)^*$ is the complex conjugate of $\mathbf{i}_{dq}^s$. Note that the computation result is remarkably simple. It follows from (6.32) and (6.33) that the stator flux of IPMSM is described in the stationary dq coordinate such that

$$\lambda_{dq}^s = L_s \mathbf{i}_{dq}^s - \frac{3}{2}L_\delta e^{j2\theta}(\mathbf{i}_{dq}^s)^* + \psi_m e^{j\theta}, \tag{6.34}$$

where $L_s = \frac{3}{2}L_{ms} + L_{ls}$. Based on (6.33), we have the inductance matrix in the stationary dq frame:

$$\mathbf{T}(0)\left[\mathbf{L}_{abcs} - \mathbf{L}_{rlc}(\theta)\right]\mathbf{T}^{-1}(0) = \mathbf{T}(0)\mathbf{L}_{abcs}\mathbf{T}^{-1}(0) - \mathbf{T}(0)\mathbf{L}_{rlc}(\theta)\mathbf{T}^{-1}(0)$$
$$= \begin{bmatrix} L_s - \frac{3}{2}L_\delta\cos 2\theta & -\frac{3}{2}L_\delta\sin 2\theta & 0 \\ -\frac{3}{2}L_\delta\sin 2\theta & L_s + \frac{3}{2}L_\delta\cos 2\theta & 0 \\ 0 & 0 & L_{ls} \end{bmatrix}.$$

Therefore, we have the following in the matrix form:

$$\begin{bmatrix} \lambda_d^s \\ \lambda_q^s \end{bmatrix} = \begin{bmatrix} L_s - \frac{3}{2}L_\delta\cos 2\theta & -\frac{3}{2}L_\delta\sin 2\theta \\ -\frac{3}{2}L_\delta\sin 2\theta & L_s + \frac{3}{2}L_\delta\cos 2\theta \end{bmatrix}\begin{bmatrix} i_d^s \\ i_q^s \end{bmatrix} + \psi_m\begin{bmatrix} \cos\theta \\ \sin\theta \end{bmatrix}. \tag{6.35}$$

By comparing (6.35) with (6.16), we can see clearly that $\frac{3}{2}L_\delta\cos 2\theta$ and $\frac{3}{2}L_\delta\sin 2\theta$ are originated from the rotor saliency.

IPMSM Dynamics in Stationary Frame

Using (6.34), the stationary IPMSM dynamic model is obtained as

$$
\begin{aligned}
\mathbf{v}_{dq}^s &= r_s \mathbf{i}_{dq}^s + \frac{d}{dt} \boldsymbol{\lambda}_{dq}^s \\
&= r_s \mathbf{i}_{dq}^s + \frac{d}{dt}(L_s \mathbf{i}_{dq}^s - \frac{3}{2} L_\delta e^{j2\theta} \mathbf{i}_{dq}^{s*}) + j\psi_m e^{j\theta}.
\end{aligned}
\tag{6.36}
$$

Rewriting (6.36) in the matrix form, it follows that

$$
\begin{aligned}
\begin{bmatrix} v_d^s \\ v_q^s \end{bmatrix} &= r_s \begin{bmatrix} i_d^s \\ i_q^s \end{bmatrix} + \begin{bmatrix} L_s - \frac{3}{2} L_\delta \cos 2\theta & -\frac{3}{2} L_\delta \sin 2\theta \\ -\frac{3}{2} L_\delta \sin 2\theta & L_s + \frac{3}{2} L_\delta \cos 2\theta \end{bmatrix} \frac{d}{dt} \begin{bmatrix} i_d^s \\ i_q^s \end{bmatrix} \\
&\quad - 3\omega L_\delta \begin{bmatrix} -\sin 2\theta & \cos 2\theta \\ \cos 2\theta & \sin 2\theta \end{bmatrix} \begin{bmatrix} i_d^s \\ i_q^s \end{bmatrix} + \omega\psi_m \begin{bmatrix} -\sin \theta \\ \cos \theta \end{bmatrix}.
\end{aligned}
\tag{6.37}
$$

This stationary model is useful in developing a high frequency model for a signal injection based sensorless algorithm.

IPMSM Dynamics in Synchronous Reference Frame

Further transforming flux (6.34) into the synchronous reference frame, we obtain

$$
\begin{aligned}
\boldsymbol{\lambda}_{dq}^e &= L_s e^{-j\theta} \mathbf{i}_{dq}^s - \frac{3}{2} L_\delta e^{j\theta} (\mathbf{i}_{dq}^s)^* + \psi_m \\
&= L_s \mathbf{i}_{dq}^e - \frac{3}{2} L_\delta (\mathbf{i}_{dq}^e)^* + \psi_m,
\end{aligned}
\tag{6.38}
$$

Letting

$$
L_d = L_s - \frac{3}{2} L_\delta
\tag{6.39}
$$

$$
L_q = L_s + \frac{3}{2} L_\delta,
\tag{6.40}
$$

we obtain

$$
\lambda_d^e = L_d i_d^e + \psi_m,
\tag{6.41}
$$

$$
\lambda_q^e = L_q i_q^e.
\tag{6.42}
$$

Note that the terms with 2θ in (6.35) disappear completely in the synchronous frame, leaving an asymmetry between L_d and L_q. Note that

$$
\mathbf{v}_{dq}^e = r_s \mathbf{i}_{dq}^e + \mathrm{p}\,\boldsymbol{\lambda}_{dq}^e + j\omega\boldsymbol{\lambda}_{dq}^e,
\tag{6.43}
$$

where

$$
j\omega\boldsymbol{\lambda}_{dq}^e = -\omega L_q i_q^e + j\omega(L_d i_d^e + \psi_m),
\tag{6.44}
$$

$$
\mathrm{p}\boldsymbol{\lambda}_{dq}^e = L_d\,\mathrm{p}i_d^e + jL_q\,\mathrm{p}\,i_q^e.
\tag{6.45}
$$

Equivalently, the voltage equation is

$$v_d^e = r_s i_d^e + L_d \frac{di_d^e}{dt} - \omega L_q i_q^e \tag{6.46}$$

$$v_q^e = r_s i_q^e + L_q \frac{di_q^e}{dt} + \omega L_d i_d^e + \omega \psi_m , \tag{6.47}$$

or

$$\begin{bmatrix} v_d^e \\ v_q^e \end{bmatrix} = \begin{bmatrix} r_s + \mathrm{p}L_d & -\omega L_q \\ \omega L_d & r_s + \mathrm{p}L_q \end{bmatrix} \begin{bmatrix} i_d^e \\ i_q^e \end{bmatrix} + \omega \begin{bmatrix} \psi_m \\ 0 \end{bmatrix} . \tag{6.48}$$

Note again that coupling terms, $-\omega L_q i_q^e$ and $\omega L_d i_d^e$, are originated from rotating the coordinate and they make an interference between d and q dynamics. In the ODE form, (6.46) and (6.47) are written as

$$\frac{d}{dt} \begin{bmatrix} i_d^e \\ i_q^e \end{bmatrix} = \begin{bmatrix} -\frac{r_s}{L_d} & \omega \frac{L_q}{L_d} \\ -\omega \frac{L_d}{L_q} & -\frac{r_s}{L_q} \end{bmatrix} \begin{bmatrix} i_d^e \\ i_q^e \end{bmatrix} - \frac{\omega \psi_m}{L_q} \begin{bmatrix} 0 \\ 1 \end{bmatrix} + \begin{bmatrix} \frac{1}{L_d} v_d^e \\ \frac{1}{L_q} v_q^e \end{bmatrix} . \tag{6.49}$$

Note that $L_\delta = \frac{1}{3}(L_q - L_d)$ causes the rotor magnetic saliency, and $L_\delta/L_{ms} = \gamma_2/\gamma_0$. If $L_\delta = 0$, then $L_d = L_q$ and (6.49) turns out to be the same as the SPMSM dynamics, (6.23).

Matrix Formalism

The same voltage equation is derived through matrix formalism. Recall that the transformation into the reference frame is achieved by multiplying by $e^{-\mathbf{J}\theta}$. For example, $\mathbf{v}_{dq}^e = e^{-\mathbf{J}\theta} \mathbf{v}_{dq}^s$. Recall from (4.38) that $e^{\mathbf{J}\theta} \frac{d}{dt}(e^{-\mathbf{J}\theta}) = -\omega \mathbf{J}$. With the use of (6.34), it follows that

$$e^{-\mathbf{J}\theta} \mathbf{v}_{dq}^s = r_s e^{-\mathbf{J}\theta} \mathbf{i}_{dq}^s + e^{-\mathbf{J}\theta} \frac{d}{dt} \left(e^{\mathbf{J}\theta} e^{-\mathbf{J}\theta} \boldsymbol{\lambda}_{dq}^s \right),$$

$$\mathbf{v}_{dq}^e = r_s \mathbf{i}_{dq}^e + e^{-\mathbf{J}\theta} \frac{d}{dt} \left(e^{\mathbf{J}\theta} \boldsymbol{\lambda}_{dq}^e \right),$$

$$= r_s \mathbf{i}_{dq}^e + e^{-\mathbf{J}\theta} \frac{d}{dt} \left(e^{\mathbf{J}\theta} \right) \boldsymbol{\lambda}_{dq}^e + \frac{d}{dt} \boldsymbol{\lambda}_{dq}^e,$$

$$= r_s \mathbf{i}_{dq}^e + \omega \mathbf{J} \boldsymbol{\lambda}_{dq}^e + \begin{bmatrix} L_d \frac{di_d^e}{dt} \\ L_q \frac{di_q^e}{dt} \end{bmatrix},$$

$$= r_s \mathbf{i}_{dq}^e + \omega \begin{bmatrix} -L_q i_q^e \\ L_d i_d^e + \psi_m \end{bmatrix} + \begin{bmatrix} L_d \frac{di_d^e}{dt} \\ L_q \frac{di_q^e}{dt} \end{bmatrix} . \tag{6.50}$$

Summary

1. In the stationary dq frame, the inductance matrix appears as

$$\begin{bmatrix} L_s - \frac{3}{2} L_\delta \cos 2\theta & -\frac{3}{2} L_\delta \sin 2\theta \\ -\frac{3}{2} L_\delta \sin 2\theta & L_s + \frac{3}{2} L_\delta \cos 2\theta \end{bmatrix} .$$

At the same time, the voltage term associated with the rotor motion is

$$-3\omega L_\delta \begin{bmatrix} -\sin 2\theta & \cos 2\theta \\ \cos 2\theta & \sin 2\theta \end{bmatrix} \begin{bmatrix} i_d^s \\ i_q^s \end{bmatrix} .$$

It should be noted that all the sine and cosine are functions of 2θ. It refers to the fact that the saliency does not discriminate the polarity.

2. In the synchronous frame, the complex equations are changed into a remarkably simple form. In the end, the IPMSM voltage equation is the same as that of SPMSM in the synchronous frame, except the fact that $L_d \neq L_q$.

Exercise 6.6   Consider a two pole IPMSM. Assume that the permanent magnets are unmagnetized. The phase flux is calculated via FEM for different rotor angles when a constant DC current, $i^s_{abc} = [100, -50, -50]^T$ A is injected. The following results are obtained:

$$
\begin{aligned}
[\lambda_a, \lambda_b, \lambda_c]^T &= [0.029, -0.0145, -0.0145]^T & \text{for } \theta = 0, \\
&= [0.0305, -0.0175, -0.0130]^T & \text{for } \theta = \frac{\pi}{6}, \\
&= [0.0350, -0.0175, -0.0170]^T & \text{for } \theta = \frac{\pi}{2}.
\end{aligned}
$$

a) Find L_{ms}, L_{lk}, and L_δ.
b) Find L_d and L_q.

Solution. $L_{ms} = 300\ \mu H$, $L_{ls} = 10\ \mu H$, and $L_\delta = 20\ \mu H$.   ∎

6.2.4   Multiple Saliency Effect

Other than the fundamental component, there can be a higher order saliency term. The next high order saliency would be described with functions of 4θ, i.e., double frequency of the fundamental saliency. The three phase inductance is described as [4]

$$\bar{\mathbf{L}}_{abcs} = \mathbf{L}_{abcs} - \mathbf{L}_{rlc}(\theta) - \mathbf{L}_{rl4}(4\theta),$$

where

$$\mathbf{L}_{rl4}(4\theta) = L_{\delta 4} \begin{bmatrix} \cos 4\theta & \cos(4\theta + 2\pi/3) & \cos(4\theta - 2\pi/3) \\ \cos(4\theta + 2\pi/3) & \cos(4\theta - 2\pi/3) & \cos 4\theta \\ \cos(4\theta - 2\pi/3) & \cos 4\theta & \cos(4\theta + 2\pi/3) \end{bmatrix}. \quad (6.51)$$

Note that $\mathbf{L}_{rl4}(4\theta) \neq \mathbf{L}_{rlc}(4\theta)$, i.e., the sequence is different from the fundamental saliency. In the stationary dq frame, we have

$$\mathbf{L}^s_{rl4} = \mathbf{T}(0)\mathbf{L}_{rl4}\mathbf{T}^{-1}(0) = -\frac{3}{2}L_{\delta 4} \begin{bmatrix} -\cos 4\theta & \sin 4\theta & 0 \\ \sin 4\theta & \cos 4\theta & 0 \\ 0 & 0 & 1 \end{bmatrix}. \quad (6.52)$$

Further, $\mathbf{L}^s_{rl4}$ is transformed into the synchronous frame as

$$\mathbf{L}^e_{rl4} = \mathbf{R}(\theta)\mathbf{L}^s_{rl4}\mathbf{R}^{-1}(\theta) = -\frac{3}{2}L_{\delta 4} \begin{bmatrix} -\cos 6\theta & \sin 6\theta & 0 \\ \sin 6\theta & \cos 6\theta & 0 \\ 0 & 0 & 1 \end{bmatrix}. \quad (6.53)$$

Then the high-frequency voltage model can be obtained such that

$$\begin{bmatrix} v_{dh}^e \\ v_{qh}^e \end{bmatrix} = \begin{bmatrix} L_d - \frac{3}{2}L_{\delta 4}\cos 6\theta & \frac{3}{2}L_{\delta 4}\sin 6\theta \\ \frac{3}{2}L_{\delta 4}\sin 6\theta & L_q + \frac{3}{2}L_{\delta 4}\cos 6\theta \end{bmatrix} \mathrm{p} \begin{bmatrix} i_{dh}^e \\ i_{qh}^e \end{bmatrix}. \tag{6.54}$$

Exercise 6.7   Consider an IPMSM with multiple saliency. Derive (6.52) and (6.52) from (6.51).

6.2.5   Multi-pole PMSM Dynamics and Vector Diagram

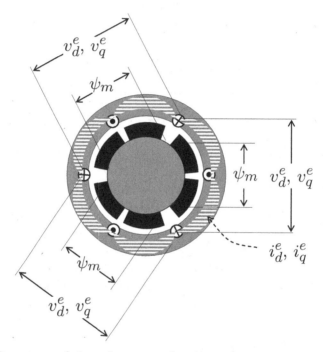

Figure 6.24: Structure of six pole motor showing a series connection of three equal sub-dynamics.

Equations (6.46) and (6.47) describe the dynamics of a two pole machine. Specifically, v_d^e and v_q^e are the voltages of a single pole pair winding, and ψ_m is the flux linkage of a single rotor pole pair. Most commonly, the P-pole system is constructed by connecting $P/2$-pole pair subsystems in series. Fig. 6.24 shows an example of six pole PMSM, in which just a-phase windings are depicted. Note that the electrical speed is equal to $\omega_e = \frac{P}{2}\omega$ in the P-pole motor, since each pole pair winding experiences $\frac{P}{2}$ periodic changes of flux linkage at each rotation.

Consider a P-pole IPMSM in which two pole pairs are repeated $P/2$-times along the air gap periphery and the unit windings are connected in series. The P-pole machine dynamics is developed on the basis of two pole dynamics. Multiplying (6.43) by $P/2$, it follows that

$$\left(\frac{P}{2}\mathbf{v}_{dq}^e\right) = \left(\frac{P}{2}r_s\right)\mathbf{i}_{dq}^e + j\left(\frac{P}{2}\omega\right)\left(\frac{P}{2}\boldsymbol{\lambda}_{dq}^e\right) + \frac{d}{dt}\left(\frac{P}{2}\boldsymbol{\lambda}_{dq}^e\right). \tag{6.55}$$

In the series connected P-pole IPMSMs, the phase voltage, resistance, and flux linkage increase by a factor of $\frac{P}{2}$, whereas the current remains the same. Let $\tilde{v}_d^e = (P/2)v_d^e$, $\tilde{v}_q^e = (P/2)v_q^e$, $\tilde{r}_s = (P/2)r_s$, $\tilde{L}_d = (P/2)L_d$, $\tilde{L}_q = (P/2)L_q$, and $\tilde{\psi}_m = (P/2)\psi_m$. Then, we have

$$\tilde{v}_d^e = \tilde{r}_s i_d^e + \tilde{L}_d \frac{di_d^e}{dt} - \omega_e \tilde{L}_q i_q^e \tag{6.56}$$

$$\tilde{v}_q^e = \tilde{r}_s i_q^e + \tilde{L}_q \frac{di_q^e}{dt} + \omega_e \tilde{L}_d i_d^e + \omega_e \tilde{\psi}_m. \tag{6.57}$$

However, for simplicity, we will abuse the notations in the following: v_d^e, v_q^e, r_s, L_d, L_q, and ψ_m will be used even in high pole machines, instead of $\tilde{v}_d^e$, $\tilde{v}_q^e$, $\tilde{r}_s$, $\tilde{L}_d$, $\tilde{L}_q$, and $\tilde{\psi}_m$. Then, the P-pole IPMSM dynamics are given as

$$v_d^e = r_s i_d^e + L_d \frac{di_d^e}{dt} - \omega_e L_q i_q^e \tag{6.58}$$

$$v_q^e = r_s i_q^e + L_q \frac{di_q^e}{dt} + \omega_e L_d i_d^e + \omega_e \psi_m. \tag{6.59}$$

That is, the same equation is used for $P-$pole motor with ω replaced by ω_e. The PMSM equations are summarized in Table 6.3.

Table 6.3: PMSM dynamic equations.

SPMSM in stationary frame
$\dfrac{d}{dt}\begin{bmatrix} i_d^s \\ i_q^s \end{bmatrix} = -\dfrac{r_s}{L_s}\begin{bmatrix} i_d^s \\ i_q^s \end{bmatrix} - \omega_e \dfrac{\psi_m}{L_s}\begin{bmatrix} -\sin\theta_e \\ \cos\theta_e \end{bmatrix} + \dfrac{1}{L_s}\begin{bmatrix} v_d^s \\ v_q^s \end{bmatrix}.$
IPMSM in stationary frame
$\begin{bmatrix} v_d^s \\ v_q^s \end{bmatrix} = r_s \begin{bmatrix} i_d^s \\ i_q^s \end{bmatrix} + \begin{bmatrix} L_s - \frac{3}{2}L_\delta \cos 2\theta_e & -\frac{3}{2}L_\delta \sin 2\theta_e \\ -\frac{3}{2}L_\delta \sin 2\theta_e & L_s + \frac{3}{2}L_\delta \cos 2\theta_e \end{bmatrix} \dfrac{d}{dt}\begin{bmatrix} i_d^s \\ i_q^s \end{bmatrix}$ $-3\omega_e L_\delta \begin{bmatrix} -\sin 2\theta_e & \cos 2\theta_e \\ \cos 2\theta_e & \sin 2\theta_e \end{bmatrix}\begin{bmatrix} i_d^s \\ i_q^s \end{bmatrix} + \omega_e \psi_m \begin{bmatrix} -\sin\theta_e \\ \cos\theta_e \end{bmatrix}$
SPMSM in synchronous frame
$\dfrac{d}{dt}\begin{bmatrix} i_d^e \\ i_q^e \end{bmatrix} = \begin{bmatrix} \frac{-r_s}{L_s} & \omega_e \\ -\omega_e & -\frac{r_s}{L_s} \end{bmatrix}\begin{bmatrix} i_d^e \\ i_q^e \end{bmatrix} - \omega_e \dfrac{\psi_m}{L_s}\begin{bmatrix} 0 \\ 1 \end{bmatrix} + \dfrac{1}{L_s}\begin{bmatrix} v_d^e \\ v_q^e \end{bmatrix}$
IPMSM in synchronous frame
$\dfrac{d}{dt}\begin{bmatrix} i_d^e \\ i_q^e \end{bmatrix} = \begin{bmatrix} -\frac{r_s}{L_d} & \omega_e \frac{L_q}{L_d} \\ -\omega_e \frac{L_d}{L_q} & -\frac{r_s}{L_q} \end{bmatrix}\begin{bmatrix} i_d^e \\ i_q^e \end{bmatrix} - \omega_e \dfrac{\psi_m}{L_q}\begin{bmatrix} 0 \\ 1 \end{bmatrix} + \begin{bmatrix} \frac{1}{L_d}v_d^e \\ \frac{1}{L_q}v_q^e \end{bmatrix}.$

Equivalent Circuit

The rotor flux linkage is equivalently expressed as a product of d-axis inductance, L_d and a virtual current, i_f, i.e.,

$$\psi_m = L_d i_f. \tag{6.60}$$

With i_f, a PMSM equivalent circuit is depicted as shown in Fig. 6.25.

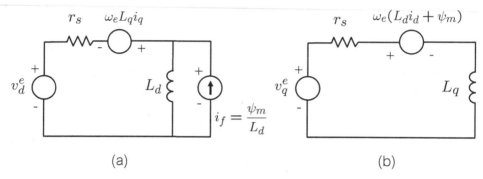

(a)   (b)

Figure 6.25: Equivalent circuit of PMSM: (a) d-axis and (b) q-axis.

Exercise 6.8   Consider a two pole IPMSM is running at 3600 rpm in the steady state. The stator coil resistance is $r_s = 0.01\ \Omega$. The operating conditions are

$$\text{phase voltage} \quad \mathbf{v}_{abc}^s = -100\left[\sin(377t + \tfrac{\pi}{3}),\ \sin(377t - \tfrac{\pi}{3}),\ \sin(377t - \pi)\right]^T,$$

$$\text{back EMF} \quad \mathbf{e}_{abc}^s = -125\left[\sin(377t),\ \sin(377t - \tfrac{2\pi}{3}),\ \sin(377t - \tfrac{4\pi}{3})\right]^T,$$

$$\text{current} \quad \mathbf{i}_{abc}^s = -50\left[\sin(377t + \tfrac{\pi}{4}),\ \sin(377t - \tfrac{5\pi}{12}),\ \sin(377t - \tfrac{13\pi}{12})\right]^T.$$

Determine L_d and L_q.

Solution. Via transformation into the stationary dq frame, it follows that

$$\mathbf{v}_{dq}^s = 100\begin{bmatrix} -\sin\left(\omega t + \tfrac{\pi}{3}\right) \\ \cos\left(\omega t + \tfrac{\pi}{3}\right) \end{bmatrix},\quad \mathbf{e}_{dq}^s = 125\begin{bmatrix} -\sin\left(\omega t\right) \\ \cos\left(\omega t\right) \end{bmatrix},\ \text{and } \mathbf{i}_{dq}^s = 50\begin{bmatrix} -\sin\left(\omega t + \tfrac{\pi}{4}\right) \\ \cos\left(\omega t + \tfrac{\pi}{4}\right) \end{bmatrix}.$$

Thus, in the synchronous reference frame,

$$\mathbf{v}_{dq}^e = 50\begin{bmatrix} -\sqrt{3} \\ 1 \end{bmatrix},\quad \mathbf{e}_{dq}^e = 125\begin{bmatrix} 0 \\ 1 \end{bmatrix},\ \text{and } \mathbf{i}_{dq}^e = 25\sqrt{2}\begin{bmatrix} -1 \\ 1 \end{bmatrix}.$$

from voltage equation in synchronous reference frame, steady-state condition and back EMF is only at q-axis

$$L_d = \frac{v_q^e - r_s i_q^e - \omega\psi_m}{\omega i_d^e} = \frac{50 - 0.01 \times 25\sqrt{2} - 125}{377 \times (-25\sqrt{2})} = 5.65\ \text{mH},$$

$$L_q = \frac{r_s i_d^e - v_d^e}{\omega i_q^e} = \frac{0.01 \times (-25\sqrt{2}) + 50\sqrt{3}}{377 \times 25\sqrt{2}} = 6.47\ \text{mH}. \qquad \blacksquare$$

6.3 PMSM Torque Equations

Using the right hand rule, torque is represented as a vector in the axial direction. With the use of orthogonal unity vectors $\overrightarrow{i}$, $\overrightarrow{j}$, and $\overrightarrow{k}$, we have

$$
\begin{aligned}
T_e &= \frac{3}{2}\frac{P}{2}\left(\boldsymbol{\lambda}^e_{dq} \times \mathbf{i}^e_{dq}\right)_k \\
&= \frac{3}{2}\frac{P}{2}\left(\begin{bmatrix}\lambda^e_{ds}\\ \lambda^e_{qs}\\ 0\end{bmatrix} \times \begin{bmatrix}i^e_d\\ i^e_q\\ 0\end{bmatrix}\right)_k \\
&= \frac{3}{2}\frac{P}{2}\begin{vmatrix}\overrightarrow{i} & \overrightarrow{j} & \overrightarrow{k}\\ \lambda^e_{ds} & \lambda^e_{qs} & 0\\ i^e_d & i^e_q & 0\end{vmatrix}_k \\
&= \frac{3}{2}\frac{P}{2}(\lambda^e_{ds}i^e_q - \lambda^e_{qs}i^e_d) \\
&= \frac{3}{2}\frac{P}{2}\left[(L_d i^e_d + \psi_m)i^e_q - L_q i^e_q i^e_d\right] \\
&= \frac{3P}{4}\left[\psi_m i^e_q - (L_q - L_d)i^e_d i^e_q\right].
\end{aligned}
\tag{6.61}
$$

Note that $\psi_m i^e_q$ is the electro-magnetic torque based on the Lorentz force, whereas $-(L_q - L_d)i^e_d i^e_q$ is the reluctance torque caused by the $L_d - L_q$ asymmetry. Note that $L_d = L_q$ in the case of SPMSM. Therefore, the reluctance torque is not utilized in SPMSM. Thereby, the PM usage per torque is larger in SPMSM than in IPMSM.

With the lossless model, $v^e_d = -\omega_e L_q i^e_q$ and $v^e_q = \omega_e L_d i^e_d + \omega_e \psi_m$, the total electric power is

$$
\begin{aligned}
P_e &= \frac{3}{2}(v^e_q i^e_q + v^e_d i^e_d) \\
&= \frac{3}{2}\omega_e(\psi_m i^e_q + (L_d - L_q)i^e_d i^e_q) \\
&= \frac{3P}{4}\omega_r(\psi_m i^e_q + (L_d - L_q)i^e_d i^e_q).
\end{aligned}
\tag{6.62}
$$

Torque is derived from the power such that $T_e = \frac{\partial P_e}{\partial \omega_r}$. Then the identical result, (6.61) follows.

Note further that the torque equation is independent of the coordinate frame.

$$
T_e = \frac{3P}{4}Im(\mathbf{i}^e_{dq} \cdot \boldsymbol{\lambda}^{e\,*}_{dq}) = \frac{3P}{4}Im(\mathbf{i}^e_{dq}e^{j\theta_e} \cdot e^{-j\theta_e}\boldsymbol{\lambda}^{e\,*}_{dq}) = \frac{3P}{4}Im(\mathbf{i}^s_{dq} \cdot \boldsymbol{\lambda}^{s\,*}_{dq}).
$$

In the stationary frame, the torque equation has the same form:

$$
T_e = \frac{3P}{4}(\lambda^s_{ds}i^s_q - \lambda^s_{qs}i^s_d),
\tag{6.63}
$$

where

$$
\begin{aligned}
\lambda^s_d &= L_s i^s_d - \frac{3}{2}L_\delta(-i^s_d \cos 2\theta_e + i^s_q \sin 2\theta_e) + \psi_m \cos\theta_e, \\
\lambda^s_q &= L_s i^s_q - \frac{3}{2}L_\delta(i^s_d \sin 2\theta_e - i^s_q \cos 2\theta_e) + \psi_m \sin\theta_e.
\end{aligned}
$$

Summary

1. The torque equation can be derived from the power relation, as the cross product of flux and current vectors. The forms are the same, independently of coordinate frame.

2. The torque consists of two parts: magnetic torque and reluctance torque. the former depends only on i_q^e, whereas the later on both i_d^e and i_q^e.

3. The SPMSMs do not have reluctance torque, since $L_d = L_q$. In case of IPMSMs, the d-axis current should have negative polarity, i.e., it is necessary to let $i_d^e < 0$ in order to utilize the reluctance torque.

Exercise 6.9   Consider a case where there are cross coupling inductances [5], L_{dq} and L_{qd} such that

$$\begin{bmatrix} \lambda_d^e \\ \lambda_q^e \end{bmatrix} = \begin{bmatrix} L_d & L_{dq} \\ L_{qd} & L_q \end{bmatrix} \begin{bmatrix} i_d^e \\ i_q^e \end{bmatrix} + \begin{bmatrix} \psi_d \\ 0 \end{bmatrix}.$$

Show that the corresponding torque equation is given as

$$T_e = \frac{3P}{4}\left(\psi_m i_q^e + (L_d - L_q)i_d^e i_q^e + L_{dq}i_q^{e2} - L_{qd}i_d^{e2}\right). \tag{6.64}$$

Exercise 6.10   Consider the same operation status of an IPMSM as written in Exercise 6.8. Show via MATLAB simulation that

$$T_e = \frac{v_a i_a + v_b i_b + v_c i_c}{\omega_r} = \frac{3}{2}\frac{v_d^e i_d^e + v_q^e i_q^e}{\omega_r} = \frac{3}{2}\frac{P}{2}(\psi_m i_q^e + (L_d - L_q)i_d^e i_q^e).$$

6.4   PMSM Block Diagram and Control

Suppose that J is the rotor inertia, and B is the friction coefficient, and T_L is a load torque. Then, the mechanical equation is given by

$$J\frac{d\omega_r}{dt} + B\omega_r = T_e - T_L. \tag{6.65}$$

Based on (6.46) and (6.47), the IPMSM dynamics can be depicted as shown in Fig. 6.26.

The typical control block diagram is shown in Fig. 6.27. To implement the current controller in the synchronous reference frame, measured currents should be transformed into the synchronous frame. For such transformation, the flux angle, θ_e is required, and the angle is obtained from the position sensor, e.g. absolute encoder or resolver. In case of resolver, resolver-to-digital converter (RDC) is required to convert the resolver signal into position and velocity data. In some home appliances

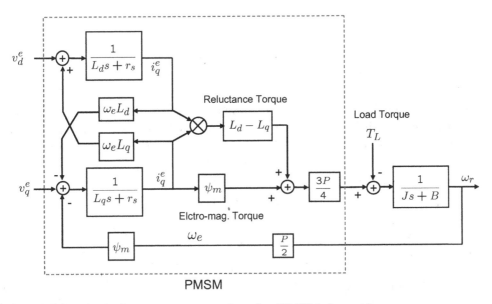

Figure 6.26: Block diagram representing the PMSM dynamics.

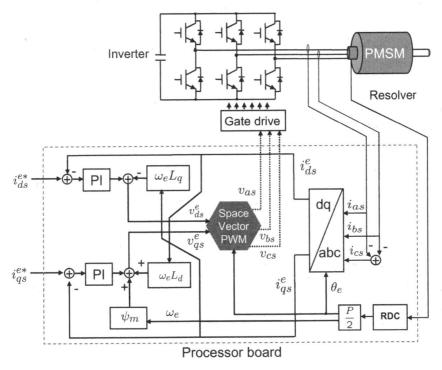

Figure 6.27: Control block diagram of PMSM using the coordinate transformation map.

where precise control is not necessary like blowers or compressors, or the system cost is critical, (speed) sensorless control is often utilized.

In general, PI controllers are used for current regulation, since the current appears DC in the synchronous frame. The current controller involves decoupling and back EMF compensation:

$$v_d^e = (PI)(i_d^{e*} - i_d^e) - \omega L_q i_q^e, \tag{6.66}$$
$$v_q^e = (PI)(i_q^{e*} - i_q^e) + \omega L_d i_d^e + \omega \psi_m. \tag{6.67}$$

The decoupling and back EMF compensations are needed in the high speed range, because they grow with speed and become dominant. For current sensing, Hall sensors are most widely utilized. Usually, two phase currents are measured by current sensors. The third one is made according to $i_c = -(i_a + i_b)$. The space vector modulation is a pulse width modulation method that converts the voltage vector (v_d^e, v_q^e) directly into switch turn-on duties.

6.4.1　MATLAB Simulation

IPMSM parameters for simulation are listed in Table 6.4, and PI gains are $K_P = 1.07$ and $K_I = 350$ for current control and $K_P = 20$ and $K_I = 40$ for speed control. Fig. 6.28 shows a speed response and the corresponding dq currents when a step load torque (212 Nm) is applied at $t = 10$ sec. Fig. 6.29 shows the phase currents when the speed changes from 500 rpm to -500 rpm.

Table 6.4: PMSM parameters for MATLAB simulation.

Number of poles	6	r_s	6.5 mΩ
L_d	0.538 mH	L_q	0.824 mH
PM flux linkage (ψ_m)	0.162 Wb		
Inertia (J)	0.1 kgm^2	Damping coeff. (B)	0

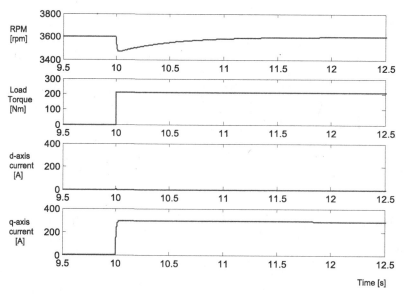

Figure 6.28: Speed and current responses to a step load applied at $t = 10$ sec.

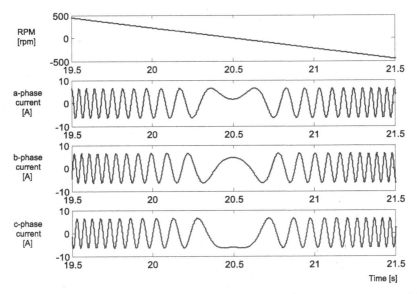

Figure 6.29: Phase currents when the speed changes from 500 rpm to -500 rpm.

Exercise 6.11   Based on the block diagram obtain abc-phase currents at zero crossing.

a) Construct a Simulink block diagram for the SPMSM model in the abc frame with the parameters: Number of poles $= 6$, $r_s = 20$ mΩ, $L_m = 0.3$ mH, $L_{ls} = 0.02$ mH, $\psi_m = 0.6$ Wb, $J = 0.1$ kgm^2, and $B = 0.0001$ Ns/m.

b) Construct a current controller and a speed controller based on Fig. 6.30. Establish current and speed controller based on the SPMSM dq model;

c) Obtain speed and current responses when the speed changes from -1000 to 1000 rpm.

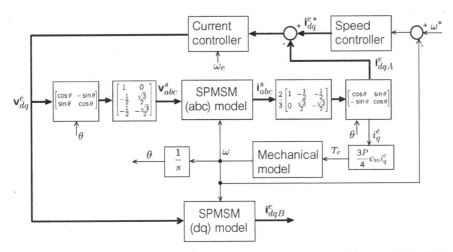

Figure 6.30: (Exercise 6.11) Speed control block diagram for the SPMSM in the abc frame.

References

[1] J. R. Hendershot and TJE Miller, *Design of Brushless Permanent-magnet Motors,* Oxford Science Publications, 1994.

[2] P. C. Krause, O. Wasynczuk, and S. D. Sudhoff, *Analysis of Electric Machinery,* IEEE Press, 1995.

[3] D. W. Novotny and T. A. Lipo, *Vector Control and Dynamics of AC Drives,* Clarendon Press, Oxford 1996.

[4] P. L. Xu and Z. Q. Zhu, " Comparison of Carrier Signal Injection Methods for Sensorless Control of PMSM Drives," Proc. of IEEE ECCE, pp. 5616−5623, 2015.

[5] I. Jeong, B.G. Gu, J. Kim, K. Nam, and Y. Kim, "Inductance Estimation of Electrically Excited Synchronous Motor via Polynomial Approximations by Least Square Method," *IEEE Trans. on Ind. Appl.,* vol. 51, no. 2, pp. 1526−1537. Mar./Apr. 2015.

[6] M. L. Woldesemayat, H. Lee, S. Won and K. Nam, "Modeling and Verification of Six-Phase Interior Permanent Magnet Synchronous Motor," *IEEE Trans. on Power Electron.,* Early Access, Dec. 2107.

[7] Z.Q. Zhu, Y. Li, D. Howe, C.M. Bingham, and D. A. Stone, "Improved rotor position estimation by signal injection in brushless AC motors, accounting for cross-coupling magnetic saturation," *IEEE Trans. Ind. Appl.,* vol. 45, No. 5, pp. 1843−1849, Sept./Oct. 2009.

Problems

6.1   Consider a two pole machine that has air gap length, g, average air gap diameter, D, and stack length, L. Suppose that the stored energy in the air gap is

$$W_{fld} = \frac{\mu_0 \pi D L}{4g} \left(\text{MMF}_s^2 + \text{MMF}_r^2 + 2\text{MMF}_s \text{MMF}_r \cos\theta \right).$$

Determine the torque, T_e.

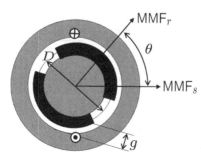

6.2   Consider a six pole IPMSM section shown below. It is assumed that the rotor
without a PM rotates when 40 A current flows through the a-phase windings,
i.e., $i_a = 40$ A and $i_b = i_c = 0$ A. The corresponding flux λ^s_{as} of a-phase
winding is plotted in the right side.

a) Assuming that the leakage inductance is equal to zero, determine L_s
and L_δ.

b) Determine L_d and L_q.

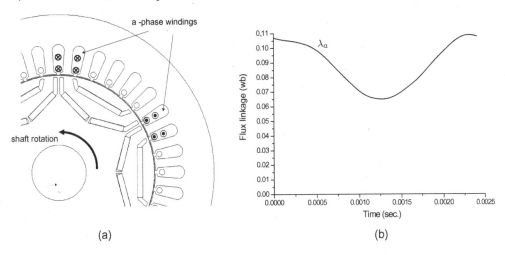

(a)   (b)

6.3   Figure below shows the back EMFs of phase windings at open terminal at
6000 rpm. The number of poles is 4, and $L_d = 1.2$ mH and $L_q = 2.3$ mH.
Suppose that the motor is running at 6000 rpm when $i^e_d = -6.6$ A and $i^e_q =$
37.5 A.

a) Estimate the back EMF constant, ψ_m.
b) Construct a voltage vector diagram for the lossless model.
c) Calculate torque.
d) Calculate power factor.

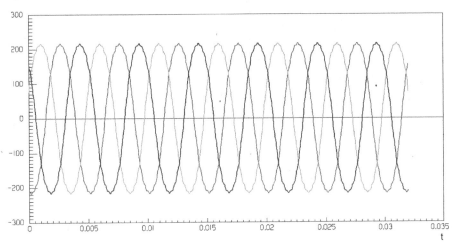

6.4 Determine whether $L_q > L_d$ or $L_d > L_q$ for the PMSMs shown in Fig. 6.12.

6.5 Consider a SPMSM depicted in (a) of the following figure. Each phase coil is concentrated in two narrow slots displaced by π. The PM coverage is θ_{PM}, and the PM field is modeled ideally as a square function as shown in (b). Let the maximum flux linkage denoted by ψ_m.

The PM flux linkage to the a-phase coil, $\psi_a(\omega t)$ is given as

$$
\psi_a(\omega t) = \begin{cases} \psi_m, & 0 \leq \omega t < \frac{\pi - \theta_{PM}}{2} \\ -\frac{2\psi_m}{\theta_{PM}}\omega t + \frac{\pi\psi_m}{\theta_{PM}}, & \frac{\pi - \theta_{PM}}{2} \leq \omega t < \frac{\theta_{PM} - \pi}{2} \\ -\psi_m, & \frac{\theta_{PM} - \pi}{2} \leq \omega t < \pi. \end{cases}
$$

(a) Show that $\psi_a(\omega t)$, $\psi_b(\omega t)$, and $\psi_c(\omega t)$ are expanded in Fourier series such that

$$
\psi_a(\omega t) = \psi'_m \sum_{\nu=1,3,5,\cdots} \frac{1}{\nu^2} \cos(\nu\omega t),
$$

$$
\psi_b(\omega t) = \psi'_m \sum_{\nu=1,3,5,\cdots} \frac{1}{\nu^2} \cos\left(\nu(\omega t - \frac{2\pi}{3})\right),
$$

$$
\psi_c(\omega t) = \psi'_m \sum_{\nu=1,3,5,\cdots} \frac{1}{\nu^2} \cos\left(\nu(\omega t - \frac{4\pi}{3})\right).
$$

where $\psi'_m = \frac{8\psi_m}{\pi\theta_{PM}}$.

(b) Transform $[\psi_a, \psi_b, \psi_c]^T$ into the stationary dq frame to obtain (ψ_d^s, ψ_q^s).

(c) Transform (ψ_d^s, ψ_q^s) further into the synchronous frame (ψ_d^e, ψ_q^e).

(d) Let $\theta_{PM} = \frac{2\pi}{3}$ and and the peak flux linkage be $\psi_m = 0.1$ Wb. Find the fundamental component of the flux linkage.

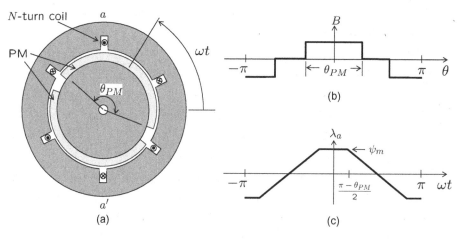

(a)   (b)   (c)

6.6 Consider an IPMSM with the following parameters: $L_s = 0.8$ mH, $L_\delta = 0.1$ mH, $r_s = 0.036$ mΩ, $\psi_m = 0.2229$ V·s. The IPMSM model is developed using the two coordinate frames as shown below: one is in the synchronous frame, the other in the stationary frame. However, the current controller is constructed in the synchronous frame with PI gains, $K_p = 30$ and $K_i = 40$, and current commands, $(i_d^{e*}, i_q^{e*}) = (-2, 10)$ A. The voltage, (v_d^s, v_q^s) of the stationary frame is obtained from (v_d^e, v_q^e) via coordinate transformation:

$$\begin{bmatrix} v_d^s \\ v_q^s \end{bmatrix} = \begin{bmatrix} \cos\theta & -\sin\theta \\ \sin\theta & \cos\theta \end{bmatrix} \begin{bmatrix} v_d^e \\ v_q^e \end{bmatrix}.$$

Let $\omega = 15$ rad/sec.

Show that mapping of (i_d^e, i_q^e) into the stationary frame yields the same vector as (i_d^s, i_q^s), i.e., the two calculation paths, ① and ② yield the same results. Utilize Rounge-Kutta fourth method with step size $h = 0.00001$ sec to solve the ODE. It is recommended to program (6.37) and (6.49) into M-file of MATLAB as follows:

```
function xx = rook(t,xy,h,w)
xytemp=xy;
kk1= ex45fc(t,xytemp,h,w);
kk2= ex45fc(t+h/2,xytemp+kk1*0.5*h,h,w);
kk3= ex45fc(t+h/2,xytemp+kk2*0.5*h,h,w);
kk4= ex45fc(t+h,xytemp+kk3*h,h,w);
xx = xytemp+h*(kk1+2*kk2+2*kk3+kk4)/6;
```

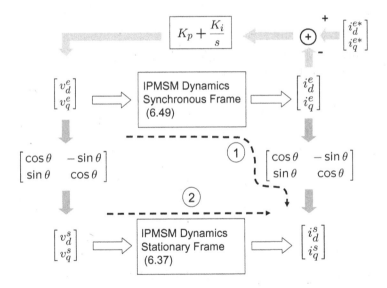

6.7 Consider a two pole IPMSM with $N = 4$, $A = 300\text{mm}^2$, $\gamma_0 = 0.01\text{mm}^{-1}$, $\gamma_2 = 0.05\text{mm}^{-1}$.

a) Calculate the L_{ms} and L_δ.

b) Using the result of a), find the L_d and L_q. Assume the $L_{ls} = 0$.

c) Suppose that the PMs are unmagnetized, and that current of magnitude, $I = \sqrt{i_d^2 + i_q^2} = 100\text{A}$ is injected. Find the maximum achievable reluctance torque.

6.8   Derive the stationary IPMSM model (6.37) from

$$\mathbf{v}_{dq}^s = r_s \mathbf{i}_{dq}^s + \frac{d}{dt}\left(L_s \mathbf{i}_{dq}^s - \frac{3}{2}L_\delta e^{j2\theta}\mathbf{i}_{dq}^{s*}\right) + j\psi_m e^{j\theta}.$$

6.9   The six phase motors considered in this paper have two sets of three phase windings which are separated by $\pi/6$ in electrical angle. The first group is named abc and the second group is xyz. The following figure shows schematic diagram of two pole six phase IPMSM. Rotor angle θ is defined by the angle between the direct axis of the rotor PM and the a phase current axis. The abc coil has flux linking with the xyz coil, as well as the rotor PM.

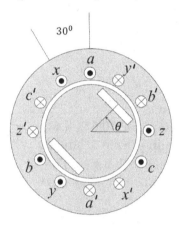

Let the current of xyz coil is denoted by $\mathbf{i}_{xyz} = \begin{bmatrix} i_x, & i_y, & i_z \end{bmatrix}^T$. Then the flux linkage of abc coil [6] is described by

$$\boldsymbol{\lambda}_{abc} = \mathbf{L}_{abcs}\mathbf{i}_{abc} + \mathbf{M}\mathbf{i}_{xyz} - L_\delta \mathbf{A}(\theta)\mathbf{i}_{abc} - L_\delta \mathbf{A}\left(\theta - \frac{\pi}{12}\right)\mathbf{i}_{xyz} + \boldsymbol{\psi}(\theta),$$

where $\boldsymbol{\psi}(\theta) = \psi_{pm}\begin{bmatrix} \cos\theta, & \cos(\theta - \frac{2\pi}{3}), & \cos(\theta + \frac{2\pi}{3}) \end{bmatrix}^T,$

$$\mathbf{A}(\theta) = \begin{bmatrix} \cos 2\theta & \cos(2\theta - \frac{2\pi}{3}) & \cos(2\theta + \frac{2\pi}{3}) \\ \cos(2\theta - \frac{2\pi}{3}) & \cos(2\theta + \frac{2\pi}{3}) & \cos 2\theta \\ \cos(2\theta + \frac{2\pi}{3}) & \cos(2\theta) & \cos(2\theta - \frac{2\pi}{3}) \end{bmatrix}.$$

(a) Find $\mathbf{M}$.

(b) Let the flux linkage of xyz coil denoted by $\boldsymbol{\lambda}_{xyz} = \begin{bmatrix} \lambda_x, & \lambda_y, & \lambda_z \end{bmatrix}^T$. Derive an expression for $\boldsymbol{\lambda}_{xyz}$ in terms of $\mathbf{i}_{xyz}$, $\mathbf{i}_{abc}$, and $\boldsymbol{\psi}(\theta - \frac{\pi}{6})$.

(c) Dual coordinates are also used in the dq stationary coordinate: (α, β) for the abc-frame, whereas (γ, σ) for the xyz-frame. Use the same transformation matrix, $\mathbf{T}$ for both abc and xyz variables, and let $\boldsymbol{\lambda}_{\gamma\sigma 0} \equiv \begin{bmatrix} \lambda_\gamma, & \lambda_\sigma, & 0 \end{bmatrix}^T \equiv \mathbf{T}\boldsymbol{\lambda}_{xyz}$ and $\mathbf{i}_{\gamma\sigma 0} \equiv \begin{bmatrix} i_\gamma, & i_\sigma, & 0 \end{bmatrix}^T = \mathbf{T}\mathbf{i}_{xyz}$. Assuming that $\mathbf{i}_{abc}$ and $\mathbf{i}_{xyz}$ are balanced, disregard the third component and let $\boldsymbol{\lambda}_{\gamma\sigma} \equiv \begin{bmatrix} \lambda_\gamma, & \lambda_\sigma \end{bmatrix}^T$ and $\mathbf{i}_{\gamma\sigma} \equiv \begin{bmatrix} i_\gamma, & i_\sigma \end{bmatrix}^T.$

Show that in the stationary rectangular coordinates

$$\boldsymbol{\lambda}_{\alpha\beta} = \mathbf{L}_g(\theta)\mathbf{i}_{\alpha\beta} + \mathbf{M}_g(\theta)\mathbf{i}_{\gamma\sigma} + \boldsymbol{\psi}_2'(\theta),$$

$$\boldsymbol{\lambda}_{\gamma\sigma} = \mathbf{L}_g(\theta - \frac{\pi}{6})\mathbf{i}_{\gamma\sigma} + \mathbf{M}_g^T(\theta)\mathbf{i}_{\alpha\beta} + \boldsymbol{\psi}_2'(\theta - \frac{\pi}{6}),$$

where

$$\mathbf{L}_g(\theta) = \begin{bmatrix} L_M - \frac{3}{2}L_\delta \cos 2\theta & -\frac{3}{2}L_\delta \sin 2\theta \\ -\frac{3}{2}L_\delta \sin 2\theta & L_M + \frac{3}{2}L_\delta \cos 2\theta \end{bmatrix} = L_M\mathbf{I} - L_\delta\mathbf{G}(\theta),$$

$$\mathbf{M}_g(\theta) = \begin{bmatrix} \frac{3\sqrt{3}L_{ms}}{4} - \frac{3}{2}L_\delta \cos(2\theta - \frac{\pi}{6}) & -\frac{3L_{ms}}{4} - \frac{3}{2}L_\delta \sin(2\theta - \frac{\pi}{6}) \\ \frac{3L_{ms}}{4} - \frac{3}{2}L_\delta \sin(2\theta - \frac{\pi}{6}) & \frac{3\sqrt{3}L_{ms}}{4} + \frac{3}{2}L_\delta \cos(2\theta - \frac{\pi}{6}) \end{bmatrix}$$

$$= \left[\mathbf{M}_s - \frac{3}{2}L_\delta\mathbf{G}\left(\theta - \frac{\pi}{12}\right) \right],$$

$$\mathbf{M}_s = \frac{3}{4}L_{ms} \begin{bmatrix} \sqrt{3} & -1 \\ 1 & \sqrt{3} \end{bmatrix} \quad \text{and} \quad \mathbf{G}(\theta) = \begin{bmatrix} \cos 2\theta & \sin 2\theta \\ \sin 2\theta & -\cos 2\theta \end{bmatrix}.$$

(d) Let $\boldsymbol{\lambda}_{dq1}^e = [\lambda_{d1}^e, \lambda_{q1}^e]^T = \mathbf{R}(\theta)\boldsymbol{\lambda}_{\alpha\beta}$, $\boldsymbol{\lambda}_{dq2}^e = [\lambda_{d2}^e, \lambda_{q2}^e]^T = \mathbf{R}(\theta)\boldsymbol{\lambda}_{\gamma\sigma}$, $\mathbf{i}_{dq1}^e = [i_{d1}^e, i_{q1}^e]^T = \mathbf{R}(\theta)\mathbf{i}_{\alpha\beta}$, $\mathbf{i}_{dq2}^e = [i_{d2}^e, i_{q2}^e]^T = \mathbf{R}(\theta)\mathbf{i}_{\gamma\sigma}$, and

$$\mathbf{R}(\theta) = \begin{bmatrix} \cos\theta & \sin\theta \\ -\sin\theta & \cos\theta \end{bmatrix}.$$

Derive the following flux linkage in the synchronous frame:

$$\boldsymbol{\lambda}_{dq1}^e = \mathbf{L}_{dq1}\mathbf{i}_{dq1}^e + \mathbf{M}_f\mathbf{i}_{dq2}^e + \boldsymbol{\psi}_{dq1}^e,$$

$$\boldsymbol{\lambda}_{dq2}^e = \mathbf{L}_{dq2}\mathbf{i}_{dq2}^e + \mathbf{M}_f^T\mathbf{i}_{dq1}^e + \boldsymbol{\psi}_{dq2}^e,$$

where $\boldsymbol{\psi}_{dq1}^e \equiv \mathbf{R}(\theta)\boldsymbol{\psi}'(\theta) = \psi_{pm}[1, 0]^T$, $\boldsymbol{\psi}_{dq2}^e \equiv \mathbf{R}(\theta)\boldsymbol{\psi}'(\theta - \frac{\pi}{6}) = \psi_{pm}[\frac{\sqrt{3}}{2}, -\frac{1}{2}]^T$, $L_M = \frac{3}{2}L_{ms} + L_{ls}$,

$$\begin{bmatrix} L_M - \frac{3}{2}L_\delta & 0 \\ 0 & L_M + \frac{3}{2}L_\delta \end{bmatrix} = \mathbf{L}_{dq1},$$

$$\begin{bmatrix} L_M - \frac{3}{4}L_\delta & \frac{3\sqrt{3}}{4}L_\delta \\ \frac{3\sqrt{3}}{4}L_\delta & L_M + \frac{3}{4}L_\delta \end{bmatrix} = \mathbf{L}_{dq2},$$

$$\frac{3}{4} \begin{bmatrix} \sqrt{3}(L_{ms} - L_\delta) & -(L_{ms} - L_\delta) \\ (L_{ms} + L_\delta) & \sqrt{3}(L_{ms} + L_\delta) \end{bmatrix} = \mathbf{M}_f.$$

6.10 Consider the PM flux linkage with the following harmonic components:

$$\boldsymbol{\lambda}_{abc} = (\mathbf{L}_{abc} - L_\delta\mathbf{L}_{rlc}(2\theta))\mathbf{i}_{abc} + \boldsymbol{\psi}_{PM}$$

where

$$\boldsymbol{\psi}_{PM} =$$
$$\psi_m \begin{bmatrix} \cos\theta \\ \cos(\theta - 2\pi/3) \\ \cos(\theta + 2\pi/3) \end{bmatrix} + \psi_{5m} \begin{bmatrix} \cos 5\theta \\ \cos 5(\theta - 2\pi/3) \\ \cos 5(\theta + 2\pi/3) \end{bmatrix} + \psi_{7m} \begin{bmatrix} \cos 7\theta \\ \cos 7(\theta - 2\pi/3) \\ \cos 7(\theta + 2\pi/3) \end{bmatrix}.$$

a) Using the transformation matrix $\mathbf{T}(0)$, derive the following expression in the stationary dq frame:

$$\begin{bmatrix} \lambda_d^s \\ \lambda_q^s \end{bmatrix} = \begin{bmatrix} L_s - \frac{3}{2}L_\delta \cos 2\theta & -\frac{3}{2}L_\delta \sin 2\theta \\ -\frac{3}{2}L_\delta \sin 2\theta & L_s + \frac{3}{2}L_\delta \cos 2\theta \end{bmatrix} \begin{bmatrix} i_d^s \\ i_q^s \end{bmatrix}$$

$$+\psi_m \begin{bmatrix} \cos\theta \\ \sin\theta \end{bmatrix} + \psi_{5m} \begin{bmatrix} \cos 5\theta \\ -\sin 5\theta \end{bmatrix} + \psi_{7m} \begin{bmatrix} \cos 7\theta \\ \sin 7\theta \end{bmatrix}.$$

b) Using the rotational transformation, $\mathbf{R}(\theta)$ derive the following equation:

$$\begin{bmatrix} \lambda_d^e \\ \lambda_q^e \end{bmatrix} = \begin{bmatrix} L_d & 0 \\ 0 & L_q \end{bmatrix} \begin{bmatrix} i_d^e \\ i_q^e \end{bmatrix} + \begin{bmatrix} \psi_m + \psi_{5m}\cos 6\theta + \psi_{7m}\cos 6\theta \\ -\psi_{5m}\sin 6\theta + \psi_{7m}\sin 6\theta \end{bmatrix}.$$

c) Using the previous result, obtain the torque equation below:

$$T_e = \frac{3}{2}\frac{P}{2}\left[\psi_m i_q^e + (L_d - L_q)i_d^e i_q^e + (\psi_{5m} + \psi_{7m})\cos 6\theta\, i_q^e + (\psi_{5m} - \psi_{7m})\sin 6\theta\, i_d^e\right].$$

d) Assume that $L_d = 0.4$ mH, $L_q = 0.8$ mH, $P = 4$, $\psi_m = 0.1$ Wb , $\psi_{5m} = 0.05$ Wb , $\psi_{7m} = 0.02$ Wb. Let $[i_d^e, i_q^e]^T = [-10, 50]$ A. Using Excel, plot the torque over one revolution period.

6.11 Assume that an IPMSM has a cross coupling inductance [8] which appears in the $abc-$frame as

$$\mathbf{L}_{cm}(2\theta) = \frac{2}{3}L_{dq} \begin{bmatrix} \sin 2\theta & \sin(2\theta - 2\pi/3) & \sin(2\theta + 2\pi/3) \\ \sin(2\theta - 2\pi/3) & \sin(2\theta + 2\pi/3) & \sin 2\theta \\ \sin(2\theta + 2\pi/3) & \sin 2\theta & \sin(2\theta - 2\pi/3) \end{bmatrix}.$$

a) Show that in the stationary frame

$$\mathbf{L}_{cm}^s = \mathbf{T}(0)\mathbf{L}_{cm}(2\theta)\mathbf{T}^{-1}(0) = -L_{dq} \begin{bmatrix} -\sin 2\theta & \cos 2\theta \\ \cos 2\theta & \sin 2\theta \end{bmatrix}.$$

b) Show that in the synchronous frame

$$\mathbf{L}_{cm}^e = \mathbf{R}(\theta)\mathbf{L}_{cm}^s\mathbf{R}^{-1}(\theta) = -\begin{bmatrix} 0 & L_{dq} \\ L_{dq} & 0 \end{bmatrix}.$$

c) The cross coupling inductance, $\mathbf{L}_{cm}(\theta)$ can be added to (6.30). Then, it follows that

$$\bar{\mathbf{L}}_{abcs} = \mathbf{L}_{abcs} - \mathbf{L}_{rlc}(\theta) - \mathbf{L}_{cm}(\theta).$$

Using the result of b), show the voltage equation is obtained in the synchronous frame such that

$$\begin{bmatrix} v_d^e \\ v_q^e \end{bmatrix} = \begin{bmatrix} r_s + \mathrm{p}\,L_d - \omega L_{dq} & -\omega L_q + \mathrm{p}\,L_{dq} \\ \omega L_d + \mathrm{p}\,L_{dq} & r_s + \mathrm{p}\,L_q + \omega L_{dq} \end{bmatrix} \begin{bmatrix} i_d^e \\ i_q^e \end{bmatrix} + \begin{bmatrix} 0 \\ \omega_e \psi_m \end{bmatrix}.$$

6.12 A multiple rotor saliency can be described by [4]

$$\bar{\mathbf{L}}_{abcs} = \mathbf{L}_{abcs} - \mathbf{L}_{rlc}(\theta) - \mathbf{L}_{mu}(\theta),$$

where

$$\mathbf{L}_{mu}(\theta) = \frac{2}{3}L_{\delta 4} \begin{bmatrix} \cos 4\theta & \cos(4\theta + 2\pi/3) & \cos(4\theta - 2\pi/3) \\ \cos(4\theta + 2\pi/3) & \cos(4\theta - 2\pi/3) & \cos 4\theta \\ \cos(4\theta - 2\pi/3) & \cos 4\theta & \cos(4\theta + 2\pi/3) \end{bmatrix}, \cdot$$

a) Show that in stationary frame

$$\mathbf{L}_{mu}^s = \mathbf{T}(0)\mathbf{L}_{mu}\mathbf{T}^{-1}(0) = L_{\delta 4} \begin{bmatrix} -\cos 4\theta & \sin 4\theta \\ \sin 4\theta & \cos 4\theta \end{bmatrix}.$$

b) Show that in synchronous frame

$$\mathbf{L}_{mu}^e = \mathbf{R}(\theta)\mathbf{L}_{mu}^s\mathbf{R}^{-1}(\theta) = L_{\delta 4} \begin{bmatrix} -\cos 6\theta & \sin 6\theta \\ \sin 6\theta & \cos 6\theta \end{bmatrix}.$$

Chapter 7

PMSM Control Methods

In the IPMSM, torque also depends on the d-axis current. Therefore, like the induction motor, there are many ways of combining dq currents for a constant torque production. But, the loss is different for each current set, (i_d^e, i_q^e). The maximum torque per ampere (MTPA) is a current minimizing control technique for a given torque, thus it is a copper loss minimizing control.

When the motor speed increases, the back-EMF approaches the maximum available terminal voltage from the inverter. To extend the speed range, negative d-axis current is applied to weaken the air gap flux. Otherwise, the torque would drop rapidly due to the voltage limit. In other words, field weakening is a high speed solution within the voltage boundary.

It is necessary to extend the constant power speed range (CPSR) for high speed applications. This chapter discusses various control methods to ensure high performance over a wide speed range under voltage and current limits.

7.1 Machine Sizing

Consider a single pole section of a P-pole PM motor shown in Fig. 7.1, where D_r is the rotor diameter, $\tau_p (= \pi D_r / P)$ is the pole pitch and l_{st} is the stack length. Assume that the rotor PM yields a sinusoidal field distribution in the air gap, and that the rotor speed is ω_r. Then air gap field density by the PM is modeled as a traveling wave, $B_m \sin\left(\frac{\pi}{\tau_p}x - \frac{P}{2}\omega_r t\right)$.

Let N_{ph} be the number of coil turns per phase and k_w be the winding factor. As shown in Fig. 7.2, the average level of $\sin x$ over π is equal to $\frac{2}{\pi}$. Note that PM flux $\frac{2}{\pi}B_m \cos\frac{P}{2}\omega_r t$ passes through $\frac{N_{ph}}{P/2}k_w$-turn coil over the area, $l_{st}\tau_p$. Therefore, the PM flux linkage to the stator coil is

$$\lambda_{pole} = \frac{N_{ph}}{P/2}k_w l_{st} \int_0^{\tau_p} B_m \sin\left(\frac{\pi}{\tau_p}x - \frac{P}{2}\omega_r t\right) dx \tag{7.1}$$

$$= \frac{N_{ph}}{P/2}k_w l_{st}\tau_p \frac{2}{\pi}B_m \cos\frac{P}{2}\omega_r t. \tag{7.2}$$

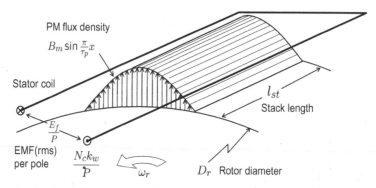

Figure 7.1: A single pole section of a PM synchronous motor.

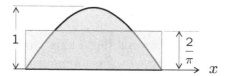

Figure 7.2: Average of a half sine wave.

Since (7.2) is the flux linkage of a phase winding per pole, the phase voltage follows from the Lenz law such that

$$e_f = -\frac{d}{dt}\frac{P}{2}\lambda_{pole} = N_{ph}k_w\Phi_f\frac{P}{2}\omega_r\sin\omega_r t, \tag{7.3}$$

where $\Phi_f = l_{st}\tau_p\frac{2}{\pi}B_m$. Therefore, the rms value of the induced voltage is equal to

$$E_f = \frac{1}{\sqrt{2}}2\pi f_e \cdot N_{ph}k_w\Phi_f = 4.44f_e N_{ph}k_w\Phi_f, \tag{7.4}$$

where $f_e \equiv \frac{P}{2}\frac{\omega_r}{2\pi}$ is the electrical frequency.

The electric loading is defined as the ampere-turn per circumferential length of the stator bore:

$$A_m = \frac{2mN_{ph}I_{max}}{P\tau_p} = \frac{2mN_{ph}\sqrt{2}I_{rms}}{\pi D_r}, \tag{7.5}$$

where m is the number of phases and I_{rms} is the rms phase current. Note that $2mN_{ph}$ is the total number of conductors in the all slots. The electric loading is a description of current sheet along the air gap periphery, and it is limited by the ability of heat dissipation from the conductor bundles. In the following sizing equation, the electric loading is utilized instead of current.

Apparent electromagnetic power that crosses air gap is equal to the product of induced voltage and current:

$$
\begin{aligned}
P_{gap} = mE_f I_{rms} &= m\sqrt{2}\pi f_e N_{ph}k_w\Phi_f\frac{A_m\pi D_r}{2\sqrt{2}mN_{ph}} \\
&= \frac{\pi}{2}k_w\frac{f_e}{P/2}(P\tau_p)l_{st}B_m A_m D_r \\
&= \frac{\pi}{4}k_w D_r^2 l_{st}B_m A_m\omega_r . \tag{7.6}
\end{aligned}
$$

Table 7.1: Magnetic and electric loadings.

Machine type	Magnetic loading B_{max}(T)	RMS electric loading $A_m/\sqrt{2}$(A/cm)
IM (air cooling, $\sim$ 10 kW)	0.6$\sim$1	100$\sim$ 350
IM (air cooling, 10 $\sim$ 1000 kW)	0.6$\sim$1	350$\sim$ 550
PMSM (air cooling, $\sim$ 10 kW)	0.8	150
PMSM (water cooling, 10 $\sim$ 100 kW)	0.8	$\sim$ 700

In deriving the third equality, $\pi D_r = P\tau_p$ and $f_e/(P/2) = \omega_r/(2\pi)$ are utilized. It is assumed here that all the air gap power is converted into mechanical power. Since torque is obtained as $T_e = P_{gap}/\omega_r$, it follows from (7.6) that

$$T_e = k_w B_m A_m \text{Vol}_{(rotor)}, \tag{7.7}$$

where $\text{Vol}_{(rotor)} = \frac{\pi}{4}D_r^2 l_{st}$ denotes the rotor volume as depicted in Fig. 7.3. Here, B_m is called the magnet loading. Torque appears as the product of electric loading, magnet loading, and the rotor volume. It is an important motor design equation. In order to maximize the torque per volume, the maximum electric and magnet loadings must be utilized. Therefore, B_m and A_m are almost fixed once the motor type and cooling method are determined. Then, the remaining factor is rotor volume. Then, torque is proportional to the rotor volume as depicted graphically in Fig. 7.3.

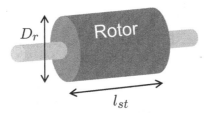

Figure 7.3: Torque is proportional to the rotor volume.

Ranges of electric and magnetic loadings are $A_m/\sqrt{2}$: 150 $\sim$ 2000 A/cm and B_{max} : 0.6 $\sim$ 1 T. High electric loading leads to a large heat generation in the coil. If the cooling method is more efficient, a higher electric loading can be used. For example, $A_m/\sqrt{2} = 150$ A/cm for a totally enclosed PM servo motor without a fan. But with water cooling, it can be increased to 700 A/cm. Typical ranges of magnetic and electric loadings are listed in Table 7.1.

7.1.1 Machine Sizes under Same Power Rating

Since $P_{gap} = T_e\omega_r$, there are two ways of increasing the motor power: to increase torque or to increase speed. For the same power, the rated torque decreases as the rated speed increases. Since the motor volume is proportional to torque, the motor volume reduces as the rated speed increases. That is, a high speed motor has a smaller volume.

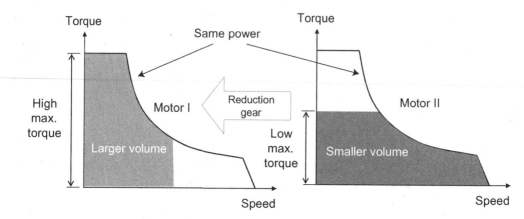

Figure 7.4: Size comparison for the same power.

Fig. 7.4 shows two motor operation regions marked by shaded areas. Note that the two parabola curves, which represent a constant power, are identical. That is, Motor I and Motor II have the same power levels. There is a trade-off between maximum torque and maximum speed. Motor I has a high maximum (starting) torque but a low maximum speed, whereas Motor II has a lower starting torque but a higher maximum speed. Based on (7.7), Motor I obviously has a larger volume than Motor II.

Note, however, that the high speed motor can be fit into a desired operation range with a reduction gear. Hence, a high speed motor along with a reduction gear could offer a reduced volume. For this lower volume, there is a tendency to utilize high speed motors in electric vehicles (EVs) [1].

7.2   Current Voltage and Speed Limits

To control the motor current, the inverter voltage should be larger than the motor terminal voltage. On the other hand, the DC link voltage is limited. The steady state model for PMSM dynamics is given as

$$v_d^e = r_s i_d^e - \omega_e L_q i_q^e \tag{7.8}$$
$$v_q^e = r_s i_q^e + \omega_e L_d i_d^e + \omega_e \psi_m. \tag{7.9}$$

Based on (7.8) and (7.9), the steady state current and voltage vectors are depicted as shown in Fig. 7.5. Two cases are compared during motoring when $i_d^e < 0$ and $i_d^e > 0$. It is observed from Fig. 7.5 (a) that the negative d-axis current reduces the motor terminal voltage since $\omega_e L_d i_d^e < 0$ is laid on the q-axis. This coupling voltage cancels out the back EMF, $\omega \psi_m$ which grows with the speed. Thereby, it extends the speed region.

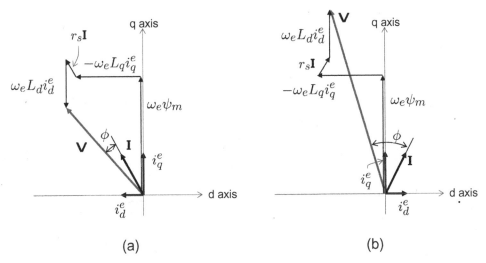

Figure 7.5: Voltage magnitude and power factor depending on i_d^e polarities during motoring: (a) $i_d^e < 0$ and (b) $i_d^e > 0$.

On the other hand, positive d-axis current increases the voltage vector. The terminal voltage cannot be larger than the inverter limit. Comparing the two, one can find a great difference in power factor. In Fig. 7.5 (a) both voltage and current vectors are in the second quadrant, and the angle difference, ϕ is relatively small. If $i_d^e > 0$, the current and voltage vectors are in different quadrants as shown in Fig. 7.5 (b). Thereby, the power factor is small.

In summary, negative d-axis current increases the power factor while limiting the voltage growth. Note that the reactive current just increases the loss while not contributing to torque generation. Therefore, a high power factor means high motor efficiency.

The benefit of field weakening control can be seen from the perspective of torque production. Here it is assumed that $L_d < L_q$. Then, a negative d-axis current gives a positive reluctance torque, i.e., it yields $\frac{3}{4}P(L_d - L_q)i_d^e i_q^e > 0$. Thus, more torque is added to the magnetic torque. In some well designed IPMSMs, the former is as large as the latter component in some high speed range.

Finally, the ohmic voltage drop, $r_s\mathbf{I}_s$ is parallel to the current vector. Thereby it increases the terminal voltage. Since it is normally small, it is often omitted for simplicity.

Fig. 7.6 shows similar vector diagrams in case of regeneration. If i_q^e is negative, then the shaft torque is negative. It absorbs external mechanical power and converts it into electrical power, i.e., it works as a generator. The two cases are compared when $i_d^e > 0$ and $i_d^e < 0$. The same result is observed with regard to the magnitude of the terminal voltage. The terminal voltage is smaller with $i_d^e < 0$. In case of regeneration, it is desired to achieve -1 as the power factor, i.e., PF $= -1$ or $\phi \approx 180°$. In this perspective, the case with $i_d^e < 0$ is better. Unlike motoring, the ohmic voltage drop helps to reduce the terminal voltage.

The benefits of the field weakening are summarized below:

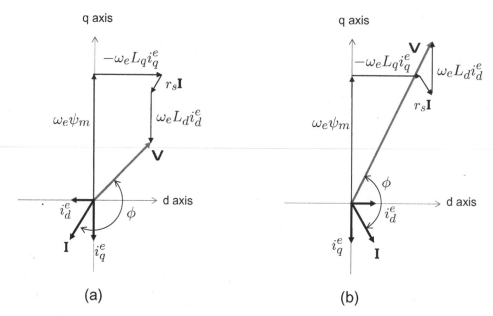

Figure 7.6: Voltage magnitude and power factor depending on i_d^e polarities during regeneration: (a) $i_d^e < 0$ and (b) $i_d^e > 0$.

Summary

1. To increase the motor speed under a voltage limit, it is necessary to use negative d-axis current. It helps to reduce the motor terminal voltage.
2. High power factor is achieved if $i_d^e < 0$.
3. When $i_d^e < 0$, the reluctance torque is added in the case that $L_d < L_q$.
4. The use of the reluctance torque is related to the increase in the power factor. High power factor is a desirable attribute for high efficiency.
5. The same goes for the regeneration. Therefore, a negative d-axis current is also necessary during generation mode.

7.2.1   Torque versus Current Angle

Fig. 7.7 shows a vector diagram of a lossless model. Note that the current angle β is measured from the q-axis by convention. With the polar description, it follows that

$$i_d^e = -I \sin \beta, \tag{7.10}$$
$$i_q^e = I \cos \beta, \tag{7.11}$$

where $I = \sqrt{i_d^{e\,2} + i_q^{e\,2}}$ and $\beta = \tan^{-1}(-i_d^e/i_q^e)$.

Substituting (7.10) and (7.11) into the torque equation, we have

$$T_e = \frac{3P}{4}\left[\psi_m I \cos \beta + \frac{1}{2}(L_q - L_d)I^2 \sin 2\beta\right]. \tag{7.12}$$

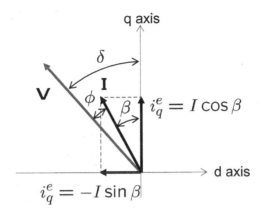

Figure 7.7: Current angle, β and power factor angle, ϕ.

Note further that the magnetic torque, $\frac{3P}{4}\psi_m I \cos\beta$ is a function of $\cos\beta$, while the reluctance torque, $\frac{3P}{4}\frac{1}{2}(L_q - L_d)I^2 \sin 2\beta$ is a function of $\sin 2\beta$. In order to make the reluctance torque positive when $L_q > L_d$, current angle should be larger than zero, i.e., $\beta > 0$. It means that the d-axis current needs to be negative. Fig. 7.8 shows torque versus current angle when $L_q > L_d$. The relative magnitude of reluctance torque is not small for a medium β value. Note that the peak torque is achieved around $\beta = 40°$.

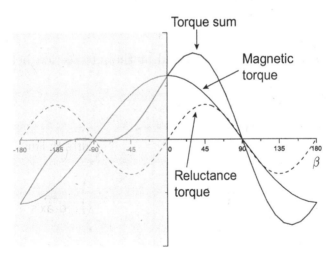

Figure 7.8: Torque against the current angle. The torque components versus current angle for $L_d < L_q$.

7.3 Extending Constant Power Speed Range

Let the magnitude of maximum voltage denoted by V_m. Then the voltage limit is described by

$$v_d^{e\,2} + v_q^{e\,2} \le V_m^2. \tag{7.13}$$

The rotor flux was described with a virtual current source i_f i.e., $\psi_m = L_d i_f$. It mimics the wound field synchronous motor. A simplified steady state IPMSM model is obtained from (7.8) and (7.9) by dropping the ohmic voltage such that

$$v_d^e = -\omega_e L_q i_q^e \tag{7.14}$$
$$v_q^e = \omega_e L_d i_d^e + \omega_e \psi_m. \tag{7.15}$$

Then, it follows from (7.13) that

$$\frac{i_q^{e\,2}}{V_m^2/(\omega_e L_q)^2} + \frac{(i_d^e + i_f)^2}{V_m^2/(\omega_e L_d)^2} \le 1. \tag{7.16}$$

It is an equation of ellipse. Note that the major and minor axes are functions of the speed. Since $L_d < L_q$, then the major axis is horizontal. It is obvious from (7.16) that the ellipses shrink to $(-i_f, 0)$ as shown in Fig. 7.9 when the speed ω increases.

Meanwhile, the current limit is a fixed circle:

$$i_d^{e\,2} + i_q^{e\,2} \le I_m^2. \tag{7.17}$$

Feasible maximum power solution is found at the intersection between an ellipse and the circle.

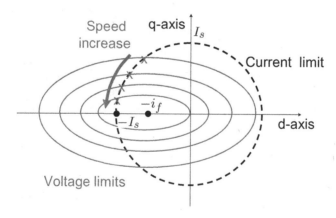

Figure 7.9: Current limit and voltage limits for various speeds in the current plane.

Fig. 7.10 shows torque-speed curve along with current contour. At the base speed, often called rated speed, the terminal voltage reaches the maximum with a rated load (current). The rated condition is marked by A on the current limit circle. As the speed increases further, the voltage limit shrinks. Thus, the maximum power

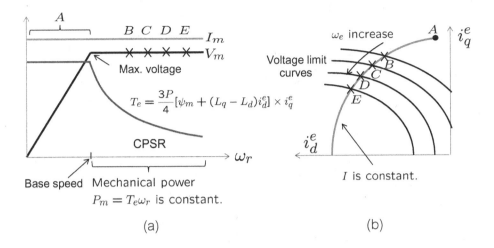

Figure 7.10: (a) Torque, voltage, and current versus speed and (b) the corresponding voltage and current contours.

solution moves down to the horizontal axis along the circle. Fig. 7.10 (b) shows a sequence of operation points B, C, D, and E when the speed increases. A larger (negative) d-axis current needs to be supplied to meet the voltage requirement. Specifically, $\omega_e L_d i_d^e$ counteracts the growing back EMF, $\omega_e \psi_m$, and a larger i_d^e (in the negative direction) is obtained at the expense of reducing the q-axis current. Those marked intersections in Fig. 7.10 (b) belong to the contour of constant (apparent) power $P_e = V_m \times I_m$, since the voltage and the current are kept constant. Most of the electrical power will be converted into the mechanical power $P_m = T_e \times \omega_m$, if the loss is small. If the power factor is almost unity, the contour, $\overline{BE}$ characterizes the constant power speed range (CPSR).

7.3.1 Torque Speed Profile

The CPSR is determined by the relative magnitudes between ψ_m and $L_d I_m$. They are the field strengths of rotor PM and stator coil. Since i_q^e approaches zero as ω_e becomes large, it is obvious from voltage limit equation (7.16) that

$$(i_d^e + i_f)^2 \leq \frac{V_m^2}{(L_d \omega_e)^2}. \tag{7.18}$$

For sufficiently large ω_e, the right hand side of (7.18) vanishes. Therefore, it should follow that $i_d^e \to -i_f$ as $\omega_e \to \infty$. As was observed in [2], the ability of producing power at infinite speed is determined by the criteria

$$\psi_m \leq L_d I_m. \tag{7.19}$$

It means that the demagnetizing flux, $L_d I_m$ by the stator coil must be larger than the PM flux, ψ_m to extend the speed range indefinitely. Fig. 7.11 shows three cases with the voltage and current limits and power plots versus speed. Specific illustrations for the three cases are summarized in the following: [3]

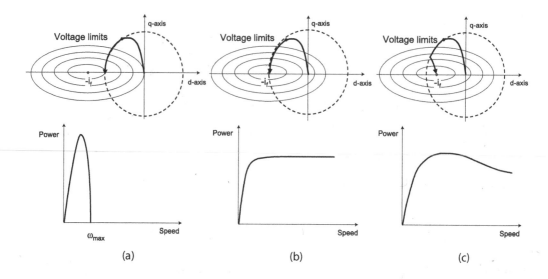

Figure 7.11: Current contours and power versus speed for three cases: (a) $\psi_m > L_d I_m$, (b) $\psi_m = L_d I_m$, and (c) $\psi_m < L_d I_m$.

i) $\psi_m > L_d I_m$: This corresponds to the case where the rotor flux is larger than the maximum field that the stator current can provide. Above the rated speed, the power drops drastically. Since the ellipses center, $-i_f$ lies out of the current limit, there will be no more intersection above the critical speed defined by

$$\omega_c \equiv \frac{V_m}{\psi_m - L_d I_m}. \qquad (7.20)$$

That is, the current limit and voltage limit curves are separated for $\omega_e > \omega_c$, so that no proper solution exists.

ii) $\psi_m = L_d I_m$: This is the case where $i_f = I_m$. Since the ellipse center lies on the current limit circle, intersection always exists for any arbitrarily large ω. Thus, the constant maximum power can be extended to the infinite speed theoretically.

iii) $\psi_m < L_d I_m$: The constant power range will be extended to the infinite speed, too. But, the output power is lowered after peaking. This design setting is most widely used.

Summary

1. The current limit appears as a fixed circle, whereas the voltage limits are represented by a group of shrinking ellipses.

2. The voltage ellipse can be defined with a virtual current, i_f. Then the voltage limits are centered at $(-i_f, 0)$ in the current plane. If $L_d < L_q$, the ellipses are pressed to the horizontal direction.

3. If $\psi_m > L_d I_m$, then no intersection between the current and voltage limits will be found above a certain speed. In such a case, the power drops quickly. To prevent it, the PM usage should be limited in accordance with the current capability, i.e., $i_f < I_m$.

7.4 Current Control Methods

Since there is no magnetic saliency in SPMSMs, the torque is determined only by q-axis current. In contrast, the reluctance torque plays an important role in IPMSMs. Thus, numerous combinations of (i_d^e, i_q^e) are feasible for a fixed torque generation.

7.4.1 Maximum Torque per Ampere Control

The motor loss consists of copper loss, iron loss, mechanical loss, and stray loss. In the low speed area, the copper loss takes about 80% of the total loss. The maximum torque per ampere (MTPA) is a control method of maximizing the torque for given magnitude of current. It can be stated equivalently as "the current minimizing solution for a given torque." Thus, the MTPA is very useful for copper loss minimization.

To obtain the desired solution, we take the differentiation of T_e with respect to β: It follows from (7.12) that

$$\frac{\partial T_e}{\partial \beta} = \frac{3P}{4}[-\psi_m I \sin \beta + (L_q - L_d)I^2 \cos 2\beta] = 0.$$

Equivalently,

$$2(L_q - L_d)I \sin^2 \beta + \psi_m \sin \beta - (L_q - L_d)I = 0.$$

Thus, we have

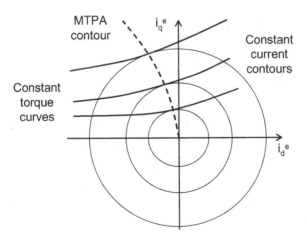

Figure 7.12: MTPA contour.

$$\beta = \sin^{-1}\left[\frac{-\psi_m + \sqrt{\psi_m^2 + 8(L_q - L_d)^2 I^2}}{4(L_q - L_d)I}\right].$$ (7.21)

Equation (7.21) gives us the loss minimizing current angle for a given current magnitude I. Further, since $i_d = -I\sin\beta$, we have

$$i_d^e = \frac{1}{4(L_q - L_d)}\left(\psi_m - \sqrt{\psi_m^2 + 8I^2(L_q - L_d)^2}\right),$$ (7.22)

$$i_q^e = \sqrt{I^2 - i_d^{e2}}.$$ (7.23)

The dotted line in Fig. 7.12 is the MTPA contour when the current magnitude increases from zero to I. Note that it is the set of points, $(-I\sin\beta, I\cos\beta)$ that a torque line intersects a (current) circle tangentially as shown in Fig. 7.12.

Different Derivation Using Lagrangian

The MTPA is viewed as the torque maximization under the current constraints. Consider power maximization problem under the current limit (7.17):

$$\max_{i_d^e, i_q^e} T_e \qquad \text{under} \quad i_d^e{}^2 + i_q^e{}^2 - I^2 \leq 0.$$

Let the Lagrangian be defined by [4]

$$\mathcal{L}(i_d^e, i_q^e) = \frac{3P}{4}\left(\psi_m i_q^e + L_d(1 - \xi)i_d^e i_q^e\right) + \mu_1\left(i_d^e{}^2 + i_q^e{}^2 - I^2\right),$$

where $\xi \equiv L_q/L_d$ is the saliency ratio and μ_1 is a Lagrange coefficient. Necessary conditions for the optimality are

$$\frac{\partial\mathcal{L}}{\partial i_d^e} = \frac{3P}{4}(1 - \xi)L_d i_q^e + 2\mu_1 i_d^e = 0,$$ (7.24)

$$\frac{\partial L}{\partial i_q^e} = \frac{3P}{4}L_d\left(i_f + (1 - \xi)i_d^e\right) + 2\mu_1 i_q^e = 0.$$ (7.25)

Then we obtain from (7.24) and (7.25) that

$$(1 - \xi)\left(i_q^e{}^2 - i_d^e{}^2\right) - i_f i_d^e = 0.$$

Since the solution is found on the boundary, we utilize the boundary equation, $i_d^e{}^2 + i_q^e{}^2 - I^2 = 0$. Then, we have the following second order equation:

$$2(i_d^e)^2 + \frac{i_f}{(1 - \xi)}i_d^e - I^2 = 0.$$

Of the two, the meaningful solution is given by

$$i_d^e = \frac{1}{4(\xi - 1)}\left(i_f - \sqrt{i_f^2 + 8I^2(\xi - 1)^2}\right)$$

$$= \frac{1}{4(\xi - 1)L_d}\left(\psi_m - \sqrt{\psi_m^2 + 8I^2(\xi - 1)^2 L_d^2}\right).$$ (7.26)

This result agrees with the previous solution (7.22).

Exercise 7.1   Consider an eight pole motor with the following parameters: $\psi_m = 0.103$ Wb, $L_d = 234$ μH, and $L_q = 562$ μH. Plot the torque versus current angle, β and the MTPA points in the current plane for $I = 100, 200, 300$, and 400 A.

Solution.   In general, the MTPA solutions are found experimentally and used

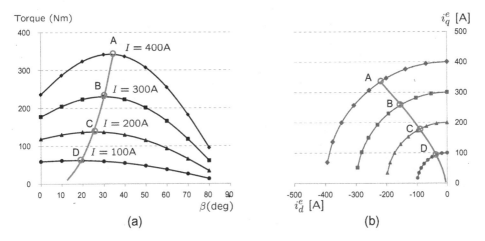

Figure 7.13: Torque versus current angle for different currents: (a) Maximum torque versus β and (b) MTPA contour in the current plane (Solution to Exercise 7.1).

after writing them into a lookup table (LUT). The main reason is that inductance changes depending on current magnitude [5]. Fig. 7.13 suggests a method of finding the MTPA points by experiments: For a given current magnitude, measure the torque while changing β's as shown in Fig. 7.13 (a). Record the peak value and the corresponding current pair in the lookup table. For example, $i_d^e = -215$ A and $i_q^e = 330$ A are recorded for $T_e = 340$ Nm. Repeat the process for different currents.

However, note that the MTPA control is not feasible in the field-weakening region due to the voltage limit. Online MTPA searching algorithms were proposed by [6], [7]. The MTPA method is also applied in the direct torque control (DTC) [8], [9].

7.4.2   Transversal Intersection with Current Limit

It is considered here a non-tangential (transversal) intersection with the current limit. Assume that a constant torque line defined by $\frac{3P}{4}\left(\psi_m i_q^e + (L_d - L_q)i_d^e i_q^e\right) = T_0$ intersects $i_d^{e\,2} + i_q^{e\,2} - I_m^2 = 0$ at two points as shown in Fig. 7.14. The two points A and B have the same current magnitude. But, B is preferable since the voltage magnitude is smaller.

Combining torque and current equations, a quartic polynomial is obtained such that

$$i_d^4 + A_1 i_d^3 + B_1 i_d^2 + C_1 i_d + D_1 = 0, \tag{7.27}$$

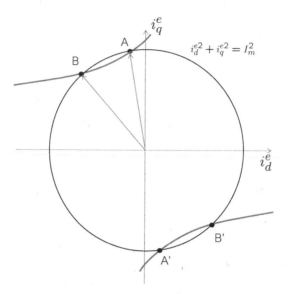

Figure 7.14: Transversal intersection with the current limit circle.

where

$$A_1 = \frac{2\psi_m}{(L_d - L_q)},$$

$$B_1 = \frac{\psi_m^2}{(L_d - L_q)^2 - I_s^2},$$

$$C_1 = \frac{2\psi_m I_s^2}{(L_d - L_q)}, \text{ and}$$

$$D_1 = \frac{1}{(L_d - L_q)^2}\left(\frac{16T_0^2}{9P^2} - I_s^2\psi_m^2\right).$$

Lodovico and Ferrari derived a general solution to quartic polynomials [10], [11], [12]:

$$i_d = -\frac{A_1}{4} \pm_s \frac{\eta_1}{2} \pm_t \frac{\mu_1}{2}, \qquad (7.28)$$

where

$$\alpha_1 = \frac{1}{3}\left(3A_1C_1 - 12D_1 - B_1^2\right) \qquad (7.29)$$

$$\beta_1 = \frac{1}{27}\left(-2B_1^3 + 9A_1B_1C_1 + 72B_1D_1 - 27C_1^2 - 27A_1^2D_1\right) \qquad (7.30)$$

$$\gamma_1 = \frac{B_1}{3} + \sqrt[3]{-\frac{\beta_1}{2} + \sqrt{\frac{\beta_1^2}{4} + \frac{\alpha_1^3}{27}}} + \sqrt[3]{-\frac{\beta_1}{2} - \sqrt{\frac{\beta_1^2}{4} + \frac{\alpha_1^3}{27}}} \qquad (7.31)$$

$$\eta_1 = \sqrt{\frac{A_1^2}{4} - B_1 + \gamma_1} \qquad (7.32)$$

$$\mu_1 = \sqrt{\frac{3}{4}A_1^2 - \eta_1^2 - 2B_1 \pm_s \frac{1}{4\eta_1}\left(4A_1B_1 - 8C_1 - A_1^3\right)}. \qquad (7.33)$$

It should be noted that two $\pm_s$ should have the same sign, while the sign of $\pm_t$ is independent. Let r_1, r_2, r_3, and r_4 be the solutions of a quartic polynomial. Then, the discriminant is defined as [14]

$$\mathcal{D} = (r_1 - r_2)^2 (r_1 - r_3)^2 (r_1 - r_4)^2 (r_2 - r_3)^2 (r_2 - r_4)^2 (r_3 - r_4)^2. \tag{7.34}$$

The discriminant tells us about the nature of roots [13]:

> if at least two of the roots coincide, then $\mathcal{D} = 0$;
> if all four roots are real or complex conjugates, then $\mathcal{D} > 0$;
> if two roots are real and two are complex conjugates, then $\mathcal{D} < 0$.

In the case of (7.27), the discriminant is equal to

$$
\begin{aligned}
\mathcal{D} = {} & 256 D_1^3 - 192 A_1 C_1 D_1^2 - 128 B_1^2 D_1^2 + 144 B_1 C_1^2 D_1 - 27 C_1^4 + 144 A_1^2 B_1 D_1^2 \\
& - 6 A_1^2 C_1^2 D_1 - 80 A_1 B_1^2 C_1 D_1 + 18 A_1 B_1 C_1^3 + 16 B_1^4 D_1 - 4 B_1^3 C_1^2 \\
& - 27 A_1^4 D_1^2 + 18 A_1^3 B_1 C_1 D_1 - 4 A_1^3 C_1^3 - 4 A_1^2 B_1^3 D_1 + A_1^2 B_1^2 C_1^2.
\end{aligned}
\tag{7.35}
$$

Fig. 7.14 shows a case of four real roots. Note that two other roots are in the fourth quadrant. Thus the discriminant is positive. Point B is obtained with the minus sign from $\pm_s$ and $\pm_t$:

$$(i_{d\alpha}, i_{q\alpha}) = \left(-\frac{A_1}{4} - \frac{\eta_1}{2} - \frac{\mu_1}{2}, \frac{\frac{4T_0}{3P}}{\psi_m + (L_d - L_q) i_{d\alpha}} \right). \tag{7.36}$$

Summary

1. In the case of IPMSM, the d-axis current plays a role in the determination of torque. Question is how to minimize the current magnitude while generating a desired torque. The MTPA yields the current minimizing solution. Thus, it is an essential technique for minimizing copper losses and widely used in practice.

2. The MTPA solution is found at a tangential intersection between torque and current curves. It is nonlinear so that it is used practically on the basis of a lookup table.

3. The MTPA is independent of the speed. However, the voltage limit adds an additional constraint above the base speed. The MTPA is, in general, not used in the field weakening range.

7.4.3   Maximum Power Control

Motors are designed, in general, to reach the maximum voltage at the rated speed under rated load. Above the rated speed, the voltage limit causes an increase in the current angle, which means more d-axis current and less q-axis current. Thereby, the airgap flux is weakened and torque decreases. With the lossless motor model (7.14) and (7.15), the power is given by

$$P_e = \frac{3}{2} \left(v_q^e i_q^e + v_d^e i_d^e \right) = \frac{3}{2} \omega_e \left(\psi_m i_q^e + (L_d - L_q) i_d^e i_q^e \right). \tag{7.37}$$

More formally, the power maximization problem in the field weakening region is stated as follows:

$$\max_{i_d^e,\, i_q^e} P_e \qquad \text{under} \quad c_1(i_d^e,\, i_q^e) \leq 0, \quad c_2(i_d^e,\, i_q^e) \leq 0,$$

where

$$c_1(i_d^e,\, i_q^e) = i_d^{e\,2} + i_q^{e\,2} - I_m^2, \tag{7.38}$$

$$c_2(i_d^e,\, i_q^e) = \xi^2 i_q^{e\,2} + (i_d^e + \frac{\psi_m}{L_d})^2 - \frac{V_m^2}{\omega_e^2 L_d^2}. \tag{7.39}$$

It is obvious from the geometric viewpoint that the maximum power is found at an intersection of the two curves, $c_1 = 0$ and $c_2 = 0$. Intersection points of the current and voltage limits make a locus of maximum power versus speed. That is, the maximum power is obtained along the current limit circle $c_1(i_d^e,\, i_q^e) = 0$. Substituting $i_q^{e\,2} = I_m^2 - i_d^{e\,2}$ into equation $c_2 = 0$, it follows that

$$0 = (1 - \xi^2)i_d^{e\,2} + 2\frac{\psi_m}{L_d}i_d^e + \frac{\psi_m^2}{L_d^2} + \xi^2 I_m^2 - \frac{V_m^2}{\omega_e^2 L_d^2}. \tag{7.40}$$

Since the intersection point in the second quadrant is meaningful, we just need to consider the negative i_d^e solution. Thus, we have

$$i_d^{e*} = \frac{1}{(\xi^2 - 1)}\left(\frac{\psi_m}{L_d} - \sqrt{\xi^2 \frac{\psi_m^2}{L_d^2} + (\xi^2 - 1)(\xi^2 I_m^2 - \frac{V_m^2}{\omega_e^2 L_d^2})}\right), \tag{7.41}$$

$$i_q^{e*} = \sqrt{I_m^2 - i_d^{e*\,2}}. \tag{7.42}$$

This characterizes the maximum CPSR, when the power factor is close to unity. However, if $I_m < i_f$, the power curve drops rapidly. See the following example.

Exercise 7.2   Consider an eight pole motor with the following parameters: $\psi_m = 0.053$ Wb, $L_d = 234$ μH, and $L_q = 562$ μH. Assume that $V_m = 180$ V and $I_m = 450$ A.
a) Using (7.22) and (7.23), calculate the MTPA point, $(i_{da}^e,\, i_{qa}^e)$ for $I_m = 450$ A.
b) Find the maximum feasible speed ω_{ea} when $V_m = 180$ V at the point, $(i_{da}^e,\, i_{qa}^e)$, i.e., determine ω_{ea} so that $V_m = \sqrt{v_d^{e\,2} + v_q^{e\,2}} = 180$ V while current is given by $(i_{da}^e,\, i_{qa}^e)$.
c) Using (7.40), find the speed ω_{ek} when the maximum power point reaches $(i_d^{e*},\, i_q^{e*}) = (-I_m,\, 0)$.
d) Using Excel, plot power and torque along the current limit circle for $\omega_{ea} \leq \omega_e \leq \omega_{ek}$.

Solution.
a) $i_d^e = \frac{1}{4(L_q - L_d)}\left(\psi_m - \sqrt{\psi_m^2 + 8I_m^2(L_q - L_d)^2}\right) = -280.4$ A,
   $i_q^e = \sqrt{I_m^2 - i_d^e} = 352$ A.

b) $\omega_e = \dfrac{V_m}{\sqrt{(L_q i_q^e)^2 - (L_d i_d + \psi_m)^2}} = 908.1$ rad/s, i.e, $\dfrac{836.2}{4 \times 2\pi} \times 60 = 2167$ rpm.

c) Inserting $i_q^e = 0$ and $i_d^e = I_m$ into (7.39), we have

$$\omega_e = \frac{V_m}{L_d I_m + \psi_m} = \frac{180}{234 \times 10^{-6} \times 450 + 0.103} = 3442 \text{ rad/s, i.e., 8221 rpm.}$$

d) See Fig. 7.15.

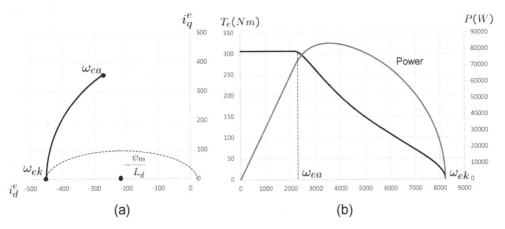

Figure 7.15: Contour along the current limit: (a) current trajectory and (b) torque and power (Solution to Exercise 7.2 (d)).

7.4.4 Maximum Torque per Voltage Control

This control strategy is valid in the high speed operation of IPMSMs when $I_m > \psi_m / L_d$. It is a case where the current limit circle includes the center of the voltage limits. Thus, infinite speed operation is feasible without reducing power to zero. But if we follows the current limit circle, the motor power drops at a finite speed as shown in Fig. 7.15.

At a certain speed, the operation point must deviate from the current limit circle and should follow another contour that converges to the center of voltage limits. In other words, there is a break point, namely ω_{eb} in the speed range as shown in Fig. 7.16. For lower speeds, $\omega_e \le \omega_{eb}$, the maximum torque is at an intersection with the current limit circle. However for higher speeds, $\omega_e \ge \omega_{eb}$, it is located at a tangential intersection with a voltage limit curve.

In the following, a method of calculating the tangential intersection is considered. Note that the flux equation follows from the voltage limit, (7.16):

$$(L_q i_q^e)^2 + \lambda_d^2 = (V_m/\omega_e)^2, \tag{7.43}$$

where $\lambda_d \equiv L_d i_d^e + \psi_m$. Substituting (7.43) into torque equation (7.12), it follows that

$$T_e^2 = \left(\frac{3P}{4}\right)^2 \left(\lambda_d - L_q \frac{\lambda_d - \psi_m}{L_d}\right)^2 \frac{(V_m/\omega_e)^2 - \lambda_d^2}{L_q^2}.$$

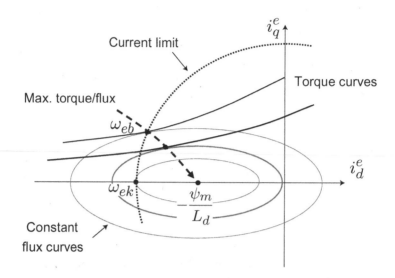

Figure 7.16: Maximum torque per flux solutions.

Taking differentiation with respect to λ_d and making it equal to zero result in [15]

$$\lambda_d = \frac{-L_q\psi_m + \sqrt{L_q^2\psi_m^2 + 8(L_d - L_q)^2(V_m/\omega_e)^2}}{4(L_d - L_q)}. \tag{7.44}$$

Using (7.44), the torque/flux maximizing current command is determined at

$$i_d^e = \frac{\lambda_d - \psi_m}{L_d} \tag{7.45}$$

$$i_q^e = \frac{\sqrt{(V_m/\omega_e)^2 - \lambda_d^2}}{L_q}. \tag{7.46}$$

It is the maximum torque solution for a given voltage. Thus, it is often called *maximum torque per voltage* (MTPV).

Exercise 7.3   Consider an eight pole motor with the following parameters: $\psi_m = 0.053$ Wb, $L_d = 234$ μH, and $L_q = 562$ μH. Assume that $V_m = 180$ V and $I_m = 450$ A.
a) Using Excel, find ω_{eb} at which the MTPV line intersects the current limit circle.
b) Using Excel, plot the current contour, torque, and power based on the MTPV for $\omega_{eb} \leq \omega_e \leq 3442$ (8221 rpm). Fill the rest interval, $0 \leq \omega_e \leq \omega_{eb}$ with the results obtained in Exercise 7.2. Compare it with the one shown in Fig. 7.15.

Solution.
a) Two i_d^e's are calculated based on (7.41) and (7.45). ω_{eb} is found when the two i_d^e's agree. The result is $\omega_{be} = 1939$, i.e., 4630 rpm.
b) Differently from the case of following the current limit circle, high power is maintained up to 8221 rpm.

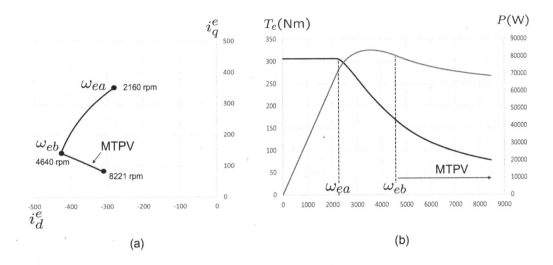

Figure 7.17: Contour along the MTPV: (a) current trajectory and (b) torque and power versus speed (Solution to Exercise 7.3).

MTPV Solution Using Ferrari Method

To find the solutions of (6.61) and (7.16), we are faced with the following quartic polynomial:

$$i_d^4 + A_2 i_d^3 + B_2 i_d^2 + C_2 i_d + D_2 = 0, \tag{7.47}$$

where

$$A_2 = \frac{2\psi_m}{L_d - L_q}\left(2 - \frac{L_q}{L_d}\right)$$

$$B_2 = \frac{\psi_m^2}{(L_d - L_q)^2} + \frac{4 \cdot \psi_m^2}{L_d(L_d - L_q)} + \frac{\psi_m^2}{L_d^2} - \frac{V_s^2}{\omega_e^2 L_d^2}$$

$$C_2 = \frac{2\psi_m}{L_d}\left(\frac{\psi_m^2}{(L_d - L_q)^2} + \frac{\psi_m^2}{L_d(L_d - L_q)} - \frac{V_s^2}{\omega_e^2 L_d(L_d - L_q)}\right)$$

$$D_2 = \frac{1}{(L_d - L_q)^2}\left(\frac{\psi_m^4}{L_d^2} + \frac{L_q^2}{L_d^2} \cdot \frac{16T_0^2}{9P^2} - \frac{V_s^2}{\omega_e^2} \cdot \frac{\psi_m^2}{L_d^2}\right).$$

Differently from the previous method, (7.44)–(7.46), it can also be found using the discriminant. The discriminant for (7.47) is

$$\mathcal{D} = \frac{256}{729} \cdot \frac{L_q^2(L_d - L_q)^4}{L_d^{10}} \cdot \frac{T_0^2}{P^6\omega_e^8}\left(\zeta T_0^4 + \delta T_0^2 + \sigma\right), \tag{7.48}$$

where

$$\zeta = 4096\omega_e^8 L_d^4 L_q^4 (L_d - L_q)^2$$

$$\delta = -144P^2\omega_e^4 L_d^2 L_q^2(-\psi_m^4\omega_e^4 L_q^4 + 20\psi_m^2\omega_e^2 V_s^2 L_d^2 L_q^2 + 8V_s^4 L_d^4 - 32V_s^4 L_d^3 L_q$$
$$+48V_s^4 L_d^2 L_q^2 - 32V_s^4 L_d L_q^3 - 40\psi_m^2\omega_e^2 V_s^2 L_d L_q^3 + 20\psi_m^2\omega_e^2 V_s^2 L_q^4 + 8V_s^4 L_q^4)$$

$$\sigma = 81P^4 V_s^2 (V_s L_d - \omega_e\psi_m L_q - V_s L_q)^3 (V_s L_d + \omega_e\psi_m L_q - V_s L_q)^3.$$

Since $\mathcal{D} = 0$ at the tangential point, we find from (7.48) that

$$T_0 = \sqrt{\frac{-\delta + \sqrt{\delta^2 - 4\zeta\sigma}}{2\zeta}} \equiv T_{0t}. \tag{7.49}$$

Equations, (7.45) and (7.46) give the location in the current plane, whereas (7.49) gives directly the torque solution, T_{0t}. Note that T_{0t} is the maximum available torque for a given speed. If (7.45) and (7.46) are substituted into the torque equation, (6.61), then the result should be the same as (7.49).

Exercise 7.4   Consider an eight pole motor with the following parameters: $\psi_m = 0.053$ Wb, $L_d = 234$ μH, and $L_q = 562$ μH. Assume that $V_m = 180$ V and $I_m = 450$ A. At 12,000 rpm, find the maximum feasible torque based on (7.45) and (7.46). Also find it based on (7.49) and check both results are the same.

Solution.   Using (7.44)–(7.46), $(i_d^e, i_q^e) = (-372, 19.8)$ at $\omega_e = 5026.5$. Correspondingly, $T_e = 51.9$ Nm. On the other hand, we have

$$\zeta = 0.053712$$
$$\delta = -117.77$$
$$\sigma = -72925.96$$

The classical MTPV method also yields the same result, $T_0 = 51.9$ Nm. ∎

7.4.5   Combination of Maximum Power Control Methods

Fig. 7.18 (a) shows a control sequence as the speed increases: Starting from an MTPA point, the current set points move along the maximum power, and maximum torque/flux contours, sequentially. Table 7.2 shows the parameters of an IPMSM. Within the base speed range, current is set at A, an MTPA point. Beyond the base speed (4550 rpm), the maximum power is tracked following a segment, $\overline{AB}$ of the (rated) current limit circle. From point B, the operation points follow the maximum torque/flux line, $\overline{BC}$. Fig. 7.18 (b) shows the corresponding torque and power plots over $\overline{AB}$ and $\overline{BC}$. As mentioned previously, the power changes in $\overline{AB}$ due to the PF variation, even though the maximum voltage and current are used.

Exercise 7.5   Consider an IPMSM with parameters in Table 7.2.
a) Determine a maximum MTPA point, A of Fig. 7.18.
b) Determine torque at A and compare it with the case of $(i_d^e, i_q^e) = (0, 40)$A.
c) Estimate the speed at K of Fig. 7.18.

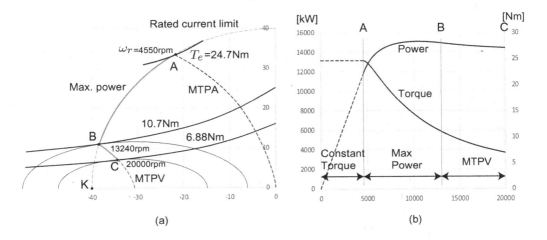

Figure 7.18: (a) Combination of control strategies for IPMSM and (b) torque and power plots.

Table 7.2: Specifications of an IPMSM.

No. of poles	6	Power (peak)	15 kW
DC link voltage (V_m)	300 V	Base speed	4550 rpm
Inductance (L_d)	3.05 mH	Rated current (I_m)	40 A
Inductance (L_q)	6.2 mH	Flux (ψ_m)	0.0948 Wb

Solution.
(a) Putting $I = 40$ into (7.22) and (7.23), it follows that $(i_d^e, i_q^e) = (-21.74, 33.57)$ A.
(b) Then
$$T_e = \tfrac{3 \times 6}{4} \times \left[0.0948 \times 33.57 + (3.05 - 6.2) \times 10^{-3} \times (-21.74) \times 33.57 \right] = 24.7 \text{ Nm}.$$
On the other hand, $T_e = \tfrac{3 \times 6}{4} \times 0.0948 \times 40 = 17.1$ Nm for $(i_d^e, i_q^e) = (0, 40)$ A.
(c) The voltage equation at K is

$$\omega_e = -\frac{V_m/L_d}{i_d^e + i_f} = \frac{V_m}{-L_d i_d^e + \psi_m} = 11052.$$

Therefore, the speed at K is 35197 rpm. ∎

Exercise 7.6   Consider an IPMSM with parameters in Table 7.2.
a) Determine the d and q axis current yielding the maximum torque under the voltage constraint at 20000 rpm.
b) Determine torque, power, and power factor at that point. In calculating the power factor, assume that the motor is lossless.

Solution.
a) $\omega_e = 20000$ rpm $\left(\frac{2\pi}{60} \right) \times 3 = 6284$. Using (7.44), it follows that $\lambda_d = -0.011$ Wb. Therefore, we obtain from (7.45) and (7.46) that $i_d^e = -34.7$ A and $i_q^e = 7.5$ A.

b) $T_e = 4.5 \times \left[0.0948 \times 7.5 + (3.05 - 6.2) \times 10^{-3} \times 7.5 \times (-34.7)\right] = 6.88$ Nm. The shaft power is equal to $P_m = T_e \times \omega_r = 6.88 \times 6284/3 = 14.4$ kW. The apparent electrical power is $P_e = \frac{3}{2} V_m I = 1.5 \times 300 \times \sqrt{34.7^2 + 7.5^2} = 16$ kW. Therefore, the power factor is equal to $\cos\phi = 14.4/16 = 0.9$.   ∎

7.4.6   Unity Power Factor Control

Depending upon the combination of i_d and i_q, the unity power factor (PF) can be obtained in the high speed region where large $-i_d^e$ is utilized. With the unity PF, the voltage magnitude can be minimized along with a minimum current. Thereby it is related to the minimization of iron and copper losses. Subsequently, the unity PF condition is derived and interpreted geometrically. From the geometry shown in Fig. 7.7, it follows that

$$\delta = \tan^{-1}\left(\frac{-v_d^e}{v_q^e}\right) = \tan^{-1}\left(\frac{\omega_e L_q i_q^e}{\omega_e \psi + \omega_e L_d i_d^e}\right) \tag{7.50}$$

$$\beta = \tan^{-1}\left(\frac{-i_d^e}{i_q^e}\right). \tag{7.51}$$

The unity PF condition is satisfied when current angle and voltage angle are the same, i.e., $\delta = \beta$. Therefore, it follows from (7.50) and (7.51) that

$$\frac{\xi i_q^e}{i_f + i_d^e} = \frac{-i_d^e}{i_q^e}. \tag{7.52}$$

Rearranging (7.52), the unity PF condition is given by

$$\frac{i_q^{e\,2}}{\left(\frac{i_f}{2\sqrt{\xi}}\right)^2} + \frac{(i_d^e + \frac{i_f}{2})^2}{\left(\frac{i_f}{2}\right)^2} = 1. \tag{7.53}$$

That is, the unity PF condition is described as an ellipse with center $(-\frac{i_f}{2}, 0)$ that passes through the origin and $(-i_f, 0)$. Since the minor axis is $\frac{i_f}{2\sqrt{\xi}}$, the shape of unity PF ellipse is pressed as the saliency ratio, ξ, increases.

The MTPA, unity PF, and max torque/flux solutions are compared in Fig. 7.20. Three points are on a same constant torque line: An MTPA point is marked with P, and it is the smallest current solution. However, the voltage magnitude is the largest. Specifically, $V_{mP} > V_{mQ} > V_{mR}$. On the other hand, the tangential solution, R is the solution with the least voltage magnitude. Thereby, R is a solution of the minimum iron loss. However, it requires the largest current magnitude, incurring the largest copper loss. Finally, Q is the solution of unity power factor. It is an intermediate solution, causing neither excessive iron nor copper loss.

Max Torque Satisfying Unity PF

The maximum torque, while satisfying the unity power factor, is found when the torque curve meets the unity PF ellipse tangentially. Substitute $T_e = (3P/4)(\psi_m + (L_d - L_q)i_d^e)i_q^e$ into (7.52) and apply the double root condition.

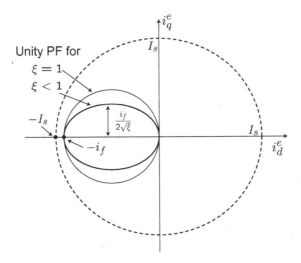

Figure 7.19: Unity PF ellipses for different saliency ratios, ξ's.

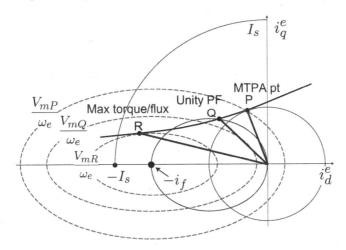

Figure 7.20: Current magnitude comparison of MTPA, unity PF, MTPV solutions.

The other approach is to utilize the gradient vectors. The level curves of unity PF and torque are characterized by $\xi i_q^{e\,2} + i_d^{e\,2} + i_f i_d^e = c_1$ and $\psi i_q^e + (L_d - L_q)i_d^e i_q^e = c_2$ for some constants, c_1 and c_2. Note that the gradient vector is orthogonal to the tangent vector at each point of the level curve.

At the tangential intersection point, the gradient vectors of the two level curves must be parallel as shown in Fig. 7.21. According to the definition in Section 1.1, the gradient vectors are obtained as

$$\nabla(\xi i_q^{e\,2} + i_d^{e\,2} + i_f i_d^e) = [2i_d^e + i_f, \ 2\xi i_q^e]^T \tag{7.54}$$

$$\nabla(\psi i_q^e + (L_d - L_q)i_d^e i_q^e) = [(L_d - L_q)i_q^e, \ \psi_m + (L_d - L_q)i_d^e]^T. \tag{7.55}$$

The parallel condition requires

$$(2i_d^e + i_f)(\psi_m + (L_d - L_q)i_d^e) = 2\xi(L_d - L_q)i_q^{e\,2}. \tag{7.56}$$

Utilizing again the unity power condition, $i_q^{e\,2} = -i_q^{e\,2} - i_f i_d^e$, it follows that

$$(2i_d^e + i_f)(if + (1 - xi)i_d^e) = 2\xi(1 - \xi)i_q^{e\,2} - 2\xi(1 - \xi)(i_q^{e\,2} + i_f i_d^e). \qquad (7.57)$$

Rearranging (7.57), the second order equation follows:

$$i_d^e + \left(\frac{1}{2(1 - \xi)} + \frac{3}{4}\right) i_f i_d^e + \frac{i_f^2}{4(1 - \xi)} = 0. \qquad (7.58)$$

Utilizing the quadratic formula, the desired solution follows.

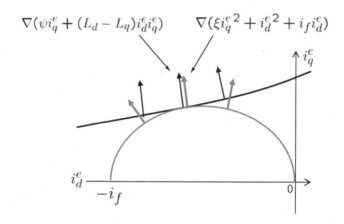

$$\nabla(\psi i_q^e + (L_d - L_q)i_d^e i_q^e) \qquad \nabla(\xi i_q^{e\,2} + i_d^{e\,2} + i_f i_d^e)$$

Figure 7.21: Gradient vectors to torque and unity PF curves.

Exercise 7.7　　Consider an eight pole motor with the following parameters: $\psi_m = 0.053$ Wb, $L_d = 234$ μH, and $L_q = 562$ μH. Assume that $V_m = 180$ V and $I_m = 450$ A.

    a) Using Excel, calculate torque while current (i_d^e, i_q^e) follows the unity PF ellipse for $-i_f \leq i_d^e \leq -i_f/2$. Find i_d^e that maximizes the torque.
    b) Using (7.57), find i_d^e of the maximum torque. Check the results of a) and b) are the same.
    c) Mark the area in the torque-speed plane where unity PF operation is feasible. Utilize the maximum torque-speed curve obtained in Fig. 7.17.

Summary

    1. The MTPA solution gives a solution of minimum copper loss at the expense of a large voltage. Thus, it could cause a large iron loss.
    2. The tangential intersection between torque and voltage equations yields a maximum torque for a given flux level. It can be stated equivalently as the minimum flux solution, i.e., the solution of minimum iron loss for a fixed torque and speed. However, it requires a large current magnitude, i.e., large copper loss.

3. The unity PF contour appears as an ellipse with end points at the origin and $(-\psi_m/L_d, 0)$. The unity PF solution gives an intermediate solution between the above two. It is a solution that minimizes both copper and iron losses at the same time.

7.4.7 Current Control Contour for SPMSM

In the case of SPMSM, the d-axis current has nothing to do with the torque and can be only used for field weakening. The MTPA locus is simply a segment of the q-axis. Fig. 7.22 shows a contour for the maximum torque operation. Above the rated speed, the maximum torque is determined at the intersection between current and voltage limits $(\overline{AB})$. Note that point 'B' is an intersection with the vertical line passing $(-\psi_m/L_d, 0)$. But after point 'B', the maximum torque is found at an intersection between a torque line and the vertical line.

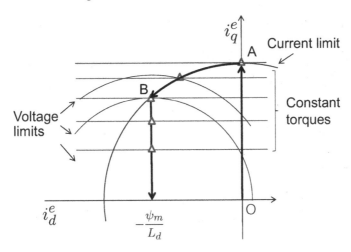

Figure 7.22: SPMSM maximum torque contour.

Exercise 7.8   Consider an SPMSM with the data: $P = 6$, $\psi_m = 0.0948$ V·s, and $L_d = L_q = 3.05 \times 10^{-3}$ H. Assume that $V_m{=}180$ V and $I_m = 40$ A. Plot the maximum torque, power, and the current contour for speed up to 12000 rpm.

Solution. See Fig. 7.23. ■

7.4.8 Properties when $\psi_m = L_d I_s$

With $\psi_m = L_d I_s$, the voltage limit is described by an ellipse with the center at $(-I_s, 0)$:

$$\frac{(i_q^e)^2}{V_s^2/(\omega_e \xi L_d)^2} + \frac{(i_d^e + I_s)^2}{V_s^2/(\omega_e L_d)^2} \leq 1. \tag{7.59}$$

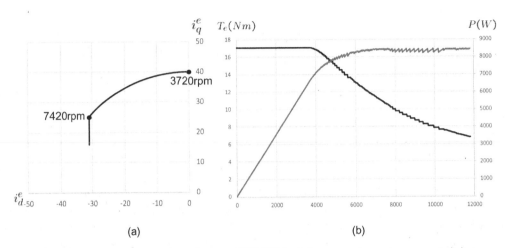

Figure 7.23: Maximum power for an SPMSM: (a) current contour and (b) torque and power (Solution to Exercise 7.8).

Thus, the ellipse shrinks down to the point $(-I_s, 0)$ as ω increases. Then, the CPSR can be extended to infinite speed theoretically. In this section, some properties are investigated theoretically. First, we consider the Kuhn-Tucker theorem, which provides necessary conditions for optimality under inequality constraints.

Kuhn-Tucker Theorem [4]. Let $f : U \to \mathbb{R}$ be a C^1 function on a certain open set $U \in \mathbb{R}^n$, and let $h_i : \mathbb{R}^n \to \mathbb{R}$, $i = 1, ..., \ell$, be C^1 functions. Suppose x^* is a local maximum of f on the set $D = U \cap \{x \in \mathbb{R}^n \mid h_i(x) \leq 0, \quad i = 1, ..., \ell\}$. Then there exist $\mu_1^*, ..., \mu_\ell^* \in \mathbb{R}$ such that

i) $\mu_i^* \leq 0$, for $i = 1, ..., \ell$;

ii) $\mu_i^* h_i(x^*) = 0$, for $i = 1, ..., \ell$;

iii) $\frac{\partial}{\partial x} f(x^*) + \sum_{i=1}^{\ell} \mu_i^* \frac{\partial}{\partial x} h_i(x_i^*) = 0$.

The coefficients $\mu_1, \cdots, \mu_\ell$ are called Kuhn-Tucker (K-T) multipliers and have the similar meaning to the Lagrangian multipliers. Note that x^* is on the boundary or interior of D. Hence, condition *ii)* states that either x^* is a point of the boundary $h_i = 0$, or an interior point of a region specified by $h_i < 0$. If $h_i(x^*) < 0$, it should follow that $\mu_i = 0$. Therefore, it is called the *complementary slackness condition*.

Since the power optimization is considered under voltage and current constraints, the Lagrangian is set as

$$\mathcal{L}(i_d^e, i_q^e, \mu_1, \mu_2) = v_d^e i_d^e + v_q^e i_q^e + \frac{\mu_1}{2}(i_d^{e\,2} + i_q^{e\,2} - I_m^2) + \frac{\mu_2}{2}(v_d^{e\,2} + v_q^{e\,2} - V_m^2)$$

$$= \omega_e \left[-L_q i_q^e i_d^e + (\psi_m + L_d i_d^e) i_q^e \right] + \frac{\mu_1}{2}(i_d^{e\,2} + i_q^{e\,2} - I_m^2)$$

$$+ \frac{\mu_2}{2} \left[(L_q i_q^e)^2 + (\psi_m + L_d i_d^e)^2 - V_m^2 \right].$$

Then it follows from Condition *iii)* that

$$\frac{\partial \mathcal{L}}{\partial i_d^{ie}} = \omega_e(-L_q i_q^{ie} + L_d i_q^{ie}) + \mu_1 i_d^{ie} + \mu_2(\psi_m + L_d i_d^{ie})L_d = 0,$$

$$\frac{\partial \mathcal{L}}{\partial i_q^{ie}} = \omega_e[(L_d - L_q)i_d^{ie} + \psi_m] + \mu_1 i_q^{ie} + \mu_2 L_q^2 i_q^{ie} = 0.$$

Rearranging, we have

$$\begin{bmatrix} i_d^{ie} & (\psi_m + L_d i_d^{ie})L_d \\ i_q^{ie} & L_q^2 i_q^{ie} \end{bmatrix} \begin{bmatrix} \mu_1 \\ \mu_2 \end{bmatrix} = \omega_e \begin{bmatrix} (L_q - L_d)i_q^{ie} \\ (L_q - L_d)i_d^{ie} - \psi_m \end{bmatrix}$$

From the Cramer's rule, it follows that

$$\mu_1 = \omega_e \frac{(L_q - L_d)L_q^2 i_d^{ie\,2} - (\psi_m + L_d i_d^{ie})[(L_q - L_d)i_d^{ie} - \psi_m]}{(L_q^2 - L_d^2)i_d^{ie} i_q^{ie} - \psi_m L_d i_q^{ie}}$$

$$\mu_2 = \omega_e \frac{(L_q - L_d)(i_d^{ie\,2} - i_q^{ie\,2}) - \psi_m i_d^{ie}}{(L_q^2 - L_d^2)i_d^{ie} i_q^{ie} - \psi_m L_d i_q^{ie}}.$$

Note that $(L_q^2 - L_d^2)i_d^{ie} i_q^{ie} - \psi_m L_d i_q^{ie} < 0$ in the second quadrant. On the other hand, $(L_q - L_d)L_q^2 i_d^{ie\,2} - (\psi_m + L_d i_d^{ie})[(L_q - L_d)i_d^{ie} - \psi_m] > 0$ is satisfied sufficiently in the second quadrant. Also, $(L_q - L_d)(i_d^{ie\,2} - i_q^{ie\,2}) - \psi_m i_d^{ie} > 0$ for at least $\beta > 45°$. Therefore, it follows that $\mu_1 < 0$ and $\mu_2 < 0$. Then from Condition *ii)*, the solution should be found on $c_1(i_d^{ie}, i_q^{ie}) = 0$ and $c_2(i_d^{ie}, i_q^{ie}) = 0$. That is, $\mu_1 < 0$ and $\mu_2 < 0$ imply that the optimal solution is found on the boundary of the current and voltage limits: Thereby, the intersection, (7.41) and (7.42) is the power maximizing solution. This proves the following theorem:

Theorem 1. Suppose that $i_d^{ie} \leq -I_m/\sqrt{2}$, $i_q^{ie} > 0$, and that $\xi \geq 1$. Then for $\omega_e > 0$ and $\beta > 45°$, the intersection, (7.41) and (7.42) of curves $c_1(i_d^{ie}, i_q^{ie}) = 0$ and $c_2(i_d^{ie}, i_q^{ie}) = 0$ maximizes $P_e = \frac{3}{2}(v_d^e i_d^e + v_q^e i_q^e)$ subject to $c_1 \leq 0$ and $c_2 \leq 0$.

Remark. Equivalently, $\beta > 45°$ is stated as $i_d^{ie} \leq -I_m/\sqrt{2}$. According to Theorem 1, the maximum power versus speed moves along the intersection of the voltage and current limits. The loci may be represented by $\{\max_{c_1, c_2 \leq 0} P_e(\omega_e) \mid \omega_b \leq \omega_e\}$, where ω_b is a base speed. As far as the operating point is on that loci, V_m and I_m remain the same. Hence, power depends only on the PF.

Power Factor

If $\psi_m = L_d I_m$, the solutions, (7.41) and (7.42) reduce to

$$i_d^{e*} = \frac{1}{\xi^2 - 1}\left(I_m - \sqrt{\xi^4 I_m^2 - \frac{V^2}{\omega^2 L_d^2}(\xi^2 - 1)}\right) \tag{7.60}$$

$$i_q^{e*} = \frac{1}{\xi^2 - 1}\sqrt{-2\xi^2 I_m^2 + \frac{V^2}{\omega^2 L_d^2}(\xi^2 - 1) + 2I_m\sqrt{\xi^4 I_m^2 - \frac{V^2}{\omega^2 L_d^2}(\xi^2 - 1)}}. \tag{7.61}$$

Solution pair (7.60) and (7.61) are the power maximizing solutions for a given speed ω.

For simplicity, only a case where $\psi_m = L_d I_m$ is considered. Then the voltage limit has the center at $(i_d^e, i_q^e) = (-I_m, 0)$.

$$\frac{(i_q^e)^2}{V_m^2/(\omega_e \xi L_d)^2} + \frac{(i_d^e + I_m)^2}{V_m^2/(\omega_e L_d)^2} \leq 1. \tag{7.62}$$

With the above polar coordinate representation, power is equal to

$$P_e = \frac{3}{2}\left(v_q^e i_q^e + v_d^e i_d^e\right) = \frac{3}{2}\left(\omega_e \psi_m I_m \cos\beta - \omega_e \frac{L_d - L_q}{2} I_m^2 \sin 2\beta\right). \tag{7.63}$$

Let $\mathbf{V}$ and $\mathbf{I}$ be the voltage vector and current vector, respectively. Note from Fig. 7.7 that

$$\mathbf{V} = V_m \angle \delta + 90° = v_d^e + jv_q^e,$$
$$\mathbf{I} = I_m \angle \beta + 90° = i_d^e + ji_q^e.$$

Utilizing the notion of inner product, (7.63) can be expressed equivalently as

$$P_e = \frac{3}{2}\mathbf{V} \cdot \mathbf{I} = \frac{3}{2}V_m I_m \cos\phi, \tag{7.64}$$

Theorem 2. Consider lossless PMSM model (7.14) and (7.15) with current limit, $c_1(i_d^e, i_q^e) \leq 0$ and voltage limit, $c_2(i_d^e, i_q^e) \leq 0$. Assume that $\psi_m = L_d I_m$. Then, the PF converges to unity as the speed increases. Correspondingly, the maximum power is obtained at infinite speed, i.e.,

$$\lim_{\omega_e \to \infty}\{\max_{c_1, c_2 \leq 0} P_e(\omega_e)\} = \frac{3}{2}V_m I_m.$$

Proof. As speed increases, the voltage ellipse shrinks to the center, $(-I_m, 0)$. Since the maximum power is given at the intersection, the solution follows the current limit circle and converges to $(-I_m, 0)$. That is,

$$\lim_{\omega_e \to \infty} \mathbf{I} = I_m \angle 180°. \tag{7.65}$$

On the other hand, it follows from (7.14) and (7.15) that

$$V_m = \sqrt{(v_q^e)^2 + (v_d^e)^2} = \omega_e L_d I_m \sqrt{(1 - \sin\beta)^2 + \xi^2 \cos^2\beta}. \tag{7.66}$$

Note that

$$\frac{V_m^2}{L_d^2 I_m^2} = \omega_e^2[(1 - \sin\beta)^2 + \xi^2 \cos^2\beta] = \omega_e^2(1 - \sin\beta)[2 + (\xi^2 - 1)(1 + \sin\beta)].$$

Note that the left hand side is a positive constant, whereas the right hand side contains variables, ω_e and β. Therefore, the right hand side converges to a constant as $\omega_e \to \infty$ and $\beta \to \pi/2$, i.e., it should follow that

$$\lim_{\substack{\omega_e \to \infty \\ \beta \to \pi/2}} \omega_e^2 (1 - \sin \beta) = \alpha, \tag{7.67}$$

where $\alpha = \frac{V_m^2}{2\xi^2 L_d^2 I_m^2}$. Utilizing (7.67), we obtain

$$\begin{aligned}
\lim_{\substack{\omega_e \to \infty \\ \beta \to \pi/2}} v_d^e &= -\lim_{\substack{\omega_e \to \infty \\ \beta \to \pi/2}} \omega_e L_q I_m \cos \beta \\
&= -\lim_{\substack{\omega_e \to \infty \\ \beta \to \pi/2}} \omega_e L_q I_m \sqrt{1 - \sin \beta} \sqrt{1 + \sin \beta} = -\sqrt{2\alpha} L_q I_m = -V_m.
\end{aligned}$$

Therefore, it follows from (7.66) that $\lim_{\omega_e \to \infty} v_q^e = 0$. This, in turn, implies that voltage angle δ converges to $90°$, i.e.,

$$\lim_{\substack{\omega_e \to \infty \\ \beta \to 90°}} \mathbf{V} = V_m \angle 180°. \tag{7.68}$$

Hence together with (7.65), the PF convergence to unity is obtained. On the other hand, it follows from (7.63) and (7.67) that

$$\begin{aligned}
\lim_{\substack{\omega_e \to \infty \\ \beta \to 90°}} P_e &= \lim_{\substack{\omega_e \to \infty \\ \beta \to 90°}} \frac{3}{2} \left[\omega_e L_d I_m^2 \cos \beta (1 - \sin \beta) + \omega_e L_q I_m^2 \sin \beta \cos \beta \right] \\
&= \lim_{\substack{\omega_e \to \infty \\ \beta \to 90°}} \frac{3}{2} \left[\frac{L_d I_m^2 \cos \beta}{\omega_e} \alpha + L_q I_m^2 \omega_e \sin \beta \sqrt{1 - \sin^2 \beta} \right] \\
&= \frac{3}{2} \sqrt{2} L_q I_m^2 \sqrt{\alpha} = \frac{3}{2} V_m I_m. \qquad \blacksquare
\end{aligned}$$

7.4.9 Per Unit Model of PMSM

For a given quantity (voltage, current, power, impedance, torque, etc.), the system quantities can be expressed as fractions of a defined base unit quantity. The per-unit value is the number related to a base quantity. The base voltage is chosen as the nominal rated voltage of the system. Once the base power and voltage are chosen, the base current and impedance are determined by the electrical laws of circuits. The per-unit system is widely used in the power system industry especially for transformers and AC machines. Calculations are simplified, and the machine characteristics are easily identified even if the unit size varies widely.

One's are assigned as per unit stator current and per unit flux at per unit speed when a rated current gives the maximum (rated) torque. The scaled values are denoted with subscript, n. The DC link voltage of the inverter is denoted by V_{dc}. The maximum voltage is reached when the rated torque is produced at the base frequency. Thus, the base voltage, V_b, is calculated as the phase voltage that can be obtained at the base speed under the rated current condition. Other base values are defined as follows:

$$\text{voltage} \quad : \quad V_b = \frac{V_{dc}}{\sqrt{3}}$$

$$\text{current} \quad : \quad I_b = \sqrt{i_{db}^2 + i_{qb}^2}$$

$$\text{flux linkage} \quad : \quad \psi_b = \sqrt{(\psi_m + L_d i_{db})^2 + (L_q i_{qb})^2}$$

$$\text{frequency} \quad : \quad \omega_b = \frac{V_b}{\psi_b}$$

$$\text{inductance} \quad : \quad L_b = \frac{\psi_b}{I_b} .$$

A normalized dq model is

$$v_{dn} = \frac{v_d}{V_b} = -\frac{\omega_e \xi L_d i_q}{\omega_b \psi_b} = -\frac{\omega_e \xi L_d i_q}{\omega_b L_b I_b} = -\omega_n \xi L_{dn} i_{qn}$$

$$v_{qn} = \frac{v_q}{V_b} = \frac{\omega_e L_d i_d + \omega_e \psi_m}{\omega_b \psi_b} = \frac{\omega_e L_d i_d}{\omega_b L_b I_b} + \frac{\omega_e \psi_m}{\omega_b \psi_b} = \omega_n L_{dn} i_{dn} + \omega_n \psi_n,$$

where

$$\omega_n = \frac{\omega_e}{\omega_b}$$

$$L_{dn} = \frac{L_d}{L_b} = \frac{\omega_b L_d I_b}{V_b}, \quad L_{qn} = \frac{L_q}{L_b} = \frac{\omega_b L_q I_b}{V_b},$$

$$v_{dn} = \frac{v_d}{V_b}, \quad v_{qn} = \frac{v_q}{V_b},$$

$$i_{dn} = \frac{i_d}{I_b}, \quad i_{qn} = \frac{i_q}{I_b},$$

$$\psi_n = \frac{\psi_m}{\psi_b}.$$

Since $P_{en} = v_{dn} i_{dn} + v_{qn} i_{qn}$ and $T_n = P_{en}/\omega_n$, the final normalized dq model turns out to be [3]

$$v_{dn} = -\omega_n \xi L_{dn} i_{qn}, \qquad (7.69)$$

$$v_{qn} = \omega_n L_{dn} i_{dn} + \omega_n \psi_n, \qquad (7.70)$$

$$T_n = \psi_n i_{qn} - (\xi - 1) L_{dn} i_{dn} i_{qn}. \qquad (7.71)$$

The normalized torque equation, (7.71), indicates that ψ_n can be traded partially with ξ. If ξ increases, then ψ should be reduced correspondingly. That is, the use of permanent magnet can be reduced as the saliency ratio increases. Since the cost of PM material, typically neodymium or samarium-cobalt, is high, it motivates us to design high saliency ratio motors. But, due to the mechanical strength requirements and manufacturing complexity, it is not easy to increase ξ arbitrarily high.

Exercise 7.9 Consider an IPMSM-A with the following parameters: six pole, $L_d = 3$ mH, $L_q = 6.2$ mH, $\psi_m = 0.146$ Wb, $\omega_b = 816.8$ rad/sec, and $i_{db} = 22$ A, $i_{qb} = 34$ A.
a) Calculate the normalized model.
b) Show that $v_{dn}^2 + v_{qn}^2 = 1$ when $\omega_n = 1$.

Solution.
a)

$$\xi = \frac{0.0062}{0.003} = 2.07$$

$$\psi_b = \sqrt{(0.146 - 0.003 \times 22)^2 + (0.0062 \times 34)^2} = 0.225$$

$$V_b = \omega_b \psi_b = 816.8 \times 0.225 = 183.78$$

$$I_b = \sqrt{22^2 + 34^2} = 40$$

$$L_{dn} = \frac{\omega_b L_d I_b}{V_b} = \frac{816.8 \times 0.003 \times 40}{183.78} = 0.533$$

$$\psi_n = \frac{\psi_m}{\psi_b} = \frac{0.146}{0.225} = 0.649$$

$$v_{dn} = -1.1\omega_n i_{qn},$$

$$v_{qn} = 0.533\omega_n i_{dn} + 0.649\omega_n,$$

$$T_n = 0.649 i_{qn} - 0.57 i_{dn} i_{qn}.$$

b) $v_{dn}^2 + v_{qn}^2 = (1.1 \times 0.85)^2 + (0.649 - 0.533 \times 0.55)^2 = 1$. ∎

7.4.10 Power-Speed Curve

If the rotor flux is strong such that $\psi_m > L_d I_b$, then the speed range cannot be extended indefinitely. The critical speed, ω_c, is defined by the speed at which power drops to zero. At ω_c, the q-axis current is equal to zero, while $i_b^e = I_b$. Therefore, it follows from (7.20) at the critical speed

$$\frac{\omega_c \psi_m}{\omega_b \psi_b} - \frac{\omega_c L_d I_b}{V_b} = \frac{V_b}{V_b} \qquad \Rightarrow \qquad \psi_n - L_{dn} = \frac{1}{\omega_{cn}}, \qquad (7.72)$$

where $\omega_{cn} \equiv \frac{\omega_c}{\omega_b}$. It should be noted that the critical speed is inversely proportional to $\psi_n - L_{dn}$.

Exercise 7.10 Consider IPMSM-A whose parameters listed in Exercise 7.9. Determine ω_c in rpm.

Solution.
$\omega_{cn} = \frac{1}{\psi_n - L_{dn}} = \frac{1}{0.649 - 0.533} = 8.62$. Thus, $\omega_c = 8.62 \times 2600 = 22412$ rpm. ∎

Exercise 7.11   Consider IPMSM-B whose parameters are the same as those of
IPMSM-A except $\psi_m = 0.16$ Wb.
a) Determine ω_{cn}.
b) Plot the power curves of IPMSM-A and IPMSM-B.

Solution. $\psi_n = \dfrac{0.16}{\sqrt{(0.16-0.003\times22)^2+(0.0062\times34)^2}} = 0.693$,

$\omega_{cn} = \dfrac{1}{\psi_n-L_{dn}} = \dfrac{1}{0.693-0.533} = 6.25$, and $\omega_c = 6.25 \times 2600 = 16250$ rpm. ∎

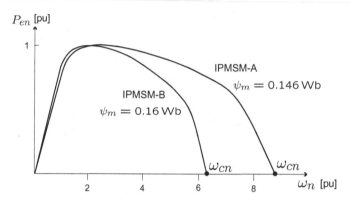

Figure 7.24: Normalized power curves for IPMSM-A and IPMSM-B.

7.4.11   Wide CPSR

High torque capability and large CPSR are the trading requirements when the
voltage and the current are limited. The power-speed profile is determined by both
saliency ratio and PM flux linkage. Their proportion determines the power profile
shape. If the proportion of PM flux linkage is dominant, the power has a sharp
peak in the low speed range and drops quickly, so that the CPSR is narrow. On the
contrary, if the reluctance proportion is large compared with PM linkage, a broad
CPSR is obtained with a sacrifice in the peak power.

   Fig. 7.25 shows a guide to balance PM flux and saliency ratio. The vertical axis
represents saliency, whereas the horizontal axis is the normalized PM flux. Each
point in the plane represents a specific motor design. The optimal combination is
indicated by a solid line [3]. It is a separatrix between high peak power and wide
CPSR. The upper region is characterized by a group of motors that have high peak
power, whereas the lower region a group of motors that have wide CPSR. It should
be noted that the power factor is close to unity along the optimal combination
line. It tells us that the unity power factor is required to extend the CPSR without
lowering the power level. On the other hand, the unity power requires a balance
between the PM flux and stator flux. While L_q is fixed, the saliency ratio is equal
to

$$\xi = \frac{L_q I_s}{L_d I_s} = \frac{L_q}{\psi_m} I_s \quad \Rightarrow \quad \xi = \frac{L_{qn}}{\psi_n} \quad \text{with normalized variables.} \qquad (7.73)$$

It looks like a parabolic curve of the form $y = c/x$, when $x = \psi_n$ changes. It supports the rationale why the separatrix has a curved decreasing function shape as shown in Fig. 7.25.

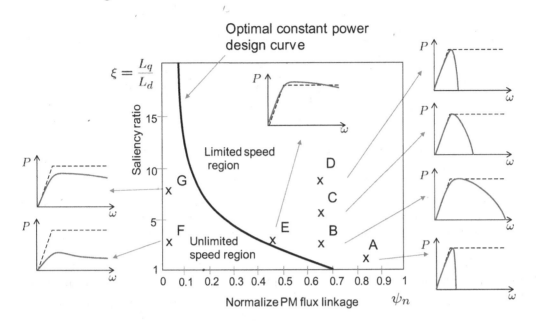

Figure 7.25: Optimal design curve for wide CPSR and power profiles.

Design B, C, and D that belong to the upper right region are the cases with different saliency for the same PM flux. Power of design D drops off quickly, because the PM flux is too high compared with the current capability. In general, the power profile becomes sharpened as the design point moves up away from the optimal design curve.

On the other hand, a motor belonging to the lower left region can maintain a constant power to a very high speed. But the torque, as well as the power, is low. Design E is a very desirable profile which has the peak power just after the rated speed and then keeps almost constant level. Such designs are found in a narrow band of the optimal design curve.

Exercise 7.12   Draw a power profile curve with the normalized variables using Excel. The voltage limit is determined by

$$ i_{qn} = \frac{1}{\xi L_{dn}} \sqrt{\frac{1}{\omega_n^2} - (L_{dn}i_{dn} + \psi_n)^2}. $$

Let $L_{dn} = 0.9$ and $\psi_n = 0.75$. For $\xi = 2.5$, find the intersections with $i_{dn}^2 + i_{qn}^2 = 1$ while increasing $\omega_n \geq 1$. For each intersection, calculate

$$ P_{en} = \omega_n T_n = \omega_n[\psi_n i_{qn} - (\xi - 1)L_{dn}i_{dn}i_{qn}]. $$

Draw a curve (ω_n, P_{en}) for $\omega_n \geq 1$. Repeat the above process for $\xi = 5$ and 10.

References

[1] Y. Mizuno et al., "Development of New Hybrid Transmission for Compact-Class Vehicles," *Proc. of SAE, Paper No. 2009-01-0726,* 2009.

[2] D. W. Novotny and T. A. Lipo, *Vector Control and Dynamics of AC Drives,* Clarendon Press, Oxford 1996.

[3] W.L. Soong and T.J.E. Miller, "Field-weakening performance of brushless synchronous AC motor drives," *IEE Proc-Electr. Power Appl.,* vol. 141, no. 6, pp. 331–340, Nov. 1994.

[4] D. G. Luenberger, *Linear and Nonlinear Programming,* 2nd Ed., Kluwer Academic Publisher, Norwell, 2004.

[5] A. Consoli, G. Scarcella, G. Scelba, and A. Testa, "Steady-state and transient operation of IPMSMs under maximum-torque-per-ampere control," *IEEE Trans. Ind. Appl.,* vol. 46, no. 1, pp. 121–129, Jan./Feb. 2010.

[6] S. Kim, Y. D. Yoon, S. K. Sul, and K. Ide, "Maximum torque per ampere (MTPA) control of an IPM machine based on signal injection considering inductance saturation," *IEEE Trans. Power Electron.,* vol. 28, no. 1, pp. 488–497, Jan. 2013.

[7] A. Ahmed, Y. Sozer, and M. Hamdan, "Maximum torque per ampere control for buried magnet PMSM based on DC-link power measurement," *IEEE Trans. Power Electron.,* vol. 32, no. 2, pp. 1299–1311, Feb. 2017.

[8] T. Inoue, Y. Inoue, S. Morimoto, and M. Sanada, "Mathematical model for MTPA control of permanent-magnet synchronous motor in stator flux linkage synchronous frame," *IEEE Trans. Ind. Appl.,* vol. 51, no. 5, pp. 3620–3628, Sep./Oct. 2015.

[9] A. Shinohara, Y. Inoue, S. Morimoto, and M.Sanada, "Direct calculation method of reference flux linkage for maximum torque per ampere control in DTC-based IPMSM drives" *IEEE Trans. Power Electron.,* vol. 32, no. 3, pp. 2114–2122, Mar. 2017.

[10] Stewart Ian, *"Galois Theory",* Chapman and Hall/CRC, 2003.

[11] Gerolamo Cardano, *"The Rules of Algebra: (Ars Magna)",* Dover Publications, 2007.

[12] Daniel Zwillinger, *"CRC Standard Mathematical Tables and Formulae",* CRC Press, 1996.

[13] Ronald S. Irving, *"Integers, Polynomials, and Rings: A Course in Algebra",* Springer, 2003.

[14] Blinn J. F., "Quartic discriminants and tensor invariants", *IEEE Computer Graphics and Applications*, vol. 22, no. 2, pp. 86–91, Mar./Apr. 2002.

[15] S. Morimoto, Y. Takeda, T. Hirasa, K. Taniguchi, "Expansion of Operating Limits for Permanent Magnet Motor by Current Vector Control Considering Inverter Capacity," *IEEE Trans. on Ind. Appl.*, vol. 26, no. 5, pp. 866–871, 1990.

[16] T. Jahns, "Getting rare-earth magnets out of EV traction machines," *IEEE Electrification mag.*, vol. 5, no. 1, pp. 6–18, Mar. 2017.

Problems

7.1 It is desired to design an IPMSM whose rated power is 36 kW at the base speed, 3000 rpm. Assume that $k_w = 0.96$, $B_m = 0.8$ T and $A_m = 550$ A/cm. Calculate T_e and the stack length, l_{st}, when the rotor diameter is $D_r = 11$ cm.

7.2 Plot the magnet and reluctance torque components for $I_m = 30$ A versus speed above 2600 rpm for the IPMSM listed in Table 7.2.
Repeat the same when $L_d = 2.2$ mH.

7.3 Consider an IPMSM with the parameters: Number of poles, $P = 8$, $\psi_m = 0.0927$ Wb, $L_d = 289$ μH, $L_q = 536$ μH, $I_m = 240$ A.

a) Draw the voltage and current vector diagram at 6000 rpm when $\beta = 70.6°$, and at 12000 rpm when $\beta = 80.5°$. Determine whether the phase voltage is acceptable when the inverter DC link voltage is 360 V. (Hint: The inverter DC link voltage is equal to the peak of the maximum line-to-line voltage.)
b) Determine the maximum current angle, β under $V_m = 360/\sqrt{3}$ when the (braking) torque is -100 Nm at 6000 rpm. Draw the voltage and current vector diagram at that moment.

7.4 Consider an IPMSM designed for integrated starter and generator. The motor parameters are; Number of poles, $P = 8$, $L_d = 20$ μH, $L_q = 60$ μH. The peak (phase) back EMF is 60 V at 16500 rpm. The inverter DC link voltage is 48 V and the maximum current is $I_m = 240$ A. Plot the maximum torque and power versus speed up to 16500 rpm.

7.5 Obtain the per-unit model for the IPMSM listed in Table 7.2. Note that the base currents are $(i_{db}, i_{qb}) = (-21.7, 33.6)$ A.

7.6 Consider an IPMSM with the following parameters: Motor power: 100 kW, rated speed: 2600 rpm, rated torque: 300 Nm $L_d = 0.1$ mH, $L_q = 0.35$ mH, $r_s = 26$ mΩ, $\psi_m = 0.052$ Wb, Rated current: 354 A_{rms}, rated voltage: 102 V_{rms}, $P=8$. The base currents are $(i_{db}, i_{qb}) = (-314, 388)$ A.

a) Determine the normalized model.

b) Show that $v_{dn}^2 + v_{qn}^2 = 1$ when $\omega_n = 1$.

c) Determine the critical speed, ω_{cn} and ω_c in rpm.

7.7 Consider an eight pole motor with the following parameters: $\psi_m = 0.053$ Wb, $L_d = 234$ μH, and $L_q = 562$ μH. Assume that $V_m = 180$ V and $I_m = 450$ A. Based on (7.44)–(7.46), find the break point ω_{eb} for the max torque/flux branch from the current limit circle. (Hint: Using Excel find ω_e that satisfies $i_d^{e\,2} + i_q^{e\,2} = I_m^2$.)

7.8 Consider an IPMSM with the following parameters: Number of poles, $P = 4$, $L_d = 4$ mH, $L_q = 11$ mH, back EMF constant $= 200$ V/krpm, maximum phase current $= 3.5$ A(rms), and maximum phase voltage $= 162$ V(rms).

a) Find an intersection with the MTPA curve and the current limit circle.

b) Find the speed, ω_{ea} at which the voltage limit passes the intersection point of a).

c) Find the break speed where the max torque/flux contour starts during max power operation.

d) Plot the maximum torque and power curves versus speed for $0 \le \omega_e \le 2\omega_{eb}$. Assume that the motor produces the maximum torque for $0 \le \omega_e \le \omega_{ea}$.

7.9 Consider an eight pole motor with the following parameters: $\psi_m = 0.103$ Wb, $L_d = 234$ μH, and $L_q = 562$ μH. Assume that $V_m = 180$ V. Based on (7.44)-(7.46), calculate the maximum torque at 14000 rpm. Using the Ferrari's method (7.49), find the maximum torque at 14000 rpm.

7.10 Consider a PMSM with parameters: $\psi_m = 0.0948$ Wb, $V_{dc} = 300$ V, $L_d = 0.00305$ H, $L_q = 0.0062$ H, $I_s = 40$ A.

a) Find an MTPA point on $i_d^{e\,2} + i_q^{e\,2} = I_s^2$.

b) Find the electrical speed, ω_e such that the voltage limit contour intersects the MTPA point.

c) Plot the current angle β for $I_s = 20$ A, 40 A, and 60 A when the PM flux ψ_m changes from 0 to 0.50 Wb.

7.11 Consider a six pole IPMSM with the following parameters: $V_{dc} = 300$ V, $L_d = 0.00305$ H, $L_q = 0.0062$ H, $\psi_m = 0.0948$ Wb.

a) Plot the voltage limit contour for ω_r=7200 rpm when $r_s = 0$ Ω.

b) Plot the voltage limit contour for ω_r=7200 rpm when $r_s = 2$ Ω.

7.12 Consider an eight pole IPMSM with the following parameters: $r_s = 0.03$ Ω, $L_d = 0.00305$ H, $L_q = 0.0062$ H, $\psi_m = 0.0948$ Wb.

a) Draw the voltage vector in the current plane and obtain the power factor angle when $i_d^e = -150$ A and $i_q^e = 180$ A (motoring).

b) Draw the voltage vector in the current plane and obtain the power factor angle when $i_d^e = -150$ A and $i_q^e = -180$ A (regeneration).

c) Discuss the role of ohmic drop, $r_s I_s$ in the magnitude of $\mathbf{V_s}$ for a) and b).

7.13 Consider an eight pole IPMSM with the following parameters: $L_d = 0.00305$ H, $L_q = 0.0062$ H, $\psi_m = 0.0948$ Wb, $V_{dc} = 360$ V, $I_s = 450$ A.

a) Using Excel, calculate MTPV points for $\omega_r = 12000, 11900, 11800, \cdots,$ 5000 rpm.

b) Find the speed at which the MTPV line intersects the current limit.

Chapter 8

Magnetism and Motor Losses

The motor loss consists mainly of copper loss and iron loss. Copper loss indicates the Joule loss of the stator coil, whereas iron loss refers to the core losses. Since there is no secondary winding in the PMSM, the copper loss of a PMSM is lower than that of the corresponding IM. For this reason, PMSMs generally have higher efficiency than IMs. The iron loss takes place in a magnetic material that works in the alternating field. It is generally separated into hysteresis and eddy current losses. The former comes from the hysteresis band of the core, while the latter is caused by magnetic induction. The hysteresis loss is higher than the eddy current loss below the base frequency. The eddy current loss, however, becomes comparable in the high frequency range. In this chapter, basics of magnetism are discussed and the loss models are developed from the perspective of electromagnetics.

8.1 Soft and Hard Ferromagnetism

In most atoms, opposite magnetic spins are paired, such that the net magnetic moment of orbiting electrons is zero. However, in the atoms of ferromagnetic materials (i.e., iron, cobalt, and nickel), such a pairing is not complete and results in non-zero magnetic moments. Furthermore, the magnetic moments are aligned in a small area. The group of aligned atoms is called domain, and it looks like a small magnet. A ferromagnetic material looks segmented into many small domains, as shown in Fig. 8.1. In an unmagnetized state, the domains are random, so that the magnetic moments are averaged to zero. When an external magnetizing field is applied, the areas of domains change depending on the degree of alignment: Aligned domains with the external field will expand, whereas opposite field domains will shrink. If the external field is very strong, the domains will stay aligned after the field is removed. That means there is a yielding point after which the state cannot be restored to its original state. It illustrates the mechanism of generation of permanent magnetism. Fig. 8.1 shows the domain variation in response to the external field.

Fig. 8.2 shows a C-core with a magnetic specimen in the core gap. An N-turn coil is wound over the C-core, and the field density, B is controlled by current change. The permeability of C-core is assumed to be infinite. Fig. 8.2 (b) shows

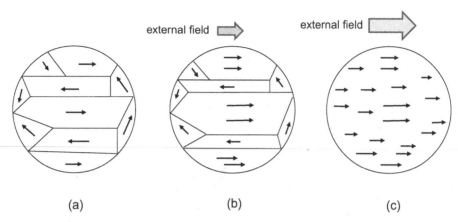

Figure 8.1: Domains of a ferromagnetic material. Each area changes under the influence of an external field: (a) unmagnetized state without an external field, (b) under the field of a medium strength, and (c) a state of complete magnetization.

a typical $B - H$ curve. An unmagnetized ferromagnetic material will follow the dashed line from the origin, o. As the field intensity, H increases, more and more domains tend to be aligned with the external field. Thus, B increases in proportion to the current, i. Its slope is defined as the permeability, μ. When H is very high, almost all magnetic domains are aligned. In such a case, an additional increase in B is hardly expected. It is a point of magnetic saturation, and represented by a point, d in the $B - H$ plane. Once the magnet material is saturated, the local (incremental) permeability is drastically reduced.

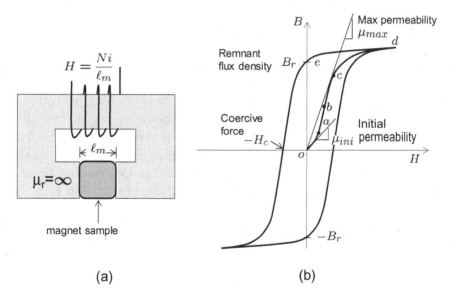

Figure 8.2: (a) C-core for $B - H$ curve measurement and (b) $B - H$ curve.

If H is reduced from d, the B-field contour does not follow the same path. Instead, it returns to an interception point, e in the B-axis. It is called remnant flux density, B_r because some fields remain even though the magnetizing force is zero. As the current increases in the negative direction, the B-field reduces to zero. The intersection of H-axis means coercive force, or coercivity of the material. It represents the force required to remove the residual magnetism from the material. A high coercive value is an indicator of intrinsic stability of the magnet material. If H continues to increase in the negative direction, the material is magnetized in the opposite direction, and will be saturated again. The complete closed loop is called a hysteresis loop. The whole contour is symmetric with respect to the origin. Three important terms are summarized in the following:

> **Remnant flux density** (B_r) - The residual flux density of a material when the magnetizing force H returns to zero.
> **Coercive force** (H_c)- Amount of reverse H-field required to make the flux density of a material zero.
> **Permeability** (μ) - The rate of field density increase with respect to the magnetizing force, H.

8.1.1 Permanent Magnet

The coercivity H_c represents a resilience against demagnetization. PMs are made of hard magnetic materials that have large coercivity force. Including natural induction, the normal induction line of PM appears as a tilted straight line in the second quadrant, and is called *demagnetization curve*. Nowadays, the remnant flux density and coercivity of a sintered Nd-Fe-B magnet reach 1.4 T and 2000 kA/m, respectively. Hard permanent magnets have a relative recoil permeability, $\mu_r = 1 \sim 1.1$. Therefore, the slope of recoil line has 45° when the horizontal axis is plotted along with $\mu_0 H$ [1].

8.1.2 Air Gap Field Determination

Now consider a C-core including a piece of PM in the middle section as shown in Fig. 8.3(a). Assume that the PM has a uniform cross sectional area of A_m, the PM height is ℓ_m. Assume also that the air gap area and height are A_g and g, respectively. Further assume that the core permeability is infinity. From the Ampere's law, it follows that

$$\oint \mathbf{H} \cdot d\ell = H_m \ell_m + \frac{B_m A_m}{\mu_0 A_g} g = 0. \tag{8.1}$$

Then we have $B_m = -\frac{\ell_m}{g}\frac{A_g}{A_m}(\mu_0 H_m)$. Thus, the permeance coefficient is equal to

$$P_c \equiv -\frac{B_m}{\mu_0 H_m} = \frac{A_g/g}{A_m/\ell_m} \approx \frac{\mathcal{R}_m}{\mathcal{R}_g} = \frac{\mathcal{P}_g}{\mathcal{P}_m}, \tag{8.2}$$

where $\mathcal{R}_m = 1/\mathcal{P}_m$ and $\mathcal{R}_g = 1/\mathcal{P}_g$ are the reluctances of magnet and air gap, respectively. The permeance coefficient is simply given as $\frac{\ell_m}{g}$ if $A_m \approx A_g$. That is, the permeance coefficient is determined by the ratio of PM thickness to air gap height.

Solving the permeance curve, $y = -\frac{\ell_m}{g}x$ with the demagnetization curve, $y = x + B_r$, we obtain the gap field, B_g from the intersection. The slope of permeance

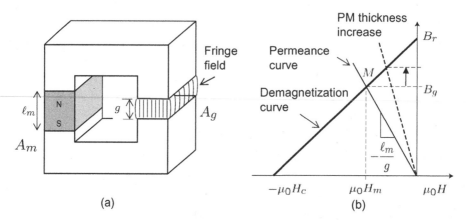

Figure 8.3: (a) C-core with a PM and (b) demagnetization and permeance curves for air gap field determination.

curve increases with the magnet length, ℓ_m as shown in Fig. 8.3(b). Then the gap field increases as the intersection moves up.

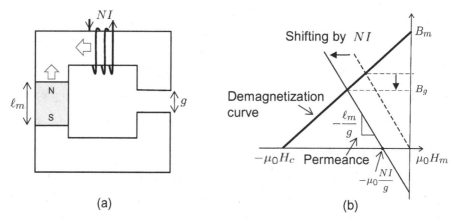

Figure 8.4: C-core with a PM and a winding, and a diagram for air gap field determination.

Now consider a C-core with a PM and an N-turn coil conducting current I, as shown in Fig. 8.4(a). Since the current enforces a reverse field, the air gap field density drops. From the Ampere's law, we have

$$H_m\ell_m + H_g g = NI. \tag{8.3}$$

Therefore,

$$B_g = -\frac{\ell_m}{g}(\mu_0 H_m) - \mu_0\frac{NI}{g}. \tag{8.4}$$

As shown in Fig. 8.4(b), the coil MMF shifts the permeance curve to the left horizontally. The MMF by ampere-turn, NI acts against MMF by PM, thereby the gap field reduces. It illustrates the gap field reduction in the field weakening operation of PMSM.

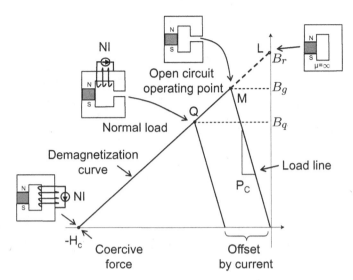

Figure 8.5: $B - H$ demagnetization curve of a PM material.

When the demagnetizing current reduces gradually, the flux density returns to the original point. The minor hysteresis loop takes place in the recovering process from recoiling. But, it can be approximated usually by a straight line. Change of operating points along the recoil line is reversible, i.e., the original state is recovered without a loss of magnetism. Fig. 8.5 shows a demagnetization curve with different magnet circuits.

Exercise 8.1   Consider a Nd-magnet with $B_r = 1.3$ T and $\ell_m = 5$ mm. Suppose that the air gap height is $g = 0.8$ mm. Determine the air gap field density.

Solution. The slope of permeance curve is $5/0.8 = 6.25$. It is needed to solve $y = x + 1.3$ and $y = -6.25x$. The solution is equal to $x = -0.179$. Therefore, the gap field is equal to $B_g = 1.12$ T.   ∎

8.1.3   Temperature Dependence and PM Demagnetization

Fig. 8.6 (a) shows the change of a demagnetization curve of a Nd-Fe-B magnet as the temperature increases. If thermal energy is added, magnetic moments fluctuate more causing field reduction. It is described as a depression of the demagnetization curve as shown in Fig. 8.6. Both the remanent flux density and coercive force decrease with temperature. The deflection point of a normal curve is called *knee point*. It resides normally in the third quadrant for temperature below 80°. However, as the temperature increases the knee point migrates upward and appears in the second quadrant.

Permanent loss of PM flux takes place in some parts under a high current loading and high temperature. It is predicted by the intersection of a load line below the

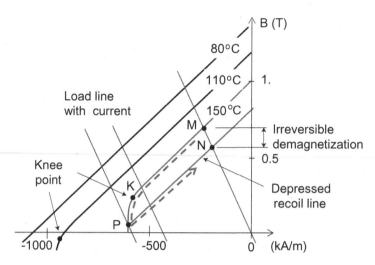

Figure 8.6: Temperature dependance of a Nd-magnet: The demagnetization curve is depressed with temperature while the knee point elevates.

knee point. Consider the $150°C$ demagnetization curve shown in Fig. 8.6. Suppose that the load line is shifted by a current injection and crosses the demagnetization curve above the knee point, K. Then the operation point recoils to M following the original curve as the current reduces. It is a reversible process during which no permanent flux loss occurs.

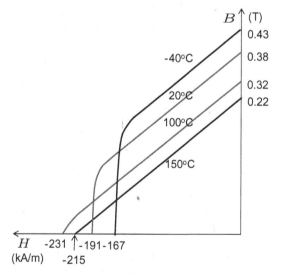

Figure 8.7: Temperature dependence of a ferrite magnet.

On the contrary, if the operation point is pushed down to P which is below K, then the operation point traces back to N following another depressed recoil line, $\overline{PN}$. The original flux density is not recovered even after the magnet is cooled down to the room temperature. The loss is irreversible, and such a perpetual damage is called *irreversible demagnetization*.

Note from Fig. 8.7 that ferrite magnets have a different temperature dependence. The coercive force reduces, as the temperature decreases. Thus, demagnetization takes place at a cold temperature.

8.1.4 Hysteresis Loss

Hysteresis loss is caused by intermolecular friction when a varying magnetic field is applied to the magnetic material. Fig. 8.8 shows a core and winding to illustrate the hysteresis loss. The area enclosed by the core flux is denoted by S, and its boundary is ∂S. On the other hand, S_c and ∂S_c denote the coil loop area and its boundary in the transverse plane of the field. Denote by $\mathbf{E}$ the electric field between the terminal. The electrical power delivered to the core via the coil is equal to

$$P = v \times i = \int_{\partial S_c} (\mathbf{E} \cdot d\boldsymbol{\ell}) \iint_S (\mathbf{J} \cdot d\mathbf{S}), \tag{8.5}$$

where v is the coil terminal voltage and i is the coil current. From the Ampere's law, it follows that

$$i = \iint_S \mathbf{J} \cdot d\mathbf{S} = \oint_{\partial S} \mathbf{H} \cdot d\boldsymbol{\ell}. \tag{8.6}$$

Also from the Faraday law, we obtain

$$v = \int_{\partial S_c} \mathbf{E} \cdot d\boldsymbol{\ell} = -\frac{\partial}{\partial t} \iint_{S_c} \mathbf{B} \cdot d\mathbf{S}. \tag{8.7}$$

Substituting (8.6) and (8.7) into (8.5), it follows that

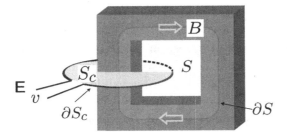

Figure 8.8: Illustration of hysteresis loss.

$$
\begin{aligned}
P &= -\oint_{\partial S} \mathbf{H} \cdot d\boldsymbol{\ell} \left(\iint_{S_c} \frac{\partial \mathbf{B}}{\partial t} \cdot d\mathbf{S} \right) \\
&= -\frac{\partial}{\partial t} \iiint_{vol} \left(\int_B \mathbf{H} \cdot d\mathbf{B} \right) d(vol),
\end{aligned} \tag{8.8}
$$

where *(vol)* is the core volume. The amount of change in the stored magnetic energy
is equal to

$$W = \iiint_{vol} \left(\int_0^B \mathbf{H} \cdot d\mathbf{B} \right) d(vol) \tag{8.9}$$

Therefore, the integral

$$\int_0^B \mathbf{H} \cdot d\mathbf{B} \tag{8.10}$$

implies the magnetic energy per volume when the flux density changes from zero to
B.

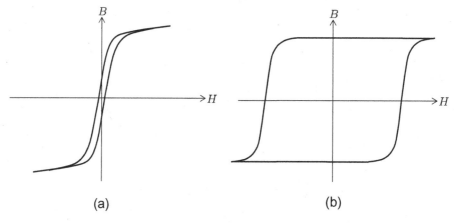

Figure 8.9: Hysteresis loops: (a) soft and (b) hard magnet materials.

　　Fig. 8.9 shows the two different types of hysteresis loop. Note that the area
signifies the power loss per a current cycle. It is called *hysteresis loss*. Therefore,
the hysteresis loss increases in proportion to the frequency. Soft magnetic materials
are characterized by the small loop area, and suited for AC applications such as
transformer, motor, AC reactor, etc. Meanwhile, hard magnetic materials have
large coercive force, and are used for making permanent magnets.
　　In general, the hysteresis loss of a soft magnetic material is approximated by

$$P_{hys} = c_h f B_m^n, \qquad \text{(W/kg)}, \tag{8.11}$$

where $n = 1.5 < n < 2.5$, c_h is a constant determined by the property of the
ferromagnetic material, f is the frequency of excitation, and B_m is the peak value
of the flux density. The exponent, n depends on the core material but is often given
as 1.6 [2].

8.1.5　Skin Depth and Eddy Current Loss

When the magnetic field that flows through a conductive material changes its
strength, circular currents are induced inside the material in such a way that their

associated field counteracts the change of the original magnetic field. This weakens the field density within the material, while causing Joule loss. The field density decreases exponentially with increasing depths in the conductor.

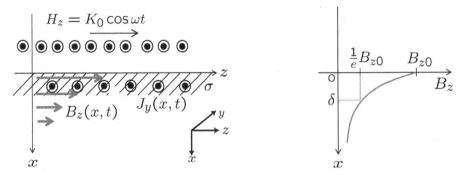

Figure 8.10: Eddy current and skin depth.

Look at a current sheet parallel over a conductive material as shown in Fig. 8.10. It forces a field, B_z inside the material. Next if the current varies then it will cause a current induction in the material and the induced current, called *eddy current*, alters the net field [3]. Since the eddy current is interactive with the net field, both current and field must be considered simultaneously to solve the problem.

The current flowing in the coil is modeled as current sheet $K_0 \cos \omega t$, and the induced eddy current is denoted by $J_y(x,t)$. Note that $J_y = \sigma E_y$, where σ is the conductance of the material. From the Faraday's law, we obtain in the steady state

$$\frac{1}{\sigma} \frac{\partial J_y}{\partial x} = -j\omega B_z, \tag{8.12}$$

Therefore,

$$B_z = \frac{j}{\omega \sigma} \frac{\partial J_y}{\partial x}. \tag{8.13}$$

Together with the Ampere's law, $-\frac{\partial B_z}{\partial x} = \mu J_y$, the following second order differential equation follows:

$$\frac{\partial^2 B_z}{\partial x^2} + j\omega \mu \sigma B_z = 0. \tag{8.14}$$

The characteristic equation for (8.14) is $s^2 + j\omega \mu \sigma = 0$. Note also that $\sqrt{j} = \frac{1+j}{\sqrt{2}}$. The roots are

$$s = \pm j\sqrt{j\omega \mu \sigma} = \pm j\sqrt{\frac{\omega \mu \sigma}{2}(1+j)} = \pm j\frac{1+j}{\delta} \tag{8.15}$$

where

$$\delta = \sqrt{\frac{2}{\omega \mu \sigma}}. \tag{8.16}$$

The general solution is obtained as

$$B_z(x) = C_1 e^{-\frac{x}{\delta}} e^{j\frac{x}{\delta}} + C_2 e^{\frac{x}{\delta}} e^{-j\frac{x}{\delta}}. \tag{8.17}$$

The second term would go infinity as $x \to \infty$. Thus, $C_2 = 0$ from the physical perspective. Therefore,

$$B_z(x) = B_{z0}e^{-\frac{x}{\delta}}e^{-j\frac{x}{\delta}}. \tag{8.18}$$

Since $|e^{-j\frac{x}{\delta}}| = 1$, the amplitude drops to $1/e = 0.368$ at $x = \delta$. At a depth, $x = \delta$ from the surface, the magnitude decreases by 63%. The parameter, δ is called the *penetration depth* or *skin depth*. Note from (8.16) that the skin depth depends on the material properties and frequency, not on the magnitude of the field.

Exercise 8.2   The copper resistivity is 0.0168 $\mu\Omega$m. Calculate the skin depth of the copper plate at 2 kHz.

Solution. Note that $\mu_0 = 4\pi \times 10^{-7}$.

$$\delta = \sqrt{\frac{2 \times 0.0168 \times 10^{-6}}{2\pi \times 2000 \times 4\pi \times 10^{-7}}} = 1.459 \text{ mm.} \qquad \blacksquare$$

Eddy Current Loss Reduction by Lamination

The eddy current refers to induced circulating currents in a conducting material when it is subjected to alternating magnetic field. The associated ohmic loss is dissipated as heat. Along with the hysteresis loss, the eddy current loss is another large loss component in transformers and motors. For a given time-varying magnetic field, a decisive measure to reduce the eddy current loss is to laminate the core block as thin as possible. Then, the cross-sectional area is reduced, and it hinders the formation of current loops.

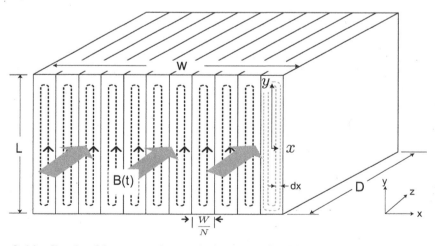

Figure 8.11: Stack of laminated metal sheets and eddy current loops.

Fig. 8.11 shows a lamination stack to which a time varying field is applied along the z-axis. It is assumed that the surface of each lamination is insulated. To see the lamination effects, consider a small loop whose area is $4xy$ in the vertical plane of a laminated piece. Then according to the Lenz's law, EMF around the loop is equal to

$$v = \frac{d\Phi}{dt} = \frac{4L}{W/N}x^2\frac{dB}{dt}.$$

Consider the loop of incremental width dx. The loop length is equal to $4(y + x) \approx 4y = 4\frac{L}{W/N}x$. Then, resistance of the loop is equal to

$$r_x = 4\frac{L}{W/N}\frac{x}{\sigma D dx},$$

where σ is the conductivity of the core material. Assume that $L \gg W/N$. Then the dissipated power in each incremental loop is

$$dP(x) = \frac{4\sigma LD}{W/N}x^3 dx \left(\frac{dB}{dt}\right)^2.$$

Therefore, the power loss, P_{sheet} in each sheet is

$$P_{sheet} = \int_0^{W/N} dP(x) = \frac{L\sigma DW^3}{16N^3}\left(\frac{dB}{dt}\right)^2.$$

Assume further that flux density is given by $B(t) = B_m \sin(2\pi ft)$. The total eddy current loss of the stack is

$$P_{st} = N \times P_{sheet} = \frac{\pi^2\sigma LDW^3}{4N^2}f^2 B_m^2 \cos^2(2\pi ft).$$

Taking the time-average, it follows that

$$P_{st} = \frac{\pi^2\sigma LDW^3}{8N^2}f^2 B_m^2. \tag{8.19}$$

It tells us that the eddy current loss is

1) proportional to the square of frequency; f^2
2) proportional to the square of field density; B_m^2
3) proportional to the conductivity; σ
4) inversely proportional to the square of lamination number; $1/N^2$.

However, (8.19) is not the exact loss estimate, since the secondary current effect is not taken into consideration. Specifically, the induced current also creates B-field and it has to be counted together with the external field, similarly to (8.12)-(8.14). Nevertheless, (8.19) gives a good loss trend. It needs to be stressed that the eddy current decreases drastically according to $1/N^2$.

8.1.6 Electrical Steel

Iron core is most widely used as a medium for the AC field. As a result, hysteresis and eddy current losses occur steadily during operation. It reduces machine efficiency and creates cooling problems. Alloy with silicon narrows the hysteresis loop of the material and increases the electrical resistivity of the steel. Thus it is good for reducing both hysteresis and eddy current losses. Silicon steel is made by adding 0.5~5% silicon (in weight) to the molten pure steel. However, as the silicon content increases, the steel becomes brittle and the saturation level decreases. The advantages of silicon steel are summarized as [4]:

(1) Increases the permeability;
(2) Reduces hysteresis loss;
(3) Increases the resistivity of the material;
(4) Prevents deterioration of magnetic properties when subjected to temperatures of about $100°C$ over long periods.

Reduction in thickness plays a crucial role in reducing the eddy loss. For transformers and motors, $0.35 \sim 0.5$ mm thickness silicon steel is most widely utilized. Further, the sheet surface is usually treated with an insulating coating using some organic or inorganic materials. Fig. 8.12 shows the B-H curve of a non-oriented electrical steel (27PNF1500) used for traction motors. Drastic saturation starts from $B = 1.5$ T.

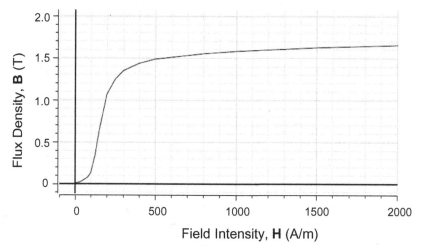

Figure 8.12: B-H curve of a non-oriented electrical steel (27PNF1500).

Various soft magnetic materials are made depending on the alloy components and the crystallization process. If electrical steel is rolled without a special processing to control the crystal orientation, magnetic properties are similar in all directions. Such a steel is called *isotropic* or *non-oriented* steel. However, in the *non-isotropic* or *grain oriented* steel, crystallization is optimized in the rolling direction. Thus, the magnetic flux density is increased by 30% in the coil rolling direction. Isotropic electrical steel is used for the motor core in which the field

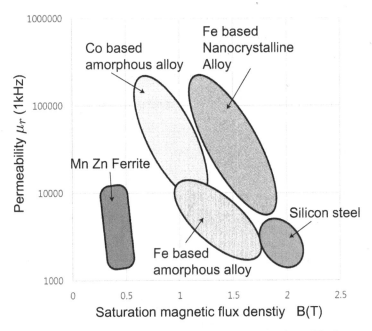

Figure 8.13: Permeability versus magnetic flux density for soft ferromagnetic materials.

rotates, whereas the grain oriented steel is used in the large power and distribution transformers.

High permeability helps to reduce the magnetizing current, which in turn reduces the use of core material, i.e., machine size. High (initial) permeability materials are good for the core of a noise filter, and high flux density materials are suited for high current inductors. Fig. 8.13 shows typical soft magnetic materials depending on the range of permeability and maximum flux density.

8.2 Motor Losses

Iron and copper losses are the main sources of the temperature increase in the motor. The proportion of two losses changes depending on the supply voltage and the operating speed. In general, the copper loss is more dominant over the iron loss.

Copper Losses

The copper loss is equal to

$$P_{cu} = \frac{3}{2}r_s I^2 = \frac{3}{2}r_s(i_d^2 + i_q^2). \tag{8.20}$$

Other than the DC loss, a significant amount of AC loss takes place especially in the bar winding. The AC loss is a high-frequency effect, which is considered to be a reduction of conduction area. It is resulted from skin and proximity effects. The skin effect is characterized by the current localization on the surface. Similarly, the proximity effect is also a current localization, but the source field is different: leakage field of a core, the field created nearby parallel conductors, or other stray field. Fig. 8.14 (b) and (c) shows the proximity effect caused by the slot leakage field. Current also interacts with the magnetic field of neighboring conductors.

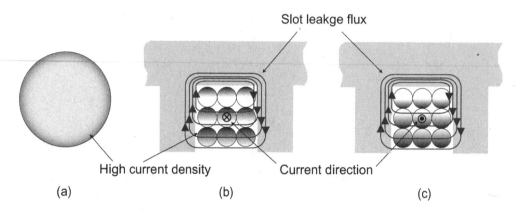

Figure 8.14: Current distribution – Dark area represents a high current density: (a) skin effect and (b), (c) proximity effect by slot leakage field.

Specifically, the stranded coils in a stator slot experience different level of leakage field. Stronger leakage field is applied to the bottom part of the coil (slot opening side), as shown in Section 3.3. Subsequently, current is distributed uneven. Dark area represents the area of higher current density. The current displacement in the rotor bar of induction motor is a similar consequence.

The decrease in effective conduction area by these nature is regarded as the AC resistance. Recall from (8.16) that the skin depth, δ decreases as the frequency increases. Thereby, the AC loss grows as the frequency increases. Nowadays, the motor speed range is wide and the number of poles is high, so that the AC loss is high in the stator windings [5]. It is reported in [6] that AC/DC loss ratio is about 2.3 at 800 Hz in a BLDCM.

Consider a slot of an electric machine which is filled with many stranded wires shown in Fig. 8.15. The wire group has z_t layers and the resistance increase is estimated differently based on the layers. The ratio of the effective AC resistance, r_{AC} to the DC resistances, r_{DC} is [7]:

$$k_{rk} = \frac{r_{AC}}{r_{DC}} = \varphi(\xi) + k(k-1)\psi(\xi) \qquad (8.21)$$

where

$$\varphi(\xi) = \xi \frac{\sinh(2\xi) + \sin(2\xi)}{\cosh(2\xi) - \cos(2\xi)}$$

$$\psi(\xi) = 2\xi \frac{\sinh(\xi) - \sin(\xi)}{\cosh(\xi) + \cos(\xi)}.$$

Note that ξ is the relative height of a conductor to the skin depth which is defined

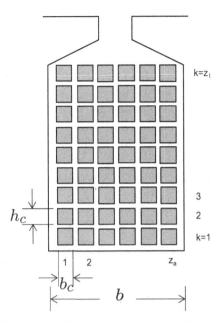

Figure 8.15: A model for the AG resistance calculation in a slot.

as

$$\xi = \frac{h_c}{\sqrt{\frac{2}{\omega\mu_0\sigma_c}\frac{b}{z_ab_c}}},$$

where σ_c is the conductivity of a conductor. The average resistance factor is approximated as for $\xi \leq 1$

$$k_{ra} = \left.\frac{r_{AC}}{r_{DC}}\right|_{avr} = \varphi(\xi) + \frac{z_t^2 - 1}{3}\psi(\xi) \approx \begin{cases} 1 + \frac{z_t^2 - 0.2}{9}\xi^4 & \text{rectangular section} \\ 1 + 0.59\frac{z_t^2 - 0.2}{9}\xi^4 & \text{circular section.} \end{cases}$$

Iron Loss

An empirical formula for iron loss based on the Steinmetz equation is given as

$$P_{fe} = c_{se}f^\alpha B_m^\beta, \tag{8.22}$$

where the coefficients, c_{se}, α and β are determined by fitting the loss model to curves from measurement data. The above loss model is modified as the generalized Steinmetz equation that utilizes a single-valued function of B and the rate of change dB/dt [8]:

$$P_{fe} = \frac{1}{T}\int_0^T c_{se}\left|\frac{dB}{dt}\right|^\alpha |B(t)|^{\beta-\alpha}dt, \tag{8.23}$$

An advantage is that the generalized equation has a DC-bias sensitivity [10].

Another approach is to separate the iron losses into static hysteresis loss and dynamic eddy current loss:

$$P_{fe} = c_h B_m^2 f + c_e B_m^2 f^2, \tag{8.24}$$

where c_e is the eddy current loss coefficient [9]. It has been proven correct for several nickel-iron (NiFe) alloys but lacks accuracy for silicon-iron (SiFe) alloys [10]. Including the excess loss, the iron loss of electrical steel is extended to

$$P_{fe} = c_h B_m^2 f + c_e B_m^2 f^2 + c_{ex}(B_m f)^{3/2}, \tag{8.25}$$

where c_{ex} is the coefficient of excess eddy current loss. The coefficients are calculated using the curve fitting of the iron loss data from manufacturers or from material test data. The above equation is based on the assumption of sinusoidal excitation.

Stray Loss

The stray losses are due to the higher winding space harmonics and slot harmonics. These losses take place in the surface layers of the stator and rotor adjacent to the air gap and in the volume of the teeth. The calculation of stray losses is difficult and does not guarantee a satisfactory accuracy. In practice, the stray losses are evaluated roughly according to

$$P_{str} = c_{str}(2\pi f)^2(i_d^2 + i_q^2), \tag{8.26}$$

where c_{str} is the stray loss coefficient [11].

Friction Loss

The friction loss refers to all mechanical drag forces when the rotor rotates. Typical motor friction losses are windage and bearing losses. A universal or brushed type DC motor includes additional losses caused by bushings or brushes. The frictional losses, in general, are proportional to the rotor speed, whereas windage losses are proportional to the cube of the rotor speed.

8.3   Loss Minimizing Control for IPMSMs

With the MTPA control, the copper loss is minimized. However, the iron loss should be also considered to improve the overall efficiency. The loss minimizing control (LMC) is an optimal control method that minimizes the total motor loss. The LMC problem becomes often complicated since it requires us to solve a quartic polynomial [8]. Furthermore, it turns out to be a constrained optimization problem especially in the field weakening range. Numerous researchers worked on the LMC problem [8]−[24]. Despite some on-line LMC methods [8], [22], look-up table (LUT) based control methods are popularly used in practice [23]. In the following, a Lagrange equation is utilized to find a loss minimizing solution. It is a summary of [23].

8.3.1   PMSM Loss Equation and Flux Saturation

In the following, the LMC is illustrated with an electric vehicle (EV) motor. Its picture and parameters are listed in Fig. 8.16 and Table 8.1.

Figure 8.16: Example EV propulsion motor.

Table 8.1: Parameters and coefficients of an IPMSM for an EV

Input DC link voltage [V]	240
Maximum output power [kW]	80
Maximum torque [Nm]	220
Maximum speed [rpm]	11000
Maximum phase current [A]	400
Rated output power [kW]	40
Rated torque [Nm]	133
Rated speed [rpm]	2600
Rated phase current [A]	216
Number of poles (P)	6
Permanent magnet flux (λ_m) [Wb]	0.07
Switching frequency [kHz]	8
Nominal d-axis inductance (L_d) [μH]	375
Nominal q-axis inductance (L_q) [μH]	835
Stator resistance (r_s) [mΩ]	29.5
Coefficient of iron loss (c_{fe})	2.1×10^{-2}

The stator flux is proportional to the stator current. However, as the current increases, core saturation develops gradually. Fig. 8.17 shows the measured values of fluxes by utilizing the steady state relations:

$$\lambda_d = \frac{1}{\omega}(v_q - r_s i_q),$$

$$\lambda_q = -\frac{1}{\omega}(v_d - r_s i_d).$$

But some values corresponding to high currents ($\sqrt{i_d^2 + i_q^2} \geq 500$ A) are extrapolated based on the data obtained from FEM calculation. Note that almost no saturation in λ_d takes place along d-axis current. However, saturation in λ_q becomes apparent as i_q increases. Based on the measured values of flux linkage, the inductances are

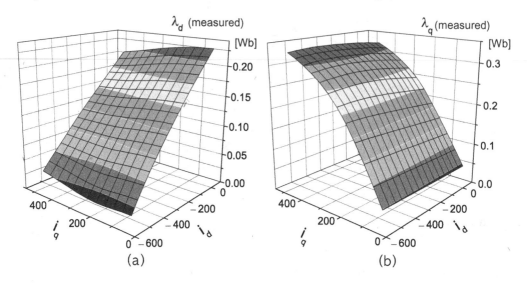

Figure 8.17: Measured stator flux linkages versus dq axis current: (a) d-axis flux linkage and (b) q-axis flux linkage.

calculated via numerical differentiation:

$$L_d = \frac{\partial \lambda_d}{\partial i_d} \approx \frac{\Delta \lambda_d}{\Delta i_d}\Big|_{i_q=const.},$$

$$L_q = \frac{\partial \lambda_q}{\partial i_q} \approx \frac{\Delta \lambda_q}{\Delta i_q}\Big|_{i_d=const.} \tag{8.27}$$

Fig. 8.18 shows the inductances versus current obtained by processing the experimental data according to (8.27). The saturation effect is noticeable in L_q as i_q increases. As indicated in Fig. 8.18 (b), the variation in L_q is approximated as a linear function in i_q such that

$$L_q = L_{q0}(1 - \alpha i_q), \tag{8.28}$$

where $\alpha > 0$ is a constant representing the slope.

Substituting $\lambda_d = L_d i_d + \psi_m$ and $\lambda_q = L_q i_q$ into the iron loss, the total loss P_t is equal to

$$
\begin{aligned}
P_t &= P_{cu} + P_{fe} + P_{str} \\
&= (\tfrac{3}{2}r_s + c_{str}\omega^2)(i_d^2 + i_q^2) + c_{fe}\omega^\gamma(\lambda_d^2 + \lambda_q^2) \\
&= k_1(\omega)i_d^2 + k_2(\omega)i_q^2 + k_3(\omega)i_d + k_4(\omega)
\end{aligned} \tag{8.29}
$$

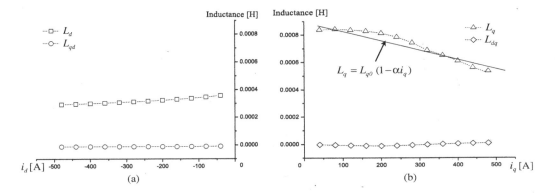

Figure 8.18: Measured values of (a) L_d and L_{qd} versus i_d when $i_q = 160$ A, (b) L_q and L_{dq} versus i_q when $i_d = -160$ A.

where $\gamma = 1.6$,

$$k_1(\omega) = \frac{3}{2}r_s + c_{fe}\omega^\gamma L_d^2 + c_{str}\omega^2,$$

$$k_2(\omega) = \frac{3}{2}r_s + c_{fe}\omega^\gamma L_q^2 + c_{str}\omega^2,$$

$$k_3(\omega) = 2c_{fe}\omega^\gamma L_d\psi_m,$$

$$k_4(\omega) = c_{fe}\omega^\gamma \psi_m^2.$$

8.3.2 Solution Search by Lagrange Equation

The goal is to find the loss minimizing current set (i_d, i_q) for a given torque value T_0 and speed ω. Further, there are voltage and current limits. The loss minimization is formulated as

$$\text{Minimize } P_t(i_d, i_q) = (\frac{3}{2}r_s + c_{str}\omega^2)(i_d^2 + i_q^2) + c_{fe}\omega^\gamma \left((L_d i_d + \psi_m)^2 + L_q^2 i_q^2\right)$$

$$\text{subject to } \quad \frac{3P}{4}(\psi_m i_q + (L_d - L_q)i_d i_q) - T_0 = 0, \tag{8.30}$$

$$(L_d i_d + \psi_m)^2\omega^2 + \omega^2(L_q i_q)^2 \le V_m^2, \tag{8.31}$$

$$i_d^2 + i_q^2 \le I_m^2. \tag{8.32}$$

Since the loss minimizing control problem is an optimization under inequality constraints, one may need to apply Kuhn-Tucker theorem in Section 7.4.8. However, the cases are separated depending on whether the optimal solution is found on the boundary or in the interior of the constraints.

First, consider the optimization in the interior. Let the Lagrangian be defined by

$$\mathcal{L}(i_d, i_q) = P_t(i_d, i_q) + \mu(T_e - T_0)$$

where μ is a Lagrange multiplier. The necessary conditions for the existence of the optimal solution are

$$
\frac{\partial \mathcal{L}(i_d, i_q)}{\partial i_d} = 3r_s i_d + 2c_{str}\omega^2 i_d + 2c_{fe}\omega^\gamma L_d^2 i_d
$$

$$
+ \; 2c_{fe}\omega^\gamma L_d \psi_m + \mu \frac{3P}{4}(L_d - L_q)i_q = 0.
$$

(8.33)

$$
\frac{\partial \mathcal{L}(i_d, i_q)}{\partial i_q} = 3r_s i_q + 2c_{str}\omega^2 i_q + 2c_{fe}\omega^\gamma L_q^2 i_q
$$

$$
+ \; \mu \frac{3P}{4}\psi_m + \mu \frac{3P}{4}(L_d - L_q)i_d = 0.
$$

(8.34)

Eliminate μ from (8.33) and (8.34) and replace i_q by $T_0/\frac{3P}{4}(\psi_m + (L_d - L_q)i_d)$. Then the following fourth order equation is derived:

$$
f_{\omega, T_0}(i_d) = A i_d^4 + B i_d^3 + C i_d^2 + D i_d + E = 0
$$

(8.35)

where

$$
A = \frac{27P^3}{64}(L_d - L_q)^3(3r_s + 2c_{str}\omega^2 + 2c_{fe}\omega^\gamma L_d^2)
$$

$$
B = \frac{27P^3}{64}\psi_m(L_d - L_q)^2(9r_s + 6c_{str}\omega^2 + 6c_{fe}\omega^\gamma L_d^2 + 2(L_d - L_q)c_{fe}\omega^\gamma L_d)
$$

$$
C = \frac{27P^3}{64}\psi_m^2(L_d - L_q)(9r_s + 6c_{str}\omega^2 + 6c_{fe}\omega^\gamma L_d^2 + 6(L_d - L_q)c_{fe}\omega^\gamma L_d)
$$

$$
D = \frac{27P^3}{64}\psi_m^3(3r_s + 2c_{str}\omega^2 + 2c_{fe}\omega^\gamma L_d^2 + 6(L_d - L_q)c_{fe}\omega^\gamma L_d)
$$

$$
E = \frac{27P^3}{32}\psi_m^4 c_{fe}\omega^\gamma L_d - \frac{9P}{4}(L_d - L_q)r_s T_0^2 - \frac{3P}{2}(L_d - L_q)c_{str}\omega^2 T_0^2
$$

$$
- \frac{3P}{2}(L_d - L_q)c_{fe}\omega^\gamma L_q^2 T_0^2
$$

Note that all the coefficients $A - E$ contains ω, and that only E includes torque T_0.

Fig. 8.19 shows the plots of (8.35) for different torques and speeds. They are close to straight lines in the region where $i_d < 0$, so that it is easy to find the zero crossing points which were marked by $\times$. The second row of Fig. 8.19 shows the curves of motor losses P_t versus i_d along the constant torque lines. The loss curves were calculated by utilizing (8.29). It should be noted that P_t has the minimum values where function f crosses zero, as predicted by the necessary conditions (8.30), (8.33), and (8.34) for optimality. That is, the minimum values of P_t are found at i_ds which satisfy $f_{\omega, T_0}(i_d) = 0$. The third row of Fig. 8.19 shows the plots of constant torque curves in the current plane with the current and voltage limits. The optimal points are also marked by $\times$. However, as the speed increases, the voltage limit curve shrinks. As a result, some solutions located out of the voltage limit, and marked by $*$. Those points should be replaced by the points on the boundary. That is, in such cases, the solution is found at an intersection between a voltage limit curve and a constant torque line [21]. The solution on the boundary was marked by $\otimes$. An algorithm to construct an LMC look-up table is shown in [23].

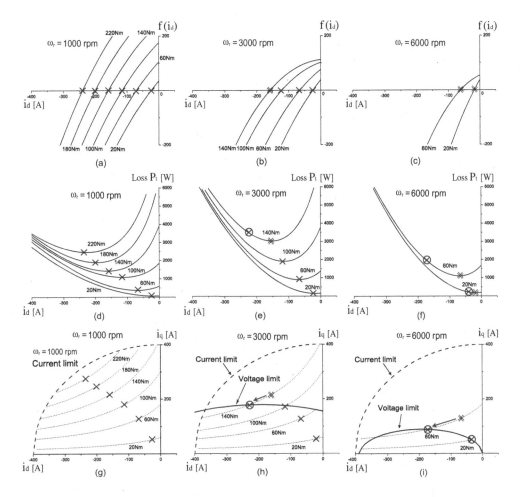

Figure 8.19: Calculated results illustrating how to find the loss minimizing (i_d, i_q) for three different speeds: (a), (b), and (c): plots of $f(i_d)$ for different torques, (d), (e), and (f): power loss curves along constant torque contours (g), (h), and (i): constant torque contours in the (i_d, i_q) plane.

8.3.3 LMC Based Controller and Experimental Setup

Fig. 8.20 shows a PMSM controller, which includes the LMC table. The LMC table requires torque and speed as the input variables, and provides an optimal current command. The LMC table is made under the largest DC link voltage. Thus, the optimal current may not be used due to a reduced voltage limit in the high speed range. Thus, a table output should be checked whether or not it is feasible under a given voltage limit. If it is not feasible, then the output value must be adjusted according to the following algorithm. It is a method of finding an optimal solution, (i_d^k, i_q^k) that satisfies the voltage limit:

i) Let $i_d^{k'} = i_d^k - \Delta i_d$;

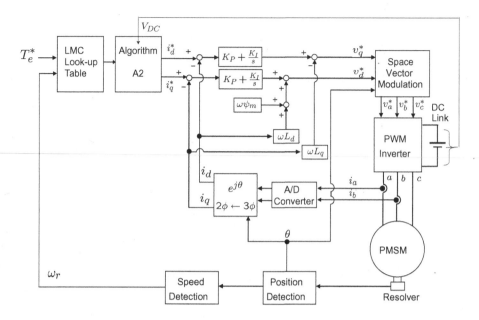

Figure 8.20: LMC structure for IPMSM.

ii) Find the corresponding q-axis current utilizing
$$i_q^{jk'} = \tfrac{4}{3P}T_0^k / \left(\psi_m + (L_d - L_q)i_d^{k'}\right);$$

iii) Check whether $(i_d^{k'}, i_d^{k'}) \in U_V^j$. If 'yes', stop. If 'no', let $(i_d^{k}, i_q^{k}) = (i_d^{k'}, i_q^{k'})$, and go to Step i).

As i)-iii) is repeated, the solution point moves to the left along the constant torque line T_0^k. Finally, the point reaches an intersection between the voltage limit and torque curves which is the desired solution.

The experimental environment is shown in Fig. 8.21. It includes the PI current controllers along with the decoupling and back-EMF compensation. The proposed LMC were implemented utilizing a floating-point DSP (MPC5554). The PWM switching frequency was selected to be 8 kHz and the dead time 2 μs. Current control routine was carried out every 125 μs, and torque command was refreshed at every 1.25 ms.

The PMSM under test was controlled in a torque control mode, and the dynamo induction motor was controlled in a speed control mode. Inverter output power was monitored by using a power analyzer. Multiplying a measured torque by the motor speed, the shaft power, $P_{out} = T_e \cdot \omega_r$ was obtained. Iron loss coefficient c_{fe} was selected to be 0.021 based on FEM calculation results around the nominal operating point, and we let $\gamma = 1.5$. It was assumed that the mechanical losses and the windage loss are small enough to be negligible. As a consequence, $P_t - P_{cu} - P_{fe}$ would be close to the stray loss, P_{str}. Based on this method, the coefficient, c_{str} was selected to be 6.5×10^{-9}. This approximation matches to a rough estimation, $P_{str} \approx 0.03 \sim 0.05 P_{out}$ [11], of the stray loss of small motors.

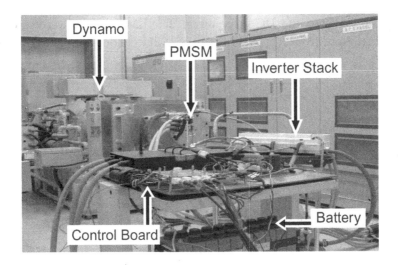

Figure 8.21: Photograph of the experimental setup.

8.3.4 Experimental Results

The parameters, as well as loss coefficients, c_{fe} and c_{str}, of the PMSM used in the experiment are summarized in Table 8.1. The loss minimizing current sets can be found by the experimental method that scans the motor losses over feasible mesh points in the current plane. Fig. 8.22 shows the loss minimizing (i_d, i_q) under fixed

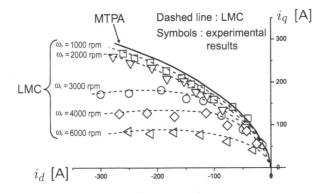

Figure 8.22: Loss minimizing (i_d, i_q) under fixed speeds 1000, 3000, and 6000 rpm, while torque is increasing.

speeds 1000~6000 rpm, while increasing the load torque. It compares computed results of the LMC (dotted line) with the results obtained from the experimental scanning method (symbol). This shows that the two methods agree in a wide area. Fig. 8.22 also shows the contour of MTPA. Note again that MTPA is independent of the speed, and thus it cannot reflect the iron loss or the stray loss, which are dependent upon the speed. The LMC results at 1000 rpm are similar to those of MTPA. Fig. 8.23 shows the plots of responses to a varying load torque at 2000 rpm. Note that both i_d and i_q change when the torque varies. It also displays the plots of measured shaft power and DC link power. The LMC results were compared with those of MTPA. Note from Fig. 8.23 (c) that LMC yielded lower loss than MTPA.

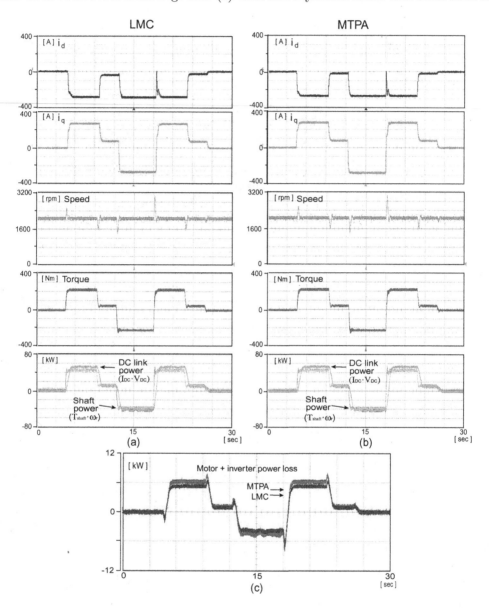

Figure 8.23: Current responses when load torque changes at 2000rpm: (a) LMC, (b) MTPA, and (c) loss comparision.

Summary

The optimal condition ended up with a fourth order polynomial in i_d. A zero crossing point of the polynomial was shown to be the loss minimizing point. The calculated minimizing solutions were compared with the values obtained from an experimental scanning method. The two results agreed well in a wide torque-speed range.

References

[1] J. R. Hendershot and TJE Miller, *Design of Brushless Permanent-magnet Motors*, Oxford Science Publications, New York, 1994.

[2] C. C. Mi, G. R. Slemon, and R. Bonert, "Minimization of iron losses of permanent magnet synchronous machines," *IEEE Trans. on Energy Convers.*, vol. 20, no. 1, pp. 121–127, Mar. 2005.

[3] J. Lammeraner and M. Stafl, *Eddy Currents*, CRC, 1966.

[4] J. McFarlane, "Silicon steel sheets – An outline of properties, applicationsa and recent developments," Symposium on Production, Properties and Applications of Alloy and Special Steels, NML, Jamshedpur, pp. 440–447, Feb. 1956.

[5] P.H. Mellor, R. Wrobel and N. McNeill, "Investigation of proximity losses in a high speed brushless permanent magnet motor", Proc. of 41st IEEE IAS Annual Meeting., vol. 3, pp. 1514–1518, Oct. 2006.

[6] M. Popescu and D. G. Dorrell, "Proximity Losses in the Windings of High Speed Brushless Permanent Magnet AC Motors with Single Tooth Windings and Parallel Paths," *IEEE Trans. Magn.*, vol. 49, no. 7, pp. 3913–3916, July 2013.

[7] J. Pyrhonen, T. Jokinen, and V. Hrabovkova, *Design of Rotating Electrical Machines*, 2nd. Ed., John, Wiley & Son. Inc., 2014.

[8] J. Li, T. Abdallah, and C. Sullivan, "Improved calculation of core loss with nonsinusoidal waveforms," in Industry Applications Conference, Thirty-Sixth IAS Annual Meeting. Conference Records, vol. 4, pp. 2203–2210, 2001.

[9] C. C. Mi, G. R. Slemon, and R. Bonert, "Modeling of Iron Losses of Permanent-Magnet Synchronous Motors," *IEEE Trans. on Ind. Appl.*, vol. 39, No. 3, pp. 734–742, May 2003.

[10] Andreas Krings et al., "Measurement and Modeling of Iron Losses in Electrical Machines," Laboratory of Electrical Energy Conversion, Royal Institute of Technology, 2014.

[11] J. F. Gieras and M. Wing, *Permanent Magnet Motor Technonology, Design Applications*, 2nd. Ed., Marcel, Dekker, Inc., New York, 2002.

[12] S. Jung, J. Hong, and K. Nam, "Current Minimizing Torque Control of the IPMSM Using Ferrari's Method," *IEEE Trans. Power Elec.*, vol. 28, no.12, pp. 5603–5617, Dec. 2013.

[13] J.C. Moreira, T.A. Lipo, and V. Blasko, "Simple efficiency maximizer for an adjustable frequency induction motor drive," *IEEE Trans. Ind. Appl.*, vol. 27, no. 5, pp. 940–946, 1991.

[14] R.D. Lorenz and S.M. Yang, "Efficiency-optimized flux trajectories for closed-cycle operation of field-orientation induction machine drives," *IEEE Trans. Ind. Appl.*, vol. 28, no. 3, pp. 574–580, 1992.

[15] S. Morimoto, Y. Tong, Y. Takeda, and T. Hirasa, "Loss minimization control of permanent magnet synchronous motor drives," *IEEE Trans. Ind. Electron.*, vol. 41, no. 5, pp. 511–517, Oct. 1994.

[16] Y. Nakamura, F. Ishibashi, and S. Hibino, "High-efficiency drive due to power factor control of a permanent magnet synchronous motor," *IEEE Trans. Power Electron.*, vol. 10, Issue 2, pp. 247–253, Mar. 1995.

[17] F. Fernandez-Bernal, A. Garcia-Cerrada, and R. Faure, "Model based loss minimization for DC and AC vector-controlled motors including core saturation," *IEEE Trans. Ind. Appl.*, vol. 36, no. 3, pp. 755–763, 2000.

[18] F. Abrahamsen, F. Blaabjerg, J.K. Pedersen, and P.B. Thoegersen, "Efficiency-optimized control of medium-size induction motor drives," *IEEE Trans. Ind. Appl.*, vol. 37, no. 6, pp. 1761–1767, 2001.

[19] S. Lim and K. Nam, "Loss-minimising control scheme for induction motors," *IEE Proc.-Electr. Power Appl.*, vol. 151, no. 4, July 2004.

[20] N. Bianchi, S. Bolognani, and M. Zigliotto, "High-performance PM synchronous motor drive for an electrical scooter," *IEEE Trans. Ind. Appl.*, vol. 19, no. 4, pp. 715–723, Dec. 2004.

[21] G. Gallegos-Lopez, F. S. Gunawan, and J. E. Walters, "Optimum torque control of permanent-magnet AC machines in the field-weakened region," *IEEE Trans. Ind. Appl.*, vol. 41, no. 4, pp. 1020–1028, Jul./Aug. 2005.

[22] C. Cavallaro, A. O. D. Tommaso, R. Miceli, A. Raciti, G. R. Galluzzo, and M. Tranpanese, "Efficiency enhancement of permanent-magnet synchronous motor drives by online loss minimization approaches," *IEEE Trans. Ind. Electron.*, vol 52, no. 4, pp. 1153–1160, Aug. 2005.

[23] J. Lee, K. Nam, S. Choi, and S. Kwon, "Loss-Minimizing Control of PMSM With the Use of Polynomial Approximations," *IEEE Trans. on Power Elec.*, vol 24, no. 4, pp. 1071–1082, Apr. 2009.

[24] M. Preindl and S. Bolognani, "Optimal State Reference Computation With Constrained MTPA Criterion for PM Motor Drives," *IEEE Trans. Power Elec.*, vol. 30, no.8, pp. 4524–4535, Aug. 2015.

Problems

8.1 Consider the toroidal core with $N = 20$-turn coil shown below. The core cross sectional area is $A = 1$ cm^2 and the mean periphery is $\ell = 30$ cm. Assume that AC current, $i(t) = 30 \sin 2\pi \times 100\, t$ A, flows in the coil. With the schematic hysteresis band shown below, estimate the hysteresis loss. Derive the energy loss, $W_h = A\ell \int_0^B H dB$ from $\int ei\, dt$.

Hint. The loss per cycle is $W_h = A\ell \int_{-1.6}^{1.6} H dB$.

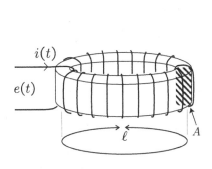
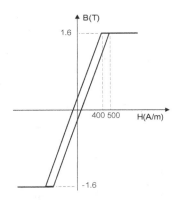

8.2   Consider a cylindrical core of radius, R with conductivity σ shown below.

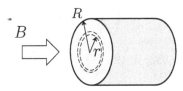
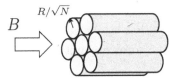

a) Suppose that a time-varying field B passes through the core from the front surface. Derive EMF around the loop of radius, r.

b) Derive the resistance of the loop.

c) Derive dissipated power, dP_r in each incremental loop.

d) Derive the total power loss when flux density is given by $B(t) = B_m \sin(2\pi f t)$.

e) Calculate the total power loss when the core is constructed with N small cylindrical core whose radius is $R/\sqrt{N}$.

8.3 Find the skin depth of pure copper and aluminum under 100 kHz time varying field.

8.4 Solve the following problem:

$$\text{Minimize } P_t(i_d, i_q) = \frac{3}{2}r_s(i_d^2 + i_q^2) + c_{fe}\omega^\gamma\left((L_d i_d + \psi_m)^2 + L_q^2 i_q^2\right)$$

$$\text{subject to } \frac{3P}{4}(\psi_m i_q + (L_d - L_q)i_d i_q) - T_0 = 0.$$

8.5 The classical eddy current loss is calculated as a sum of harmonic contributions:

$$P_{ec} = c_{ec}\sum_{n=1}^{\infty} B_n^2(nf)^2,$$

where B_n is the peak value of the magnetic flux density for the harmonic order n and f the fundamental frequency. Neglect the skin effect of the eddy current. Show that c_{ec} is written as:

$$c_{ec} = \frac{\pi^2 d^2}{6\rho\rho_\epsilon},$$

where d is the sheet thickness, ρ the sheet density and ρ_ϵ the specific electrical resistance of the steel.

8.6 Consider a PM block shown below. Suppose that $J_z(x) = \frac{x}{\rho_e}\frac{dB}{dt}$, where ρ_m is the specific electrical resistance of the PM. Then the eddy current loss per unit volume is equal to

$$k_m = \frac{1}{b_m}\int_{-\frac{b_m}{2}}^{\frac{b_m}{2}} \rho_m J_z^2(x)dx = \frac{b_m^2}{12\rho_m}\left(\frac{dB}{dt}\right)^2$$

Derive the total magnet loss when $B = B_m\sin(2\pi ft)$.

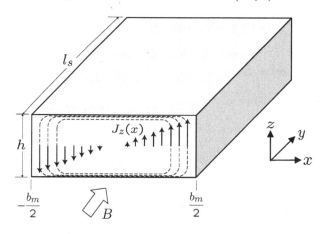

Chapter 9

PMSM Sensorless Control

Rotor position information is essential for the field orientation control of PMSMs. The position sensors are, in general, expensive and fragile under mechanical shock, increasing the drive cost. The sensorless control is a signal processing technique that aims at eliminating the position sensor. It utilizes only measured electrical quantities such as motor currents and voltages. Since it is an estimation algorithm set in a processor, it reduces the system cost. Various sensorless control techniques are widely used in home appliances, transportation system, cranes, etc., where the cost pressure is immense.

Numerous sensorless techniques have been proposed over two decades: They are categorized into back-EMF based methods and saliency based methods. Matsui pioneered the back-EMF based method for SPMSM [1]. Tomita et al.introduced a disturbance observer for an EMF estimation [2]. Ortega et al. studied several types of nonlinear observer for SPMSM [3]-[6]. The back-EMF based methods are advantageous in the medium and high speed regions, and the implementation is relatively simple. However performance is poor at low speed, and full torque cannot be handled at zero speed. Further, they are sensitive to the environment change such as temperature variation and noise.

In the saliency-based methods, a high-frequency signal has to be injected to detect the rotor position. Corley and Lorenz applied the heterodyne filtering technique in extracting a desired signal out of the modulated carrier [7]. Zhu et al. considered the cross-coupling magnetic saturation [8]. Aihara et al. combined a signal injection technique with a back-EMF based position estimation method [9]. A sliding mode current observer was utilized to find out the position and velocity estimates [10]. The influence of measurement errors and inverter irregularities in the performance of the sensorless control was studied by Nahid-Mobarakeh et al.[11]. Bolognani et al. applied the extended Kalman filter and suggested a guideline for choosing the noise covariance matrices [12]. Recently, PWM based voltage injection methods are studied widely for fast response and high signal to noise ratio [13]-[16].

9.1   IPMSM Dynamics in a Misaligned Frame

Since the exact rotor angle is not known, the field oriented control must be constructed based on an angle estimate. The PMSM dynamics in a misaligned coordinate frame carries additional terms caused by the angle estimation error. The additional terms look like unknown nonlinear disturbances to the system. Denote by $\bar{\theta}_e$ an angle estimate. The variables based on $\bar{\theta}_e$ are marked by overline. Let

$$\Delta\theta_e = \bar{\theta}_e - \theta_e, \tag{9.1}$$

$$\Delta\omega_e = \bar{\omega}_e - \omega_e. \tag{9.2}$$

Fig. 9.1 shows aligned and misaligned axes, and their difference.

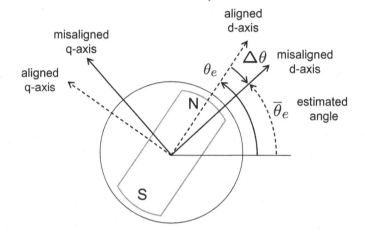

Figure 9.1: Misaligned (tilted) dq frame.

An IPMSM model based on $\bar{\theta}_e$ is developed in the following. Recall from (6.34) that the flux in the stationary frame is equal to

$$\lambda_{dq}^s = L_s i_{dq}^s - \frac{3}{2} L_\delta e^{j2\theta_e} i_{dq}^{s*} + \psi_m e^{j\theta_e}, \tag{9.3}$$

where $L_s = \frac{3}{2} L_{ss} + L_{ls}$. Transforming the flux (9.3) to the frame of $e^{j\bar{\theta}_e}$, we obtain

$$\overline{\lambda}_{dq}^e \equiv e^{-j\bar{\theta}_e} \lambda_{dq}^s = L_s e^{-j\bar{\theta}_e} i_{dq}^s - \frac{3}{2} L_\delta e^{j(2\theta - \bar{\theta}_e)} (i_{dq}^s)^* + \psi_m e^{-j\Delta\theta_e}.$$

Defining the current in the reference frame by $\overline{i}_{dq}^e = e^{-j\bar{\theta}_e} i_{dq}$, we have

$$\overline{\lambda}_{dq}^e = L_s \overline{i}_{dq}^e - \frac{3}{2} L_\delta (\overline{i}_{dq}^e)^* e^{-j2\Delta\theta_e} + \psi_m e^{-j\Delta\theta_e}.$$

Therefore,

$$j\bar{\omega}_e \overline{\lambda}_{dq}^e = j\bar{\omega}_e L_s \overline{i}_{dq}^e - j\bar{\omega}_e \frac{3}{2} L_\delta (\overline{i}_{dq}^e)^* e^{-j2\Delta\theta_e} + j\bar{\omega}_e \psi_m e^{-j\Delta\theta_e} \tag{9.4}$$

$$p\overline{\lambda}_{dq}^e = L_s p\overline{i}_{dq}^e - \frac{3}{2} L_\delta (p\overline{i}_{dq}^e)^* e^{-j2\Delta\theta_e} + 3L_\delta j\Delta\omega_e (\overline{i}_{dq}^e)^* e^{-j2\Delta\theta_e} - j\Delta\omega_e \psi_m e^{-j\Delta\theta_e}, \tag{9.5}$$

where $\mathbf{p} \equiv \frac{d}{dt}$ is the differential operator. Substituting (9.4) and (9.5) into (6.20), it follows that

$$
\begin{bmatrix} \overline{v}_d^e \\ \overline{v}_q^e \end{bmatrix} = r_s \begin{bmatrix} \overline{i}_d^e \\ \overline{i}_q^e \end{bmatrix} + \begin{bmatrix} L_s - \frac{3}{2}L_\delta \cos 2\Delta\theta_e & \frac{3}{2}L_\delta \sin 2\Delta\theta_e \\ \frac{3}{2}L_\delta \sin 2\Delta\theta_e & L_s + \frac{3}{2}L_\delta \cos 2\Delta\theta_e \end{bmatrix} \frac{d}{dt} \begin{bmatrix} \overline{i}_d^e \\ \overline{i}_q^e \end{bmatrix}
$$
$$
+ (3\Delta\omega_e - \frac{3}{2}\overline{\omega}_e)L_\delta \begin{bmatrix} \sin 2\Delta\theta_e & \cos 2\Delta\theta_e \\ \cos 2\Delta\theta_e & -\sin 2\Delta\theta_e \end{bmatrix} \begin{bmatrix} \overline{i}_d^e \\ \overline{i}_q^e \end{bmatrix}
$$
$$
+ \overline{\omega}_e L_s \begin{bmatrix} -\overline{i}_q^e \\ \overline{i}_d^e \end{bmatrix} + \omega_e \psi_m \begin{bmatrix} \sin \Delta\theta_e \\ \cos \Delta\theta_e \end{bmatrix} \tag{9.6}
$$

Note that if $\Delta\theta_e = -\theta_e$, $\Delta\omega_e = -\omega_e$, and $\overline{\omega}_e = 0$, then the misaligned equation (9.6) reduces to

$$
\mathbf{v}_{dq}^s = r_s \mathbf{i}_{dq}^s + \begin{bmatrix} L_s - \frac{3}{2}L_\delta \cos 2\theta_e & -\frac{3}{2}L_\delta \sin 2\theta_e \\ -\frac{3}{2}L_\delta \sin 2\theta_e & L_s + \frac{3}{2}L_\delta \cos 2\theta_e \end{bmatrix} \frac{d}{dt} \mathbf{i}_{dq}^s
$$
$$
- 3\omega_e L_\delta \begin{bmatrix} -\sin 2\theta_e & \cos 2\theta_e \\ \cos 2\theta_e & \sin 2\theta_e \end{bmatrix} \mathbf{i}_{dq}^s + \omega_e \psi_m \begin{bmatrix} -\sin \theta_e \\ \cos \theta_e \end{bmatrix} \tag{9.7}
$$

which is the same as the stationary IPMSM model, (6.37). That is, if estimated angle stays at zero, then the synchronous IPMSM model turns out to be stationary IPMSM model.

On the other hand, if we let $\Delta\theta_e = 0$, then (9.6) turns out to be

$$
\begin{bmatrix} \overline{v}_d^e \\ \overline{v}_q^e \end{bmatrix} = r_s \begin{bmatrix} \overline{i}_d^e \\ \overline{i}_q^e \end{bmatrix} + \begin{bmatrix} L_s - \frac{3}{2}L_\delta & 0 \\ 0 & L_s + \frac{3}{2}L_\delta \end{bmatrix} \frac{d}{dt} \begin{bmatrix} \overline{i}_d^e \\ \overline{i}_q^e \end{bmatrix}
$$
$$
- \frac{3}{2}\overline{\omega}_e L_\delta \begin{bmatrix} 0 & 1 \\ 1 & 0 \end{bmatrix} \begin{bmatrix} \overline{i}_d^e \\ \overline{i}_q^e \end{bmatrix} + \overline{\omega}_e L_s \begin{bmatrix} -\overline{i}_q^e \\ \overline{i}_d^e \end{bmatrix} + \omega_e \psi_m \begin{bmatrix} 0 \\ 1 \end{bmatrix}. \tag{9.8}
$$

Noting $L_d = L_s - \frac{3}{2}L_\delta$ and $L_q = L_s + \frac{3}{2}L_\delta$, (9.8) is equal to the IPMSM dynamic model.

9.1.1 Different Derivation Using Matrices

Dynamic model description in a new frame which is shifted angularly by $-\Delta\theta_e$, is achieved by multiplying

$$
e^{\mathbf{J}\Delta\theta_e} = \begin{bmatrix} \cos \Delta\theta_e & -\sin \Delta\theta_e \\ \sin \Delta\theta_e & \cos \Delta\theta_e \end{bmatrix} \quad \text{for} \quad \mathbf{J} = \begin{bmatrix} 0 & -1 \\ 1 & 0 \end{bmatrix}.
$$

to the voltage equation. The voltage and current vectors in the misaligned frame are given by $\overline{\mathbf{v}}_{dq}^e = e^{-\mathbf{J}\Delta\theta_e} \mathbf{v}_{dq}^e$ and $\overline{\mathbf{i}}_{dq}^e = e^{-\mathbf{J}\Delta\theta_e} \mathbf{i}_{dq}^e$. Then, the IPMSM dynamics in the misaligned coordinate is written as

$$
\overline{\mathbf{v}}_{dq}^e = e^{-\mathbf{J}\Delta\theta_e} \begin{bmatrix} r_s + \mathbf{p}L_d & -\omega_e L_q \\ \omega_e L_d & r_s + \mathbf{p}L_q \end{bmatrix} e^{\mathbf{J}\Delta\theta_e} \overline{\mathbf{i}}_{dq}^e + e^{-\mathbf{J}\Delta\theta_e} \begin{bmatrix} 0 \\ \omega_e \psi_m \end{bmatrix}. \tag{9.9}
$$

Note that the autonomous part has a form of the similarity transformation. In some literature, the following notations are used:

$$L_{av} \equiv \frac{L_d + L_q}{2} = L_s,$$

$$L_{df} \equiv \frac{L_q - L_d}{2} = \frac{3}{2}L_\delta.$$

Then,

$$e^{-\mathbf{J}\Delta\theta_e} \begin{bmatrix} pL_d & 0 \\ 0 & pL_q \end{bmatrix} e^{\mathbf{J}\Delta\theta_e} \bar{\mathbf{i}}_{dq}^e$$

$$= \Delta\omega_e \begin{bmatrix} L_{df}\sin 2\Delta\theta_e & -L_{av} + L_{df}\cos 2\Delta\theta_e \\ L_{av} + L_{df}\cos 2\Delta\theta_e & -L_{df}\sin 2\Delta\theta_e \end{bmatrix} \begin{bmatrix} \bar{i}_d^e \\ \bar{i}_q^e \end{bmatrix}$$

$$+ \begin{bmatrix} L_{av} - L_{df}\cos 2\Delta\theta_e & L_{df}\sin 2\Delta\theta_e \\ L_{df}\sin 2\Delta\theta_e & L_{av} + L_{df}\cos 2\Delta\theta_e \end{bmatrix} p \begin{bmatrix} \bar{i}_d^e \\ \bar{i}_q^e \end{bmatrix}. \tag{9.10}$$

Similarly, it follows that

$$e^{-\mathbf{J}\Delta\theta_e} \begin{bmatrix} 0 & -\omega_e L_q \\ \omega_e L_d & 0 \end{bmatrix} e^{\mathbf{J}\Delta\theta_e} \bar{\mathbf{i}}_{dq}^e$$

$$= (\bar{\omega}_e - \Delta\omega_e) \begin{bmatrix} -L_{df}\sin 2\Delta\theta_e & -L_{av} - L_{df}\cos 2\Delta\theta_e \\ L_{av} - L_{df}\cos 2\Delta\theta_e & L_{df}\sin 2\Delta\theta_e \end{bmatrix} \begin{bmatrix} \bar{i}_d^e \\ \bar{i}_q^e \end{bmatrix}. \tag{9.11}$$

Combining the first part of (9.10) with (9.11), it follows that

$$\Delta\omega_e \begin{bmatrix} L_{df}\sin 2\Delta\theta_e & -L_{av} + L_{df}\cos 2\Delta\theta_e \\ L_{av} + L_{df}\cos 2\Delta\theta_e & -L_{df}\sin 2\Delta\theta_e \end{bmatrix} \begin{bmatrix} \bar{i}_d^e \\ \bar{i}_q^e \end{bmatrix}$$

$$+ (\bar{\omega}_e - \Delta\omega_e) \begin{bmatrix} -L_{df}\sin 2\Delta\theta_e & -L_{av} - L_{df}\cos 2\Delta\theta_e \\ L_{av} - L_{df}\cos 2\Delta\theta_e & L_{df}\sin 2\Delta\theta_e \end{bmatrix} \begin{bmatrix} \bar{i}_d^e \\ \bar{i}_q^e \end{bmatrix}$$

$$= \underbrace{L_{df}(2\Delta\omega_e - \bar{\omega}_e)}_{(3\Delta\omega_e - \frac{3}{2}\bar{\omega}_e)L_\delta} \begin{bmatrix} \sin 2\Delta\theta_e & \cos 2\Delta\theta_e \\ \cos 2\Delta\theta_e & -\sin 2\Delta\theta_e \end{bmatrix} \begin{bmatrix} \bar{i}_d^e \\ \bar{i}_q^e \end{bmatrix} + \bar{\omega}_e L_{av} \begin{bmatrix} -\bar{i}_q^e \\ \bar{i}_d^e \end{bmatrix}. \tag{9.12}$$

Then, the same equation as (9.6) follows from (9.9)–(9.12).

Exercise 9.1   Show the following identity:

$$e^{-\mathbf{J}\Delta\theta_e} \begin{bmatrix} 0 & -\omega_e L_q \\ \omega_e L_d & 0 \end{bmatrix} e^{\mathbf{J}\Delta\theta_e} = \omega_e \begin{bmatrix} -L_{df}\sin 2\Delta\theta_e & -L_{av} - L_{df}\cos 2\Delta\theta_e \\ L_{av} - L_{df}\cos 2\Delta\theta_e & L_{df}\sin 2\Delta\theta_e \end{bmatrix}.$$

9.1.2   Dynamic Model for Sensorless Algorithm

Most sensorless algorithms are based on the estimated reference frame. Angle error, $\Delta\theta_e$ is estimated at each step, and the reference frame is updated incrementally such that $\hat{\theta}_{e(k+1)} = \hat{\theta}_{e(k)} + \widehat{\Delta\theta}_{e(k)}$, where $\widehat{\Delta\theta}_{e(k)}$ is an estimate of angle error at step k.

Angle estimation time is relatively short compared with the speed change. Thereby, it is possible to let $\Delta\omega_e = 0$. The dynamic model (9.6) in the estimated frame is given by

$$\begin{bmatrix} \bar{v}_d^e \\ \bar{v}_q^e \end{bmatrix} = \begin{bmatrix} L_{av} - L_{df}\cos 2\Delta\theta_e & L_{df}\sin 2\Delta\theta_e \\ L_{df}\sin 2\Delta\theta_e & L_{av} + L_{df}\cos 2\Delta\theta_e \end{bmatrix} \frac{d}{dt}\begin{bmatrix} \bar{i}_d^e \\ \bar{i}_q^e \end{bmatrix} + \omega_e\psi_m \begin{bmatrix} \sin\Delta\theta_e \\ \cos\Delta\theta_e \end{bmatrix}$$

$$+ r_s \begin{bmatrix} \bar{i}_d^e \\ \bar{i}_q^e \end{bmatrix} - \bar{\omega}_e L_{df} \begin{bmatrix} \sin 2\Delta\theta_e & \cos 2\Delta\theta_e \\ \cos 2\Delta\theta_e & -\sin 2\Delta\theta_e \end{bmatrix}\begin{bmatrix} \bar{i}_d^e \\ \bar{i}_q^e \end{bmatrix} + \bar{\omega}_e L_s \begin{bmatrix} -\bar{i}_q^e \\ \bar{i}_d^e \end{bmatrix}. \quad (9.13)$$

Assume that $\Delta\theta$ is small. Then, $\sin 2\Delta\theta_e \approx 0$ and $\cos 2\Delta\theta_e \approx 1$. With such an approximation,

$$-\bar{\omega}_e L_{df} \begin{bmatrix} \sin 2\Delta\theta_e & \cos 2\Delta\theta_e \\ \cos 2\Delta\theta_e & -\sin 2\Delta\theta_e \end{bmatrix}\begin{bmatrix} \bar{i}_d^e \\ \bar{i}_q^e \end{bmatrix} + \bar{\omega}_e L_s \begin{bmatrix} -\bar{i}_q^e \\ \bar{i}_d^e \end{bmatrix}$$

$$= \left(L_{df} \begin{bmatrix} \cos 2\Delta\theta_e & -\sin 2\Delta\theta_e \\ -\sin 2\Delta\theta_e & -\cos 2\Delta\theta_e \end{bmatrix} + L_s \mathbf{I} \right) \bar{\omega}_e \begin{bmatrix} -\bar{i}_q^e \\ \bar{i}_d^e \end{bmatrix} \approx \begin{bmatrix} -\omega_e L_q \bar{i}_q^e \\ \omega_e L_d \bar{i}_d^e \end{bmatrix}. \quad (9.14)$$

The approximation results in the familiar coupling voltage that appears in the coordinate transformation into a rotating frame. Note that $\Delta\theta_e$ is hardly obtained from (9.14) since $L_s \gg L_{df} = \frac{3}{2}L_\delta$. Neglecting the terms which are independent of $\Delta\theta_e$, the voltage equation has the following two $\Delta\theta_e$-dependent terms:

$$\underbrace{\begin{bmatrix} L_{av} - L_{df}\cos 2\Delta\theta_e & L_{df}\sin 2\Delta\theta_e \\ L_{df}\sin 2\Delta\theta_e & L_{av} + L_{df}\cos 2\Delta\theta_e \end{bmatrix} \frac{d}{dt}\begin{bmatrix} \bar{i}_d^e \\ \bar{i}_q^e \end{bmatrix}}_{\text{signal injection}} + \underbrace{\omega_e\psi_m \begin{bmatrix} \sin\Delta\theta_e \\ \cos\Delta\theta_e \end{bmatrix}}_{\text{back EMF}} \quad (9.15)$$

Sensorless algorithms are classified into two categories depending which terms are used: The back EMF based algorithms are the methods of utilizing $\omega_e\psi_m[\sin\Delta\theta_e, \cos\Delta\theta_e]$ for angle estimation. Since it is dependant on the speed, it cannot be used in the low speed region. To overcome such a limit, signal injection methods were developed that extract an estimate of $\Delta\theta_e$ from the inductance matrix. It should be noted that the signal injection method cannot be used in the SPMSM where $L_{df} = 0$. Often, some sensorless algorithms are based on the stationary model.

9.2 Back-EMF Based Angle Estimation

Many back EMF based sensorless algorithms were developed in the estimated rotating frame [1], [2] or the stationary frame [3], [4], [5], [6].

9.2.1 Morimoto's Extended EMF Observer

Morimoto et al. developed an IPMSM model with an extended back-EMF [18], and used disturbance observers to estimate the back-EMF. The extended EMF observer is constructed in the estimated frame which rotates synchronously. Consider the following voltage equation in the aligned synchronous frame:

$$\begin{bmatrix} v_d^e \\ v_q^e \end{bmatrix} = \begin{bmatrix} r_s + \mathrm{p}L_d & -\omega_e L_q \\ \omega_e L_d & r_s + \mathrm{p}L_q \end{bmatrix} \begin{bmatrix} i_d^e \\ i_q^e \end{bmatrix} + \begin{bmatrix} 0 \\ \omega_e \psi_m \end{bmatrix}. \tag{9.16}$$

Both L_q and L_d appear with the differential operator, p. It can be changed equivalently into the following form:

$$\begin{bmatrix} v_d^e \\ v_q^e \end{bmatrix} = \begin{bmatrix} r_s + \mathrm{p}L_d & -\omega_e L_q \\ \omega_e L_q & r_s + \mathrm{p}L_d \end{bmatrix} \begin{bmatrix} i_d^e \\ i_q^e \end{bmatrix} + \begin{bmatrix} 0 \\ E_{ex} \end{bmatrix} \tag{9.17}$$

where

$$E_{ex} = \omega_e[(L_d - L_q)i_d^e + \psi_m] - (L_d - L_q)(\mathrm{p}i_q^e).$$

Note that only L_d appears in the diagonal with p's, whereas L_q in the off-diagonal: On the other hand, E_{ex}, called extended EMF, contains the differential operator. In the vector form, (9.17) is rewritten as

$$\mathbf{v}_{dq}^e = [(r_s + \mathrm{p}L_d)\mathbf{I} + \omega_e L_q \mathbf{J}]\, \mathbf{i}_{dq}^e + \boldsymbol{\xi} \tag{9.18}$$

where

$$\mathbf{v}_{dq}^e = \begin{bmatrix} v_d^e \\ v_q^e \end{bmatrix}, \qquad \mathbf{i}_{dq}^e = \begin{bmatrix} i_d^e \\ i_q^e \end{bmatrix}, \qquad \boldsymbol{\xi} = \begin{bmatrix} 0 \\ E_{ex} \end{bmatrix}.$$

Note that $\mathbf{I}$ and $\mathbf{J}$ can be handled easily with the rotational transformation. In order to change the vectors into the misaligned coordinate $(\bar{\theta}_e)$, it is necessary to multiply $e^{-\mathbf{J}\Delta\theta_e}$, where $\Delta\theta_e = \bar{\theta}_e - \theta_e$. Then the dynamics in the misaligned coordinate is given as

$$\bar{\mathbf{v}}_{dq}^e = e^{-\mathbf{J}\Delta\theta_e}\left[(r_s + \mathrm{p}L_d)\mathbf{I} + \omega_e L_q \mathbf{J}\right] e^{\mathbf{J}\Delta\theta_e}\bar{\mathbf{i}}_{dq}^e + e^{-\mathbf{J}\Delta\theta_e}\boldsymbol{\xi}, \tag{9.19}$$

where $\bar{\mathbf{v}}_{dq}^e = e^{-\mathbf{J}\Delta\theta_e}\mathbf{v}_{dq}^e$ and $\bar{\mathbf{i}}_{dq}^e = e^{-\mathbf{J}\Delta\theta_e}\mathbf{i}_{dq}^e$. Rearranging (9.19), it follows that

$$\begin{bmatrix} \bar{v}_d^e \\ \bar{v}_q^e \end{bmatrix} = \begin{bmatrix} r_s + \mathrm{p}L_d & -\omega_e L_q \\ \omega_e L_q & r_s + \mathrm{p}L_d \end{bmatrix} \begin{bmatrix} \bar{i}_d^e \\ \bar{i}_q^e \end{bmatrix} + \begin{bmatrix} \xi_d \\ \xi_q \end{bmatrix} \tag{9.20}$$

where

$$\begin{bmatrix} \xi_d \\ \xi_q \end{bmatrix} = E_{ex}\begin{bmatrix} \sin\Delta\theta_e \\ \cos\Delta\theta_e \end{bmatrix} + (\bar{\omega}_e - \omega_e)L_d\begin{bmatrix} -\bar{i}_q^e \\ \bar{i}_d^e \end{bmatrix}, \tag{9.21}$$

and $\bar{\omega}_e = \frac{d}{dt}\bar{\theta}_e$. It is interesting to see that the dynamic model is also simple in misaligned frame with the extended EMF. The dynamic model (9.20) can be seen as a linear system, but disturbed by unknown disturbance, $[\xi_d, \xi_q]^T$. Normally, there is almost no speed variation in the short period of angle estimation. Then, we let $\bar{\omega}_e = \omega_e$ without loss of generality. It is necessary to estimate the disturbance, since it contain angle information. Suppose that we obtained estimates, $\hat{\xi}_d$ and $\hat{\xi}_q$. Then, the angle error can be estimated such that

$$\widehat{\Delta\theta}_e = \bar{\theta}_e - \hat{\theta}_e = \tan^{-1}\left(\frac{\hat{\xi}_d}{\hat{\xi}_q}\right). \tag{9.22}$$

It motivates us to use disturbance observers. Note further that it is not necessary to estimate E_{ex} since it is canceled in (9.22).

Disturbance Observer

The disturbance observer was introduced by Ohnishi, and refined later by Umeno and Hori [19]. It is often used to estimate unknown load torque in the speed loop of many electro-mechanical systems. Fig. 9.2 shows a block diagram illustrating a typical disturbance compensator. The plant is denoted by $G(s)$. The compensator includes the inverse model, $1/G(s)$, thereby the system $G(s)$ must be nonminimum phase, i.e., $G(s)$ should have zeros in the left half plane of $\mathbb{C}$. Furthermore, $1/G(s)$ is an improper system, i.e., the degree of numerator is larger than that of the denominator. Thereby, it includes differentiators which are unacceptable practically. Therefore, a low pass filter, $Q(s)$ must be used and as a consequence, a band limited solution is obtained. Fig. 9.2 (b) shows a case of practical realization in which $Q(s)$ and $1/G(s)$ are combined. Obviously, $Q(s)/G(s)$ should be proper. In general, the disturbance observer performs satisfactorily for a DC like disturbance.

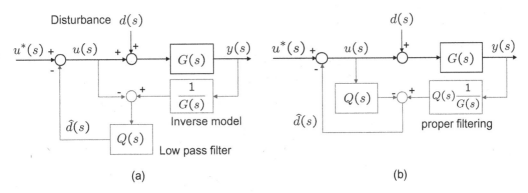

Figure 9.2: Disturbance compensator: (a) conceptual diagram and (b) equivalent realization.

Fig. 9.3 shows an angle error estimation scheme utilizing two disturbance observers for ξ_d and ξ_q. In this particular example, the disturbance observers are applied to the current loop, and ξ_d and ξ_q are assumed to vary slowly. Since the model is a first order system, a first order low pass filter, $\frac{\alpha}{s+\alpha}$ is used with the unity DC gain.

Exercise 9.2 Derive the impulse response for the filter, $\frac{\alpha}{s+\alpha}$ and show that it converges to the delta function, $\delta(t)$ as $\alpha \to \infty$.

Solution. $\mathcal{L}^{-1}\left(\frac{\alpha}{s+\alpha}\right) = \alpha e^{-\alpha t}u_s(t)$, where $u_s(t)$ is the unity step function. Also, $\int_{-\infty}^{\infty} \alpha e^{-\alpha t}u_s(t)\,dt = 1$ independently of α. ∎

Angle and Speed Estimation

When an observer is constructed in an estimated frame, $\widehat{\Delta\theta_e}$ is obtained instead of $\hat{\theta}_e$. To obtain an angle estimate, one may add $\bar{\theta}_e$ to $\widehat{\Delta\theta_e}$, i.e., $\hat{\theta}_e = \widehat{\Delta\theta_e} + \bar{\theta}_e$. However, errors and noise are involved naturally in the estimation process, thereby

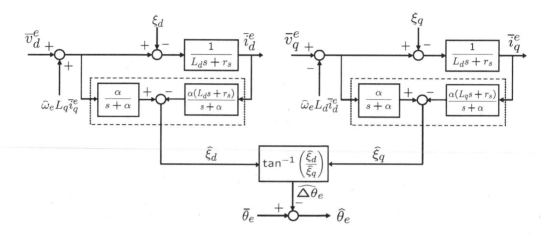

Figure 9.3: Angle error estimation utilizing disturbance observers for the extended EMF.

it is necessary to update $\hat{\theta}_e$ more conservatively. That is, a certain type of low pass filter is needed to reduce the noise effects.

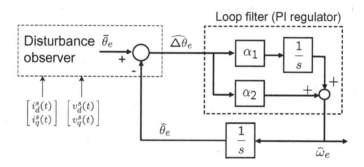

Figure 9.4: Angle estimation from $\widehat{\Delta\theta}_e$ that utilizes a tracking filter.

A tracking filter is often used to extract an estimate $\hat{\theta}_e$ from $\theta_e - \bar{\theta}_e$. Fig. 9.4 shows the block diagram of a tracking filter [20], [21], which consists of a PI regulator and an integrator. The estimator is described by

$$\dot{\hat{\omega}} = \alpha_1(\hat{\theta}_e - \bar{\theta}_e) + \alpha_2(\hat{\omega}_e - \bar{\omega}_e).$$
$$\dot{\hat{\theta}}_e = \hat{\omega}_e. \tag{9.23}$$

Note that the denominator of the closed loop system is equal to $s^2 + \alpha_2 s + \alpha_1 = 0$. The PI gains (α_2, α_1) are selected such that the filter bandwidth is larger than the signal bandwidth. Due to inherent tracking capability, $\hat{\theta}_e$ will track θ_e, as far as $\theta_e - \hat{\theta}_e$ is correct. For a high gain, the PI controller behaves like a bang-bang controller, and Jang et al. [22],[23] utilized a bang-bang controller instead of the PI controller.

The node prior to the integrator signifies a speed estimate. That is, $\hat{\omega}_e$ is obtained without a differentiation. As shown in (9.4), the disturbance observer works with the tracking filter to yield an angle estimate.

Experimental Results

Experiments were performed with an IPMSM developed for an air conditioner compressor. The specifications were listed in Table 9.1. The control algorithms were implemented in a TMS320vc33 DSP board. The PWM switching frequency was set to be 5 kHz and the dead time 2 μs. The current control and speed/position estimation were carried out every 0.2 ms, and the speed control loop was activated every 2 ms. Fig. 9.5 shows the responses of θ_e, $\hat{\theta}_e$, $\theta_e - \hat{\theta}_e$, and $\hat{\omega}_r$ at 500 and

Table 9.1: Motor parameters of an IPMSM used for experiments.

Parameters [Unit]	Values	Parameters [Unit]	Values
Input DC link voltage [V]	540	Inductance (L_d) [mH]	3
Rated output power [kW]	11	Inductance (L_q) [mH]	6.2
Rated speed [r/min]	5000	Stator resistance (r_s) [Ω]	0.151
Rated torque [Nm]	21	Rotor flux (ψ_m) [Wb]	0.09486
Rated current [Arms]	25	Number of poles (P)	6
Rated voltage [Vrms]	220		

6000 rpm (field weakening region). Note that the angle error is less than $\pm 5°$ at 500 rpm and $\pm 2°$ at 6000 rpm. Fig. 9.6 shows a transient response to a step speed command from 3000 to 4000 rpm. At the point of abrupt speed change, the angle error is developed, but settles down within 0.15 sec. To make the transient response better, α needs to be increased as much as possible. In this experiment, $\alpha = 3450$. Fig. 9.7 shows speed control responses: 500 rpm $\rightarrow$ 6000 rpm $\rightarrow$ 500 rpm. This shows that the field weakening control with the Morimoto's algorithm is also stable.

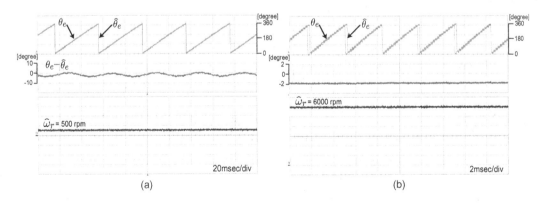

Figure 9.5: Angle estimation with the disturbance observer: θ_e, $\hat{\theta}_e$, $\theta_e - \hat{\theta}_e$, and $\hat{\omega}_r$ at (a) 500 rpm and (b) 6000 rpm.

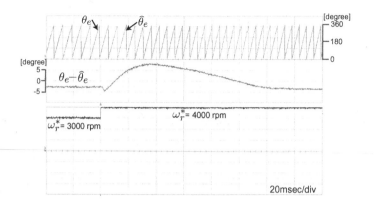

Figure 9.6: Transient response for a step speed change from 3000 rpm to 4000 rpm.

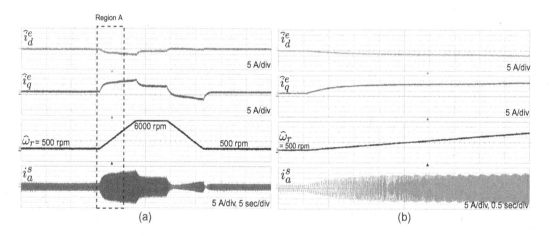

Figure 9.7: Speed control responses with the Morimoto's sensorless algorithm: (a) 500 rpm → 6000 rpm → 500 rpm and (b) a magnified plot of region A.

Summary

1. The Morimoto's sensorless algorithm was developed for IPMSMs. The essential part is to make L_d appear in the diagonal part with the differential operator and L_q in the off-diagonal part of the voltage equation, while driving out the remaining terms into the extended EMF term. **2.** The extended EMF based model has a simple structure in the misaligned coordinate. Utilizing the disturbance observer, voltage disturbances are detected which carry angle information.

9.2.2   Ortega's Nonlinear Observer for Sensorless Control

Ortega et al.[3] developed a nonlinear observer for SPMSM in the stationary frame. Later its performance was demonstrated through experiments [4]. With $L_\delta = 0$, the

P-pole SPMSM stationary model is derived from (9.7) such that

$$L_s \frac{d}{dt} \begin{bmatrix} i_d^s \\ i_q^s \end{bmatrix} = -r_s \begin{bmatrix} i_d^s \\ i_q^s \end{bmatrix} - \omega_e \psi_m \begin{bmatrix} -\sin\theta_e \\ \cos\theta_e \end{bmatrix} + \begin{bmatrix} v_d^s \\ v_q^s \end{bmatrix} \tag{9.24}$$

$$T_e = \frac{3P}{4} \psi_m (i_q^s \cos\theta_e - i_d^s \sin\theta_e). \tag{9.25}$$

The goal is to estimate $[-\sin\theta_e, \cos\theta_e]^T$ using $\mathbf{i}_{dq}^s$ and $\mathbf{v}_{dq}^s$. The flux in the stationary frame is given as

$$\boldsymbol{\lambda}_{dq}^s = L_s \begin{bmatrix} i_d^s \\ i_q^s \end{bmatrix} + \psi_m \begin{bmatrix} \cos\theta_e \\ \sin\theta_e \end{bmatrix}. \tag{9.26}$$

It is an air gap flux which is the sum of stator flux activated by current and the rotor (PM) flux. It can be obtained by integrating $\mathbf{v}_{dq}^s - r_s \mathbf{i}_{dq}^s$. Letting $\mathbf{y} \equiv \mathbf{v}_{dq}^s - r_s \mathbf{i}_{dq}^s$, it follows from (9.24) and (9.26) that

$$\dot{\boldsymbol{\lambda}}_{dq}^s = \mathbf{y}. \tag{9.27}$$

That is, integration of $\mathbf{v}^s - r_s\mathbf{i}^s$ gives flux. Define a function $\boldsymbol{\eta} : \mathbb{R}^2 \to \mathbb{R}^2$ by

$$\boldsymbol{\eta}(\boldsymbol{\lambda}_{dq}^s) \equiv \boldsymbol{\lambda}_{dq}^s - L_s \begin{bmatrix} i_d^s \\ i_q^s \end{bmatrix} = \psi_m \begin{bmatrix} \cos\theta_e \\ \sin\theta_e \end{bmatrix}. \tag{9.28}$$

It is the rotor flux and the Euclidean norm is equal to

$$\|\boldsymbol{\eta}(\boldsymbol{\lambda}_{dq}^s)\|^2 = \psi_m^2. \tag{9.29}$$

Let $\hat{\boldsymbol{\lambda}}_{dq}^s$ be an estimate of $\boldsymbol{\lambda}_{dq}^s$. It can be obtained by integrating (9.27). Nevertheless, the pure integrator is marginally stable, since it is unbounded in the presence of a DC offset.

Consider a cost

$$J = \frac{1}{2}(\psi_m^2 - \|\boldsymbol{\eta}(\hat{\boldsymbol{\lambda}}_{dq}^s)\|^2)^2.$$

Then the gradient is given by

$$\frac{\partial J}{\partial \hat{\boldsymbol{\lambda}}_{dq}^s} = -2\boldsymbol{\eta}(\hat{\boldsymbol{\lambda}}_{dq}^s)(\psi_m^2 - \|\boldsymbol{\eta}(\hat{\boldsymbol{\lambda}}_{dq}^s)\|^2). \tag{9.30}$$

Utilizing the gradient, Ortega proposed a nonlinear observer with an error injection:

$$\begin{aligned} \dot{\hat{\boldsymbol{\lambda}}}_{dq}^s &= \mathbf{y} + \frac{\gamma}{2}\boldsymbol{\eta}[\psi_m^2 - \|\boldsymbol{\eta}\|^2], \\ &= \mathbf{v}_{dq}^s - r_s\mathbf{i}_{dq}^s + \frac{\gamma}{2}\boldsymbol{\eta}(\hat{\boldsymbol{\lambda}}_{dq}^s)[\psi_m^2 - \|\boldsymbol{\eta}(\hat{\boldsymbol{\lambda}}_{dq}^s)\|^2], \end{aligned} \tag{9.31}$$

where $\hat{\boldsymbol{\lambda}}_{dq}^s \in \mathbb{R}^2$ is the observer state variable and $\gamma > 0$ is the observer gain. If an estimate, $\hat{\boldsymbol{\lambda}}_{dq}^s$ is correct, it should satisfy (9.29). It states that the locus of rotor flux is a circle of radius, ψ_m. Thus, the difference, $\|\boldsymbol{\eta}(\hat{\boldsymbol{\lambda}}_{dq}^s)\|^2 - \psi_m^2$ is an error measure, thus can be used for steering $\hat{\boldsymbol{\lambda}}_{dq}^s$.

Theorem 9.1 [3]
Consider the SPMSM model, (9.26) and (9.27), and a nonlinear observer defined by (9.28) and (9.31). The observation error $\boldsymbol{\chi}$ defined by

$$\boldsymbol{\chi} \equiv -\frac{1}{\psi_m} e^{-J\theta_e}(\hat{\boldsymbol{\lambda}}_{dq}^s - \boldsymbol{\lambda}_{dq}^s). \tag{9.32}$$

converges asymptotically to the disk $\{\boldsymbol{\chi} \in \mathbb{R}^2 | \, \|\boldsymbol{\chi}\| \leq 2\}$ as $t \to \infty$.

Note that $\boldsymbol{\chi}$ is a normalized flux error in the synchronous frame. The theorem states that $\boldsymbol{\chi}$ is bounded in a disk of radius 2 centered at $[\cos\theta_e, \sin\theta_e]^T$. We may rewrite

$$\hat{\boldsymbol{\lambda}}_{dq}^s = L_s \begin{bmatrix} i_d^s \\ i_q^s \end{bmatrix} + \psi_m \begin{bmatrix} \cos\hat{\theta}_e \\ \sin\hat{\theta}_e \end{bmatrix}. \tag{9.33}$$

Then, an angle estimate is obtained from $\hat{\boldsymbol{\lambda}}_{dq}^s$ such that

$$\hat{\theta}_e = \tan^{-1}\left(\frac{\hat{\lambda}_q^s - L_s i_q^s}{\hat{\lambda}_d^s - L_s i_d^s}\right). \tag{9.34}$$

Remark.
Since the above nonlinear observer is constructed based on the stationary model, it does not require speed information in the construction of an observer. This is an advantage when it is compared with the observers constructed on a synchronous frame. See for example [1],[11].

Control Block Diagram

The sensorless control block for a SPMSM, which includes the nonlinear observer, is shown in Fig. 9.8. The nonlinear observer outputs angle estimate $\hat{\theta}$, based on which the field orientation control is synthesized. The conventional PI controllers are utilized for d and q axis current control along with the decoupling and back-EMF compensation. The IP type speed controller utilizes $\hat{\omega}_e$ that comes out from a speed estimator.

Jansson et al. [17] pointed that the injection of d-axis current enhanced the robustness of sensorless system against r_s variation. However, d-axis current pulses are more effective as the persistent exciting condition is required for parameter convergence. Fig. 9.8 shows a voltage pulse train applied to the d-axis [4]. In the experiment, the pulse frequency was 200 Hz, the peak level was 50 V, and the pulse duty was 0.2 msec. However, such d-axis current was not injected if $|\omega_r| > 100$ rpm.

Experimental Results

Experiments were performed with a motor-generator set that was made with two SPMSMs as shown in Fig. 9.9. All the nonlinear observer and control algorithms

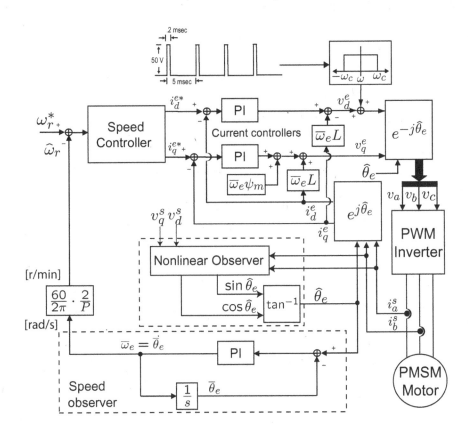

Figure 9.8: The overall sensorless control block diagram with the nonlinear observer and the speed estimator.

were implemented in a TMS320vc33 DSP board. The PWM switching frequency was set to be 8 kHz and the dead time 2 μs. The current control algorithm were carried out every 125 μs, and the speed control loop was activated every 1.25 ms.

Figure 9.9: Photo of the experiment setup.

Table 9.2: Motor parameters of a SPMSM used for experiments

Parameters [Unit]	Values
Input DC link voltage [V]	200
Rated output power [kW]	0.3
Rated torque [Nm]	3.0
Rated speed [r/min]	1000
Rated phase current [A]	3.0
Number of poles (P)	8
Rotor flux (ψ_m) [Wb]	0.11
Switching frequency [kHz]	8
Stator inductance (L) [mH]	1.14
Stator resistance (R_s) [Ω]	0.675

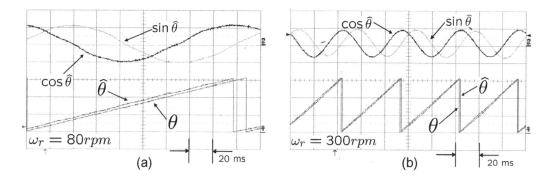

Figure 9.10: Comparison between the real and the estimated position data under no-load condition at (a) 80 rpm and (b) 300 rpm.

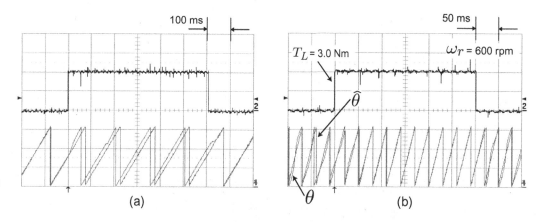

Figure 9.11: Comparison between the real and the estimated position data under a full step load at (a) 100 rpm and (b) 600 rpm.

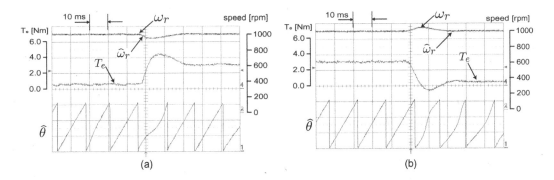

Figure 9.12: Speed and the corresponding torque responses at 1000 rpm when a full load torque is (a) applied and (b) removed.

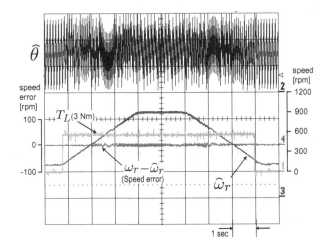

Figure 9.13: Speed control response to the full step load torque.

Fig. 9.10 shows $\sin\hat{\theta}$, $\cos\hat{\theta}$, and $\hat{\theta}$, along with real position θ measured by a 6000 pulses/rev encoder under no-load when (a) $\omega_r = 80$ rpm and (b) 300 rpm, respectively. Note that the position errors at 300 rpm are smaller than those at 80 rpm. Fig. 9.11 (a) and (b) show behaviors of the position estimates when a full step load was applied when $\omega_r = 100$ rpm and 600 rpm, respectively. One can also notice that the steady state position errors at a higher speed are smaller. Fig. 9.12 (a) and (b) show the responses of the speed estimates and the corresponding torque at the time of full load loading and removal when $\omega_r = 1000$ rpm, respectively. Fig. 9.13 shows a macroscopic view of the behaviors of speed and angle estimates when the speed changes from $\omega_r = 100$ rpm to 900 rpm with a full load. Fig. 9.14 shows a stable performance at 10 rpm (0.01 pu) with a 1.5 Nm (0.5 pu) load. Fig. 9.14 (b) is an expanded plot of real and estimated angles shown in Fig. 9.14 (a). Note that the d-axis current has a shape of pulse train.

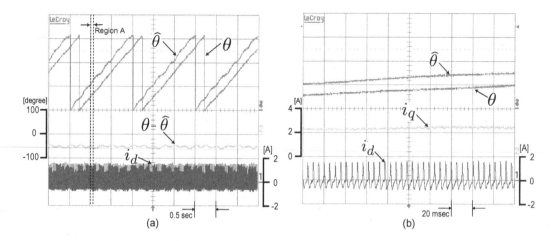

Figure 9.14: (a) Experimental results under half load (1.5 Nm) at 10 rpm and (b) an expanded plot of Region A in (a).

Summary

1. The nonlinear observer is developed for SPMSMs in the stationary frame. One advantage is that it does not require the speed information, so that a speed estimator can be constructed separately.
2. Robustness can be enhanced with a d-axis current injection in the low speed region.
3. It has a simple structure and performs well in the practical system.

9.2.3   Bobtsov's Initial Parameter Estimator

The initial rotor position estimation is essential for sensorless control of synchronous machines. If the initial angle is known, then the motor starting performance would be improved greatly. Using the gradient method, Bobtsov et al. [5] developed an observer that estimated the initial value of rotor flux. Later, a modified algorithm that intended to avoid the pure integrator was proposed by Choi et al. [6].

The SPMSM dynamics are modeled in the stationary frame as:

$$\dot{\boldsymbol{\lambda}}^s_{dq} = \mathbf{v}^s_{dq} - r_s \mathbf{i}^s_{dq} \tag{9.35}$$

$$\boldsymbol{\lambda}^s_{dq} = L_s \mathbf{i}^s_{dq} + \psi_m \mathbf{c}(\theta_e) \tag{9.36}$$

$$\dot{\theta}_e = \omega_e. \tag{9.37}$$

where $\mathbf{c}(\theta_e) = [\cos \theta_e, \sin \theta_e]^T$. Let the rotor flux be denoted by

$$\mathbf{x} \equiv \boldsymbol{\lambda}^s_{dq} - L_s \mathbf{i}^s_{dq} = \psi_m \mathbf{c}(\theta_e), \tag{9.38}$$

where $\mathbf{x} = [x_d, x_q]^T$. Differentiating both sides of (9.38), it follows that

$$\dot{\mathbf{x}} = \mathbf{v}^s_{dq} - r_s \mathbf{i}^s_{dq} - L_s \mathbf{p} \, \mathbf{i}^s_{dq} = \psi_m \omega_e \mathbf{J} \mathbf{c}(\theta_e). \tag{9.39}$$

Bobtsov's Initial Parameter Estimation

Consider the same form of equation as (9.39):

$$\dot{\mathbf{q}}(t) = \mathbf{v}_{dq}^s - r_s \mathbf{i}_{dq}^s - L_s p\, \mathbf{i}_{dq}^s, \qquad \mathbf{q}(0) = [0,0]^T, \qquad (9.40)$$

where $\mathbf{q} \in \mathbb{R}^2$. Note that $\mathbf{x}$ and $\mathbf{q}$ have the same dynamics, but the initial values are different. For notational convenience, we let $\boldsymbol{\eta}_0 \equiv \mathbf{x}(0) = \boldsymbol{\lambda}_{dq}^s(0) - L_s \mathbf{i}_{dq}^s(0)$. Then,

$$\mathbf{x}(t) = \mathbf{q}(t) + \boldsymbol{\eta}_0. \qquad (9.41)$$

Fig. 9.15 shows differences between $\mathbf{x}$ and $\mathbf{q}$. Note that both of them have circular contours, but the centers are different due to the initial value, $\boldsymbol{\eta}_0$.

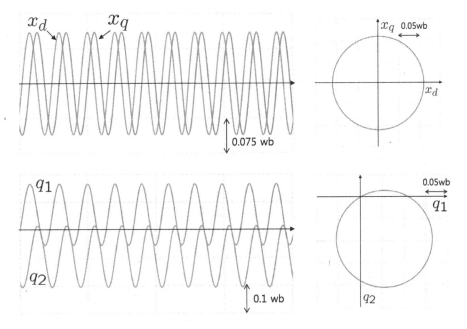

Figure 9.15: Differences between $\mathbf{x}$ and $\mathbf{q}$: (a) Time plot and (b) Lissajous plot.

An initial parameter estimator is developed from $\|\mathbf{q} + \boldsymbol{\eta}_0\|^2 = \|\mathbf{x}\|^2$. Noting that $\|\mathbf{x}\|^2 = \psi_m^2$, we obtain

$$-\|\mathbf{q}\|^2 = 2\mathbf{q}^T \boldsymbol{\eta}_0 + \|\boldsymbol{\eta}_0\|^2 - \psi_m^2 = \begin{bmatrix} 2\mathbf{q}^T & 1 \end{bmatrix} \boldsymbol{\beta}, \qquad (9.42)$$

where

$$\boldsymbol{\beta} \equiv \begin{bmatrix} \boldsymbol{\eta}_0 \\ \|\boldsymbol{\eta}_0\|^2 - \psi_m^2 \end{bmatrix} \in \mathbb{R}^3. \qquad (9.43)$$

A filter $\frac{\alpha p}{p+\alpha}$ is applied to both sides of (9.42). In fact, it functions as a differentiator although it is combined with a low pass filter. Thus, it nullifies the effect of constant term, $\|\boldsymbol{\eta}_0\|^2 - \psi_m^2$ in (9.42). Then, we have

$$y(\mathbf{q}) = \boldsymbol{\Omega}(\mathbf{q})^T \boldsymbol{\eta}_0 + \epsilon(t) \qquad (9.44)$$

where $y : \mathbb{R}^2 \to \mathbb{R}$ and $\Omega : \mathbb{R}^2 \to \mathbb{R}^2$ are defined by

$$y(\mathbf{q}) = \frac{-\alpha \mathrm{p}}{\mathrm{p} + \alpha}(\|\mathbf{q}\|^2), \tag{9.45}$$

$$\Omega(\mathbf{q}) = \frac{2\alpha \mathrm{p}}{\mathrm{p} + \alpha}(\mathbf{q}), \tag{9.46}$$

$$\text{and} \quad \epsilon(t) = \frac{\mathrm{p}\alpha}{\mathrm{p} + \alpha}\{\|\boldsymbol{\eta}_0\|^2 - \psi_m^2\} = \alpha c e^{-\alpha t} \tag{9.47}$$

for some constant $c > 0$. It should be noted that $y(\mathbf{q})$ and $\Omega(\mathbf{q})$ are available terms, because $\mathbf{q}$ is known. Note also that (9.44) allows a linear regression since it has a product form; known vector, $\Omega(\mathbf{q})$ multiplied by an unknown parameter vector, $\boldsymbol{\eta}_0$. A parameter estimator can be easily derived by applying the gradient algorithm to the error $(y(\mathbf{q}) - \Omega(\mathbf{q})^T\boldsymbol{\eta}_0)^2$ such that

$$\dot{\hat{\boldsymbol{\eta}}}_0 = \boldsymbol{\Gamma}\Omega(\mathbf{q})\big(y(\mathbf{q}) - \Omega(\mathbf{q})^T\hat{\boldsymbol{\eta}}_0\big), \tag{9.48}$$

where $\boldsymbol{\Gamma}$ is a gain matrix. Fig. 9.16 shows a block diagram of the Bobtsov's observer.

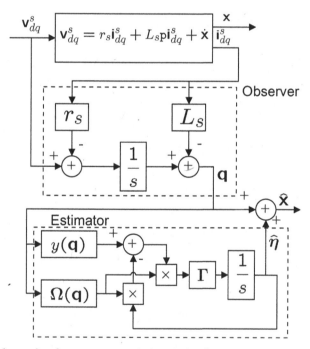

Figure 9.16: Bobtsov's observer containing initial parameter estimator.

Stability Analysis

Define the parameter error by $\tilde{\boldsymbol{\eta}}_0 = \hat{\boldsymbol{\eta}}_0 - \boldsymbol{\eta}_0$: Then,

$$\dot{\tilde{\boldsymbol{\eta}}}_0 = -\boldsymbol{\Gamma}\Omega(\mathbf{q})\Omega(\mathbf{q})^T\tilde{\boldsymbol{\eta}}_0 + \boldsymbol{\Gamma}\Omega(\mathbf{q})\epsilon(t). \tag{9.49}$$

Let a Lyapunov function be defined by

$$V(\tilde{\boldsymbol{\eta}}_0, t) = \tilde{\boldsymbol{\eta}}_0^T \Gamma^{-1} \tilde{\boldsymbol{\eta}}_0 + \int_t^\infty |\epsilon(\tau)| d\tau, \tag{9.50}$$

where $\Gamma \in \mathbb{R}^{2 \times 2}$ is assumed to be positive definite. Differentiating V along the trajectories of (9.49) yields

$$\dot{V} = -|\Omega^T \tilde{\boldsymbol{\eta}}_0|^2 - |\Omega^T \tilde{\boldsymbol{\eta}}_0 - \epsilon|^2 \le 0. \tag{9.51}$$

Thus, we conclude that $V(\tilde{\boldsymbol{\eta}}_0, t)$ is a non-increasing function. ∎

9.2.4 Comparison between Back EMF and Rotor Flux Estimate

At zero speed, the back EMF is zero. Therefore, back EMF based methods do not work properly unless the speed goes above a certain level where the back EMF is greater than the noise level. Fig. 9.17 shows the back EMF and the rotor flux when the speed changes from 100 rpm to -100 rpm. Note that the back EMF is zero during zero speed. In contrast, the rotor flux remains nonzero constant. Thus, an angle estimate, $\hat{\theta}_e = \tan^{-1}(\hat{\lambda}_{dr}^s / \hat{\lambda}_{qr}^s)$ based on the rotor flux would be better than $\hat{\theta}_e = \tan^{-1}(\hat{e}_d^s / \hat{e}_q^s)$ based on the back EMF.

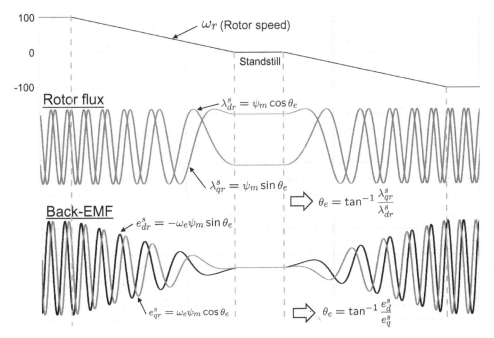

Figure 9.17: Comparison between back EMF and rotor flux estimate near zero speed.

However, the rotor flux estimation methods also yield poor performance in the low speed region. Because they are based on (9.27) and (9.40) which are the pure integrator of $\mathbf{v}_{dq}^s - r_s \mathbf{i}_{dq}^s$. Note that there are errors in $\mathbf{v}_{dq}^s$ and r_s. Voltage command is usually utilized instead of $\mathbf{v}_{dq}^s$ measurements. However, the actual voltage applied to the machine is different due to switch on-drop voltage, dead time, and other transient effects. Also, the exact value of r_s is hardly known since it is affected by temperature and AC effects. These errors are not so influential in the high speed operation. But they critically impact on the accuracy of estimation in the low speed range, since they behave like DC offsets. Fig. 9.18 shows angle estimation error and flux estimate during zero speed at the presence of 2% r_s error.

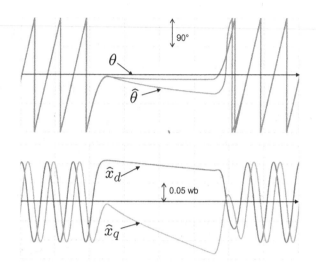

Figure 9.18: (a) Angle estimation error and (b) flux estimate during zero speed at the presence of 2% r_s error.

9.2.5   Starting Algorithm by Signal Injection

The Morimoto's extended EMF observer is based on a synchronous frame, and requires the angle update and current control successively as the rotor rotates. An incremental algorithm does not work well if the initial error is large. For example, if $\Delta\theta_e \geq \frac{\pi}{6}$, then the sensorless algorithm fails. It is because the algorithm is based on the assumption that $\Delta\theta_e$ is small. To utilize the incremental type algorithm for PMSM control, the rotor angle, although rough, must be known initially. Therefore, another method is necessary to find an initial angle estimate. Since it is required once before starting, it is called *starting algorithm*. Mizutani et al. [24] proposed a starting and low speed operation method for IPMSM. It is like an early stage signal injection method: To identify the initial position, A probing signal, which is a periodic voltage pulse, is applied to the motor in the stationary state. Since the IPMSM has a rotor saliency, its inductance differs depending on the rotor position. By analyzing the current responses, the rotor position is estimated roughly.

Position Error Estimation

In the low speed region, it can be assumed that $\bar{\omega}_e \approx 0$ and $\omega_e \approx 0$. Then, the stationary model (9.7) is reduced to

$$\mathbf{v}_{dq}^s = r_s \mathbf{i}_{dq}^s + \begin{bmatrix} L_\alpha & L_\gamma \\ L_\gamma & L_\beta \end{bmatrix} \frac{d}{dt} \mathbf{i}_{dq}^s, \qquad (9.52)$$

where $L_\alpha = L_s - \frac{3}{2} L_\delta \cos 2\Delta\theta_e$, $L_\beta = L_s + \frac{3}{2} L_\delta \cos 2\Delta\theta_e$, and $L_\gamma = \frac{3}{2} L_\delta \sin 2\Delta\theta_e$. The following voltage pulse is injected intermittently only in the (stationary) d-axis:

$$\Delta\mathbf{v} = \begin{cases} \begin{bmatrix} V_p \\ 0 \end{bmatrix}, & t_k \leq t < t_k + T_p \\ \text{zero}, & t_k + T_p \leq t < t_{k+1}, \end{cases} \qquad (9.53)$$

where V_p is a constant. Note that $L_\gamma = 0$ if there is no angle error. In other words, the d-axis voltage pulse would cause a change only in the d-axis current if the rotor is aligned to the d-axis. Otherwise, the d-axis voltage pulse will also affect the q-axis current.

Let $\Delta\mathbf{i}$ be the response to $\Delta\mathbf{v}$. The current response follows from (9.52) such that

$$\Delta\mathbf{i}(s) \; = \; \frac{1}{(L_d s + r_s)(L_q s + r_s)} \begin{bmatrix} r_s + L_\beta s & -L_\gamma s \\ -L_\gamma s & r_s + L_\alpha s \end{bmatrix} \Delta\mathbf{v}(s). \qquad (9.54)$$

Solving (9.54) for $\Delta\mathbf{v}(s) = \left[\frac{V_s}{s}, 0 \right]^T$, it follows that

$$\begin{bmatrix} \Delta i_d \\ \Delta i_q \end{bmatrix} (T_p) = \begin{bmatrix} I_0 + I_1 \cos 2\Delta\theta_e \\ -I_1 \sin 2\Delta\theta_e \end{bmatrix}, \qquad (9.55)$$

where

$$I_0 \; = \; \frac{V_p}{r_s} \left[1 - \frac{1}{2} \left(e^{-\frac{r_s}{L_d} T_p} + e^{-\frac{r_s}{L_q} T_p} \right) \right]$$

$$I_1 \; = \; \frac{1}{2} \frac{V_p}{r_s} \left(e^{-\frac{r_s}{L_d} T_p} - e^{-\frac{r_s}{L_q} T_p} \right).$$

Using (9.55), an angle error is estimated as

$$\Delta\theta_e = \frac{1}{2} \tan^{-1} \left(\frac{-\Delta i_q}{\Delta i_d - I_0} \right). \qquad (9.56)$$

Depending on the sector of angular position, the polarity of Δi_q changes. For example, see Δi_q depending on $\Delta\theta_e = \frac{\pi}{4}$ and $\frac{3\pi}{4}$ in Fig. 9.19. It should be noted that $I_1 = 0$ if there is no saliency. Thus, (9.56) cannot be used for SPMSMs. The probing voltage is superposed on the reference voltage to keep angle detection during low speed operation. The probing is repeated, for example, every 20 or 30 sampling periods. Normally, this starting algorithm is switched to the other back-EMF based sensorless method after the speed increases above a certain point.

Experimental Results

Experiments were performed with the IPMSM whose specifications are listed in Table 9.1. The PWM switching frequency was set to be 5 kHz. The probing voltage (50V, 125 μs) was injected every 20 current samplings. Fig. 9.19 shows Δi_d and Δi_q for repeated d-axis rectangular voltage pulses, while the motor shaft was turned from zero to 360°. As predicted by (9.55), Δi_d is a cosine function with a DC offset, and Δi_q is a sine function with negative sign. Fig. 9.19 also shows zoomed-in plots for $\Delta \theta_e = 0$, $\frac{\pi}{4}$, $\frac{\pi}{2}$, and $\frac{3\pi}{4}$. Note that $\Delta \theta_e = \frac{\pi}{4}$ and $\frac{3\pi}{4}$ are differentiated by the polarity of Δi_q.

Fig. 9.20 (a) shows an angle estimate and the real angle when the motor speed was increased from standstill to 100 rpm. Fig. 9.20 (b) shows a moment of transition from the starting algorithm to a back-EMF based sensorless algorithm at 350 rpm. The motor was started initially by the starting algorithm. It was switched to a back-EMF based sensorless algorithm after the back-EMF increased to a certain level.

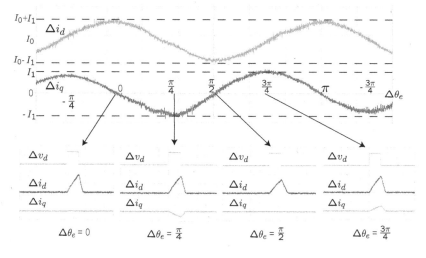

Figure 9.19: Δi_d and Δi_q for a d-axis rectangular voltage pulse, (50 V, 125 μs), when $\Delta \theta_e$ changes from zero to 360°.

Summary

1. If incremental angle is obtained by an observer, an additional method is required to estimate the initial angle.

2. Voltage pulses are injected periodically to the d-axis for IPMSMs. If the reference frame is misaligned, then a current response is also monitored in the q-axis. By analyzing the relative magnitudes of the dq currents, the real rotor flux angle is estimated.

3. Although the angle estimation error is large, the above method works satisfactorily in starting up the motor.

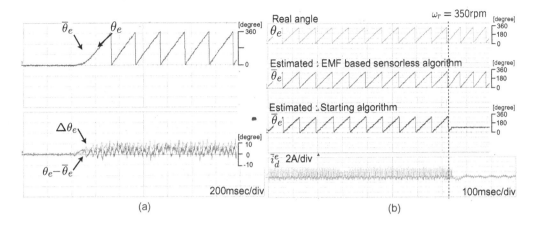

Figure 9.20: Experimental results of the starting algorithm: (a) angle estimate and real angle when the speed increases from 0 rpm to 100 rpm and (b) algorithm transition from the starting algorithm to a back-EMF based sensorless algorithm at 350 rpm.

9.3 Sensorless Control by Signal Injection

Since the signal injection methods do not depend on the back-EMF, it can be used in the low frequency region or at zero speed. High frequency probing voltage is injected into the motor terminal together with the main control voltage. Frequency being high, the probing signal does not affect the motor performance such as torque and speed. But it can increase the eddy current loss and cause some audible noise. If the rotor has a magnetic anisotropy, the responses are different in the d and q axes. The angle detection relies on $\sin 2\theta_e$ and $\cos 2\theta_e$ in the autonomous part of the high frequency model. The signal injection methods are classified according to the types of carrier signal:

1) Rotating sinusoidal signal injection in the stationary frame [7], [27] - [30];
2) Stationary sinusoidal signal injection in the (estimated) synchronous reference frame [8], [22], [23];
3) Square-wave signal injection in the stationary frame [14], [15];
4) Square-wave signal injection in the estimated reference frame [13], [14], [16].

Methods 1) and 2) require a demodulation process to extract a desired information from the high frequency responses. The demodulation is carried out via synchronous (heterodyne) rectification consisting of multiplying the carrier and passing through a low pass filter. Correspondingly, the process is rather complicated and the response is sluggish due to the low pass filter. On the other hand, the demodulation is not necessary in methods 3) and 4). The voltage pulse is synthesized each PWM interval, and the angle information is retrieved from the magnitudes of the current ripple. Therefore, these methods are advantageous in processing simplicity and high bandwidth. Three types of injection signals, 1), 2), and 4) are reviewed in the following sections.

9.3.1   Rotating Signal Injection in Stationary Frame

Recall from (6.36) that the IPMSM model in the stationary frame is obtained as

$$\mathbf{v}_{dq}^s = r_s \mathbf{i}_{dq}^s + \frac{d}{dt}(L_{av}\mathbf{i}_{dq}^s - L_{df}e^{j2\theta_e}\mathbf{i}_{dq}^{s*}) + j\omega_e\psi_m e^{j\theta_e},$$

where $L_{av} = (L_d + L_q)/2$ and $L_{df} = (L_q - L_d)/2$. Note also that the saliency effect is condensed in the term, $L_{df}e^{j2\theta_e}\mathbf{i}_{dq}^{s*}$. In the matrix formalism, it is written as

$$L_{df}\begin{bmatrix} \cos 2\theta_e & \sin 2\theta_e \\ \sin 2\theta_e & -\cos 2\theta_e \end{bmatrix}\begin{bmatrix} i_d^s \\ i_q^s \end{bmatrix}.$$

One may neglect the resistive voltage drop and the back-EMF in the high frequency model:

$$\mathbf{v}_{dq}^s = \frac{d}{dt}(L_{av}\mathbf{i}_{dq}^s - L_{df}e^{j2\theta_e}\mathbf{i}_{dq}^{s*}). \tag{9.57}$$

Assume that a high frequency rotating voltage vector is injected into the IPMSM:

$$\mathbf{v}_{dq}^s = V_h e^{j(\omega_h t + \frac{\pi}{2})}, \tag{9.58}$$

where $\omega_h \gg \omega_e$. Let the solution be of the form [25]

$$\mathbf{i}_{dq}^s = I_{cp}e^{j\omega_h t} + I_{cn}e^{-j(\omega_h t - 2\theta_e)}. \tag{9.59}$$

Substituting (9.58) into (9.57) and grouping the terms of positive and negative sequences separately, we can find the coefficients

$$I_{cp} = \frac{V_h}{\omega_h}\frac{L_{av}}{L_{av}^2 - L_{df}^2} \quad \text{and} \quad I_{cn} = \frac{V_h}{\omega_h}\frac{L_{df}}{L_{av}^2 - L_{df}^2}.$$

Equivalently,

$$\begin{bmatrix} i_d^s \\ i_q^s \end{bmatrix} = I_{cp}\begin{bmatrix} \cos\omega_h t \\ \sin\omega_h t \end{bmatrix} + I_{cn}\begin{bmatrix} \cos(2\theta_e - \omega_h t) \\ \sin(2\theta_e - \omega_h t) \end{bmatrix} \tag{9.60}$$

It is emphasized here that θ_e appears only in the negative sequence. Lorenz et al. [7],[25],[26] applied the heterodyne signal processing technique as shown in Fig. 9.22: i_d^s and i_q^s are multiplied by $\sin(2\hat{\theta}_e - \omega_h t)$ and $\cos(2\hat{\theta}_e - \omega_h t)$ which are based on an estimate $\hat{\theta}_e$. Then,

$$i_q^s \cos(2\hat{\theta}_e - \omega_h t) - i_d^s \sin(2\hat{\theta}_e - \omega_h t) = I_{cn}\sin 2(\theta_e - \hat{\theta}_e) + I_{cp}\sin 2(\omega_h t - \hat{\theta}_e), \tag{9.61}$$

where the first term is a DC, whereas the second is a high frequency term. Applying a low pass filter, the high frequency carrier is removed. Further if $\theta_e - \hat{\theta}_e$ is small, then

$$LPF\left(i_q^s \cos(2\hat{\theta}_e - \omega_h t) - i_d^s \sin(2\hat{\theta}_e - \omega_h t)\right) = I_{cn}\sin 2(\theta_e - \hat{\theta}_e) \approx 2I_{cn}(\theta_e - \hat{\theta}_e),$$

where $LPF(\cdot)$ signifies a low pass filter. Spectral description of the demodulation process is shown in Fig. 9.21.

The output of the demodulation is angle error, $\theta_e - \hat{\theta}_e$, and it is combined with a tracking filter. Due to inherent tracking capability, $\hat{\theta}_e$ will track θ_e. Fig. 9.22 shows a whole signal processing block diagram.

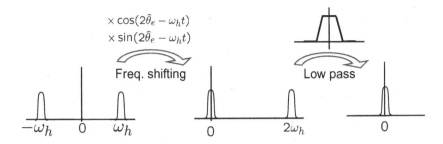

Figure 9.21: Spectral description of the demodulation.

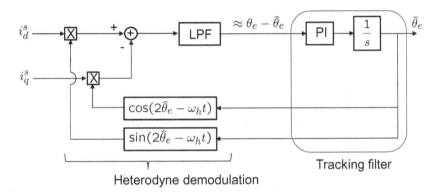

Figure 9.22: Angle detection algorithm based on signal injection utilizing the heterodyne demodulation and a tracking filter.

9.3.2 Signal Injection in a Synchronous Frame

In this part, a sinusoidal signal is injected only into the d-axis of the synchronous frame which is based on an angle estimate, and the current solution is obtained from the high frequency model of IPMSM.

High Frequency Model of IPMSM

Suppose that a high frequency signal, $[v_{dh}^e, v_{qh}^e]^T$ is added to the main driving voltage, $[v_d^e, v_q^e]^T$, in the synchronous reference frame which is aligned to the real PM flux. Then by the superposition law, the current responses are obtained such that

$$\begin{bmatrix} v_d^e + v_{dh}^e \\ v_q^e + v_{qh}^e \end{bmatrix} = \begin{bmatrix} r_s + pL_d & -\omega_e L_q \\ \omega_e L_d & r_s + pL_q \end{bmatrix} \begin{bmatrix} i_d^e + i_{dh}^e \\ i_q^e + i_{qh}^e \end{bmatrix} + \begin{bmatrix} 0 \\ \omega_e \psi_m \end{bmatrix}, \quad (9.62)$$

where i_{dh}^e, i_{qh}^e are the high frequency current responses. The high frequency part can be separated such that

$$\begin{bmatrix} v_{dh}^e \\ v_{qh}^e \end{bmatrix} = \begin{bmatrix} r_s + pL_d & -\omega_e L_q \\ \omega_e L_d & r_s + pL_q \end{bmatrix} \begin{bmatrix} i_{dh}^e \\ i_{qh}^e \end{bmatrix} \approx \begin{bmatrix} r_s + pL_d & 0 \\ 0 & r_s + pL_q \end{bmatrix} \begin{bmatrix} i_{dh}^e \\ i_{qh}^e \end{bmatrix}. \quad (9.63)$$

The last approximation in (9.63) is based on the observation that $\omega_h L_d \gg \omega_e L_q$ since $\omega_h \gg \omega_e$. Writing (9.63) in favor of the current vector, we get

$$\begin{bmatrix} i_{dh}^e \\ i_{qh}^e \end{bmatrix} = \begin{bmatrix} \frac{1}{r_s + pL_d} & 0 \\ 0 & \frac{1}{r_s + pL_q} \end{bmatrix} \begin{bmatrix} v_{dh}^e \\ v_{qh}^e \end{bmatrix}. \tag{9.64}$$

Suppose that the estimated angle $\bar{\theta}_e$ differs from the real angle by $\Delta\theta_e = \theta_e - \bar{\theta}_e$. Using the method shown in (9.9), the dynamics in the misaligned coordinate are given as

$$\begin{bmatrix} \bar{i}_{dh}^e \\ \bar{i}_{qh}^e \end{bmatrix} = \begin{bmatrix} \cos\Delta\theta_e & -\sin\Delta\theta_e \\ \sin\Delta\theta_e & \cos\Delta\theta_e \end{bmatrix} \begin{bmatrix} \frac{1}{r_s+pL_d} & 0 \\ 0 & \frac{1}{r_s+pL_q} \end{bmatrix} \begin{bmatrix} \cos\Delta\theta_e & \sin\Delta\theta_e \\ -\sin\Delta\theta_e & \cos\Delta\theta_e \end{bmatrix} \begin{bmatrix} \bar{v}_{dh}^e \\ \bar{v}_{qh}^e \end{bmatrix}$$

$$= \begin{bmatrix} \frac{\cos^2\Delta\theta_e}{r_s+pL_d} + \frac{\sin^2\Delta\theta_e}{r_s+pL_q} & \frac{\cos\Delta\theta_e\sin\Delta\theta_e}{r_s+pL_d} - \frac{\cos\Delta\theta_e\sin\Delta\theta_e}{r_s+pL_q} \\ \frac{\cos\Delta\theta_e\sin\Delta\theta_e}{r_s+pL_d} - \frac{\cos\Delta\theta_e\sin\Delta\theta_e}{r_s+pL_q} & \frac{\sin^2\Delta\theta_e}{r_s+pL_d} + \frac{\cos^2\Delta\theta_e}{r_s+pL_q} \end{bmatrix} \begin{bmatrix} \bar{v}_{dh}^e \\ \bar{v}_{qh}^e \end{bmatrix}.$$

Suppose that a high frequency signal is injected into the d-axis of the misaligned frame:

$$\begin{bmatrix} \bar{v}_{dh}^e \\ \bar{v}_{qh}^e \end{bmatrix} = \begin{bmatrix} V_h \cos\omega_h t \\ 0 \end{bmatrix}. \tag{9.65}$$

Then, it follows that

$$\begin{bmatrix} \bar{i}_{dh}^e \\ \bar{i}_{qh}^e \end{bmatrix} = \begin{bmatrix} \frac{\cos^2\Delta\theta_e}{r_s+pL_d} + \frac{\sin^2\Delta\theta_e}{r_s+pL_q} \\ \frac{\cos\Delta\theta_e\sin\Delta\theta_e}{r_s+pL_d} - \frac{\cos\Delta\theta_e\sin\Delta\theta_e}{r_s+pL_q} \end{bmatrix} V_h \cos\omega_h t. \tag{9.66}$$

Note that (9.64) and (9.66) are the same equations, but described in the coordinates that differ by $\Delta\theta_e$.

Simplified Angle Error Estimation Method

Let

$$A \equiv \frac{V_h}{r_s + pL_d} \cos\omega_h t \quad \text{and} \quad B \equiv \frac{V_h}{r_s + pL_q} \cos\omega_h t.$$

The steady state solutions of A and B are

$$A = \frac{V_h}{\sqrt{r_s^2 + \omega_h^2 L_d^2}} \cos\left(\omega_h t - \tan^{-1}\left(\frac{\omega_h L_d}{r_s}\right)\right), \tag{9.67}$$

$$B = \frac{V_h}{\sqrt{r_s^2 + \omega_h^2 L_q^2}} \cos\left(\omega_h t - \tan^{-1}\left(\frac{\omega_h L_q}{r_s}\right)\right). \tag{9.68}$$

Now consider a limiting case where $\omega_h L_d \gg r_s$ for a high $\omega_h > 0$. Then,

$$A \approx \frac{V_h}{\omega_h L_d} \sin\omega_h t \quad \text{and} \quad B \approx \frac{V_h}{\omega_h L_q} \sin\omega_h t.$$

Noting that $L_d = L_{av} - L_{df}$ and $L_q = L_{av} + L_{df}$, the solution for (9.66) reduces to

$$\begin{bmatrix} \tilde{i}^e_{dh} \\ \tilde{i}^e_{qh} \end{bmatrix} = \frac{V_h}{\omega_h L_d L_q} \begin{bmatrix} L_{av} + L_{df} \cos 2\Delta\theta_e \\ L_{df} \sin 2\Delta\theta_e \end{bmatrix} \sin(\omega_h t). \tag{9.69}$$

To obtain an estimate of $\Delta\theta_e$, $\sin(\omega_h t)$ should be eliminated. For this purpose, a synchronous rectification method is utilized: First, multiply the modulated signal by $\sin(\omega_h t)$. Then, remove the high frequency term by using a low pass filter (LPF). We let

$$X_{dh} = \frac{2\omega_h L_d L_q}{V_h} \times \mathrm{LPF}(\tilde{i}^e_{dh} \times \sin\omega_h t) \tag{9.70}$$

$$X_{qh} = \frac{2\omega_h L_d L_q}{V_h} \times \mathrm{LPF}(\tilde{i}^e_{qh} \times \sin\omega_h t), \tag{9.71}$$

where $\mathrm{LPF}(r)$ represents the filtered signal of $r(t)$. Note that

$$X_{dh} \approx L_{av} + L_{df} \cos 2\Delta\theta_e$$
$$X_{qh} \approx L_{df} \sin 2\Delta\theta_e.$$

Therefore, an angle error estimate is obtained such that

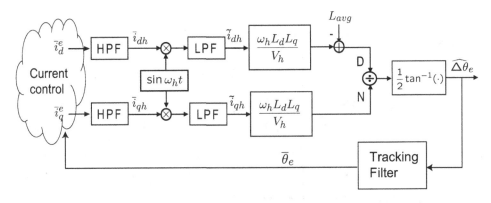

Figure 9.23: Angle estimation block diagram: Signal injection into the d-axis of an estimated frame.

$$\widehat{\Delta\theta_e} = \frac{1}{2} \tan^{-1} \left(\frac{X_{qh}}{X_{dh} - L_{av}} \right).$$

Fig. 9.23 shows the signal processing block diagram. If the rotor position estimation error is small, then (9.71) is approximated as $X_{qh} \approx 2L_{df}\Delta\theta_e$, and it can be used solely for $\Delta\theta_e$ estimation. The role and structure of the tracking filter are the same as in Fig. 9.22.

Instead of $V_h \cos\omega_h t$, pulsating voltage vector was utilized in [31], [32], [33]. It yielded better results for SPMSMs, where the machine saliency is given by the magnetic saturation and not by the structural rotor geometry.

9.3.3   PWM Level Square-Voltage Injection in Estimated Frame

Performances of sinusoidal wave injection methods are usually insufficient in some applications, since the angle estimation bandwidth is not wide enough. The frequency of injected voltage is never too high due to the limitation of PWM frequency.

To improve the bandwidth of rotor position estimation, a square-wave pulsating signal is synthesized in each PWM period, while providing the normal control voltage. This method utilizes the available maximum bandwidth. Of course, the pulse average is zero and injected into the d-axis in order not to disturb the main operation. The rotor position is estimated utilizing the magnitudes of current ripple. This estimation process is simple and straightforward since it does not require demodulation. High signal bandwidth gives a better signal to noise ratio. Due to these merits, this approach performs better than the other injection algorithms.

Current Ripple to Voltage Pulse

A square voltage pulse is designed in the synchronous estimated frame. Fig. 9.24 shows high frequency voltage components, $\bar{v}_{dh}^{e}, \bar{v}_{qh}^{e}$. Note that a square wave is synthesized only in the d-axis every PWM period [14]. The sequence of obtaining the the current ripple consists of:

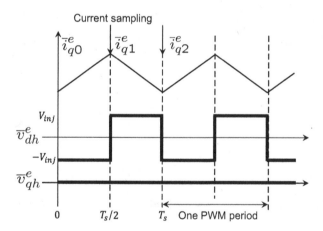

Figure 9.24: Square wave voltage in the estimated rotor reference frame and current sampling points.

1) Transform the injection voltage into the synchronous frame:

$$\mathbf{v}_{dqh}^{e} = e^{\mathbf{J}\Delta\theta}\bar{\mathbf{v}}_{dqh}^{e} = \begin{bmatrix} \cos\Delta\theta_{e} & \sin\Delta\theta_{e} \\ -\sin\Delta\theta_{e} & \cos\Delta\theta_{e} \end{bmatrix} \begin{bmatrix} \pm V_{inj} \\ 0 \end{bmatrix} = \pm V_{inj} \begin{bmatrix} \cos\Delta\theta_{e} \\ -\sin\Delta\theta_{e} \end{bmatrix}, \quad (9.72)$$

where V_{inj} is the amplitude of the injection signal and $e^{\mathbf{J}\Delta\theta_{e}}$ stands for the transformation from the estimated rotor reference frame to the real rotor reference frame.

2) High frequency current dynamics in the synchronous frame is given by

$$\mathbf{v}_{dqh}^e = \pm V_{inj} \begin{bmatrix} \cos \Delta\theta_e \\ -\sin \Delta\theta_e \end{bmatrix} = \begin{bmatrix} L_d & 0 \\ 0 & L_q \end{bmatrix} \frac{1}{T_s} \begin{bmatrix} \Delta i_{dh}^e \\ \Delta i_{qh}^e \end{bmatrix}, \tag{9.73}$$

where T_s is the PWM period, and the following approximation is utilized:

$$L_d \frac{di_{dh}^e}{dt} \approx L_d \frac{i_{dh}^e(t_i + T_s) - i_{dh}^e(t_i)}{T_s} = L_d \frac{\Delta i_{dh}^e}{T_s}.$$

3) The current ripple is transformed back into the estimated frame:

$$\begin{bmatrix} \Delta \bar{i}_{dh}^e \\ \Delta \bar{i}_{qh}^e \end{bmatrix} = e^{-\mathbf{J}\Delta\theta_e} \begin{bmatrix} \Delta i_{dh}^e \\ \Delta i_{qh}^e \end{bmatrix} = \pm T_s V_{inj} \begin{bmatrix} \cos \Delta\theta_e & -\sin \Delta\theta_e \\ \sin \Delta\theta_e & \cos \Delta\theta_e \end{bmatrix} \begin{bmatrix} \frac{\cos \Delta\theta_e}{L_d} \\ -\frac{\sin \Delta\theta_e}{L_q} \end{bmatrix}$$

$$= \pm T_s \cdot V_{inj} \begin{bmatrix} \frac{\cos^2 \Delta\theta_e}{L_d} + \frac{\sin^2 \Delta\theta_e}{L_q} \\ \frac{1}{2}\left(\frac{1}{L_d} - \frac{1}{L_q}\right) \sin 2\Delta\theta_e \end{bmatrix} \tag{9.74}$$

Like the previous method in Section 9.2.5, the desired angle information appears in the q-axis current ripple.

Angle Detection from Current Ripple

Fig. 9.24 shows the d-axis pulsating voltage and the corresponding current ripple. It depicts current and voltage in the estimated frame: The high frequency voltage component is

$$\bar{v}_{dh}^e = \begin{cases} V_{inj}, & 0 \le t < T_s/2, \\ -V_{inj}, & T_s/2 \le t < T_s. \end{cases}$$

In Fig. 9.24, two current components are added: The current increases steadily by the (normal) control voltage component, whereas the high frequency current has a triangular shape.

Let the absolute current magnitudes denoted by $\bar{i}_{d0}^e = \bar{i}_d^e(0)$, $\bar{i}_{d1}^e = \bar{i}_d^e(T_s/2)$, and $\bar{i}_{d2}^e = \bar{i}_d^e(T_s)$. Define the current increments by

$$\Delta \bar{i}_{d10}^e = \bar{i}_{d1}^e - \bar{i}_{d0}^e, \tag{9.75}$$

$$\Delta \bar{i}_{d21}^e = \bar{i}_{d2}^e - \bar{i}_{d1}^e. \tag{9.76}$$

Also, denote by $\bar{i}_{d\Delta}^e$ the control current increment over a half interval. It is an increment of the average current driven by the the current controller. Since it is assumed to have a constant slope, the amount of variation is $2\bar{i}_{d\Delta}^e$ in a single PWM period. From the geometry shown in Fig. 9.24, it follows that $\bar{i}_{dh}^e = \Delta\bar{i}_{d10}^e - \bar{i}_{d\Delta}^e$ at $t = T_s/2$, and also $\bar{i}_{dh}^e = -\Delta\bar{i}_{d21}^e + \bar{i}_{d\Delta}^e$ at $t = T_s$. The similar notation and the same definition are applied to the q-axis. It follows from (9.75) and (9.76) that

$$\Delta\bar{i}_{d10}^e - \Delta\bar{i}_{d21}^e = 2\bar{i}_{dh}^e$$

$$\Delta\bar{i}_{q10}^e - \Delta\bar{i}_{q21}^e = 2\bar{i}_{qh}^e \tag{9.77}$$

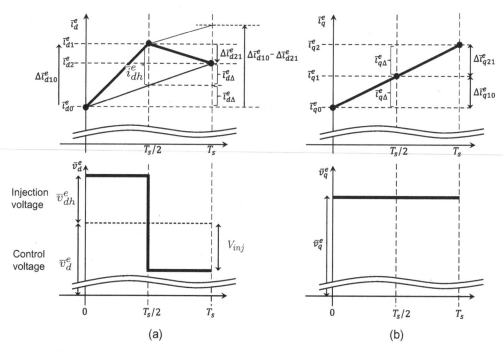

Figure 9.25: Voltage and current waves in the estimated rotor reference frame over a single PWM period: (a) d-axis current and voltage and (b) q-axis current and voltage.

Current ripples caused by the high frequency voltage are $\tilde{i}_{dh}^e$ and $\tilde{i}_{qh}^e$. Consequently, we obtain from (9.74) that

$$\begin{bmatrix} \Delta \tilde{i}_{d10}^e - \Delta \tilde{i}_{d21}^e \\ \Delta \tilde{i}_{q10}^e - \Delta \tilde{i}_{q21}^e \end{bmatrix} = 2 \begin{bmatrix} \tilde{i}_{dh}^e \\ \tilde{i}_{qh}^e \end{bmatrix} = 2T_s \cdot V_{inj} \begin{bmatrix} \frac{\cos^2 \Delta \theta_e}{L_d} + \frac{\sin^2 \Delta \theta_e}{L_q} \\ \frac{1}{2}\left(\frac{1}{L_d} - \frac{1}{L_q}\right) \sin 2\Delta \theta_e \end{bmatrix} \tag{9.78}$$

The q-axis component contains $\Delta \theta_e$ and $\sin 2\Delta \theta_e \approx 2\Delta \theta_e$. Thereby, $\Delta \theta_e$ is expressed as a function of current ripple:

$$\Delta \theta_e = 2K\tilde{i}_{qh}^e = K(\Delta \tilde{i}_{q10}^e - \Delta \tilde{i}_{q21}^e) \tag{9.79}$$

$$K = \frac{L_d L_q}{2(L_q - L_d)T_s \cdot V_{inj}}. \tag{9.80}$$

Note that this PWM level sensorless method does not need a demodulation process [14]. Simply, currents are measured twice every PWM period. Fig. 9.25 shows voltage and current waves in the estimated rotor reference frame over a PWM period. Fig. 9.26 shows the block diagram for obtaining an angle estimate using the magnitudes of q-axis current ripple.

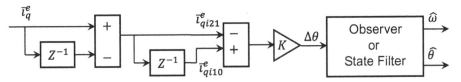

Figure 9.26: Angle estimation using the q-axis current ripple in the estimated rotor reference frame.

Summary

1. A high frequency probing signal is injected to detect the rotor angle by exploiting the magnet saliency of IPMSMs. With the signal injection method, IPMSMs can be controlled at zero speed without a position sensor.

2. For angle detection, the synchronous rectification is utilized. The process consists of carrier multiplication and low pass filtering. The low pass filtering causes phase delay, so that the response is sluggish and even instability occurs when the carrier bandwidth is not sufficiently high.

3. Considering the limits in increasing the frequency via PWM, square wave is better than sinusoidal waves. As a typical reference, note that the square wave injection frequency is 1 kHz, whereas the sinusoidal wave injection frequency is 330 Hz. It was reported in [29] that the pulsating injection method was more accurate and less sensitive to the dead-time effect.

4. The high frequency signal is injected either in the estimated frame or stationary frame. In both cases, an estimate of angle error, $\Delta\theta_e$ is obtained. Thereby, a PLL filter is naturally used to derive an absolute angle from the angle error. The strength of filtering effect can be adjusted by the PI gain of the filter.

5. Square pulses can be made by setting positive and negative levels for each half PWM period. From the magnitude of current ripple, the angle error, $\Delta\theta_e$ can be detected. It makes the demodulation process obsolete, so that it does not suffer the phase delay problem. The largest bandwidth is expected since it utilizes the maximum injection frequency and avoids the use of a low pass filter.

6. Higher signal magnitude is favorable for a larger signal to noise ratio. However, it would reduce the control voltage margin while inducing extra copper and eddy losses.

7. The high frequency signal makes audible noise. The amplitude and frequency of injected voltage need to be optimized considering the bandwidth, signal to noise ratio, loss, and acoustic noise.

8. The sensorless control algorithm is widely adopted in general purpose IPMSM drives. However, the performance is still insufficient in some applications. Typical problems are long convergence time, potential start-up failure, and limited system stability [15].

References

[1] N. Matsui, "Sensorless PM brushless DC motor drives," *IEEE Trans. on Ind. Electron.*, vol. 43, no. 2, pp. 300−308, Apr. 1996.

[2] M. Tomita, T. Senjyu, S. Doki, and S. Okuma, "New sensorless controls for brushless dc motors using disturbance observers and adaptive velocity estimations," *IEEE Trans. on Ind. Electron.*, vol. 45, no.2, pp. 274−282, Apr. 1998.

[3] R. Ortega, L. Praly, A. Astolfi, J. Lee, and K. Nam, "A Simple observer for permanent magnet synchronous motors with guaranteed stability properties," *LSS Internal Report*, Dec. 2008.

[4] J. Lee, J. Hong, K. Nam, R. Ortega, L. Praly, L., and A. Astolfi, "Sensorless control of surface-mount permanent magnet synchronous motors based on a nonlinear observer," *IEEE Trans. on Power Elec.*, vol. 25, no. 2, pp. 290−297, 2010.

[5] A. Bobtsov, A. Pyrkin, and R. Ortega, "A Robust Globally Convergent Position Observer for the Permanent Magnet Synchronous Motor," *Automatica*, vol. 61, pp. 47−54, Nov. 2015.

[6] J. Choi, K. Nam, A. A. Bobtsov, A. Pyrkin and R. Ortega, "Robust Adaptive Sensorless Control for Permanent-Magnet Synchronous Motors," *IEEE Trans. Power Electron.*, vol. 32, no. 5, pp. 3989−3997, May. 2017.

[7] M. J. Corley and R. D. Lorenz, "Rotor position and velocity estimation for a salient-pole permanent magnet synchronous machine at standstill and high speeds," *IEEE Trans. on Ind. Appl.*, vol. 34, no 4, pp. 784−789, Jul./Aug. 1998.

[8] Y. Li, Z. Q. Zhu, D. Howe, C. M. Bingham and D. A. Stone, "Improved rotor position estimation by signal injection in brushless AC motors, accounting for cross-coupling magnetic saturation," *IEEE Trans. Ind. Appl.*, vol. 45, no. 5, pp. 1843−1849, Sept./Oct. 2009.

[9] T. Aihara, A. Toba, T. Yanase, A. Mashimo, and K. Endo, "Sensorless torque control of salient-pole synchronous motor at zero-speed operation," *IEEE Trans. on Power Elec.*, vol 14, no 1, pp. 202−208, Jan. 1999.

[10] Z. Chen, M. Tomita, S. Doki, and S. Okuma, "New adaptive sliding observers for position- and velocity-sensorless controls of brushless DC motors," *IEEE Trans. on Ind. Electron.*, vol. 47, no. 3, pp. 582−591, Jun. 2000.

[11] B. Nahid-Mobarakeh, F. Meibody-Tabar, and F. M. Sargos, "Back EMF estimation-based sensorless control of PMSM: robustness with respect to measurement errors and inverter irregularities," *IEEE Trans. on Ind. Appl.*, vol. 43, no. 2, pp. 485−494, Mar./Apr. 2007.

[12] S. Bolognani, L. Tubiana, and M. Zigliotto, "Extended Kalman filter tuning in sensorless PMSM drives," *IEEE Trans. on Ind. Appl.*, vol. 39, no. 6, pp. 1741−1747, Nov./Dec. 2003.

[13] Y. D. Yoon, S. K. Sul, S. Morimoto, and K. Ide, High bandwidth sensorless algorithm for AC machines based on square-wave type voltage injection," *IEEE Trans. Ind. Appl.*, vol. 47, no. 3, pp. 1361−1370, May/Jun. 2011.

[14] S. Kim, J. I.Ha, and S.K. Sul, PWM switching frequency signal injection sensorless method in IPMSM," *IEEE Trans. Ind. Appl.*, vol. 48, no. 5, pp. 1576−1587, Sep./Oct. 2012.

[15] J. Liu, and Z. Q. Zhu, "Sensorless Control Strategy by Square-Waveform High-Frequency Pulsating Signal Injection Into Stationary Reference Frame," *IEEE J. of Emerging and Selected Topics in Power Elec.*, vol. 2, no. 2, pp. 171−180, Jun. 2014.

[16] Y. Kwon, and S. Sul, " Reduction of Injection Voltage in Signal Injection Sensorless Drives Using a Capacitor-Integrated Inverter," *IEEE Trans. on Power Elec.*, vol. 32, no. 8, pp. 6261−6273, Aug. 2017.

[17] M. Jansson, L. Harnefors, O. Wallmark, M. Leksell, "Synchronization at startup and stable rotation reversal of sensorless nonsalient PMSM drives," *IEEE Trans. on Ind. Elec.*, vol. 53, no. 2, pp. 379−387, Apr. 2006.

[18] S. Morimoto, K. Kawamoto, M. Sanada, and Y. Takeda, "Sensorless control strategy for salient-pole PMSM based on extended EMF in rotating reference frame," *IEEE Trans. on Ind. Appl.*, vol. 38, no 4, pp. 1054−1061, Jul./Aug. 2002.

[19] T. Umeno and Y. Hori, "Robust Speed Control of DC Servomotors Using Modern Two Degrees-of- Freedom Controller Design," *IEEE Trans. on Ind. Elect.*, vol. 38, no. 5, pp. 363−368, Oct. 1991.

[20] "L. Harnefors and H.-P. Nee, A general algorithm for speed and position estimation of ac motors," *IEEE Trans. Ind. Elect.*, vol. 47, no. 1, pp. 77−83, Feb. 2000.

[21] O. Wallmark and L. Harnefors, "Sensorless control of salient PMSM drives in the transition region," *IEEE Trans. Ind. Elect.*, vol. 53, no. 4, pp. 1179−1187, Aug. 2006.

[22] J. H. Jang, S. K. Sul, J. I. Ha, K. Ide, and M. Sawamura, "Sensorless drive of surface-mounted permanent-magnet motor by high-frequency signal injection based on magnetic saliency," *IEEE Trans. Ind. Appl.*, vol. 39, no. 4, pp. 1031−1039, Jul./Aug. 2003.

[23] J. H. Jang, J. I. Ha, M. Ohto, K. Ide, and S. K. Sul, "Analysis of permanent-magnet machine for sensorless control based on high-frequency signal injection," *IEEE Trans. Ind. Appl.*, vol. 40, no. 6, pp. 1595−1604, Nov./Dec. 2004.

[24] R. Mizutani, T. Takeshita, and N. Matsui, "Current model-based sensorless drives of salient-pole PMSM at low speed and standstill," *IEEE Trans. Ind. Appl.*, vol. 34, no. 4, pp. 841−846, Jul./Aug. 1998.

[25] Y.S. Jeong, R. D. Lorenz, T. M. Jahns, and S. K. Sul, "Initial rotor position estimation of an interior permanent-magnet synchronous machine using carrier-frequency injection methods," *IEEE Trans. Ind. Appl.*, vol. 41, no. 1, pp. 38−45, Jan./Feb. 2005.

[26] S. Wu, D. D. Reigosa, Y. Shibukawa, M. A. Leetmaa, R. D. Lorenz, and Y. Li, "Interior permanent-magnet synchronous motor design for improving self-sensing performance at very low speed," *IEEE Trans. Ind. Appl.*, vol. 45, no. 6, pp. 1939−1945, Nov./Dec. 2009.

[27] M. W. Degner and R. D. Lorenz, "Using multiple saliencies for the estimation of flux, position, and velocity in AC machines," *IEEE Trans. Ind. Appl.*, vol. 34, no. 5, pp. 1097−1104, Sep./Oct. 1998.

[28] P. Garcia, F. Briz, M. W. Degner, and D. Diaz-Reigosa, "Accuracy, bandwidth, and stability limits of carrier-signal-injection-based sensorless control methods," *IEEE Trans. Ind. Appl.*, vol. 43, no. 4, pp. 990−1000, Jul./Aug. 2007.

[29] D. Raca, P. Garcia, D. Reigosa, F. Briz, and R. D. Lorenz, "A comparative analysis of pulsating vs. rotating vector carrier signal injection-based sensorless control," *Proc. IEEE Appl. Power Electron. Conf. Expo.*, Feb. 2008, vol. 879, pp. 24−28.

[30] D. Raca, P. Garcia, D. D. Reigosa, F. Briz, and R. D. Lorenz, "Carrier signal selection for sensorless control of PM synchronous machines at zero and very low speeds," *IEEE Trans. Ind. Appl.*, vol. 46, no. 1, pp. 167−178, Jan./Feb. 2010.

[31] M. Linke, R. Kennel, and J. Holtz, "Sensorless position control of Permanent Magnet Synchronous Machines without Limitation at Zero Speed," *Proc. 2002 IEEE IECON Conf.*, pp. 674−679.

[32] M. Linke, R. Kennel, and J. Holtz, "Sensorless speed and position control of synchronous machines using alternating carrier injection," *Proc. Int. Elec. Mach. Drives Conf.*, Madison, pp. 1211−1217, Jun. 2003.

[33] H. Holtz, "Acquisition of Position Error and Magnet Polarity for Sensorless Control of PM Synchronous Machines," *IEEE Tran. Ind. Appl.*, vol. 44, no. 4, pp. 1172−1180, July/August 2008.

Problems

9.1 Define the observation error by $\tilde{\mathbf{x}} \equiv \hat{\mathbf{x}} - \mathbf{x}$. Show that the error dynamics directly follows from (9.26)-(9.31) such that

$$\dot{\tilde{\mathbf{x}}} = -\gamma a(\tilde{\mathbf{x}}, t) \left\{ \tilde{\mathbf{x}} + \psi_m \begin{bmatrix} \cos \theta_e(t) \\ \sin \theta_e(t) \end{bmatrix} \right\}$$

$$a(\tilde{\mathbf{x}}, t) \equiv \frac{1}{2} \|\tilde{\mathbf{x}}\|^2 + \psi_m [\tilde{x}_1 \cos \theta_e(t) + \tilde{x}_2 \sin \theta_e(t)].$$

9.2 Using the MATLAB Simulink, construct a speed controller based on the Ortega's sensorless algorithm for a SPMSM listed in Table 9.2. Follow the instructions below:

a) Use a PLL type speed estimator:

$$\dot{z}_1 = K_p(\hat{\theta}_e - z_1) + K_i z_2$$
$$\dot{z}_2 = \hat{\theta}_e - z_1$$
$$\hat{\omega} = K_p(\hat{\theta}_e - z_1) + K_i z_2,$$

where K_p and K_i are proportional and integral gains, respectively.

b) Use PI controllers for current control and compensate the cross coupling voltages, $\omega L i_d$ and $\omega L i_q$.

c) Use a PI controller for the speed control.

d) Refer to the control block diagram shown in Fig. 9.8. But neglect the d-axis voltage pulse injection shown on the top.

Obtain a comparable result (300 rpm, no load) with the experimental one shown in Fig. 9.10 (b).

9.3 Derive

$$\dot{\varepsilon}(t) = (\mathbf{A}_s - \mathbf{Kc})\varepsilon(t) - \left(\int_0^t e^{(\mathbf{A}_s - \mathbf{Kc})(t-\tau)} \mathbf{b}_s(\tau) d\tau \right) \dot{\hat{p}}(t)$$

from

$$\dot{\beta}(t) = (\mathbf{A}_s - \mathbf{Kc})\beta(t) + \mathbf{b}_s(t),$$
$$\varepsilon(t) = -\beta(t)\hat{p}(t) + \int_0^t e^{(\mathbf{A}_s - \mathbf{Kc})(t-\tau)} \mathbf{b}_s(\tau)\hat{p}(\tau) d\tau.$$

9.4 Derive from (9.19) the following:

$$\begin{bmatrix} \overline{v}_d^e \\ \overline{v}_q^e \end{bmatrix} = \begin{bmatrix} r_s + pL_d & -\omega_e L_q \\ \omega_e L_q & r_s + pL_d \end{bmatrix} \begin{bmatrix} \overline{i}_d^e \\ \overline{i}_q^e \end{bmatrix} + \begin{bmatrix} \xi_d^e \\ \xi_q^e \end{bmatrix}$$

where
$$\begin{bmatrix} \xi_d^e \\ \xi_q^e \end{bmatrix} = E_{ex} \begin{bmatrix} \sin(\bar{\theta} - \theta) \\ \cos(\bar{\theta} - \theta) \end{bmatrix} + (\bar{\omega}_e - \omega_e) L_d \begin{bmatrix} -i_q^{-e} \\ i_d^{-e} \end{bmatrix}.$$

9.5   Show that if $\bar{\omega}_e \neq \omega_e$ then

$$e^{-J\Delta\theta_e} \begin{bmatrix} \mathrm{p}L_d & 0 \\ 0 & \mathrm{p}L_q \end{bmatrix} e^{J\Delta\theta_e} \bar{i}_{dq}^{-e}$$

$$= (\bar{\omega}_e - \omega_e) \begin{bmatrix} L_{df} \sin 2\Delta\theta_e & -L_{av} + L_{df} \cos 2\Delta\theta_e \\ L_{av} + L_{df} \cos 2\Delta\theta_e & -L_{df} \sin 2\Delta\theta_e \end{bmatrix} \begin{bmatrix} i_d^{-e} \\ i_q^{-e} \end{bmatrix}$$

$$+ \begin{bmatrix} L_{av} - L_{df} \cos 2\Delta\theta_e & L_{df} \sin 2\Delta\theta_e \\ L_{df} \sin 2\Delta\theta_e & L_{av} + L_{df} \cos 2\Delta\theta_e \end{bmatrix} \begin{bmatrix} \mathrm{p}i_d^{-e} \\ \mathrm{p}i_q^{-e} \end{bmatrix}.$$

9.6   Derive (9.55) for $\Delta\mathbf{v} = [\frac{V_p}{s}, 0]^T$.

9.7   Using (9.48), show the error dynamic (9.49). Note that the parameter error is defined by $\tilde{\eta}_0 = \hat{\eta}_0 - \eta_0$.

9.8   Consider a voltage equation with the cross coupling inductance L_{dq} [8]:

$$\begin{bmatrix} v_d^e \\ v_q^e \end{bmatrix} = \begin{bmatrix} r_s + \mathrm{p}L_d - \omega_e L_{dq} & -\omega_e L_q + \mathrm{p}L_{dq} \\ \omega_e L_d + \mathrm{p}L_{dq} & r_s + \mathrm{p}L_q + \omega_e L_{dq} \end{bmatrix} \begin{bmatrix} i_d^e \\ i_q^e \end{bmatrix} + \begin{bmatrix} 0 \\ \omega_e \psi_m \end{bmatrix}.$$

a) Denote by i_d^{-e}, i_q^{-e}, and v_d^{-e}, v_q^{-e} the high frequency current and voltage components in a misaligned frame. Assuming $r_s = 0$ and $\omega_e = 0$, show that high frequency model is obtained in the misaligned synchronous frame such that

$$\begin{bmatrix} i_d^{-e} \\ i_q^{-e} \end{bmatrix} = \frac{1}{L_{av}^2 - \hat{L}_{df}^2} \begin{bmatrix} L_{av} + \hat{L}_{df} \cos(2\Delta\theta_e + \theta_c) & -\hat{L}_{df} \sin(2\Delta\theta_e + \theta_c) \\ -\hat{L}_{df} \sin(2\Delta\theta_e + \theta_c) & L_{av} - \hat{L}_{df} \cos(2\Delta\theta_e + \theta_c) \end{bmatrix} \frac{1}{\mathrm{p}} \begin{bmatrix} v_d^{-e} \\ v_q^{-e} \end{bmatrix}$$

where $\Delta\theta = \bar{\theta} - \theta$ is the error angle, $L_{av} = (L_q + L_d)/2$, $L_{df} = (L_q - L_d)/2$, $\theta_c = \tan^{-1}(L_{dq}/L_{df})$, and $\hat{L}_{df} = \sqrt{L_{df}^2 + L_{dq}^2}$. Assume that $\Delta\omega_e = 0$.

b) A sinusoidal signal is injected only into the d-axis of the misaligned frame, i.e., $[\bar{v}_d^e, \bar{v}_{qh}^e]^T = [V_h \cos\omega_h t, 0]^T$. When i_q^{-e} in a) is forced to zero, show that the steady-state position error caused by the cross coupling inductance is obtained such that

$$\Delta\theta_e = -\frac{1}{2} \tan^{-1} \frac{2L_{dq}}{L_q - L_d} = -\frac{1}{2}\theta_c.$$

9.9   In the case of multiple saliency, the high-frequency model is obtained as

$$\begin{bmatrix} v_{dh}^e \\ v_{qh}^e \end{bmatrix} = \begin{bmatrix} L_d - L_{\delta 4} \cos 6\theta_e & L_{\delta 4} \sin 6\theta_e \\ L_{\delta 4} \sin 6\theta_e & L_q + L_{\delta 4} \cos 6\theta_e \end{bmatrix} \mathrm{p} \begin{bmatrix} i_{dh}^e \\ i_{qh}^e \end{bmatrix},$$

where $L_{\delta 4}$ is the multiple saliency inductance. When a high-frequency injection into d-axis of misaligned frame, show that the high-frequency q-axis current appears as

$$\bar{i}^e_{qh} = -\frac{\bar{v}^e_{dh}}{L^2_{av} - L^2_{df} - 2L_{\delta 4}\cos 6\theta_e L_{df} - L^2_{\delta 4}}\frac{1}{\mathrm{p}}[L_{df}\sin 2\Delta\theta_e + L_{\delta 4}\sin(2\Delta\theta + 6\theta_e)].$$

where $L_{av} = (L_q + L_d)/2$ and $L_{df} = (L_q - L_d)/2$. Note that $\Delta\theta = \bar{\theta} - \theta$.

9.10 Derive from (9.7) the following voltage equation:

$$\begin{bmatrix} v^s_d \\ v^s_q \end{bmatrix} = \begin{bmatrix} r_s + \mathrm{p}L_q & 0 \\ 0 & r_s + \mathrm{p}L_q \end{bmatrix}\begin{bmatrix} i^s_d \\ i^s_q \end{bmatrix} + \omega_e\left(\psi_m + (L_d - L_q)i^e_d\right)\begin{bmatrix} -\sin\theta_e \\ \cos\theta_e \end{bmatrix}$$
$$+(L_d - L_q)(\mathrm{p}\ i^e_d)\begin{bmatrix} \cos\theta_e \\ \sin\theta_e \end{bmatrix}.$$

9.11 Derive from (9.7) the following extended EMF model in the stationary frame:

$$\begin{bmatrix} v^s_d \\ v^s_q \end{bmatrix} = \begin{bmatrix} r_s + \mathrm{p}L_d & \omega_e(L_d - L_q) \\ -\omega_e(L_d - L_q) & r_s + \mathrm{p}L_d \end{bmatrix}\begin{bmatrix} i^s_d \\ i^s_q \end{bmatrix} + E_{ex}\begin{bmatrix} -\sin\theta_e \\ \cos\theta_e \end{bmatrix}$$

where $E_{ex} = \omega_e[(L_d - L_q)i^e_d + \psi_m] - (L_d - L_q)(\mathrm{p}\ i^e_q)$ is extended EMF.

Chapter 10

Pulse Width Modulation and Inverter

Ideal switches do not consume power since no current flows in the cut-off mode and no voltage drop takes place in the conduction mode. In the pulse width modulation (PWM), only on and off modes are utilized. By adjusting on-off duties (pulse widths), the average fits into a desired voltage wave form.

A three phase voltage source inverter consisting of six switches and a common DC rail is shown in Fig. 10.1. It consists of three switching arms: (S_a, S'_a), (S_b, S'_b), and (S_c, S'_c). As the physical switching device, power MOSFETs are suitable for low voltage and high frequency applications. They behave like an ideal switch, though they consume power during transition and on-state. However, the loss portion is relatively small. For medium and high power inverters, IGBTs are widely used. For low voltage or high frequency applications, FETs are utilized. Note that each switch has an anti-parallel diode in the voltage source inverter to provide commutation paths for inductive loads.

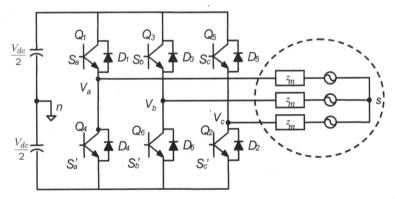

Figure 10.1: Voltage source inverter with six switching devices.

Fig. 10.2 shows typical voltage and current switching pattern, and the corresponding voltage-current contour during turn on and off. During the turn on transient, the (collector-emitter) voltage drops only after the switch (collector) current

347

builds up. Likewise, current begins to drop after the collector-emitter voltage rises in the turning off process.

The load is assumed to be an inductive load, which is treated here as a current sink. State ii) shows the state after U starts conduction. The collect-to-emitter voltage, v_{ce} drops rapidly after current increases. In state iii), the reverse recovery current of diode $\bar{U}$ flows just before the diode turning off. Since the reverse recovery current is added to the load current, a current spike takes place. Microscopically the reverse current is interpreted as homing of excess carriers to the DC link side as the depletion layer is developed [1]. In Fig. 10.2 (b), turn on and off contours are depicted in the current-voltage plane. The enclosed areas represent turn on and off losses.

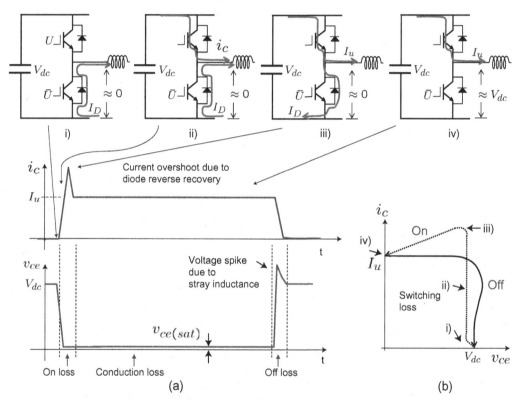

Figure 10.2: Four states during upper switch turn on: i) turn-off state (lower diode free wheeling), ii) conduction start, iii) diode reverse recovery current flow, and iv) turn-on completion: (a) current and voltage profiles of U and (b) turn on and off contours in the voltage-current plane.

Since both current and voltage are not zero, power loss takes place during transients. In Fig. 10.2 (b), turn on and off contours are depicted in the current-voltage plane. Usually, turn off loss is larger than turn on loss in IGBTs. The switching loss is directly proportional to the switching frequency. Besides the switching loss, conduction loss also takes place in the conduction stage since voltage on-drop exists in the physical switches. The on-drop voltage is normally 2~4 V and it increases as the voltage blocking capability increases. These losses cause heat in the switch, but the junction temperature must be controlled below 150°C. Thus, heat sink together with air or water cooling devices must be equipped for power semiconductors.

10.1 Switching Function and Six Step Operation

We also denote by $S_a = 1$ if switch S_a is turned on, otherwise $S_a = 0$. The same rule is applied to S_b and S_c. Consider the switching pattern for $n = 0, 1, 2, \cdots$:

$$\begin{cases} S_a = 1, & n\pi \leq \omega t < (n+1)\pi \\ S_a = 0, & (n+1)\pi \leq \omega t < (n+2)\pi, \end{cases}$$

$$\begin{cases} S_b = 1, & (n+2/3)\pi \leq \omega t < (n+5/3)\pi \\ S_b = 0, & (n+5/3)\pi \leq \omega t < (n+8/3)\pi, \end{cases}$$

$$\begin{cases} S_c = 1, & (n-2/3)\pi \leq \omega t < (n+1/3)\pi \\ S_c = 0, & (n+1/3)\pi \leq \omega t < (n+4/3)\pi. \end{cases}$$

This switching pattern is called *six step operation*, since it utilizes six switchings per period. The six inverter switching status can be mapped into the vertices of a regular hexagon. Note that $V_0 \equiv (0,0,0)$ is the case when all the upper level switches are turned on, and $V_7 \equiv (1,1,1)$ all the lower level switches are turned on. In both cases, the motor terminal voltage is equal to zero. The non-trivial six vectors are defined by

$$\begin{aligned} V_1 &: (S_a, S_b, S_c) = (1,\ 0,\ 0); \\ V_2 &: (S_a, S_b, S_c) = (1,\ 1,\ 0); \\ V_3 &: (S_a, S_b, S_c) = (0,\ 1,\ 0); \\ V_4 &: (S_a, S_b, S_c) = (0,\ 1,\ 1); \\ V_5 &: (S_a, S_b, S_c) = (0,\ 0,\ 1); \\ V_6 &: (S_a, S_b, S_c) = (1,\ 0,\ 1). \end{aligned}$$

If the vectors $V_1, \cdots, V_6$ are mapped into the stationary dq frame, we will have six vectors as shown in Fig. 10.3. Note that V_0 and V_7 represent zero vectors located at the origin. They are called space vectors, and their directions are identical to those of the resulting flux in the DC state. For example, the direction of $V_1 = (1, 0, 0)$ is the same as that of the flux (or current) vector that would result from the switching state, $(S_a, S_b, S_c) = (1, 0, 0)$. Fig. 10.4 shows this interpretation along with motor flux.

The pole voltage is defined as the motor terminal voltage with respect to the neutral point, n of the DC link. Then the pole voltage values are either $\frac{V_{dc}}{2}$ or $-\frac{V_{dc}}{2}$. Using the switching function, the pole voltages are given as

$$v_{an} = \frac{V_{dc}}{2}(2S_a - 1), \tag{10.1}$$

$$v_{bn} = \frac{V_{dc}}{2}(2S_b - 1), \tag{10.2}$$

$$v_{cn} = \frac{V_{dc}}{2}(2S_c - 1). \tag{10.3}$$

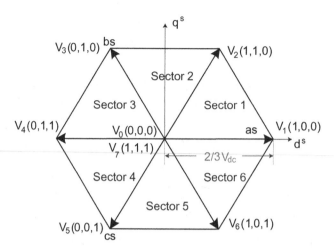

Figure 10.3: Space vector diagram.

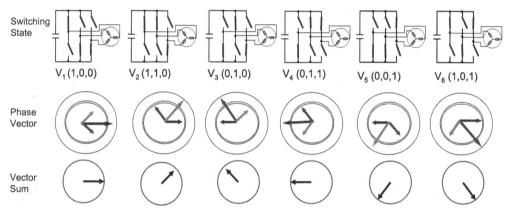

Figure 10.4: Directions of the flux vectors for the switching vectors $V_1, \cdots, V_6$.

Then, the line-to-line voltage is equal to

$$
\begin{aligned}
v_{ab} &= v_{an} - v_{bn}, \\
v_{bc} &= v_{bn} - v_{cn}, \\
v_{ca} &= v_{cn} - v_{an}.
\end{aligned}
$$

The pole voltages and line-to-line voltages versus time are shown in Fig. 10.5. Note that the peak value of the line-to-line voltage is V_{dc}. The fundamental component of the line-to-line voltage is calculated as

$$
v_{ab[1]} = \frac{2}{\pi} \int_0^\pi v_{ab} \sin \omega t \ d(\omega t) = \frac{2}{\pi} \int_{\pi/6}^{5\pi/6} V_{dc} \sin \omega t \ d(\omega t) = \frac{2\sqrt{3}V_{dc}}{\pi}. \quad (10.4)
$$

This is the maximum output voltage that an inverter can provide under a given DC link voltage, V_{dc}.

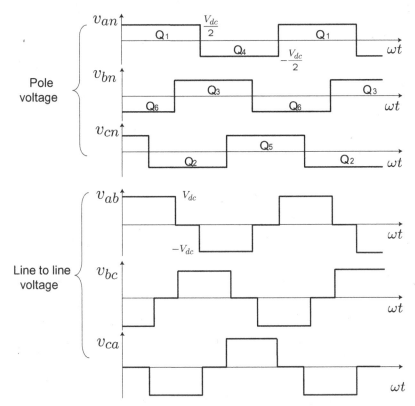

Figure 10.5: Pole and line-to-line voltages in the six step operation.

Note that v_{sn} is the voltage between between the neutral point, s of the motor and the DC link center tap, n. It is obvious that

$$v_{an} = v_{as} + v_{sn}, \tag{10.5}$$

$$v_{bn} = v_{bs} + v_{sn}, \tag{10.6}$$

$$v_{cn} = v_{cs} + v_{sn}. \tag{10.7}$$

Since v_{sn} is common to all phases, it is called common mode voltage. With the assumption that $v_{as} + v_{bs} + v_{cs} = 0$, it should follow that

$$v_{sn} = \frac{v_{an} + v_{bn} + v_{cn}}{3} = \frac{V_{dc}}{6}(2S_a + 2S_b + 2S_c - 3). \tag{10.8}$$

Then, the motor phase voltages follow from (10.5)–(10.8) as:

$$v_{as} = \frac{V_{dc}}{3}(2S_a - S_b - S_c), \tag{10.9}$$

$$v_{bs} = \frac{V_{dc}}{3}(2S_b - S_c - S_a), \tag{10.10}$$

$$v_{cs} = \frac{V_{dc}}{3}(2S_c - S_a - S_b). \tag{10.11}$$

Fig. 10.6 shows the typical phase voltage and the common mode voltage, v_{sn} when the six step mode operates. Obviously, the sum of (10.9), (10.10), and (10.11) is

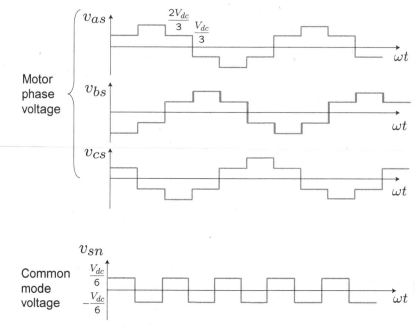

Figure 10.6: Phase voltages and the motor neutral voltage with the six step operation.

equal to zero. The peak of the fundamental component of the phase voltage is equal to

$$v_{as[1]} = \frac{2}{\pi} V_{dc}. \qquad (10.12)$$

It can be easily obtained from (10.4) by dividing $\sqrt{3}$. The six step operation uses the minimum number of switchings, whereas it produces the maximum voltage under a given DC link voltage. Therefore, it is utilized in some high power or high speed area where high voltage utilization is necessary. Also, it can be utilized when the minimum switching frequency is desired because of the switching loss. But, a major disadvantage is large current harmonics, as shown in Fig. 10.7.

Exercise 10.1   Using the Fourier series expansion, show that the fundamental component of the six step phase voltage is equal to $v_{as[1]} = \frac{2}{\pi} V_{dc}$.

10.2   PWM Methods

Numerous PWM methods were investigated by many researchers [2], [3], [4], but sinusoidal PWM and space vector PWM are the two most popular methods. The sinusoidal PWM can be realized by analog circuits, whereas the space vector PWM is usually implemented in a DSP.

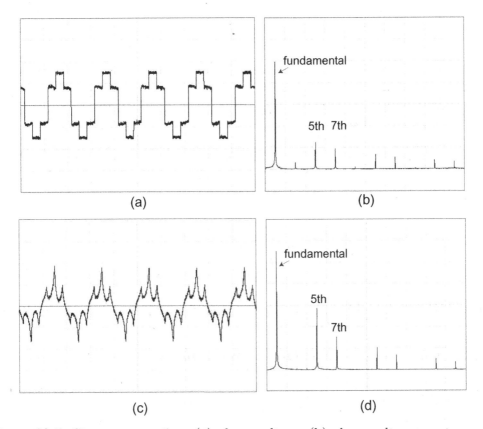

Figure 10.7: Six step operation: (a) phase voltage, (b) phase voltage spectrum, (c) current, and (d) current spectrum.

10.2.1 Sinusoidal PWM

The sinusoidal PWM (SPWM) is obtained by comparing a triangular wave with a sinusoidal wave, as shown in Fig. 10.8: Turn-on signal of the high side switch is determined according to

$$S_a = \begin{cases} 1, & \text{if } V_{sine} \geq V_{tri} \\ 0, & \text{if } V_{sine} < V_{tri} \end{cases},$$

where V_{sine} and V_{tri} are sinusoidal and triangular waves, respectively. The principle of the SPWM and the corresponding output voltage is shown in Fig. 10.8. The maximum peak voltage is equal to $V_{dc}/2$. To see the DC link voltage utilization of a PWM method, its fundamental component is compared with the feasible maximum, $\frac{2}{\pi}$ that is the fundamental component of the six step. Their ratio is defined as the modulation index, and in the case of SPWM

$$M_i = \frac{V_{[1]}(\text{sinusoidal})}{V_{[1]}(\text{six step})} = \frac{\frac{1}{2}V_{dc}}{\frac{2}{\pi}V_{dc}} = 0.785. \tag{10.13}$$

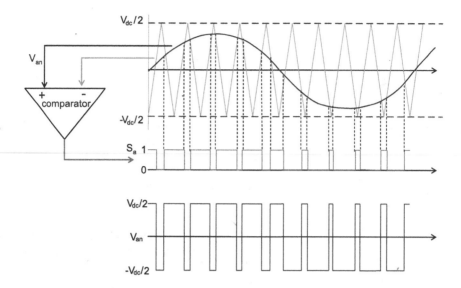

Figure 10.8: Sinusoidal PWM method by comparing with triangular wave.

10.2.2 Injection of Third Order Harmonics

The maximum modulation index differs depending on the PWM strategy. One concern is how to increase the maximum without incurring large harmonics. In the three phase system, we have a freedom to add a third order harmonic wave to all phase voltages. Note that third order harmonic waves are the same for all phases since they overlap perfectly after 120° and 240° phase shifts. Therefore, they do not alter the line-to-line voltage and the phase currents.

Note here that a third order harmonic wave can reduce the peak of the sum. Fig. 10.9 (a) shows an example: The peak of the fundamental component was one. After adding a third order component, the peak of the sum drops to 0.87. Thus, we have a room to increase the fundamental component further if the allowable peak is one. Fig. 10.9 (b) shows the case: The fundamental is increased to 1.15. However, the peak of the sum is less than 1. The voltage gain obtained through the third order harmonic component is about 1.15.

10.2.3 Space Vector Modulation

Fig. 10.3 is a space vector hexagon with six vertices, $\mathbf{V}_0, \cdots, \mathbf{V}_6$. The six sub-triangles are named Sector 1, $\cdots$, Sector 6. An arbitrary space vector is represented as a weighted sum of $\mathbf{V}_0, \cdots, \mathbf{V}_6$. Consider a vector $\mathbf{V}_a$ in Sector 1 shown in Fig. 10.10. Note that $\mathbf{V}_a = \mathbf{V}_{a1} + \mathbf{V}_{a2}$, where $\mathbf{V}_{a1}$ and $\mathbf{V}_{a2}$ are the decomposed vectors into $\mathbf{V}_1$ and $\mathbf{V}_2$, respectively.

Let T_s be the PWM switching period. Decomposed vector, $\mathbf{V}_{a1}$, is synthesized by turning on $\mathbf{V}_1(1,0,0)$ during a fractional time, T_1, which is proportional to the

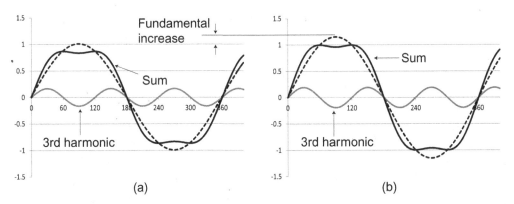

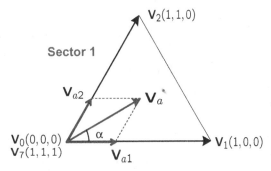

Figure 10.9: Reduction of peak voltage by adding a third order harmonics: (a) fundamental peak = 1 and (b) fundamental peak =1.15.

Figure 10.10: Calculation of on-times for v_1 and v_2 (partitioning by parallelogram).

magnitude of $\mathbf{V}_{a1}$. Specifically, T_1 and T_2 are determined as

$$\frac{|\mathbf{V}_{a1}|}{|\mathbf{V}_1|} = \frac{|\mathbf{V}_{a1}|}{(2/3)V_{dc}} = \frac{T_1}{T_s}, \tag{10.14}$$

$$\frac{|\mathbf{V}_{a2}|}{|\mathbf{V}_2|} = \frac{|\mathbf{V}_{a2}|}{(2/3)V_{dc}} = \frac{T_2}{T_s}. \tag{10.15}$$

Obviously, $T_1 + T_2 < T_s$ and the zero vector, $\mathbf{V}_0$ or $\mathbf{V}_7$ is assigned during $T_0 = T_s - (T_1 + T_2)$. Zero voltage can be applied to the terminals by shorting either lower or upper switches. Note that all inverter output terminals are tied to the negative side of the DC link in the case of $\mathbf{V}_0$, while in case of $\mathbf{V}_7$ they are shorted to the positive rail. From the geometrical relationship, it follows that

$$\begin{bmatrix} v_d^s \\ v_q^s \end{bmatrix} \equiv |\mathbf{V}_a| \begin{bmatrix} \cos(\alpha) \\ \sin(\alpha) \end{bmatrix} = \frac{T_1}{T_s}\frac{2}{3}V_{dc} \begin{bmatrix} 1 \\ 0 \end{bmatrix} + \frac{T_2}{T_s}\frac{2}{3}V_{dc} \begin{bmatrix} \cos\frac{\pi}{3} \\ \sin\frac{\pi}{3} \end{bmatrix}$$

$$= \frac{2}{3}\frac{V_{dc}}{T_s} \begin{bmatrix} 1 & \cos\frac{\pi}{3} \\ 0 & \sin\frac{\pi}{3} \end{bmatrix} \begin{bmatrix} T_1 \\ T_2 \end{bmatrix}. \tag{10.16}$$

Therefore,

$$\begin{bmatrix} T_1 \\ T_2 \end{bmatrix} = \frac{\sqrt{3}T_s}{V_{dc}} \begin{bmatrix} \sin\frac{\pi}{3} & -\cos\frac{\pi}{3} \\ 0 & 1 \end{bmatrix} \begin{bmatrix} v_d^s \\ v_q^s \end{bmatrix}. \tag{10.17}$$

This gives a relation between d, q voltage components and PWM duties in Sector 1. Generalizing (10.17) in the other sectors, we obtain

$$T_1 = \frac{\sqrt{3}T_s}{V_{dc}} \left[\sin(\frac{\pi}{3}m)v_d^s - \cos(\frac{\pi}{3}m)v_q^s \right], \tag{10.18}$$

$$T_2 = \frac{\sqrt{3}T_s}{V_{dc}} \left[-\sin(\frac{\pi}{3}(m-1))v_d^s + \cos(\frac{\pi}{3}(m-1))v_q^s \right], \tag{10.19}$$

where $m = 1, 2, \cdots, 6$ denote sector numbers. Equations (10.18) and (10.19) enable us to calculate directly the switching duties from the d, q voltages. An arbitrary vector, $\mathbf{V} = v_d^s + jv_q^s$ in Sector m can be obtained such that

$$\mathbf{V} = \mathbf{V}_m \frac{T_1}{T_s} + \mathbf{V}_{m+1} \frac{T_2}{T_s} \qquad \text{for} \quad m = 1, 2, ..., 6.$$

The above process illustrates, namely, the space vector modulation (SVM). Fig. 10.11 shows a method of creating PWM patterns for the upper switches, S_a, S_b, and S_c based on T_1 and T_2. To reduce the switching numbers and make the PWM pattern symmetric, the zero vector intervals are partitioned equally between $\mathbf{V}_0$ and $\mathbf{V}_7$ such that

$$T_0 = T_7 = \frac{T_s - (T_1 + T_2)}{2}.$$

In Sector 1, the on-duties of the phase voltage, (v_a, v_b, v_c) are set according to

$$\begin{aligned} S_a &= T_1 + T_2 + (T_s - T_1 - T_2)/2, \\ S_b &= T_2 + (T_s - T_1 - T_2)/2, \\ S_c &= (T_s - T_1 - T_2)/2. \end{aligned}$$

Then, (1,0,0) vector is turned on for T_1 and (1,1,0) vector for T_2.

Exercise 10.2   It is desired to create a voltage vector $|V_a|e^{j\alpha}$ which belongs to Sector 2. Calculate T_1 and T_2.

Solution. $T_1 = \frac{\sqrt{3}|V_a|T_s}{V_{dc}} \sin(\frac{2\pi}{3} - \alpha)$, $T_2 = \frac{\sqrt{3}|V_a|T_s}{V_{dc}} \sin(\alpha - \frac{\pi}{3})$, and $T_0 = T_s - (T_1 + T_2)$.   ■

10.2.4   Sector Finding Algorithm

The duty computing equations, (10.18) and (10.19) vary depending on the sectors. To compute the on-duties of a given vector, $\mathbf{V}^* = (v_d^s, v_q^s)$, the first step is to find a sector to which $\mathbf{V}^*$ belongs. In finding a sector, the signs of v_d^s and v_q^s are utilized firstly. Note that $v_q^s > 0$ for Sectors 1, 2, and 3, and $v_q^s < 0$ for Sectors 4, 5, and 6. Further to distinguish Sector 1 and Sector 2, $|v_q|$ and $\sqrt{3}|v_d|$ are compared. Similarly, we have

$$\begin{cases} \text{Sector 2, 5,} & \text{if } |v_q^s| > \sqrt{3}|v_d^s| \\ \text{Sector 1, 3, 4, 6,} & \text{if } |v_q^s| < \sqrt{3}|v_d^s|. \end{cases}$$

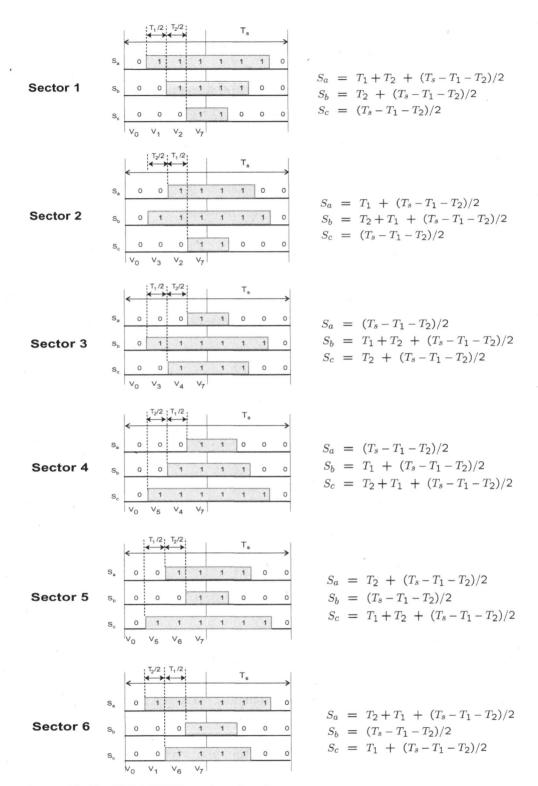

$$S_a = T_1 + T_2 + (T_s - T_1 - T_2)/2$$
$$S_b = T_2 + (T_s - T_1 - T_2)/2$$
$$S_c = (T_s - T_1 - T_2)/2$$

$$S_a = T_1 + (T_s - T_1 - T_2)/2$$
$$S_b = T_2 + T_1 + (T_s - T_1 - T_2)/2$$
$$S_c = (T_s - T_1 - T_2)/2$$

$$S_a = (T_s - T_1 - T_2)/2$$
$$S_b = T_1 + T_2 + (T_s - T_1 - T_2)/2$$
$$S_c = T_2 + (T_s - T_1 - T_2)/2$$

$$S_a = (T_s - T_1 - T_2)/2$$
$$S_b = T_1 + (T_s - T_1 - T_2)/2$$
$$S_c = T_2 + T_1 + (T_s - T_1 - T_2)/2$$

$$S_a = T_2 + (T_s - T_1 - T_2)/2$$
$$S_b = (T_s - T_1 - T_2)/2$$
$$S_c = T_1 + T_2 + (T_s - T_1 - T_2)/2$$

$$S_a = T_2 + T_1 + (T_s - T_1 - T_2)/2$$
$$S_b = (T_s - T_1 - T_2)/2$$
$$S_c = T_1 + (T_s - T_1 - T_2)/2$$

Figure 10.11: SVM PWM gating signals.

Fig. 10.12 shows the sector finding rule from (v_d^s, v_q^s), and Fig. 10.13 shows a complete sector finding algorithm.

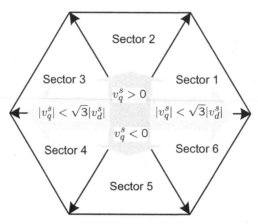

Figure 10.12: Finding sectors from (v_d^s, v_q^s).

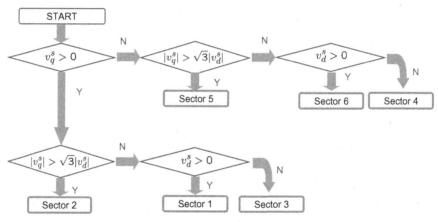

Figure 10.13: Sector finding flow chart.

10.2.5   Space Vector Waveform

Fig. 10.14 (a) shows the phase voltages represented by S_a, S_b, and S_c. As shown in Fig. 10.15, the SVM phase voltage can be decomposed as the sum of a perfect sinusoidal wave and a triangular wave. Note that the period of the triangular wave is one third of the sine wave period. The triangular wave is a sum of many triplen harmonics of order 3, 9, 15, 21, $\cdots$. Observe that the fundamental peak is reduced by the third order wave.

The largest SVM voltage is obtained when the space vector follows the inscribed circle in the hexagon. The maximum rms voltage is

$$\max\{v_a,\ v_b,\ v_c\} = \frac{1}{\sqrt{2}}\frac{\sqrt{3}}{2}\frac{2}{3}V_{dc} = \frac{1}{\sqrt{6}}V_{dc}. \tag{10.20}$$

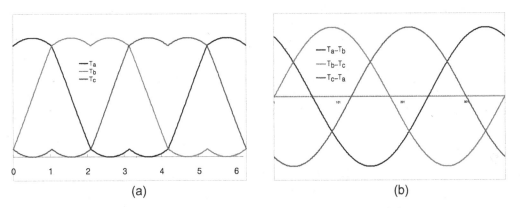

Figure 10.14: (a) Space vector PWM output (S_a, S_b, S_c) and (b) line to line voltage of the space vector PWM $(S_a - S_b, S_b - S_c, S_c - S_a)$.

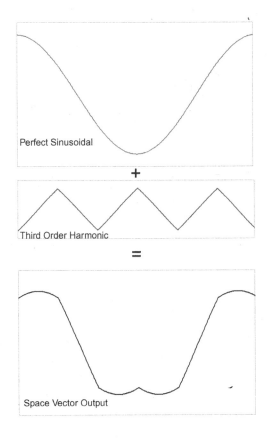

Figure 10.15: Space vector wave decomposition.

The maximum modulation index is equal to

$$M_i = \frac{\frac{1}{\sqrt{3}}V_{dc}}{\frac{2}{\pi}V_{dc}} = \frac{\pi}{2\sqrt{3}} = 0.906. \qquad (10.21)$$

It should be noted that the modulation index of SVM is larger that that of SPWM by $(0.906 - 0.785) \times 100 = 12\%$.

Now we consider another method of deriving SVM wave form. Example 10.3 shows how the third order triangular wave is made from the three phase sinusoidal waves. It is very interesting that the SVM wave, (v'_a, v'_b, v'_c) is constructed systematically without using the sector finding algorithm.

Exercise 10.3   Using Excel, follow the instructions below:
1) Draw $v_a = \sin t$, $v_b = \sin(t - 120°)$, and $v_c = \sin(t - 240°)$.
2) Find $v_{min} = \min\{v_a, v_b, v_c\}$. and $v_{max} = \max\{v_a, v_b, v_c\}$.
3) Obtain $v_{3rd} = v_{max} + v_{min}$.
4) Let $v_{sn} = -0.5v_{3rd}$, and $v'_a = v_a + v_{sn}$, $v'_b = v_b + v_{sn}$, and $v'_c = v_c + v_{sn}$.
5) Increase the amplitude of v_a, v_b, and v_c so that the peak of v'_a, v'_b, and v'_c is equal to one.

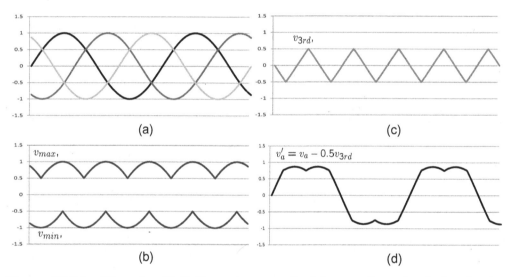

Figure 10.16: (Exercise 10.3) Procedure synthesizing a triangular 3rd order harmonic wave from the three phase sinusoidal waves: (a) v_a, v_b, v_c, (b) v_{max}, v_{min}, (c) $v_{3rd} = v_{max} + v_{min}$, (d) $v'_a = v_a - 0.5v_{3rd}$.

Switch Utilizing Factor

We denote by v_o and i_o the rms values of line to line voltage and load current, respectively. The inverter switch utilizing factor (SUF) is defined by

$$SUF = \frac{\sqrt{3}v_o i_o}{6V_T I_T}, \tag{10.22}$$

where V_T and I_T are the maximum voltage and current applied to the switches, respectively [1]. Note that the maximum rms voltages are

$$v_o = \begin{cases} \frac{\sqrt{3}}{2\sqrt{2}}V_{dc}, & \text{Sinusoidal PWM} \\ \frac{1}{\sqrt{2}}V_{dc}, & \text{Space vector PWM.} \end{cases}$$

Note also that $V_T = V_{dc}$ and $I_T = \sqrt{2}I_0$. Therefore,

$$SUF = \begin{cases} \frac{1}{8} = 0.125, & \text{Sinusoidal PWM} \\ \frac{\sqrt{3}}{12} = 0.144, & \text{Space vector PWM.} \end{cases}$$

Note that the SUF for the space vector PWM is higher than that of the SPWM, since the former utilizes third order harmonics.

10.2.6   Discrete PWM

Fig. 10.17 shows a typical discrete PWM (DPWM) which is characterized by the flat peak over certain periods. It should be noted that the DPWM pattern is obtained by adding a third order harmonic wave. Therefore, the line-to-line voltage remains purely sinusoidal, i.e., no additional current harmonics are induced.

During the peak periods, a switch stays on. Thus, switching on and off are not necessary. According to Fig. 10.17, the non-switching interval takes about 30%. Thereby, one-third of switching loss is saved. It cuts a fair amount of heat generation by IGBTs or FETs, and is useful especially when the switching frequency is high. The discrete PWM is utilized often in home-appliances such as refrigerator, washing machine, etc., where the PWM frequency is as high as 16 kHz due to audible noise.

The previous DPWM shown in Fig. 10.17 is named DPWM1. It should be noted however that the switching loss is minimized when the current reaches a maximum in the non-switching period. In this perspective, DPWM1 is preferred in applications where the power factor is close to unity. Other types of DPWM can be made by inserting flat periods in different places as shown in Fig. 10.18. The DPWM2 is good for lagging loads near 30° such as induction motor drives. The region with $\Psi=0$ is suitable for an induction machine that operates as a generator [5].

The quality of wave form worsens due to the harmonic contents. The harmonic factor for a voltage wave form, $v = \sum_{k=1}^{\infty} a_k \cos(2\pi f kt + \phi)$ is defined by

$$HDF = \frac{1}{a_1}\sqrt{\sum_{k=2}^{\infty} a_k^2}. \tag{10.23}$$

The HDF is nothing but the rms value of harmonics normalized by the fundamental component. Note that the HDF is obtained as a polynomial of the modulation index.

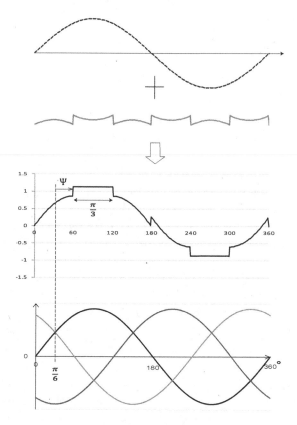

Figure 10.17: DPWM pattern obtained by a third order harmonic wave.

For example [6],

$$HDF_{SPWM} \;=\; \frac{3}{2}\left(\frac{4}{\pi}M_i\right)^2 - \frac{4\sqrt{3}}{\pi}\left(\frac{4}{\pi}M_i\right)^3 + \frac{9}{8}\left(\frac{4}{\pi}M_i\right)^4 \qquad (10.24)$$

$$HDF_{SVM} \;=\; \frac{3}{2}\left(\frac{4}{\pi}M_i\right)^2 - \frac{4\sqrt{3}}{\pi}\left(\frac{4}{\pi}M_i\right)^3 + \frac{9}{8}\left(\frac{3}{2}-\frac{9}{8}\frac{\sqrt{3}}{\pi}\right)\left(\frac{4}{\pi}M_i\right)^4 \quad (10.25)$$

The HDFs of several PWM waves are compared for $0 \le M_i \le 1$ in Fig. 10.19. In the range of low modulation index, the SPWM and SVPWM are superior to DPWM methods, whereas in the higher modulation range, the opposite is true.

10.2.7   Overmodulation Methods

The inscribed circle to the voltage hexagon represents the maximum voltage contour that can be made without a nonlinearity [7], [8], [9]. The shaded area in Fig. 10.20 represents the overmodulation region. If a trajectory follows a contour in the shaded area, the fundamental voltage can be made greater than the linear maximum. The extreme is the six step operation in which only the vertices are utilized. The increased voltage, however, accompanies unwanted harmonics. These

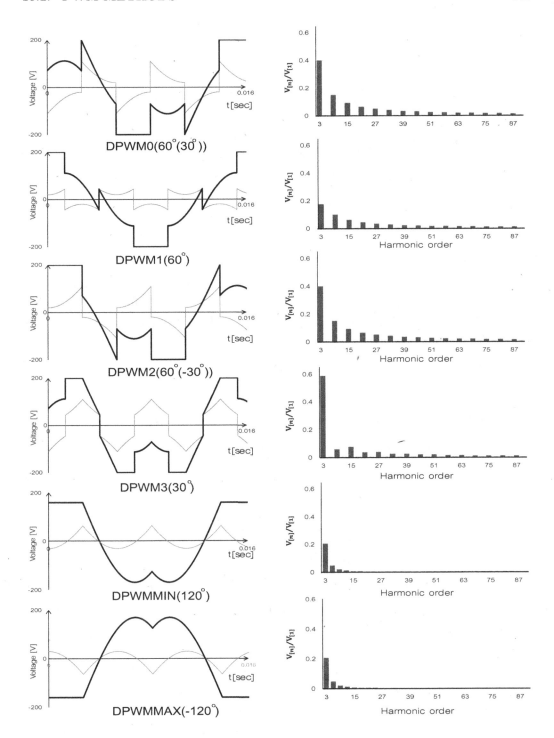

Figure 10.18: Various types of DPWM waveforms for $M_i = 0.9$.

harmonics are different from the those in the SVM and DPWMs. They include non-triplen harmonics, in addition to the triplen, that incur current harmonics.

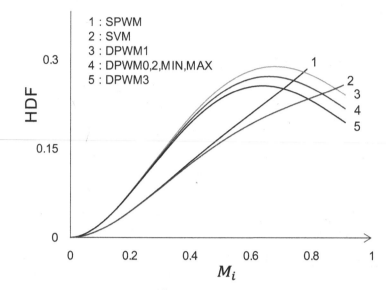

Figure 10.19: Harmonic distortion factor versus modulation index for typical PWM methods.

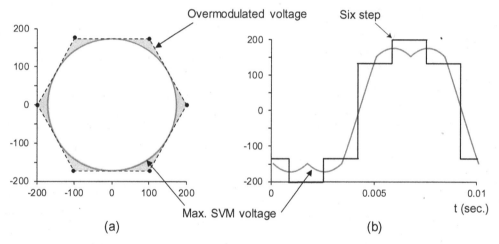

Figure 10.20: (a) Region of overmodulation between six step and maximum SVM and (b) voltage plots versus time.

Three conventional overmodulation methods are illustrated in Fig. 10.21. The minimum-phase-error PWM (MPEPWM) is the simplest method that limits the reference voltage, V^* to the intersection, a with a side of the hexagon [10]. This overmodulation strategy keeps the dq component ratio. The minimum-magnitude-error PWM (MMEPWM) is to select the voltage vector at the intersection, b with a vertical line from V^* so that two vectors have a minimum magnitude difference [11]. However, these two methods yield sluggish responses when they are used with a current controller. As a method of improving the current response in the motor control, back EMF was considered in the dynamic field-weakening PWM (DFWPWM) [12]:

Point c in Fig. 10.21 is an intersection with the line between the reference voltage, V^* and the back EMF, E.

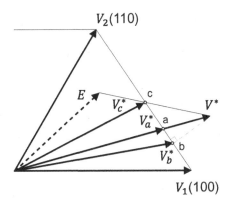

Figure 10.21: Conventional overmodulation methods.

Fig. 10.22 shows Bolognani's overmodulation technique [13]: Draw firstly a circle with a radius of $|V^*|$. Find an intersection with a side of the hexagon, and choose V_m^* as the modulated vector as shown in Fig. 10.22 (a). If $|V^*|$ is large, then no intersection is found. In such a case, the nearest vertex of the hexagon is selected as shown in Fig. 10.22 (b). It offers a better linearity between the modulation index, M_i and the reference voltage, $|V^*|$ in the overmodulation range [14]. It also provides a seamless transition to the six step operation, which is impossible with the previous three overmodulation methods.

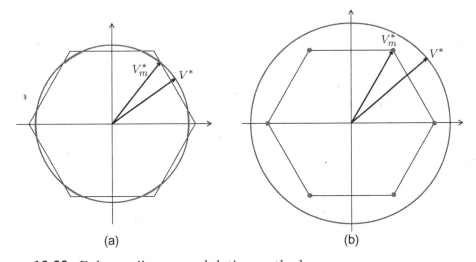

(a)   (b)

Figure 10.22: Bolognani's overmodulation method.

Fig. 10.23 shows the relation between the inverter output voltage, V_{inv} and the reference voltage, V^*. Both of them are normalized by $V_{dc}/2$. Note that the linear region of the SPWM is $[0, 1]$ and that of SVM is $[0, 2/\sqrt{3}]$. Therefore, the overmodulation starts from $2/\sqrt{3} = 1.15$. All PWM methods are bounded above by the six step operation. Thus, the upper limit is $\frac{2}{\pi}V_{dc}/(V_{dc}/2) = \frac{4}{\pi} = 1.27$. It shows

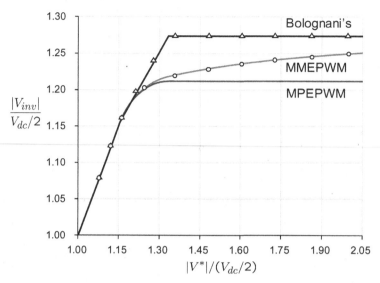

Figure 10.23: Proportionality comparision among different overmodulation methods.

that the Bolognani's overmodulation method is superior in the proportionality. In addition, it reaches the six step at a relatively low V^*.

10.3   Common Mode Current and Countermeasures

The harmonic energy fraction is described by the harmonic distortion factor, (10.23) and it increases as the modulation index increases as shown in Fig. 10.19. A third order harmonic voltage is exploited to achieve a better voltage utilization in the SVM. Especially in DPWMs, various types of harmonic voltages are used as shown in Fig. 10.18. Observe that they have many sharp edges that represent rich harmonic contents. Since they are common to all phases, they are called common mode voltage.

Common mode voltage is also generated even in the SPWM. Fig. 10.24 shows a sinusoidal PWM and the corresponding common mode voltage v_{sn} between AC and DC neutral points, where the DC link voltage is 300 V, PWM frequency is 2 kHz, modulation index is 0.8, and the frequency of reference voltage is 100 Hz. Up-to-date IGBTs have high switching speed, $dV/dt = 200{\sim}1000$ V/μs. High dV/dt of the common mode voltage is the major driving source for common mode noise. It appears in the form of conducted current (1$\sim$30 MHz), and/or radiated emission.

The common mode voltage does not incur any current if inverter and motor are floated. However, the motor and the inverter housings should be grounded to provide a common voltage reference for various sensors and communication. Furthermore, frame grounding is necessary for human safety. The frame grounding makes a closed loop with the capacitive couplings. In the case of inverter, the capacitive link is found between IGBT wafer and heat sinking plate. Meanwhile, coil, slot liner, air gap, rotor core, shaft, lubricant film, bearing ball, bearing rail,

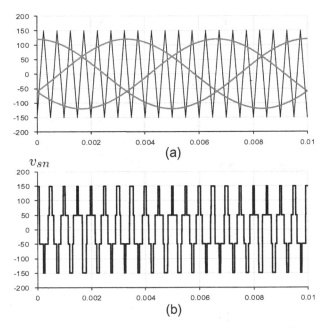

v_{sn}

Figure 10.24: SPWM: (a) comparison method with triangular wave and (b) zero sequence voltage v_{sn} between AC and DC neutral points.

motor frame etc. form a conductive path inside a motor. Fig. 10.25 shows the loop of common mode current which conducts a big loop consisting of inverter, power cable, motor, and ground system.

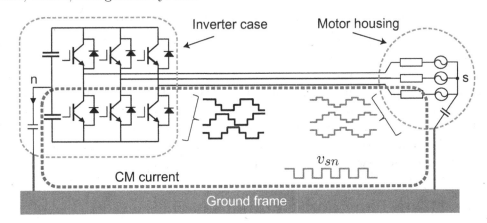

Figure 10.25: Common mode current path during six step operation.

Common mode noise limits the control bandwidth and degrades the overall quality of control. Fig. 10.26 shows common mode current flow in motor and its equivalent circuit. Apart from noise interference, the common mode voltage is critically harmful to the motor bearing. Once voltage is sufficiently high to overcome the resistance of the oil film in the bearing, the shaft (common mode) current discharges. It causes pits and fusion craters in the race wall and ball bearings. This phenomenon continues until excessive fluting noise and failure occur [15]. The common mode noise levels are strongly regulated by the rules of Federal Communications Commission (FCC) in the U.S.

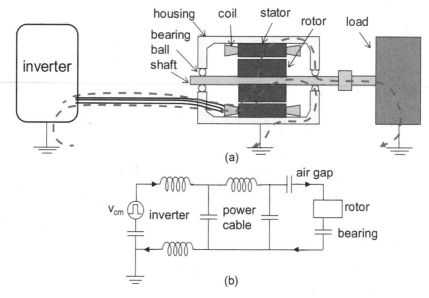

Figure 10.26: (a) Common mode current path and (b) an equivalent circuit.

The common mode current can be suppressed by the common mode noise filter as shown in Fig. 10.27 (a) [16]. Note that the coils of common mode inductor are wound with the same orientation for the three power lines. As a result, no field is induced by the differential mode, whereas a big inductance is developed to the common mode current. So, it does not inhibit the flow of differential currents, while causing an impedance to the common mode current. It is a method of impeding the common mode current.

Another way of avoiding the possible damage is to provide a low impedance path as shown in Fig. 10.27 (b). It consists of coaxial shield, RF connectors, metal cases, and the ground system. Then, the common mode current will be bypassed through coaxial shield and metal conduit, and to the ground. Impedance matching is made via RF connectors between the coaxial shield and inverter case/motor housing.

10.4 Dead Time and Compensation

It is shown in Fig. 10.2 (a) that the switch current does not fall down to zero immediately after the gating signal becomes low. This is because IGBTs have a relatively long tail current (about 1 μs.) and the switching circuit has stray inductances. Therefore, the complementary switch should be turned on after the current tail disappears. Otherwise, a shoot-through phenomenon takes place. To prevent it, a dead interval, in which both upper and lower switches are low, should be provided for each signal transition. The dead time can be set by putting off all rising edges of the gating signals. The dead time period depends on current tail,

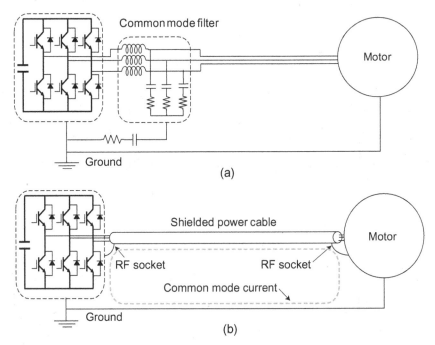

Figure 10.27: Methods of attenuating common mode noise: (a) common mode filter and (b) coaxial cable.

dV/dt capability of IGBT, stray inductance of the switching loop, and the DC link voltage. In the case of IGBTs, the typical value is $2 \sim 4$ μs.

Fig. 10.28 shows a method of generating the dead time and the voltage error caused by the dead-time. When the load is inductive, the load current continues to flow through an anti-parallel (free wheeling) diode during the dead time: Therefore, the pole voltage (inverter terminal voltage) is determined to be either $\frac{V_{dc}}{2}$ or $-\frac{V_{dc}}{2}$ depending on the current direction: If the current flows into the load $(i_{as} > 0)$, the load current flows through the anti-parallel diode of lower arm. Thus, $-\frac{V_{dc}}{2}$ appears at the terminal during the dead time, i.e., a negative error voltage is made, as shown in Fig. 10.28 (a). On the other hand, if current flows out from the load $(i_{as} < 0)$, the load current flows through the anti-parallel diode of upper arm. Thus, $\frac{V_{dc}}{2}$ appears at the terminal during the dead time, i.e., a positive error voltage is made as shown in Fig. 10.28 (b).

When the current level is low, the pole voltage during the dead time is not determined definitely. Distorted voltage and current wave forms are depicted in Fig. 10.29. It is obvious from Fig. 10.29 that the percentage of distortion is large when the current level is low. The distortion caused by dead time can be corrected by adding or subtracting the dead time interval, T_d, depending on the current polarities: To compensate the dead time error, on-duty of high side switch is increased by T_d, if $i_a < 0$. Conversely, if $i_a < 0$, then on-duty is decreased by T_d:

$$\begin{cases} T_a + T_d, & \text{if } i_a > 0, \\ T_a - T_d, & \text{if } i_a < 0, \end{cases}$$

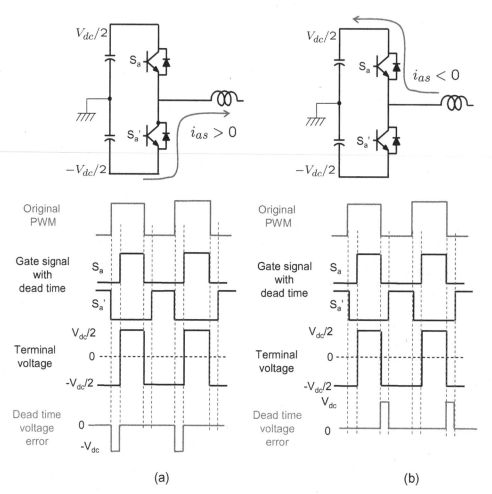

Figure 10.28: Dead-time effects depending on current direction: (a) $i_{as} > 0$ and (b) $i_{as} < 0$.

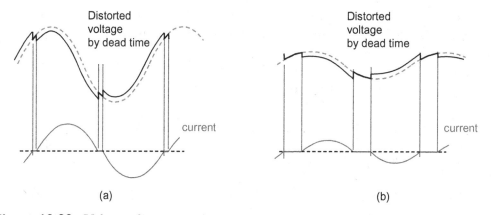

Figure 10.29: Voltage distortion due to the dead time when the current level is (a) high and (b) low.

where T_a is on-duty of the upper switch before compensation. Fig. 10.30 shows the dead time compensation method.

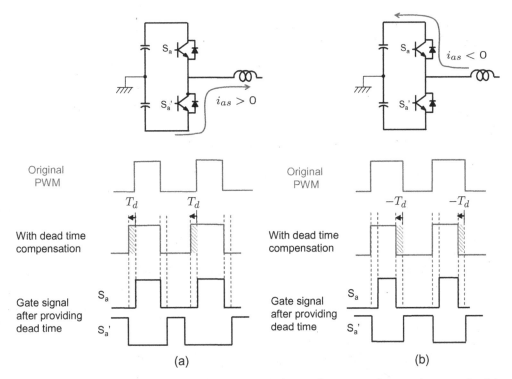

Figure 10.30: Dead time compensation depending on the current polarities: (a) $i_a < 0$ and (b) $i_a > 0$.

Fig. 10.31 shows experimental current wave forms before and after the dead time correction. It is observed that the dead time also causes a reduction in the fundamental component of the output voltage, apart from distortion. In the extreme case, the distorted output voltage generates subharmonics, which result in torque pulsation and instability at low-speed and light-load operation [17], [18]. In the sensorless vector control, the dead time voltage error causes a negative impact on the low-speed performance [19]. Some dead time compensation schemes are found in [17],[18],[20],[21].

10.5 Position and Current Sensors

In order to construct a current controller in the reference frame, the measured current vector should be mapped into the reference frame. Therefore, the rotor angular position and phase currents need to be measured. For position sensing, the absolute encoder or resolver is utilized. Since absolute encoders are normally expensive, resolvers are popularly utilized for PMSMs. However, an incremental encoder can be used with some starting methods. For current sensing, Hall effect sensors are widely used.

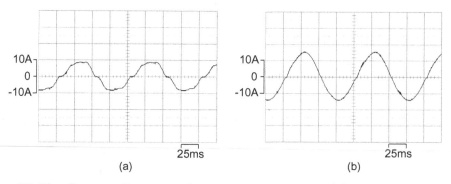

Figure 10.31: Current distortion due to the dead time: (a) before compensation and (b) after compensation.

10.5.1   Encoder

An incremental rotary encoder generates electrical pulses when the shaft rotates. By counting the pulses, angular position or speed is measured. Fig. 10.32 shows the structure of an encoder consisting of LEDs, photo transistors, and a disk with slits. Normally, the disk is made of glass and the slits are created by etching. When the disk rotates, flickering light is converted into electrical pulses by the photo transistors. The voltage pulses are transmitted outside after being conditioned. Incremental encoders output three pairs of signals: A, $\overline{A}$, B, $\overline{B}$, and Z, $\overline{Z}$. The A and B phase signals are $90°$ out of phase, as shown in Fig. 10.32. The signal states are summarized in Table 10.1. The direction is determined based on which signal comes first after resetting, $(A, B) = (0, 0)$. If B changes from 0 to 1 while $A = 0$, then the device rotates in the clockwise (CW) direction. Alternatively if B changes from 0 to 1 while $A = 0$, then it rotates in the counterclockwise (CCW) direction. The 'exclusive OR' operation, A$\oplus$B is used to increase the resolution by a factor of 4. The disk has one slot for Z-phase, so that the Z-phase pulse arises once

Table 10.1: State diagram for encoder pulses.

Direction	CW			CCW		
	Phase	A	B	A$\oplus$B	Phase	A	B	A$\oplus$B
	1	0	0	0	1	0	0	0
	2	0	1	1	2	1	0	1
	3	1	1	0	3	1	1	0
	4	1	0	1	4	0	1	1

per revolution. The Z-phase signal is used for resetting the data with an absolute position.

MT Methods

There are two speed measuring methods. With the M-method, the speed is measured by the number of encoder pulses, m_m for a given fixed time interval, T_s.

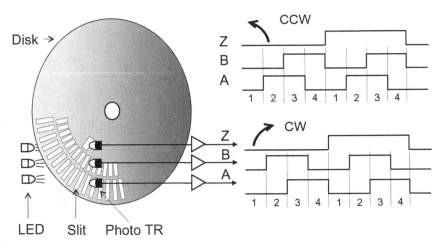

Figure 10.32: Incremental encoder and its signal patterns.

Suppose that an encoder resolution is P_{pr} pulses per revolution (PPR). Then, the number of pulses per second is $\frac{m_m}{T_s}$. Therefore, the shaft speed, N_f, in rpm is calculated as

$$N_f = \frac{60}{T_s}\frac{m_m}{P_{pr}} \text{ (rpm)}.$$

Note however that the encoder pulse width becomes wider and wider as the speed decreases, while T_s is fixed. Finally, m_m will be a fractional number at a low speed. The truncation error of m_m/T_s becomes serious at a low speed. In short, the M-method error increases as the speed decreases.

On the other hand, the T-method utilizes high frequency reference pulses to measure the encoder pulse width. It counts the number of pulses captured between the rising edges of encoder pulses. Let f_t be the frequency of reference pulses, and m_t be the number of reference pulses captured in an encoder period. Then, m_t/f_t is the encoder period in seconds, and f_t/m_t is the number of encoder pulses per second. Therefore, the shaft speed, N_f in rpm is calculated as

$$N_f = \frac{60 f_t}{P_{pr} m_t} \text{ (rpm)}.$$

The T-method is more accurate in a low speed region, but becomes less accurate with increasing speed, because m_t is decreasing. Thus, it is not appropriate for high speed measurements.

Fig. 10.34 compares the measurement errors of the M-method and T-method. An integrated method that combines both is called the M/T-method. A synchronous speed measurement method was proposed in [22].

10.5.2 Resolver and R/D Converter

Resolver is a type of angular position sensor that provides an absolute position [23]. Thus, it is suitable for PMSM control. Most resolvers consist of two parts, stator

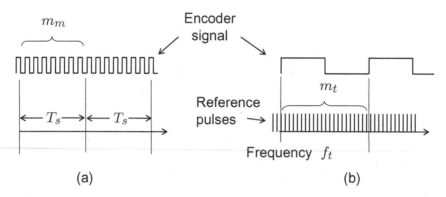

Figure 10.33: Speed measuring methods: (a) M-method and (b) T-method.

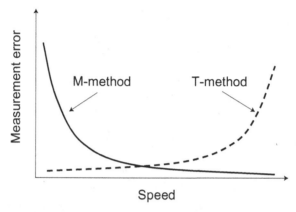

Figure 10.34: Measurement errors versus speed.

and rotor without a housing. Furthermore, the rotor of such resolver is mounted directly on the motor shaft, thereby it requires neither bearing nor shaft coupler. Since the resolver is mechanically robust, it is used in the harsh environments such as in electric vehicles.

Resolver is basically a transformer with a moving object, rotor. It has three sets of windings: one is for field excitation and the other two are for signal detection. The variable reluctance type has the exciting coil on the stator, as well as the detection coils. The rotor of reluctance resolver has a curved surface without any winding. On the other hand, the synchro type has an exciting coil on the rotor. Fig. 10.35 shows a eight pole reluctance type resolver. This type of resolver is favored for its simplicity in structure and use. Note that two sets of sensing coils are placed at different positions, and they are coupled to the exciting coils via a rotor as shown in Fig. 10.35 (b). Then the amount of coupling between the sensing coil and the exciting coil varies when the curved rotor surface rotates. In the following, we assume for convenience that the pole number is 2. Suppose that a high frequency carrier signal, $E_0 \sin \omega_{rs} t$ is injected to the exciting coil and the rotor axis is offset by θ_r from the magnet axis of a sensing coil, namely coil X. Then the induced voltage is $X = E \sin \omega_{rs} t \cos \theta_r$. The other sensing coil, namely coil Y is shifted by $\frac{\pi}{2}$ from coil X. The induced voltage of coil Y is $Y = E \sin \omega_{rs} t \sin \theta_r$.

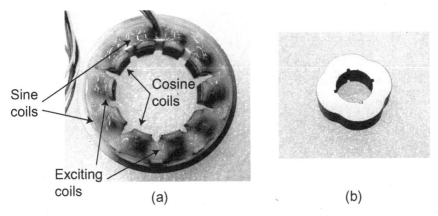

Figure 10.35: Reluctance type resolver: (a) stator and (b) rotor.

Fig. 10.36 illustrates the principle of the synchro-type resolver, and Fig. 10.37 shows a sample photo. It has a ring-type rotary transformer through which carrier signal is delivered to the rotor winding without a contact. Normally an AC source (2~8 V, 10~20 kHz) is applied as a carrier to the stationary primary coil.

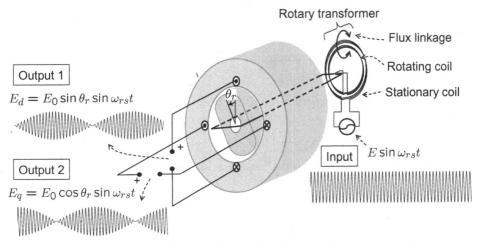

Figure 10.36: Schematic diagram and signals of a synchro-type resolver.

A block diagram of a resolver-to-digital converter (RDC) is shown in Fig. 10.38. The induced voltages of the two sensing coils have the same form as in the case of reluctance type. Fig. 10.38 shows a signal processing block diagram of a resolver to digital converter (RDC). Let φ be an internal variable of RDC which is set to track θ_r, i.e., φ is a digital value which is transmitted to a digital signal processor (DSP) as an angle estimate. As a first step, the above two signals are multiplied by $\cos\varphi$ and $\sin\varphi$, respectively. Then, we have

$$E_0 \sin\theta_r \cos\varphi \sin\omega_{rs}t - E_0 \cos\theta_r \sin\varphi \sin\omega_{rs}t = E_0 \sin(\theta_r - \varphi)\sin\omega_{rs}t.$$

Second, the carrier signal is eliminated by the synchronous rectification: The signal

Figure 10.37: Synchro-type resolver.

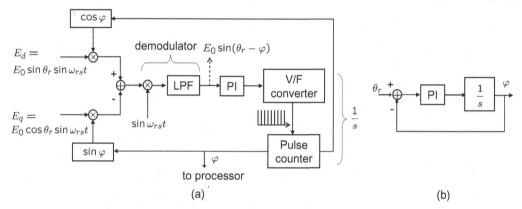

Figure 10.38: RDC block diagram: (a) functional blocks and (b) simplified equivalent block diagram.

is multiplied by $\sin \omega_{rs} t$ and passed through a low pass filter:

$$E_0 \sin(\theta_r - \varphi) \sin^2 \omega_{rs} t = E_0 \sin(\theta_r - \varphi) \frac{1 - \cos 2\omega_{rs} t}{2} \quad \stackrel{\text{LPF}}{\Rightarrow} \quad \frac{E_0}{2} \sin(\theta_r - \varphi).$$

The voltage-to-frequency (V/F) converter generates pulses in proportion to the input voltage level. Thus, a pulse counter yields digital values. Note that the V/F converter together with the pulse counter functions as an integrator, and these two are symbolized by $1/s$ in the block diagram. Then, a closed loop is completed with the resolver, as shown in Fig. 10.38 (b).

The PI controller is needed for loop tracking performance. Note that the input to the PI controller will be regulated ultimately due to the integral action. In other words, the PI controller will force $\sin(\theta_r - \varphi) \approx (\theta_r - \varphi) \to 0$. Therefore, the convergence, $\varphi \to \theta_r$ is achieved.

10.5.3   Hall Current Sensor and Current Sampling

When the current flows through a semiconductor piece that is laid in a magnetic field, the carriers (holes and electrons) experience a force in an orthogonal direction. That force makes a carrier concentration gradient, so that a voltage appears across the device width. The voltage is proportional to the magnet field, and this effect is called the Hall effect. Hall effect sensors are most widely used for current measurements, since they provide natural isolation between the measured line and the sensing circuit.

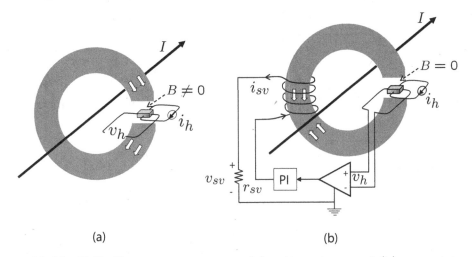

(a) (b)

Figure 10.39: Hall effect current sensors: (a) voltage type and (b) current type.

Fig. 10.39 shows two types of current sensors. The voltage type directly utilizes the Hall sensor voltage, v_h as a measured current. But, the core material shows nonlinear characteristics as the B field increases. Further, this method contains an offset error due to the hysteresis loop. Thus, the voltage type current sensors have a relatively large error (typically $\pm 1\%$).

The current type Hall current sensor employs an extra winding and a current servo amplifier. The Hall sensor is used for detecting the air gap flux. If the gap field is not equal to zero, then the servo amplifier forces current, i_{sv}, to flow in the opposite direction until the air gap field is nullified. Since the closed loop is formed with a PI controller, the air gap field remains zero. In this case, i_{sv} is proportional to I. Here, resistor, r_{sv}, is inserted to measure i_{sv}. That is, I is detected by measuring $v_{sv} = r_{sv} i_{sv}$. Since the current type sensor maintains the core field equal to zero, it is less affected by the material properties of the core. Therefore, it yields more accurate measurements (0.1% offset error).

To implement the vector control, it is necessary to sample the current for feedback. However, the current changes widely even in a single PWM interval. Therefore, the sampled value is desired to be the average value of the PWM interval.

Fig. 10.40 shows currents of symmetric and asymmetric PWMs with a current sampling point in the middle. In case of the symmetric PWM, the center value represents the average value. However, the middle point sampling does not give an average value in the asymmetric PWM. Therefore, it is better to use the symmetric PWM with the center point sampling.

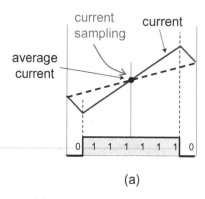
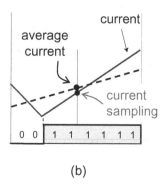

(a)   (b)

Figure 10.40: Current sampling points and the average value: (a) symmetric PWM and (b) asymmetric PWM.

References

[1] N. Mohan, T. M. Undeland, and W.P. Robbins, *Power Electronics, Converters, Applications, and Design*, John Wiley & Sons. Inc., 1995.

[2] J. Holtz, "Pulsewidth modulation a survey," *IEEE Trans. Ind. Electron.*, vol. 39, no. 5, pp. 410−420, Oct. 1992.

[3] J. Holtz, "Pulsewidth modulation for electronic power conversion," *Proc. IEEE*, vol. 82, no. 8, pp. 1194−1214, Aug. 1994.

[4] D. G. Holmes and T. A. Lipo, *Pulse Width Modulation for Power Converters: Principles and Practice*, Wiley, 2003.

[5] D. Grahame Holmes and Thomas A. Lipo, *Pulse Width Modulation for power converters: Principles and Practice*, Wiley-IEEE Press, Sep. 2003.

[6] A. M. Hava, R. J.Kerkman, and T. A. Lipo, "A high performance generalized discontinuous PWM algorithm," *IEEE Trans. on Ind. Appl.*, vol. 34, no. 5, pp. 1059−1071, Sep. 1998.

[7] J. Holtz, W. Lotzkat, and A. Khambadkone, "On continuous control of PWM inverters in the overmodulation range including the six-step mode," *IEEE Trans. Power Electron.*, vol. 8, no. 4, pp. 546−553, Oct. 1993.

[8] A. M. Hava, R. J. Kerkman, and T. A. Lipo, "Carrier-based PWM-VSI overmodulation strategies: Analysis, comparison, and design," *IEEE Trans. Power Electron.*, vol. 13, no. 4. pp. 674−689, July 1998.

[9] G. Narayanan and V. T. Ranganathan, "Extension of operation of space vector PWM strategies with low switching frequencies using different overmodulation algorithms," *IEEE Trans. Power Electron.*, vol. 17, no. 5, pp. 788−798, Sep. 2002.

[10] T. G. Habetler, F. Profumo, M. Pastorelli, and L. M. Tolbert, "Direct torque control of induction machines using space vector modulation," *IEEE Trans. Ind. Appl.*, vol. 28, pp. 1045−1053, Sep./Oct. 1992.

[11] H. Mochikawa, T. Hirose, and T. Umemoto, "Overmodulation of voltage source PWM inverter," *Conf. Rec. JIEE-IAS Conf.*, pp. 466−471, 1991.

[12] J.-K. Seok and S. Sul, "A new overmodulation strategy for induction motor drive using space vector PWM," *Conf. Rec. IEEE APEC95*, pp. 211−216, Mar. 1995.

[13] S. Bolognani and M. Zigliotto, "Novel digital continuous control of SVM inverters in the overmodulation range," *IEEE Trans. Ind. Appl.*, vol. 33, no. 2, pp. 525−530, Mar./Apr. 1997.

[14] Y. Kwon, et al., "Six-Step Operation of PMSM With Instantaneous Current Control," *IEEE Trans. on Ind. Appl.*, vol. 50, no. 4, pp. 2614−2625, Jul./Aug. 2014.

[15] W. Oh, "Preventing VFD/AC Drive Induced Electrical Damage to AC Motor Bearings," Electro Static Technology, ITW, 2006.

[16] H. Akagi et al., "Design and Performance of a Passive EMI Filter for Use With a Voltage-Source PWM Inverter Having Sinusoidal Output Voltage and Zero Common-Mode Voltage," *IEEE Trans. Power Electron.*, vol. 19, no. 4, pp. 1069−1076, July 2004.

[17] R. C. Dodson, P. D. Evans, H. T. Yazdi, and S. C. Harley, "Compensating for dead time degradation of PWM inverter waveforms," *IEE Proc. B, Electr. Power Appl.*, vol. 137, no. 2, pp. 73−81, Mar. 1990.

[18] D. Leggate and R. J. Kerkman, "Pulse-based dead-time compensator for PWM voltage inverters," *IEEE Trans. Ind. Electron.*, vol. 44, no. 2, pp. 191−197, Apr. 1997.

[19] J. Lee, T. Takeshita, and N. Matsui, "Optimized stator-flux-oriented sensorless drives of IM in low-speed performance," *in Conf. Rec. IEEE IAS Annu. Meeting*, pp. 250−256, 1996.

[20] N. Hur, K. Nam, and S. Won, "A two degrees of freedom current control scheme for dead-time compensation," *IEEE Trans. Ind. Electron.*, vol. 47, no. 3, pp. 557−564, June 2000.

[21] G. Liu, D. Wang, Y. Jin, M. Wang, and P. Zhang, "Current-detection-independent dead-time compensation method based on terminal voltage A/D conversion for PWM VSI," *IEEE Trans. on Ind. Electron.*, vol. 64, no. 10, pp. 7689−7699, 2017.

[22] T. Tsuji, T. Hashimoto, H. Kobayashi, M. Mizuochi, and K Ohnishi, "A wide-range velocity measurement method for motion control," *IEEE Trans. on Ind. Electron.*, vol. 56, no. 2, pp. 510–519, 2009.

[23] C.W de Silva, *Sensors and Actuators: Control Systems Intrumentation*, CRC Press, 2007.

Problems

10.1 Calculate $v_{as[5]}/v_{as[1]}$ and $v_{as[7]}/v_{as[1]}$ for six step inverter operation.

10.2 Suppose that the DC link voltage is $V_{dc} = 300$ V and that the switching frequency is 8 kHz. Calculate T_1 and T_2 in the symmetric PWM to synthesis a voltage vector, $100\angle 30°$ V.

10.3 Construct a DPWM1 generator using MATLAB Simulink.
a) Construct a block module for the third order harmonic wave of DPWM1 referring to the block code, $Third_S VM$ in the following figure:

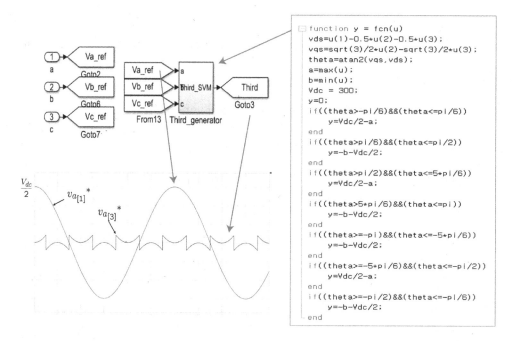

b) Construct a PWM generation block and a low pass filter, $\frac{1}{\tau_0 s+1}$, where $\tau = 7.96 \times 10^{-5}$. Obtain the filtered PWM output and its spectrum as shown below:

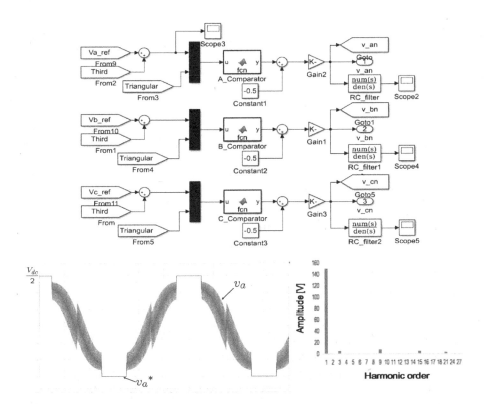

c) Find the common mode voltage and its filtered output. Obtain the filtered PWM output and its spectrum as shown below: Plot the spectrum of the common mode voltage.

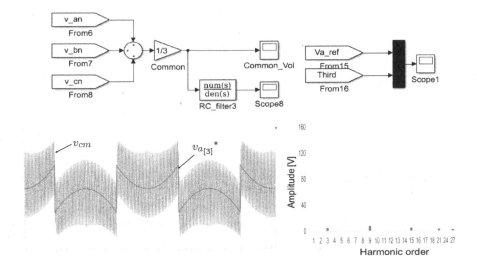

10.4 Consider a rectangular wave with zero steps.

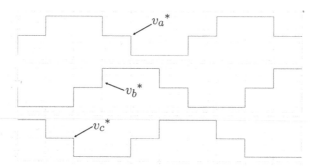

Using MATLAB Simulink, obtain the filtered PWM output and its spectrum. Obtain also the spectrum of the common mode voltage.

10.5 Determine a sector to which the voltage vector $(v_d^s, v_q^s) = (-100, -77)$ belongs according to the sector finding algorithm.

10.6 Consider an inverter with DC link voltage 300 V. A bipolar PWM method is used with 8 kHz carrier frequency. Assume further that the load is a three phase balanced inductive load and that the dead time interval is 2 μs. Calculate the (maximum) relative voltage errors, $\Delta V/V$ when the phase voltage is 10 V and 100 V, and the current flows into the load.

10.7 a) Using MATLAB, construct a triangular wave by integrating a rectangular wave of amplitude ± 0.5 and period $T_s = 125$ μs. Find spectrum of the triangular wave.

b) Note that the square wave is expressed as

$$\Pi(t) = \frac{2}{\pi} \sum_{n=1}^{\infty} \frac{1}{n} \sin\left(\frac{\pi}{2}n\right) \sin\left(n\frac{2\pi}{T_s}t\right).$$

Find the Fourier coefficients of

$$\Xi(t) = \int_0^t \Pi(t)dt$$

and plot them for $1 \leq n \leq 30$ with $T_s = 125$ μs. Check whether it is the same as the spectrum obtained in a).

10.8 Consider a 1000 pulse/rev encoder. Calculate the maximum speed error $(\Delta N_f/N_f)$ when the M-method is utilized with 100 Hz sampling $(T_s = 20$ ms) for $N_f = 10$ and 100 rpm.

10.9 Consider a resolver that has two inputs and one output as shown below. Suppose that $E_1 = E\sin\omega_{rs}t$ and $E_2 = E\cos\omega_{rs}t$. Referring to the rotor angle, θ_r defined in the figure, determine the output function.

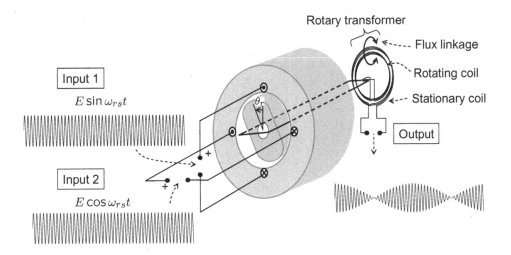

Chapter 11

Basics of Motor Design

Quality factors of motor include high torque and power density, low torque ripple, high efficiency, low noise and vibration, robustness, low cost, etc. It is not easy to meet all of the quality factors, because many of them are incompatible with each other. Motor design is a very complex and time-consuming task, because the designer should consider electro-magnetics, material properties, thermal behavior, noise and vibration, insulation material, motor control issues, limitations of power electronics, etc.

The motor generates an adjustable traveling magnetic wave. To create a complete traveling wave, the stator and rotor MMFs must be sinusoidal in time and space. However, these ideals are hardly fulfilled in practice due to manufacturing complexity, power density, material cost, etc. Thereby, motor designers compromise incessantly between the required specifications and actual limit.

11.1 Winding Methods

Multiple slot structures are utilized to create a sinusoidal MMF. By using a large number of slots, the MMF can be brought closer to a sine wave. Fig. 11.1 shows a case of four slots per pole per phase. Specifically, the a-phase winding consists of four pairs (a_1, a_1'), (a_2, a_2'), (a_3, a_3'), and (a_4, a_4') with full pitch. By the superposition law, the resulting MMF increases and decreases gradually with steps.

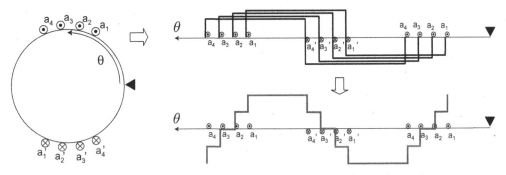

Figure 11.1: Superposition of MMFs of distributed windings ($q = 4$).

It is desired to maximize the fundamental component while minimizing MMF harmonics. Distributed and concentrated windings are differentiated by the number of slots per pole per phase: Specifically, let

$$q = \frac{N_s}{m \times 2p},$$ (11.1)

where m is the number of phases, p is the number of pole pairs, and N_s is the number of stator slots. If $q \geq 1$, then the coil span is larger than one slot pitch, and the winding method is called distributed winding. On the contrary if $q < 1$, each coil wounds a single tooth. It is called non-overlapping concentrated winding, or simply *concentrated winding.* Fig. 11.2 shows photos of distributed and concentrated windings.

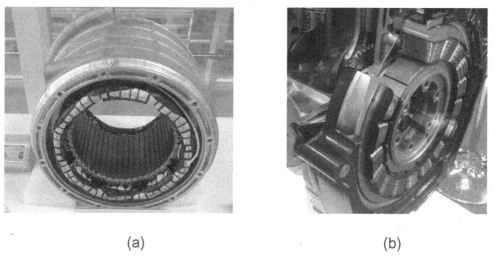

(a)   (b)

Figure 11.2: Photos of stator windings. (a) distributed winding and (b) concentrated winding.

According to the end turn, there are two types of the distributed windings: Fig. 11.3 shows top views of lap and concentric over-lapping windings. In the lap winding, coil loops are identical with a regular coil head. But in the concentric winding, coil spans are different. However, MMFs are the same in both cases, since the MMF is not affected by the end turn but by the active side of the coil in the slots. Nevertheless, the coil resistance is proportional to the total length of the coils. Evaluation points are

1. mass of copper which affects the cost;
2. ohmic voltage drops;
3. cooling,
4. simplicity in coil insertion and forming.

Fig. 11.4 (a-1) and (a-2) show two six slot machines: Note that the motor shown in (a-1) has $N_s = 6$ and $2p = 2$, thereby $q = 1$ leading to a distributed winding. On the other hand, the motor shown in (a-2) has $N_s = 6$ and $2p = 4$. It has

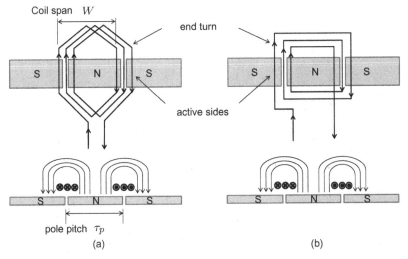

Figure 11.3: Top view of flattened state windings ($q = 3$): (a) consequent-pole lap winding and (b) consequent-pole concentric winding.

the conventional 3 teeth/2 pole structure. The coil winding is concentrated since $q = 0.5 < 1$. Fig. 11.4 (c-1) and (c-2) show the MMFs when only the a-phase current is flowing. However, these are unrealistic since it is not possible to conduct only a single phase in the three phase machine. On the other hand, it is necessary to consider $(i_a, i_b, i_c) = (I, -\frac{1}{2}I, -\frac{1}{2}I)$, $(\frac{1}{2}I, \frac{1}{2}I, -I)$, $(-\frac{1}{2}I, I, -\frac{1}{2}I)$, $(-I, \frac{1}{2}I, \frac{1}{2}I)$, $(-\frac{1}{2}I, -\frac{1}{2}I, I)$, and $(\frac{1}{2}I, -I, \frac{1}{2}I)$, where the sum of the phase currents is zero. Fig. 11.4 (d-1) and (d-2) show the MMF sums at a time when a three phase current flows, $(i_a, i_b, i_c) = (I, -I/2, -I/2)$. Fig. 11.4 (e-1) and (e-2) show the spectra of the MMF sums, from which it is necessary to see that multiples of third order harmonics disappear. This observation can be generalized to all three phase machines, i.e., no triplen harmonics appear in the three phase system.

Note further that if there is a half symmetry, even harmonics do not appear. Otherwise, even harmonics are present. The half symmetry of a periodic function is defined by $F(x) = -F(\frac{\tau_p}{2} - x)$. Note that the MMF of the concentrated winding shown in Fig. 11.4 (d-2) does not have the half symmetry. Thus, even harmonics are present as shown in Fig. 11.4 (e-2). Harmonic components are:

$$\nu = 1, 2, 4, -5, 7, 8, 10, \cdots \quad \text{for concentred winding,}$$
$$\nu = 1, -5, 7, -11, 13, \cdots \quad \text{for distributed winding.}$$

Note that $q = 0.5$ for motors of $N_s/2p = 6/4$, 9/6, 12/8, 15/10, 18/12, 21/14, etc. Fig. 11.5 shows a photo of a concentric four pole - six slot motor.

11.1.1 Full and Short Pitch Windings

The pole pitch is defined by $\tau_p = \pi D_r/2p$, where D_r is the rotor (air gap) diameter. The coil span is often called coil pitch and denoted by W. If $W = \tau_p$, the winding

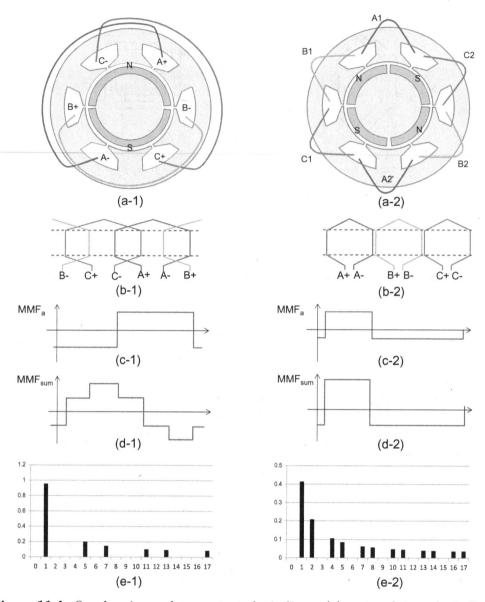

Figure 11.4: Overlapping and concentrated windings: (a) motor slots and windings, (b) winding diagrams, (c) MMFs when only a-phase current flows, (d) MMF sums for $(i_a,\ i_b,\ i_c) = (I,\ -\frac{1}{2}I,\ -\frac{1}{2}I)$, and (e) harmonics for MMFs in (d).

is full pitch. Otherwise, the winding is called *short* or *fractional pitch winding*. The flux linkage can be maximized with the full pitch winding.

Coil layers are classified according to the number of coil sides in a slot. If only one coil is inserted in each slot, it is called single layer. With the double layer, both short and full pitched windings are feasible. Fig. 11.6 (a) and (b) show short $(W/\tau_p = 5/6)$ and full pitch windings when $q = 2$, respectively. The MMF of each loop is different, but the sums are the same as shown in Fig. 11.6 (c). The reason is that the MMF depends only on the active sides, not on end-winding connections.

Figure 11.5: Photo of a four pole - six slot motor with concentric windings having $q = 0.5$.

Note that the sum has a step (zero interval) at each transition. Such steps help to reduce the harmonics. However, the improvement is offset by a decrease in the fundamental component.

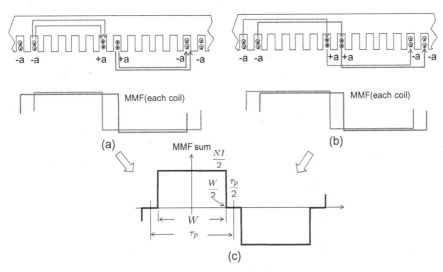

Figure 11.6: Two different windings for $q = 2$: (a) short pitch $(W/\tau_p = 5/6)$, (b) full pitch, and (c) MMF.

Pitch Factor

Consider a case where $W < \tau_p$ shown in Fig. 11.6 (c). The pitch factor is defined for each harmonic component by comparing it with the corresponding harmonic coefficient of the square wave. Specifically, the νth pitch factor, $k_{p(\nu)}$ is defined by the Fourier series expansion:

$$F(x) = \frac{NI}{2} \frac{4}{\pi} \sum_{\nu=1} \frac{1}{\nu} k_{p(\nu)} \cos \frac{\nu\pi}{\tau_p} x. \tag{11.2}$$

The Fourier coefficient for the MMF shown in Fig. 11.6 is equal to

$$\frac{4}{\pi}\frac{1}{\nu}k_{p(\nu)} \equiv 2 \times \frac{1}{\tau_p}\int_0^{\tau_p} F(x)\cos\left(\nu\frac{\pi}{\tau_p}x\right)dx$$

$$= 4 \times \frac{1}{\tau_p}\int_0^{\frac{W}{2}}\frac{NI}{2}\cos\left(\nu\frac{\pi}{\tau_p}x\right)dx = \frac{4}{\pi}\frac{1}{\nu}\sin\left(\nu\frac{\pi}{2}\frac{W}{\tau_p}\right)$$

Therefore,

$$k_{p(\nu)} = \sin\left(\nu\frac{W}{\tau_p}\frac{\pi}{2}\right). \tag{11.3}$$

Note also that $k_{p(\nu)} = \pm 1$ in the full pitch case, (2.4). Table 11.1 shows the pitch factors for $W/\tau_p = 1$, $2/3$, and $5/6$.

Table 11.1: Pitch factors for $W/\tau_p = 1$, $2/3$ and $5/6$.

ν	1	5	7	11	13
$k_{p(\nu)}$ for $W/\tau_p = 1$	1	1	-1	-1	1
$k_{p(\nu)}$ for $W/\tau_p = 2/3$	0.866	-0.866	0.866	-0.866	0.866
$k_{p(\nu)}$ for $W/\tau_p = 5/6$	0.966	0.259	-0.259	-0.966	-0.966

MMF spectra can be obtained with the discrete-time Fourier series, (2.37) and (2.38). The MMF spectra of full pitch and $W/\tau_p = 5/6$ are compared in Fig. 11.7. Observe that the 5th and 7th components are reduced significantly with the fractional pitch. Further note that the harmonic decrease is achieved at the expense of reducing the fundamental component.

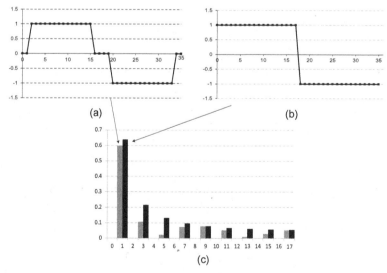

Figure 11.7: Harmonic analysis of two square MMFs: (a) 5/6 pitch winding, (b) full pitch winding, and (c) harmonics.

The same Fourier coefficient is derived differently using an MMF sum with a shifted one as shown in Fig. 11.6 (b). Two full pitch windings are displaced by $\pm\frac{\tau_p - W_c}{2}$. Then, we can derive the same pitch factor such that

$$\frac{1}{2}\cos\left(\nu\frac{\pi}{\tau_p}(x - \frac{\tau_p - W_c}{2})\right) + \frac{1}{2}\cos\left(\nu\frac{\pi}{\tau_p}(x + \frac{\tau_p - W_c}{2})\right)$$
$$= \frac{1}{2}\sin\left(\nu\pi(\frac{x}{\tau_p} + \frac{W_c}{2\tau_p})\right) - \frac{1}{2}\sin\left(\nu\pi(\frac{x}{\tau_p} - \frac{W_c}{2\tau_p})\right)$$
$$= \cos\left(\nu\pi\frac{x}{\tau_p}\right)\sin\left(\nu\frac{\pi}{2}\frac{W_c}{\tau_p}\right).$$

The harmonic leakage factor σ_h is defined as

$$\sigma_h = \frac{1}{k_{p1}^2}\sum_{\nu=2}^{\infty}\frac{k_{p(\nu)}^2}{\nu^2}. \tag{11.4}$$

It is the ratio of the high order components to the fundamental in power.

Exercise 11.1   Determine the pole number, the number of slots per pole per phase, and W/τ_p for the motors shown in Fig. 11.4 (a-1) and (a-2).

Exercise 11.2   Consider the short pitch $(W/\tau_p = 5/6)$, multi-layer winding shown in Fig. 11.8. Obtain the pitch factor using the discrete Fourier series expansion (2.37) and (2.38).

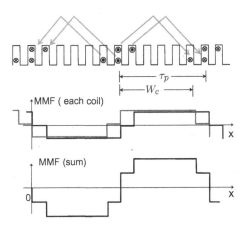

Figure 11.8: (Exercise 11.2) MMF for a short pitch winding for $q = 2$ and $W/\tau_p = 5/6$.

Exercise 11.3   Calculate the pitch factor, k_{p1} for the MMF for $q = 1/2$ shown in Fig. 11.4 (d-2).

Sol. The fundamental component is equal to

$$\frac{4}{\pi} k_{p1} \frac{NI}{2} = \frac{2}{\pi} \int_0^{\frac{\pi}{3}} \frac{4}{3} \frac{NI}{2} \cos\theta \, d\theta - \frac{2}{\pi} \int_{\frac{\pi}{3}}^{2\pi} \frac{2}{3} \frac{NI}{2} \cos\theta \, d\theta = \frac{2}{\pi} \frac{NI}{2} \left(\frac{4}{3} \frac{\sqrt{3}}{2} + \frac{2}{3} \frac{\sqrt{3}}{2} \right).$$

Thus, $k_{p1} = \frac{\sqrt{3}}{2} = 0.866$. ■

Exercise 11.4   Show that $\sigma_h = 0.46$ for a machine with $q = \frac{1}{2}$.

Exercise 11.5   Consider the MMF shown in Fig. 11.4 (d-1). Let $I = 100$ A and $N = 6$. Calculate the peak air gap field density, B when the gap height is $g = 0.5$ mm.

Distribution Factor

The slots are distributed over a curved section in the cylindrical machines. In a machine with $q > 1$, MMF vectors of individual coils are not aligned. Thus, the total MMF is not a simple algebraic sum. Fig. 11.9 shows a group of coils and their vector sum for $q = 3$. The electrical angle difference between the coils is

$$\frac{2p}{2} \frac{2\pi}{2pmq} = \frac{\pi}{mq}. \tag{11.5}$$

It is seen from the diagram that the vector sum is smaller than the sum of individual

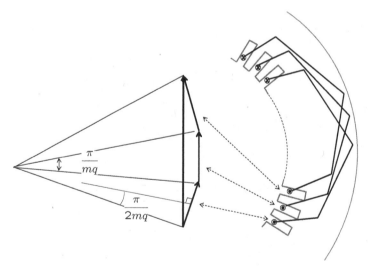

Figure 11.9: Coil distributed along an arc for $q = 3$.

vectors. The reduction is represented by a distribution factor,

$$k_d = \frac{\sin\left(\frac{\pi}{2m}\right)}{q \cdot \sin\left(\frac{\pi}{2mq}\right)}$$

For the νth order harmonic,

$$k_{d(\nu)} = \frac{\sin(\frac{\nu\pi}{2m})}{q \cdot \sin(\frac{\nu\pi}{2mq})}$$

For example, the distribution factor for a motor of $2p = 8$, $q = 2$, and $m = 3$ is equal to

$$k_d = \frac{\sin(\frac{\pi}{6})}{2 \cdot \sin(\frac{\pi}{12})} = 0.966 .$$

The winding factor is defined as a product of the pitch and winding factors:

$$k_{w(\nu)} = k_{p(\nu)} \cdot k_{d(\nu)} .$$

The distribution factors for $q = 3$ are listed in Table 11.2.

Table 11.2: The distribution factors for $q = 3$.

ν	1	5	7	11	13
$k_{d(\nu)}$ for $q = 3$	0.960	0.218	-0.177	-0.177	0.218

Exercise 11.6   Calculate the winding factors of an eight pole - 48 slot motor for harmonic numbers, $\nu = 1, 5, 7$.

11.2   MMF with Slot Openings

Previously, the MMF are calculated based on the assumption that coils are located at discrete points on the inner surface of stator core. Thus, the current is modeled as a delta function in the space. In the normal machine, coils are inserted in the slots of stator core. To give a more realistic modeling, it is assumed here that the coils are concentrated at the location of slot openings, and that the stator core has a uniform cylindrical shape. Then, the current is modeled as multiple current sheets attached on the inner surface of cylindrical core. This modeling approach is valid if the core permeability is infinite.

Fig. 11.10 (a) and (b) shows a stator with slot opening width, b_0 and an approximation with current sheets. The core is assumed to be infinitely permeable. Let R_s be the inner bore radius of stator, N_{ph} the number of turns per phase, p the number of pole pairs, and i a phase current. Then, $\frac{N_{ph}i}{p}$ is the ampere-turns per pole per phase.

The current sheet can be modeled as a square function with a narrow width, b_0 and height, $\pm\frac{N_{ph}i}{2pb_0}$ such that

$$K(x) = \begin{cases} -\frac{N_{ph}i}{2pb_0}, & \frac{\tau_p}{2} - \frac{b_0}{2} \le \theta \le \frac{\tau_p}{2} + \frac{b_0}{2} \\ \frac{N_{ph}i}{2pb_0}, & -\frac{\tau_p}{2} - \frac{b_0}{2} \le \theta \le -\frac{\tau_p}{2} + \frac{b_0}{2} \\ 0, & \text{otherwise}, \end{cases} \tag{11.6}$$

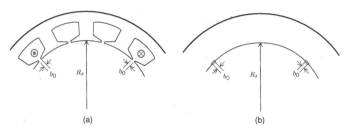

Figure 11.10: (a) Stator with slot openings and (b) an approximation with current sheets.

where x is the arc length along the stator bore. The MMF is obtained by

$$F(x) = \int K(x)dx. \tag{11.7}$$

The current sheet is depicted in Fig. 11.11 (a), and its corresponding MMF is shown in Fig. 11.11 (c). It is different from the delta-function modeling that results in a square MMF. Note that the MMF has slanted slopes in the slot opening region. It needs to be compared with a square MMF caused by delta functions shown in Fig. 11.11 (b) and (d). Using (11.6), the MMF with slots is expanded as a Fourier

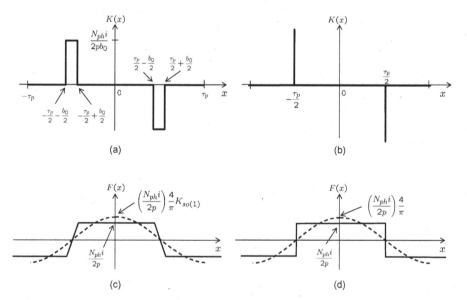

Figure 11.11: (a) Current sheet with slot opening, (b) current sheet as $b_0 \to 0$, (c) MMF for (a), and (d) MMF for (b).

series [1], [2]

$$F(x) = \sum_{\nu}^{\infty} a_\nu \cos(\nu \frac{\pi}{\tau_p} x), \tag{11.8}$$

where

$$a_\nu = 2 \times 2 \times \frac{1}{\tau_p} \int_0^{\frac{\tau_p}{2}} F(x) \cos(\nu \frac{\pi}{\tau_p} x) dx$$

$$= \frac{4}{\tau_p} \times \frac{N_{ph} i}{2p} \left[\int_0^{\frac{\tau_p}{2} - \frac{b_0}{2}} \cos(\nu \frac{\pi}{\tau_p} x) dx - \frac{2}{b_0} \int_{\frac{\tau_p}{2} - \frac{b_0}{2}}^{\frac{\tau_p}{2}} \left(x - \frac{\tau_p}{2} \right) \cos(\nu \frac{\pi}{\tau_p} x) dx \right]$$

$$= \frac{4}{\tau_p} \frac{N_{ph} i}{2p} \times \frac{2}{b_0} \left(\frac{\tau}{\nu \pi} \right)^2 \cos \left(\nu \frac{b_0}{2\tau_p} \pi - \nu \frac{\pi}{2} \right)$$

$$= \frac{4}{\pi} \frac{N_{ph} i}{2p} \frac{1}{\nu} \left(\sin \nu \frac{\pi}{2} \right) \left(\frac{1}{\nu} \frac{\tau_p}{\pi} \frac{2}{b_0} \right) \sin \left(\nu \frac{b_0}{2} \frac{\pi}{\tau_p} \right)$$

Let the slot opening factor defined by

$$K_{so(\nu)} = \left(\sin \nu \frac{\pi}{2} \right) \frac{\sin \nu \frac{\pi}{\tau_p} \frac{b_0}{2}}{\nu \frac{\pi}{\tau_p} \frac{b_0}{2}}. \tag{11.9}$$

Note that $\sin \nu \frac{\pi}{2} = 0$ or ± 1. Especially, it is equal to zero whenever ν is even number, i.e., even harmonics do not appear for MMF with the half symmetry. Then,

$$F(x) = \frac{N_{ph} i}{2p} \frac{4}{\pi} \sum_{\nu=1,3,5\cdots} \frac{1}{\nu} K_{so(\nu)} \cos(\nu \frac{\pi}{\tau_p} x). \tag{11.10}$$

Note that $K_{so(\nu)}$ is a sinc function and $\lim_{b_0 \to 0} |K_{so(\nu)}| = 1$. Therefore, as the slot opening reduces to zero, then the MMF converges to the square MMF. Comparing (11.2) with (11.10), it is observed that the general pitch factor, $k_{p(\nu)}$ is just replaced by the slot opening factor, $K_{so(\nu)}$. As the opening becomes large, the peak of MMF is reduced by the factor of $|K_{so(\nu)}|$. The fundamental component is shown in Fig. 11.11 (c) as a dotted cosine curve with amplitude, $\frac{N_{ph} i}{2p} \frac{4}{\pi} K_{so(1)}$. With the stator bore radius, R_s, it is expressed equivalently as

$$K_{so(\nu)} = \frac{\sin \nu \frac{p}{R_s} \frac{b_0}{2}}{\nu \frac{p}{R_s} \frac{b_0}{2}} \tag{11.11}$$

since $\tau_p = \frac{\pi R_s}{p}$. Note also that $\nu \frac{\pi}{\tau_p} \frac{b_0}{2} = \nu p \Delta_\alpha$, where $\Delta_\alpha = \frac{b_0}{2R_s}$. In terms of angle, $\alpha \equiv R_s x$ the MMF is expressed differently as

$$F(\alpha) = \frac{\tau_p}{\pi R_s} \frac{N_{ph} i}{2} \frac{4}{\pi} \sum_{\nu=1} \frac{1}{\nu} \left(\sin \nu \frac{\pi}{2} \right) \frac{\sin \nu p \Delta_\alpha}{\nu p \Delta_\alpha} \cos(\nu p \alpha). \tag{11.12}$$

Let the winding factor with slot opening defined by

$$\bar{k}_{w(\nu)} \equiv K_{so(\nu)} k_{d(\nu)}. \tag{11.13}$$

The MMFs of phase coils are equal to

$$F_a(x) = \frac{N_{ph}i_a}{2p}\frac{4}{\pi}\sum_\nu^\infty \frac{1}{\nu}\bar{k}_{w(\nu)}\cos(\nu\frac{\pi}{\tau_p}x), \tag{11.14}$$

$$F_b(x) = \frac{N_{ph}i_b}{2p}\frac{4}{\pi}\sum_\nu^\infty \frac{1}{\nu}\bar{k}_{w(\nu)}\cos\nu(\frac{\pi}{\tau_p}x - \frac{2\pi}{3}), \tag{11.15}$$

$$F_c(x) = \frac{N_{ph}i_c}{2p}\frac{4}{\pi}\sum_\nu^\infty \frac{1}{\nu}\bar{k}_{w(\nu)}\cos\nu(\frac{\pi}{\tau_p}x - \frac{4\pi}{3}). \tag{11.16}$$

11.2.1   MMF with Current Harmonics

Since the current is controlled by the pulse width modulation (PWM), the current contains PWM harmonics. Furthermore, inverter nonlinearities such as switch on-drop and dead-time cause additional harmonics. The three phase currents carrying harmonic components are represented with the harmonic number, n as

$$i_a = \sum_n^\infty \sqrt{2}I_{rms(n)}\cos n(p\omega_r t), \tag{11.17}$$

$$i_b = \sum_n^\infty \sqrt{2}I_{rms(n)}\cos n\left(p\omega_r t - \frac{2\pi}{3}\right), \tag{11.18}$$

$$i_c = \sum_n^\infty \sqrt{2}I_{rms(n)}\cos n\left(p\omega_r t - \frac{4\pi}{3}\right), \tag{11.19}$$

where ω_r is the rotor mechanical speed. The time harmonic order is denoted by n. In contrast, the space harmonic order is denoted by ν. It is assumed that all the harmonics contained in i_a, i_b, and i_c, are shifted equally by 120° in electrical angle.

Then, the MMF's are

$$F_a(x) = \frac{4}{\pi}\frac{N_{ph}}{2p}\sqrt{2}\sum_n^\infty I_{rms(n)}\sum_\nu^\infty \frac{\bar{k}_{w(\nu)}}{\nu}\cos n(p\omega_r t)\cos\left(\nu\frac{\pi}{\tau_p}x\right), \tag{11.20}$$

$$F_b(x) = \frac{4}{\pi}\frac{N_{ph}}{2p}\sqrt{2}\sum_n^\infty I_{rms(n)}\sum_\nu^\infty \frac{\bar{k}_{w(\nu)}}{\nu}\cos n\left(p\omega_r t - \frac{2\pi}{3}\right)\cos\nu\left(\frac{\pi}{\tau_p}x - \frac{2\pi}{3}\right), \tag{11.21}$$

$$F_c(x) = \frac{4}{\pi}\frac{N_{ph}}{2p}\sqrt{2}\sum_n^\infty I_{rms(n)}\sum_\nu^\infty \frac{\bar{k}_{w(\nu)}}{\nu}\cos n\left(p\omega_r t - \frac{4\pi}{3}\right)\cos\nu\left(\frac{\pi}{\tau_p}x - \frac{4\pi}{3}\right). \tag{11.22}$$

Note that

$$\frac{4}{\pi}\frac{N_{ph}}{2}\sqrt{2} \approx 0.9N_{ph}. \tag{11.23}$$

From the cosine law it follows that

$$\cos(np\omega_r t)\cos\left(\nu\frac{\pi}{\tau_p}x\right) = \frac{1}{2}\cos\left(np\omega_r t + \nu\frac{\pi}{\tau_p}x\right) + \frac{1}{2}\cos\left(np\omega_r t - \nu\frac{\pi}{\tau_p}x\right),$$

$$\cos\left(np\omega_r t - n\frac{2\pi}{3}\right)\cos\left(\nu\frac{\pi}{\tau_p}x - \nu\frac{2\pi}{3}\right)$$

$$= \frac{1}{2}\cos\left(np\omega_r t + \nu\frac{\pi}{\tau_p}x - (\nu+n)\frac{2\pi}{3}\right) + \frac{1}{2}\cos\left(np\omega_r t - \nu\frac{\pi}{\tau_p}x + (\nu-n)\frac{2\pi}{3}\right),$$

$$\cos\left(np\omega_r t - n\frac{4\pi}{3}\right)\cos\left(\nu\frac{\pi}{\tau_p}x - \nu\frac{4\pi}{3}\right)$$

$$= \frac{1}{2}\cos\left(np\omega_r t + \nu\frac{\pi}{\tau_p}x - (\nu+n)\frac{4\pi}{3}\right) + \frac{1}{2}\cos\left(np\omega_r t - \nu\frac{\pi}{\tau_p}x + (\nu-n)\frac{4\pi}{3}\right).$$

The MMF sum, $F_{sum}(x) \equiv F_a+F_b+F_c$ is separated into two groups: the positive and negative sequences. Let us define $M_{\nu n}^+(x)$ and $M_{\nu n}^-(x)$ by for each harmonic order, ν and n:

(Positive sequence)

$$M_{\nu n}^+(x) \equiv \frac{1}{2}\left\{1 + \cos(\nu-n)\frac{2\pi}{3} + \cos(\nu-n)\frac{4\pi}{3}\right\}\cos\left(np\omega_r t - \nu\frac{\pi}{\tau_p}x\right); \quad (11.24)$$

(Negative sequence)

$$M_{\nu n}^-(x) \equiv \frac{1}{2}\left\{1 + \cos(\nu+n)\frac{2\pi}{3} + \cos(\nu+n)\frac{4\pi}{3}\right\}\cos\left(np\omega_r t + \nu\frac{\pi}{\tau_p}x\right). \quad (11.25)$$

Therefore,

$$M_{\nu n}^+ = \begin{cases} \frac{3}{2}\cos\left(np\omega_r t - \nu\frac{\pi}{\tau_p}x\right), & \nu-n = 3m \\ 0, & \nu-n \neq 3m. \end{cases} \quad (11.26)$$

Likewise,

$$M_{\nu n}^- = \begin{cases} \frac{3}{2}\cos\left(np\omega_r t + \nu\frac{\pi}{\tau_p}x\right), & \nu+n = 3m' \\ 0, & \nu+n \neq 3m'. \end{cases} \quad (11.27)$$

Summarizing the above, it follows that

Harmonic numbers	Positive seq: $M_{\nu n}^+$	Negative seq: $M_{\nu n}^-$
$n \pm \nu \neq 3m$	0	0
$n - \nu = 3m$	$\frac{3}{2}\cos(np\omega_r t - \nu\frac{\pi}{\tau_p}x)$	0
$n + \nu = 3m'$	0	$\frac{3}{2}\cos(np\omega_r t + \nu\frac{\pi}{\tau_p}x)$

Excluding even and triplen order harmonics, it follows that $\nu = 1, 5, 7, 11, 13, \cdots$. The same is assumed to be true for the time harmonics. Note also that $n-\nu = 3m$ and $n+\nu = 3m'$ cannot be satisfied simultaneously. It can be proved by way of contradiction: Assume that they are satisfied at the same time. Then, $2n = 3(m+m')$ or $2\nu = 3(m'-m)$, i.e., either n or ν is a multiple of 3. It contradicts the fact that neither ν nor n is a multiple of 3. Therefore, either positive or negative sequence takes place exclusively for each (ν, n).

Now, let $n = 2k \pm 1$ and $\nu = 2k' \pm 1$. Then, in order to give nontrivial solution it should follow that $n + \nu = (2k \pm 1) + (2k' \pm 1) = 2(k + k') \pm 2 = 3m$, or $2(k + k') = 3m$. Also, $n - \nu = (2k \pm 1) - (2k' \pm 1) = 2(k - k') = 3m$ or $2(k - k') \pm 2 = 3m$. This means that m is a multiple of 2, and $n \pm \nu$'s are multiples of 6. Thus, we may let $n + \nu = 6c'$ and $n - \nu = 6c''$ for some integer, c' and c''. Table 11.3 shows harmonic components constituting nonzero positive and negative sequences for $\nu, n < 30$. The corresponding grid of harmonic numbers is shown in Fig. 11.12. Note also that the lines of positive and negative sequences do not intersect at integer points.

Table 11.3: Harmonic components constituting positive and negative sequences for $\nu, n < 30$.

ν	Negative sequences for $(\nu + n = 6c')$ (ν, n)	Positive sequences for $(\nu - n = 6c'')$ (ν, n)
1	(1,5), (1,11), (1,17), (1,23), (1,29),	(1,1), (1,7), (1,13), (1,19), (1,25),
5	(5,1), (5,7), (5,13), (5,19), (5,25),	(5,5), (5,11), (5,17), (5,23), (5,29),
7	(7,5), (7,11), (7,17), (7,23),	(7,1), (7,7), (7,13), (7,19), (7,25),
11	(11,1), (11,7), (11,13), (11,19),	(11,5), (11,11), (11,17), (11,23), (11,29),
13	(13,5), (13,11), (13,17),	(13,1), (13,7), (13,13), (13,19), (13,25),
17	(17,1), (17,7), (17,13),	(17,5), (17,11), (17,17), (17,23), (17,29),
19	(19,5), (19,11),	(19,1), (19,7), (19,13), (19,19), (19,25),
23	(23,1), (23,7),	(23,5), (23,11), (23,17), (23,23), (23,29),
25	(25,5),	(25,1), (25,7), (25,13), (25,19), (25,25),
29	(29,1),	(29,5), (29,11), (29,17), (29,23), (29,29),

Therefore, it follows from (11.20)−(11.22) that the MMF excited by the harmonic current is expressed as

$$
F_{sum}(x) = 0.9 \frac{N_{ph}}{p} \sum_n^\infty I_{rms(n)} \sum_{\nu \neq 2k, 3k} \frac{\bar{k}_{w(\nu)}}{\nu} (M_{\nu n}^+(x) + M_{\nu n}^-(x))
$$

$$
= 1.35 \frac{N_{ph}}{p} \sum_{\nu - n = 6c''}^\infty \frac{I_{rms(n)} \bar{k}_{w(\nu)}}{\nu} \cos\left(np\omega_r t - \nu \frac{\pi}{\tau_p} x \right)
$$

$$
+ 1.35 \frac{N_{ph}}{p} \sum_{\nu + n = 6c'}^\infty \frac{I_{rms(n)} \bar{k}_{w(\nu)}}{\nu} \cos\left(np\omega_r t + \nu \frac{\pi}{\tau_p} x \right). \quad (11.28)
$$

Summary

- There are number of winding methods in the distributed winding: short and full pitch, single and double layers. However, the MMF is determined only by the arrangement of coil active sides, independently of end-turn configuration.

- The pitch factor is the ratio of the MMF harmonic coefficient to that of the square MMF.

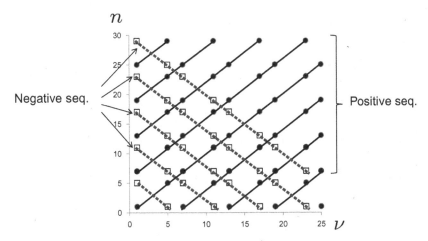

Figure 11.12: Harmonic number combinations, $\{(\nu, n)\}$ for positive and negative sequences.

- If there is a half symmetry in the MMF shape, no even harmonic component exists. No triplen harmonics exist in the three phase system. Therefore, the MMF harmonics exists for $\nu = 6k \pm 1$: The harmonic components of $\nu = 6k+1$ create positive sequences, whereas those of $\nu = 6k - 1$ negative sequences.

- The MMF harmonic traveling speed reduces as the space harmonic order, ν increases.

- The slot opening effect can be counted by replacing normal pitch factor, k_p by the slot opening factor, K_{so}, which is a sinc function. K_{so} converges to k_p as the slot width decreases.

- When there are time harmonics in the current, both positive and negative sequences take place.

Exercise 11.7   Find the MMF with the space and current harmonics for $\nu, n < 19$.

Solution.

$$F_{sum} = 1.35\frac{N_{ph}}{p}\left[I_1\bar{k}_{w(1)}\cos p(\theta - \omega_r t) + I_7\bar{k}_{w(1)}\cos p(\theta - 7\omega_r t)\right.$$

$$+ I_{13}\bar{k}_{w(1)}\cos p(\theta - 13\omega_r t) + I_5\bar{k}_{w(5)}\cos p(5\theta - 5\omega_r t)$$

$$+ I_{11}\bar{k}_{w(5)}\cos p(5\theta - 11\omega_r t) + I_1\bar{k}_{w(7)}\cos p(7\theta - \omega_r t)+$$

$$\cdots$$

$$+ I_5\bar{k}_{w(1)}\cos p(\theta + 5\omega_r t) + I_{11}\bar{k}_{w(1)}\cos p(\theta + 11\omega_r t) + I_1\bar{k}_{w(5)}\cos p(5p\theta + \omega_r t)$$

$$+ I_7\bar{k}_{w(5)}\cos p(5\theta + 7\omega_r t) + I_{13}\bar{k}_{w(5)}\cos p(5\theta + 13\omega_r t)+$$

$$\cdots$$

Table 11.4: Harmonic components constituting positive and negative sequences for $\nu, n < 19$.

ν	$M_{\nu n}^{-} \neq 0$ for $(\nu + n = 6\ell')$ (ν, n)	$M_{\nu n}^{+} \neq 0$ for $(\nu - n = 6\ell'')$ (ν, n)
1	(1,5), (1,11),	(1,1), (1,7), (1,13),
5	(5,1), (5,7), (5,13),	(5,5), (5,11),
7	(7,5), (7,11),	(7,1), (7,7), (7,13),
11	(11,1), (11,7), (11,13),	(11,5), (11,11),
13	(13,5), (13,11),	(13,1), (13,7), (13,13),

■

11.3 Fractional Slot Machines

Fractional slot machine refers to the machine with lower slot number such as $q < 1$. Then, the winding turns out to be a concentrated winding around a single tooth. It is a very attractive feature in the manufacturing, since the winding can be done with a simple needle winding machine. Further, high slot fill factor can be achieved by orderly winding. Alternatively, the stator can be easily manufactured by assembling tooth segments after winding each piece separately. The fractional slot design is widely used in high pole high torque machines [4]–[8].

11.3.1 Concentrated Winding on Segmented Core

If the coil span is equal to one tooth pitch, each winding will be concentrated over one tooth. The required condition for concentrated (non-overlapping) winding is [3]

$$\frac{N_{ph}}{GCD(N_{ph}, 2p)} = km, \tag{11.29}$$

where N_{ph} is the number of coil turns per phase, m is the number of phases, and k is an integer.

Fig. 11.13 shows a schematic drawing of concentrated winding over a core segment and a photo of the stator assembly. Each segment has a single tooth and a dovetail in the back iron, so that it can be assembled to be a circular stator. Since the coil can be wound separately before assembly, it can be neatly wrapped in a free space with a linear winding technology. Thus, the slot fill factor is higher than the case of normal coil insertion. In addition, the winding equipment cost can be reduced.

The slot liner and slot key are normally used to strengthen the isolation and protect the coils from mechanical abrasion. Further, the slot liner serves as a guide for inserting the coil. The lubricious surface facilitates coils to slide well into the slot. However, it reduces the slot fill factor, which is defined as the ratio of copper section to the entire slot area:

$$f_s = \frac{\frac{\pi}{4} d^2 n}{S}, \tag{11.30}$$

where d is the coil diameter without the varnish layer, n is the number of coils in the slot, and S is the slot area. In general, the fill factor depends on the coil type and coil arrangements. Normally, random wiring with round wire results in a poor fill factor in the range of $35 \sim 45\%$. In order to achieve higher fill factors, rectangular or flat wire needs to be used.

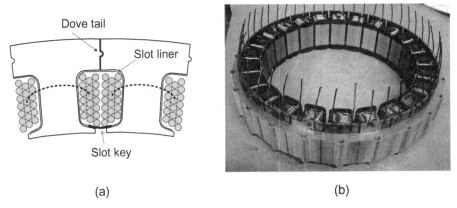

(a) (b)

Figure 11.13: (a) Concentrated winding over a core segment and (b) a stator assembly for 16 pole motor (Hyundai Motor Company).

A high slot fill factor is a key element in the realization of high power density motor. Another big advantage is that the concentrated winding has a very short end-turn length. The end turn only increases the coil resistance and leakage inductance, and does not contribute to torque production. Consequently, motor efficiency can be improved by shortening the end turn. Core segmentation is suitable for pancake type motors that have short stacking lengths with a large bore radius. Due to the non-overlapping nature, the mutual inductance is almost zero. Thereby, the concentrated winding increases the fault tolerance if the inverter fails.

But the motors of concentrated winding have some disadvantages: Large MMF harmonics cause eddy current losses in the rotor PMs [9], which often lead to demagnetization. Specifically, a great danger is the sub-order harmonics that penetrate deeply into the PMs. In general, the thicker magnet is utilized to avoid demagnetization. Further, a large effective air gap is required to reduce the torque ripple. Therefore, concentrated winding works with surface mounted PM rotor, not with the embedded type. As a result, the concentrated winding motors have generally a low power density since the reluctance torque is not utilized effectively. Finally, the segments generate a high audible noise. To reduce it, the assembled core should be pressed firmly through an outer ring. Some pros and cons of distributed and concentrated windings are summarized in Table 11.5.

11.3.2 Feasible Slot-Pole Number Combination

A large winding factor is desired in order to maximize the torque density. Thus, the coil-pitch needs to be close to the pole-pitch as much as possible. Theoretically, $N_s = 2p \pm 1$ yields the highest winding factor. Typical examples for three phase

Table 11.5: Comparison of concentrated and distributed windings

	Concentrated Windings	Distributed Windings
End turn length	short non-overlapping	long overlapping
Slot fill factor	high $(0.5 \sim 0.65)$	low $(0.35 \sim 0.45)$
Copper loss	small	large
Working harmonic	higher order	fundamental
MMF harmonics	large	small
Even harmonics	present	absent
Saliency ratio	small	large
Power density	relatively small	large
PM loss	large	small
PM usage	large	small
Suitable PM location	surface or slightly imbedded	surface or interior
Convenience in winding	easy	complex
Mutual inductance	small	large
Fault tolerance	high	low

machines are $N_s/2p = 3/2,\ 3/4,\ 9/8,\ 9/10,\ 15/14,\ 15/16,\ 21/20,\ 21/22$, etc. Those slot-pole combinations are good for low cogging torque, since $LCM(N_s, 2p)$ is large.

On the other hand, the greatest common denominator, $GCD(N_s, 2p)$ indicates a structural periodicity between stator and rotor. A large GCD is desired for low noise and vibration. However, the machines with $N_s = 2p \pm 1$ have the lowest $GCD(N_s, 2p) = 1$. In such a case, unbalanced magnetic pull takes place, i.e., the rotor experiences fluctuating magnetic pull during rotation [10], [11]. It should be avoided since the unbalanced magnetic force causes excessive acoustic noise and vibration, reducing the bearing life.

Hence, the second alternative goes with $N_s = 2p \pm 2$; In three phase machines, $N_s/2p$ combinations are 6/4, 6/8; 12/10, 12/14; 18/16, 18/20; 24/22, 24/26, etc [12]. Of course, those machines do not have the unbalanced magnetic force since $GCD(N_s, 2p) > 1$. For a balanced three phase machine, typical $N_s/2p$ combinations are listed in Table 11.6.

Star and Winding Diagrams

The star of slots is a vector representation of the EMFs that could be induced in the coil. The star diagram was developed for coil design of fractional-slot motors [5]. It provides a systematic method to determine coil in and out and shows how the EMF vectors are constituted from each coil sides. Further, it can be used to calculate the distribution factor. A vector is assigned to each coil in a slot, and is called a *slot side vector.* The slot side vectors form a spoke of the star diagram, and the EMF of phase winding.

Let

$$t = GCD(N_s, p). \tag{11.31}$$

Table 11.6: Slot-pole combinations for three phase motors [13].

	Distributed Windings								
Number of slots (N_s)	6	9	12	15	18	21	24	36	48
	2	2	2	2	2	2	2	2	2
Number of Poles ($2p$)		4	4	4	4	4	4	4	4
				10	6	8	8	6	8
					8		10	8	10

	Concentrated Windings								
Number of slots (N_s)	6	9	12	15	18	21	24	36	48
	4	6	8		12		20	30	38
Number of Poles ($2p$)	8	8	10		14		22	32	40
		12	14		16		26	38	44
			16		20			42	52

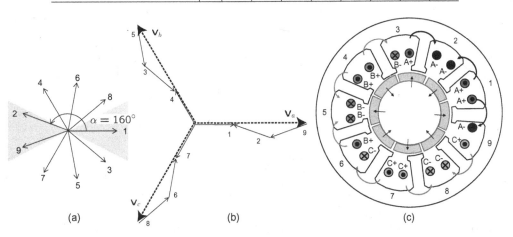

Figure 11.14: Star diagram and coil arrangements of 9 slot-8 pole motor: (a) star of coil sides, (b) phase EMF vectors, and (c) coil arrangements.

Then the star diagram is characterized by N_s/t spokes, and each spoke is composed of t slot side vectors. The motor winding is feasible only when $\frac{N_s}{mt}$ is an integer, where m is the number of phases. The electrical angle between two adjacent slots is $\alpha \equiv 2\pi p/N_s$.

Let us consider two 8 pole motors with different slot numbers, $N_s = 9$ and $N_s = 12$. Note that $t = GCD(9,4) = 1$ for the 9 slot motor, and $t = GCD(12,4) = 4$ for the 12 slot motor. Therefore, the numbers of spokes are $9/1=9$ and $12/4=3$, and the electrical angle increase over a slot is $\alpha = 2\pi \times 4/9 = 160°$ and $2\pi \times 4/12 = 120°$, respectively. Thereby, the 9 slot-8 pole motor has 9 vectors:

$$\mathbf{E}_1 = 1, \quad \mathbf{E}_2 = e^{j\frac{8\pi}{9}}, \quad \cdots, \quad \mathbf{E}_9 = e^{j8\times\frac{8\pi}{9}}.$$

The phase EMF vectors are formed by grouping the spokes in the opposite sectors after dividing the complex plane into $2m$ sectors. Fig. 11.14 shows spokes, phase

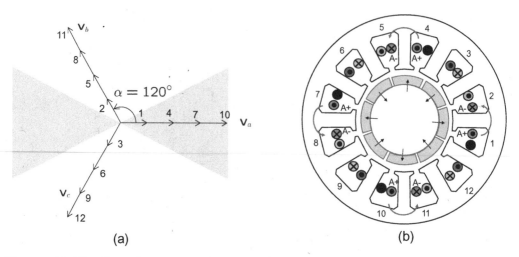

(a)   (b)

Figure 11.15: Star diagram and coil arrangements of 12 slot-8 pole motor: (a) phase EMF vectors, and (b) coil arrangements.

EMFs, and coil allocation of 9 slot-8 pole motor. In summing the vectors polarity should be considered with the signs. For example, the a phase EMF is constituted by adding the spokes belonging to the shaded area shown in Fig. 11.14 (a):

$$V_a = E_1 - E_2 - E_9.$$

Note that the plane should be divided by $2m = 6$ sectors in the case of three phase machines, and that the vectors, E_2 and E_9, belonging to the opposite sector, are added with minus sign. Fig. 11.15 shows phase EMFs and coil allocation of 12 slot-8 pole motor. It has 3 spokes, 120° advance, and all components are added with positive signs. The phase EMFs are aligned naturally with the spokes.

Steps of drawing the star and the corresponding winding diagrams are summarized in the following.

1. Assign numbers, $i = 1, 2, 3, \cdots N_s$ to slots sequentially.
2. Define the slot side vectors such that

$$E_i = e^{j\alpha(i-1)} \qquad \text{for} \quad i = 1, \cdots, S.$$

3. Divide the complex plane into $2m$ sectors of equal angle. Form a phase EMF by adding up the slot side vectors in a pair of two opposite sectors. Assign signs to each slot side vector in a way that the EMF magnitude is maximized.
4. For each slot choose coil 'in' and 'out' depending on the sign of slot side vector in forming the phase EMF. Draw the coil layout.

Fig. 11.16 shows star diagram, winding arrangements, and coil layout for a 12 slot-10 pole motor. Note that the electrical angle between two adjacent slots is $\alpha = \frac{360 \times 10}{12} = 150°$. The vectors in opposite sectors are grouped into a phase

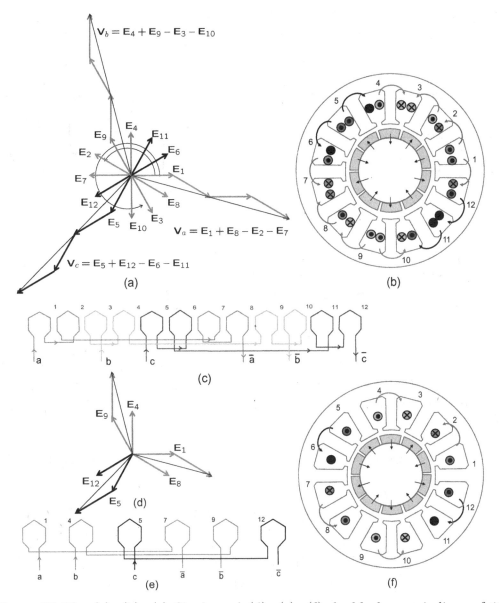

Figure 11.16: (a), (b), (c) Single and (d), (e), (f) double layer windings of 12 slot−10pole motor: (a),(d) star diagram and phase EMFs, (b),(e) winding arrangements, and (c),(f) coil layout.

EMF. For example, the a-phase EMF is made as $V_a = E_1 + E_8 - E_2 - E_7$ by assigning outward direction to slots 1 and 8 and inward direction to slots 2 and 7. From the vector sum shown in Fig. 11.16 (a), a pitch factor can be evaluated as $k_p = 4\cos 15°/4 = 0.966$. The 12 slot 10 pole motor can also have a single layer winding as shown in Fig. 11.16 (e). Based on it, the coil layout can be drawn as shown in Fig. 11.16 (d). Not that the teeth are wound alternately. With such a winding, the mutual inductance between the phases is almost zero [14], meaning that each phase is essentially magnetically isolated. The zero mutual inductance is good for fault-tolerance [15], since a failure in a phase affects little to the other phases.

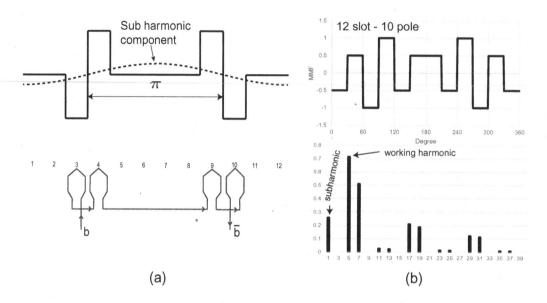

(a)   (b)

Figure 11.17: MMFs of 12 slot$-$10 pole motor: (a) single phase conduction and (b) three phase current $(i_a, i_b, i_c) = (I, -\frac{I}{2}, -\frac{I}{2})$ conduction and the MMF spectrum.

Fig. 11.17 shows MMF of a 12 slot$-$10 pole machine for a single phase conduction and a three phase conduction, $(i_a, i_b, i_c) = (I, -\frac{I}{2}, -\frac{I}{2})$. It should be noted that the 5th is the main (fundamental) component. There also exists a first order component, appearing as dotted line in Fig. 11.17 (a).

MMFs of other fractional motors are shown in Fig. 11.18. Each of them displays MMF for a single phase conduction and its harmonic components. Therefore, triplen harmonics exist. Note that even harmonics exist for odd numbered slots. Note that $q = 3/8$ for 9 slot$-$8 pole motors, in which the 4^{th} order harmonic is a working harmonic and the first and second ones are the two sub-harmonic field waves. Note also that 18 slot$-$16 pole machine is a duplicate of 9 slot$-$10 pole machine. However, magnetic pull is not developed in 18 slot$-$16 pole machine due to even slot number. The working harmonic is 8th, while the 2nd and 4th are the subharmonics.

Exercise 11.8   Draw the MMF, $F(x)$ waveform for a 18 slot$-$16 pole machine when $(i_a, i_b, i_c) = (I, -\frac{I}{2}, -\frac{I}{2})$. Find the spectrum using (2.37) and (2.38). Check that the MMF and spectrum are the same as those in Fig. 11.19:

11.3.3   Torque-Producing Harmonic and Subharmonics

The first order component is normally the torque producing (working) harmonic component in the distributed windings. The other high order harmonics, being

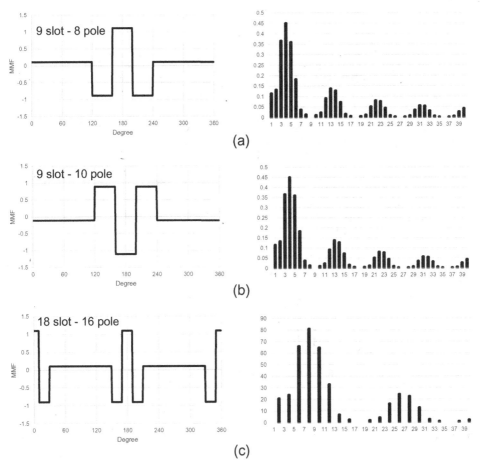

Figure 11.18: MMFs for single phase conduction and its spectrum: (a) 9 slot−8 pole, (b) 9 slot−10 pole, and (c) 18 slot−16 pole.

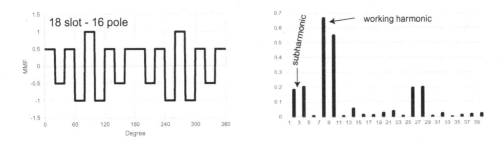

Figure 11.19: (Exercise 11.8) MMF and spectrum of a 18 slot−16 pole motor.

asynchronous to the fundamental, induce torque ripple and eddy current losses in the rotor PM, as well as in the iron core.

But in the fractional motors, working harmonic is a high order component. For example, the 5th harmonic is the working harmonic in the 12 slot−10 pole motors. In this case, the first order component, being a subharmonic, acts as a parasitic

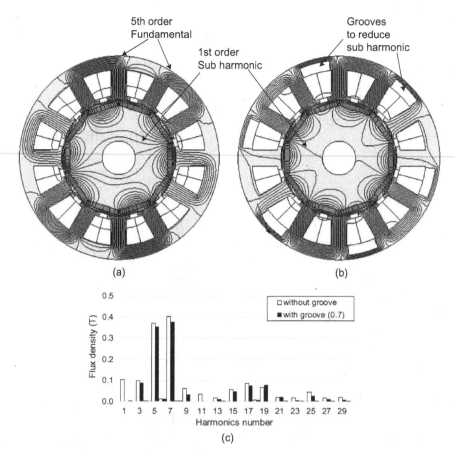

Figure 11.20: Flux lines of 12 slot−10 pole motor: (a) without groove, (b) with grooves and (c) Harmonics.

one. Fig. 11.20 shows flux lines obtained by FEM analysis, where the first order component is identified by the big loops. This subharmonic component has a long wavelength and transverses the rotor core. Those suborder harmonics are particularly harmful to the rotor PMs, since they penetrate deeply into the rotor PM and induce large eddy-current losses [16], [17]. They also generate a big radial attracting force, which can deform the rotor [18]. Fig. 11.20 (b) shows the flux lines when the stator core is modified with many grooves on the back iron. The grooves are set to hinder the passage of the first order component. As a result, suborder harmonics are reduced significantly. The groove shape optimization was studied in [19].

Fig. 11.21 show a real back EMF measurement of a 10 pole−12 slot motor with an exterior rotor. Note that the PMs are made with ferrites. It is worth noting that the back EMF looks perfect sinusoidal, which is a typical attribute of fractional slot motors. As mentioned in the above, the gap field has lots of harmonics, thereby PMs of fractional slot motors are more vulnerable to demagnetization. Thus, it is somewhat paradoxical that the harmonic-rich motors have pure sinusoidal EMFs. It is because contributions of local space harmonics are summed out to be zero on the shaft.

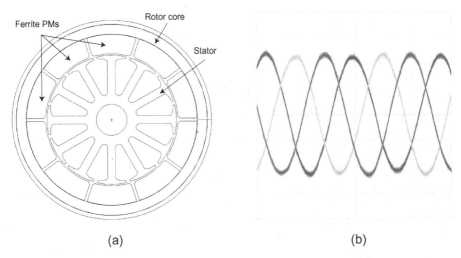

(a) (b)

Figure 11.21: 10 pole−12 slot motor with the exterior rotor: (a) sectional view and (b) back EMF.

11.4 Demagnetization Analysis

Coil Joule loss, core iron loss, and PM eddy current loss are the sources of the PM temperature rise. Nonetheless, the most dangerous threat is the PM eddy current loss, since the heat source is inside the PM itself. NdFeB magnet is basically a metal alloy and thus the field harmonics induce a considerable amount of eddy current. Its conductivity $(0.67 \times 10^{-6}$ S/m) is about 1/90 of copper or 1/15 of iron. The causes of current induction include: winding space harmonics, current harmonics, and slot harmonics. Meanwhile, ferrite magnets, being oxide ceramic, do not suffer eddy current loss.

11.4.1 PM Loss Influential Factors

Pole-slot combination is the most influential factor on the PM loss. Additionally, slot opening, PM position, PM arrangement, cavity area, and bridge thickness are other influential factors. The fundamental or a working harmonic MMF rotates asynchronously with the rotor. Therefore, the synchronous component does not induce eddy current. But, high or low order MMF harmonics seem to be AC fields to the PM, thus cause PM loss.

The MMF is a product of winding function, $N(\theta)$ and current, $I(t)$: The former governs the space harmonics, while the latter determines time harmonics. In the perspective of the space harmonics, distributed windings are better than the concentrated winding. Furthermore, higher the number of slots per pole per phase, the less space harmonics result. Similarly, multi-layer windings help reduce the space harmonics. However, such solutions increase often manufacturing complexity, as well as the cost. To reduce the demagnetization risk while minimizing the PM usage, most EV propulsion motors utilize a moderate number of slots per pole per phase, for example, $q = 2$ or 3.

a) **PWM harmonics**

The inverter output voltage contains harmonics inherently by PWM. Thereby, current harmonics are induced in a wide frequency range. If the PWM frequency is high, then the magnitude current ripple and penetration depth reduce. The PWM-related losses are noticeable on low inductance machines. In such a case, the LC filter is used at the inverter output.

b) **Slot harmonics**

The stator slot openings make a sequence of dents in the air-gap flux density, i.e., a narrow and deep field drop is made at each slot opening as shown in Fig. 11.22. This abrupt field change causes eddy current loss in the PMs. It is intensified when the slot opening is large and the air gap is small.

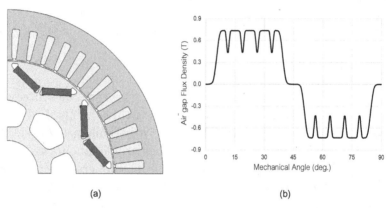

(a)   (b)

Figure 11.22: Slot effects: (a) motor model and (b) gap field density.

11.4.2   PM Loss and Demagnetization Analysis

The PM loss can be estimated via 3D FEM analysis. Fig. 11.23 shows a mesh model. Fig. 11.24 shows FEM simulation results with the PM (N39UH, resistivity $\rho = 1.4 \times 10^{-6}$ Ω·m) when the phase current is 320 A_{rms} and 200 Hz. The loss profiles are different depending on the PM segmentation. Note that the total PM loss reduces significantly as the segmentation number increases. The total PM loss after segmentation is less than 50W. It is a small number compared with other losses.

Fig. 11.25 shows that the slope of permeance curve varies as the PM thickness changes. It becomes steeper with a greater PM thickness ($\ell_{m1} > \ell_{m2}$). Then, the intersection point with the demagnetization curve moves away from the knee point (See Section 8.1.3). Therefore, it is necessary to use a sufficiently thick PM to avoid demagnetization. However, it is a costly method.

In the motor design stage, thermal analysis should be performed under some reasonable boundary conditions to guarantee that the PM temperature does not exceed above a certain level under all feasible operation modes.

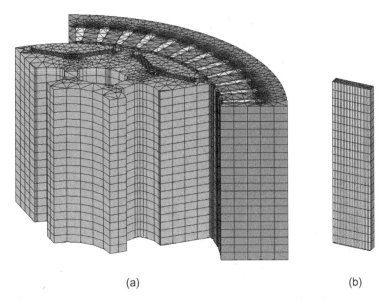

Figure 11.23: 3D mesh model for demagnetization analysis via finite element analysis (FEA): (a) core and (b) PM (single piece).

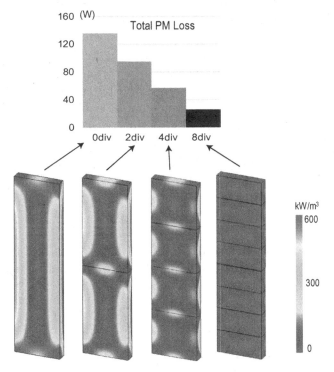

Figure 11.24: FEA results of the PM loss depending on segmentation.

11.4.3 Armature Reaction

If the stator winding conducts, the stator field is superimposed on the rotor PM field in the air gap. Consider two pole periods shown in Fig. 11.26. The symbols on

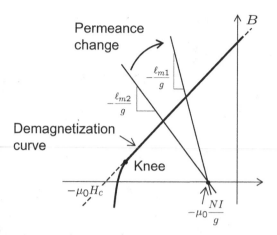

Figure 11.25: Intersection drifts away from the knee point as the PM thickens $(\ell_{m1} > \ell_{m2})$.

the stator represent the q-axis current. Assume that the corresponding field has a sinusoidal shape. Note that the peak is made in the middle of two PMs. The PMs cause DC offsets: one upward and the other downward. Then the gap field density is the highest at the leading edge of each PM, and the lowest in the tailing edge. At the leading edge, the rotor core and stator teeth are saturated normally. Thus, the resultant magnetic field is distorted, so that the field center lags behind. Such a field distortion is called armature reaction.

On the other hand, the gap field suppression by current, $\int H d\ell$ is so high at the tailing edge that a partial PM demagnetization would occur if current is great. Fig. 11.27 is a FEM simulation result of armature reaction. Dark area indicates the region of high the field density, and light area the region of low field density. Note that the PM field is almost nullified by the stator current in the tailing edge.

The demagnetization limit of the current loading is calculated from the Ampere's law:

$$A_m \frac{2\pi R_r}{P} k_w \leq H_c h_m, \tag{11.32}$$

where A_m (A/m) is the electric loading, R_r is the rotor radius, and H_c is the coercive force of PM. Therefore, the electric loading should be limited under

$$A_m \leq P \frac{h_m H_c}{2\pi R_r k_w}. \tag{11.33}$$

11.5   Torque Analysis

Torque production is the primary purpose of the electrical motors. High torque density, linearity against current magnitude, and low torque ripple are the desired attributes of the motor. However, practical machines include many non-idealities such as current harmonics, field harmonics, core saturation, etc. They result in torque ripple, cogging torque, noise, and vibration.

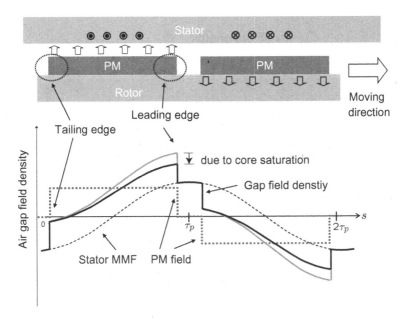

Figure 11.26: Stator current effect (armature reaction) on the leading and tailing edges of PM.

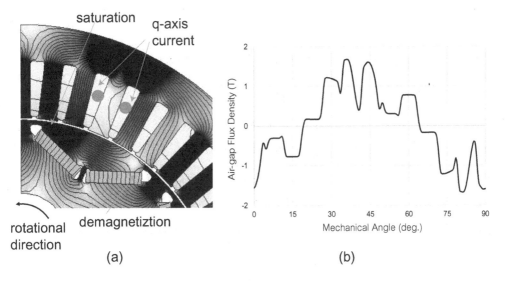

Figure 11.27: FEM results: (a) armature reaction by the stator current and (b) gap flux density.

General Torque Equation

Note from Fig. 11.28 that the energy is stored in the form of magnetic field can be depicted as the upper triangular region, $W_f = \int i\,d\lambda$. The complementary region is called co-energy, $W_c = \int \lambda\,di$. On the other hand, the mechanical energy is equal to $W_m = \int T_e\,d\theta_r$, where θ_r is the rotor mechanical (shaft) angle.

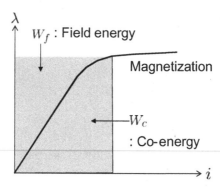

Figure 11.28: Energy and co-energy.

Note that an applied electrical energy is either stored as the field energy or converted into the mechanical energy:

$$id\lambda = dW_m + dW_f. \tag{11.34}$$

Therefore,

$$iv = i\frac{d\lambda}{dt} = T_e\omega_r + \frac{dW_f}{dt}. \tag{11.35}$$

Utilizing $\frac{1}{\omega_r} = \frac{dt}{d\theta_r}$, it follows that

$$T_e(\theta_r, \lambda) = i(\theta_r, \lambda)\frac{d\lambda}{d\theta_r} - \frac{dW_f(\theta_r, \lambda)}{d\theta_r}. \tag{11.36}$$

However, it is better to express torque in terms of co-energy, W_c, instead of W_f, since the current is a familiar variable to be handled. From the geometry shown in Fig. 11.28, we have $W_c + W_f = \lambda i$. Thereby,

$$dW_c + dW_f = \lambda di + id\lambda. \tag{11.37}$$

Hence, it follows from (11.36) and (11.37) that

$$T_e = \frac{id\lambda - (\lambda di + id\lambda - dW_c(\theta_r, i))}{d\theta_r} = \frac{\partial W_c(\theta_r, i)}{\partial\theta_r}. \tag{11.38}$$

Note here that $\lambda\frac{di}{d\theta_r} = 0$, since the current is not a function of rotor angle.

Torque Derivation from Gap Energy

The electromagnetic torque is given as the gradient of co-energy with respect to the angle. The co-energy of a motor air gap is equal to

$$W_c = \frac{1}{2\mu_0}\int_V B(\alpha, \theta_r, t)^2 dV, \tag{11.39}$$

where α is the angular position (space variable) and V represents the air gap volume.

A familiar torque equation will be derived from (11.39) in the following: The gap field is the sum of stator and rotor contributions:

$$B(\theta_r, \alpha, t) = F_s(\alpha, t)\Lambda_g(\alpha, \theta_r) + B_r(\alpha, \theta_r), \tag{11.40}$$

where $\Lambda_g(\theta_r, \alpha)$ is a permeance function of the gap per unit area, $F_s(\alpha, t)$ is the stator MMF. Assume for convenience the motor has two poles, i.e., $p = 1$. Let the d and q axis currents be $i_d(t) = I_d \cos \omega_e t$ and $i_q(t) = I_q \sin \omega_e t$. Assume further that the following traveling MMF is made via a distributed winding:

$$F_s(\alpha, \theta_r) = \frac{3}{2} N_{ph} I_d \cos(\alpha - \omega_e t) + \frac{3}{2} N_{ph} I_q \sin(\alpha - \omega_e t), \qquad (11.41)$$

where N_{ph} is the total number of turns per phase. Meanwhile, it is assumed that the PMs are imbedded in the rotor, so that the permeance is modeled as

$$\Lambda_g(\theta_r, \alpha) = \frac{\mu_0}{g(\theta_r, \alpha)} = \mu_0(\gamma_0 - \gamma_2 \cos 2(\alpha - \theta_r)), \qquad (11.42)$$

where γ_0, γ_2 are constants. Equation (11.42) states that the permeance seen by the stator coil changes depending on the rotor position, θ_r. Recall that the permeance is an inverse function of reluctance, i.e., its basic nature is μ_0 times the inverse of gap function. See Section 6.2.3 for the detail. Since the d-axis is aligned to the rotor flux, the rotor flux density is modeled as

$$B_r(\theta_r, \alpha) = \overline{B}_r \cos(\alpha - \theta_r), \qquad (11.43)$$

where $\overline{B}_r$ denotes the maximum value of B_r.

Let l_{st} be the stack length, and R_r and R_s the rotor outer and stator inner radii, respectively. Note also that $R_s = R_r + g$. Then, it follows from (11.38) and (11.39) that

$$
\begin{aligned}
T_e &= \frac{1}{2\mu_0} \frac{\partial}{\partial \theta_r} \int_V [F_s(\alpha, \theta_r)\Lambda_g(\alpha, \theta_r) + B_r(\alpha, \theta_r)]^2 dV \\
&= \frac{l_{st}}{2\mu_0} \frac{\partial}{\partial \theta_r} \int_{R_r}^{R_s} \int_0^{2\pi} [F_s(\alpha, \theta_r)\Lambda_g(\alpha, \theta_r) + B_r(\alpha, \theta_r)]^2 r \, dr \, d\alpha \\
&= \frac{l_{st}(R_s^2 - R_r^2)}{4\mu_0} \int_0^{2\pi} [F_s\Lambda_g + B_r]\left(\frac{\partial F_s}{\partial \theta_r}\Lambda_g + F_s\frac{\partial \Lambda_g}{\partial \theta_r} + \frac{\partial B_r}{\partial \theta_r}\right) d\alpha . (11.44)
\end{aligned}
$$

Note that $(F_s\Lambda_g + B_r)\frac{\partial F_s}{\partial \theta_r}\Lambda_g = 0$ since $\frac{\partial F_s}{\partial \theta_r} = 0$, i.e, a change of rotor angle does not affect the stator MMF. On the other hand,

$$
\begin{aligned}
(F_s\Lambda_g + B_r)\frac{\partial B_r}{\partial \theta_r} &= \left(\frac{3}{2} N_{ph} I_d \cos(\alpha - \theta_r)\Lambda_g + \frac{3}{2} N_{ph} I_q \sin(\alpha - \theta_r)\Lambda_g \right. \\
&\qquad \left. + \overline{B}_r \cos(\alpha - \theta_r)\right) \overline{B}_r \sin(\alpha - \theta_r) \\
&\Rightarrow \frac{3}{2} N_{ph} I_q \overline{B}_r \sin^2(\alpha - \theta_r)\Lambda_g \\
&= \frac{3}{2} N_{ph} I_q \overline{B}_r \frac{1}{2}(1 - \cos 2(\alpha - \theta_r))\mu_0(\gamma_0 - \gamma_2 \cos 2(\alpha - \theta_r)) \\
&\Rightarrow \frac{3}{2} N_{ph} I_q \overline{B}_r \frac{\mu_0}{2}(\gamma_0 + \gamma_2 \cos^2 2(\alpha - \theta_r)) \\
&\Rightarrow \frac{3}{2} N_{ph} I_q \overline{B}_r \frac{\mu_0}{2}(\gamma_0 + \frac{\gamma_2}{2}) \qquad (11.45)
\end{aligned}
$$

Now consider

$$
\begin{aligned}
B_r F_s \frac{\partial \Lambda_g}{\partial \theta_r} &= \overline{B}_r \cos(\alpha - \theta_r)\left(\frac{3}{2}N_{ph}I_d\cos(\alpha - \theta_r)\Lambda_g + \frac{3}{2}N_{ph}I_q\sin(\alpha - \theta_r)\right) \\
&\quad \times (-2\mu_0\gamma_2\sin 2(\alpha - \theta_r)) \\
&\Rightarrow -\frac{3}{2}N_{ph}I_q\overline{B}_r\frac{\mu_0}{2}\frac{\gamma_2}{2}.
\end{aligned}
\tag{11.46}
$$

Finally,

$$
\begin{aligned}
F_s^2\Lambda_g\frac{\partial \Lambda_g}{\partial \theta_r} &= -\frac{\mu_0}{g}\left(\frac{3}{2}N_{ph}I_d\cos(\alpha - \theta_r) + \frac{3}{2}N_{ph}I_q\sin(\alpha - \theta_r)\right)^2 2\mu_0\gamma_2\sin 2(\alpha - \theta_r) \\
&\Rightarrow -2\mu_0^2\frac{\gamma_2}{g}\left(\frac{3N_{ph}}{2}\right)^2 I_dI_q\sin^2 2(\alpha - \theta_r) \\
&\Rightarrow -2\mu_0^2\gamma_2\left(\frac{3N_{ph}}{2}\right)^2 I_dI_q\sin^2 2(\alpha - \theta_r)(\gamma_0 - \gamma_2\cos 2(\alpha - \theta_r)) \\
&\Rightarrow -2\mu_0^2\gamma_2\left(\frac{3N_{ph}}{2}\right)^2 I_dI_q\frac{\gamma_0}{2}.
\end{aligned}
\tag{11.47}
$$

Note that $R_s^2 - R_r^2 \approx (R_r + g_0)^2 - R_r^2 = 2g_0R_r$, where $g_0 = \frac{4}{\pi}\frac{1}{\gamma_0}$. The effective air gap seems to be larger than $\frac{1}{\gamma_0}$ by a factor of $\frac{4}{\pi} = 1.273$. Thus, it follows that

$$
\begin{aligned}
\frac{l_{st}(R_s^2 - R_r^2)}{4\mu_0}\int_0^{2\pi}\left[(F_s\Lambda_g + B_r)\frac{\partial B_r}{\partial \theta_r} + B_rF_s\frac{\partial \Lambda_g}{\partial \theta_r}\right]d\alpha &= l_{st}\frac{R_rg_0}{2}\frac{3}{2}N_{ph}I_q\overline{B}_r\pi\gamma_0 \\
&= \frac{3}{2}\underbrace{\frac{2}{\pi}\overline{B}_rl_{st}\pi R_rN_{ph}}_{=\psi_m} I_qg_0\gamma_0\frac{\pi}{4} \\
&= \frac{3}{2}I_q\psi_m.
\end{aligned}
\tag{11.48}
$$

In the above, $\frac{1}{\pi}\int_{-\pi/2}^{\pi/2}\overline{B}_r\cos\alpha\,d\alpha = \frac{2}{\pi}\overline{B}_r$ is used. On the other hand, it follows from (11.47) that

$$
\begin{aligned}
\frac{l_{st}(R_s^2 - R_r^2)}{4\mu_0}\int_0^{2\pi}F_s^2\Lambda_g\frac{\partial \Lambda_g}{\partial \theta_r}d\alpha &= -\frac{l_{st}2R_rg_0}{4\mu_0}2\pi\times 2\mu_0^2\gamma_2\left(\frac{3N_{ph}}{2}\right)^2 I_dI_q\frac{\gamma_0}{2} \\
&= -\frac{3}{2}3(\mu_0l_{st}\frac{R_r}{2}\gamma_2N_{ph}^2)I_dI_q\underbrace{g_0\gamma_0}_{=\frac{4}{\pi}} \\
&= -\frac{3}{2}3(\frac{4}{\pi^2}\mu_0l_{st}\frac{\pi R_r}{2}\gamma_2N_{ph}^2)I_dI_q \\
&= -\frac{3}{2}3\underbrace{\frac{4}{\pi^2}\mu_0\frac{l_{st}\tau_p}{p}\gamma_2N_{ph}^2}_{=L_\delta} I_dI_q \\
&= -\frac{3}{2}(L_q - L_d)I_dI_q.
\end{aligned}
\tag{11.49}
$$

Note that the winding factor is assumed to be $k_w = 1$ in (6.28). Recall from (6.39) and (6.40) that $L_q - L_d = \frac{3}{2}L_\delta$. Finally, it follows from (11.44), (11.48), and (11.49) that

$$T_e = \frac{3}{2}(\psi_m I_q + (L_d - L_q)I_d I_q). \tag{11.50}$$

Observe that (11.50) is identical to the classical torque equation (6.61) for $P = 2$.

11.5.1   Torque Ripple

Previously, only the fundamental components of MMF and saliency were utilized to derive the average torque. In reality, most motors show a periodic change of shaft torque during the operation. It is triggered by the interaction between the MMF and the air gap flux harmonics. Also, it can be excited by mechanical unbalance or eccentricity of the rotor. The torque ripple causes unwanted noise and vibration in the electric machine. The crucial factors that affect the ripple include the numbers of stator slots and poles, the magnet angle, and the slot opening width. In the case of PMSM rotor, affecting factors are the PM arrangement, dimension, angle, width of the bridge, and flux barriers around the magnets. So, extensive design studies are repeated to achieve the ripple specification.

The gap field can be written as a product of the air gap permeance and MMFs:

$$B(\theta_r, \alpha, t) = \Lambda_g(\alpha, \theta_r)\left[F_N(\alpha)I(t) + F_r(\alpha, \theta_r)\right], \tag{11.51}$$

where F_N is the stator MMF per ampere and F_r is the rotor MMF. Fig. 11.29 depicts

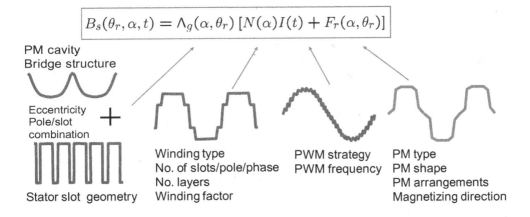

Figure 11.29: Sources for field harmonics.

the sources of field harmonics: The air gap permeance, Λ_g reflects all reluctance elements in stator and rotor. It includes the effects of stator slotting, PM cavity, eccentricity, bridge structure, and pole-slot combination. Fig. 11.30 shows bridge and cavity of a PMSM rotor. The rotor surface looks smooth with the naked eye. But the bridges are saturated by the PM field, so that the rotor surface is seen grooved in the perspective of field flow. By the same reasoning, PM cavity and other cavities affect the gap permeance. Including all reluctance components, a general expression is given by

$$\Lambda_g(\alpha) = \Lambda_{g0}\left[1 + \sum_{m=1,2,3,\cdots}^{\infty} A_m \cos(mN_s\alpha)\right],$$
(11.52)

where $\Lambda_{g0} = \frac{\mu_0}{g}$ and N_s is the number of slots.

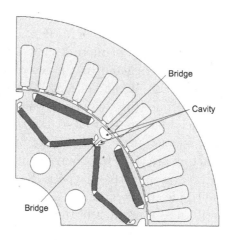

Figure 11.30: Cavity and bridge of a IPMSM.

The stator MMF, F_N depends on the winding types; the number of slots per pole per phase, the number of layers, and the winding factor. Current, I contains the PWM ripple, or other harmonics caused by inverter nonlinearity.

The rotor MMF, $F_r(\alpha, \theta_r)$ depends on magnet shape, magnetizing direction, and arrangement. Differently from the stator MMF, it is a function of rotor angle, θ_r.

$$F_r(\alpha, \theta_r) = \sum_{k=1,3,5,\cdots}^{\infty} F_{rk} \cos(kp(\alpha - \theta_r) + \phi_k),$$
(11.53)

where p is the number of pole pairs.

To minimize the torque ripple, some machines are designed in a slotless type. Most machines utilize skewing in the stator or rotor slots. Multi-step rotor skewing is used generally for PM machines [20]. Torque ripple at the peak torque is required to be less than 5% in EV applications [21]. Fig. 11.31 shows FEM simulation results of torque ripple before and after half slot step skewing. The ripple was reduced from 23% to 4.5% via skewing.

11.5.2 Cogging Torque

Cogging torque in PMSM is caused by the permeance change seen by the PM field. In particular, the PM field jumps between the stator slots when the rotor is rotating.

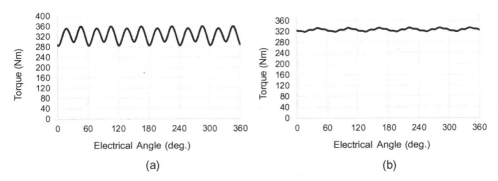

Figure 11.31: Torque ripple of an 8 pole, 48 slot, 100kW IPMSM at 3000 rpm, $I_s = 452$ A, $\beta = 45°$: (a) before and (b) after a half slot step skewing.

In other words, the PM flux is modulated as the rotor PM field slides over stator teeth. It is felt as periodic pulsating load when we turn the motor shaft by hand. Therefore, it is nothing to do with the stator current.

It will be shown later that the least common multiple of poles and slot numbers determines the minimum frequency of cogging torque. To prove this, B_r is described as a product of a rotor MMF and the permeance function:

$$B_r(\alpha, \theta_r) = F_r(\alpha, \theta_r)\Lambda_g(\alpha), \qquad (11.54)$$

where $\Lambda_g(\alpha)$ is already defined in (11.52). Note that the permeance function does not vary with respect to the rotor angle, i.e., it is not a function of θ_r, which is different from the case of stator field.

Let

$$F_r(\alpha, \theta_r) = \frac{g}{\mu_0} \sum_{k=1}^{\infty} B_{rk} \cos(pk(\alpha - \theta_r) + \phi_{rk}). \qquad (11.55)$$

The cogging torque is equal to

$$
\begin{aligned}
T_{cog} &= \frac{1}{2\mu_0} \frac{\partial}{\partial \theta_r} \int_V F_r^2(\alpha, \theta_r)\Lambda_g^2(\alpha)dV \\
&= \frac{l_{st}}{2\mu_0} \frac{\partial}{\partial \theta_r} \int_0^{2\pi} \int_{R_r}^{R_s} F_r^2(\alpha, \theta_r)\Lambda_g^2(\alpha)rdrd\alpha \\
&= \frac{l_{st}(R_s^2 - R_r^2)}{4\mu_0} \int_0^{2\pi} 2F_r(\alpha, \theta_r)\frac{\partial F_r}{\partial \theta_r}\Lambda_g^2(\alpha)d\alpha \\
&= \frac{l_{st}R_r g^2}{2\mu_0^2} \int_0^{2\pi} 2\left(\sum_{k=1}^{\infty} B_{rk} \cos(pk(\alpha - \theta_r) + \phi_{rk})\right) \\
&\quad \times \left(\sum_{l=1}^{\infty} B_{rl}pl \sin(pl(\alpha - \theta_r) + \phi_{rl})\right)\left[1 + \sum_m A_m \cos(mN_s\alpha)\right]^2 d\alpha
\end{aligned}
$$
$$\qquad (11.56)$$

After taking the integration over one complete period, many terms will vanish. Let $m' = LCM(2p, N_s)$. Then for $k = m'/2p$ and $m = m'/N_s$, some terms of (11.56) result in

$$\int_0^{2\pi} \left(\sum_k B_{rk}^2 pk \sin(2pk(\alpha - \theta_r) + 2\phi_{rk}) \right) \left(2 \sum_m A_m \cos(mN_s\alpha) \right) d\alpha$$

$$\Rightarrow \int_0^{2\pi} \sum_{m'} m' B_{r\frac{m'}{2p}}^2 A_{\frac{m'}{N_s}} \sin(m'(\alpha - \theta_r) + 2\phi_{rk}) \cos(m'\alpha) d\alpha$$

$$= \int_0^{2\pi} \sum_{m'} m' B_{r\frac{m'}{2p}}^2 A_{\frac{m'}{N_s}} \frac{1}{2} \left[\sin(m'(2\alpha - \theta_r) + 2\phi_{rk}) + \sin(-m'\theta_r + 2\phi_{rk}) \right] d\alpha$$

$$= \pi \sum_{m'} m' B_{r\frac{m'}{2p}}^2 A_{\frac{m'}{N_s}} \sin(-m'\theta_r + 2\phi_{rk}). \tag{11.57}$$

Note that the last term (11.57) is a nonzero function that alternates m' times per rotor revolution. It proves that $LCM(N_s, P)$ determines the lowest order cogging torque. When $P/GCD(N_s, P) \geq 1$ and f is the input frequency, the period and the fundamental frequency of cogging torque are specified as

$$\alpha_c = \frac{360°}{LCM(N_s, P)} \quad \text{and} \quad f_c = 2\frac{LCM(N_s, P)}{P} f. \tag{11.58}$$

There are many ways to reduce the cogging torque. See for example [21]–[24]. But the basic principle is to increase the LCM. Then, the amplitude of cogging torque is minimized. Fig. 11.32 shows a FEM simulation of the cogging torque before skewing. Since $P = 8$ and $N_s = 48$, it follows that $LCM(N_s, P) = 48$ and $\alpha_c = 7.5°$. The corresponding ripple period is 30° in electrical angle as shown in Fig. 11.32 (a). Fig. 11.32 (b) and (c) show the flux vector change in the stator teeth when the cogging torque is peak and zero, respectively.

11.5.3   Radial Force Analysis

Noise and vibration can be produced by the ripple component in the air gap flux density. It is caused by the slotted structure of the stator. The traveling electromagnetic force excites structural vibration of the stator yoke or teeth, and results in acoustic noise. This noise can be amplified highly if it resonates with the structural modes of the machine. Electromagnetically induced acoustic noise is characterized by the distinct tonalities due to the strong harmonic nature of magnetic forces in both time and space domains. Thus, the electromagnetic noise sounds unpleasant unlike broad-band noises. The noise can be reduced by designing the machine robust. However, it leads to a large volume or heavy weight. The acoustic noise and vibration are an important design issue in the optimization of electrical machines.

According to the Maxwell stress tensor method, the force densities are

$$P_r = \frac{1}{2\mu_0}(B_r^2 - B_\alpha^2), \tag{11.59}$$

$$P_\alpha = \frac{1}{\mu_0} B_r B_\alpha, \tag{11.60}$$

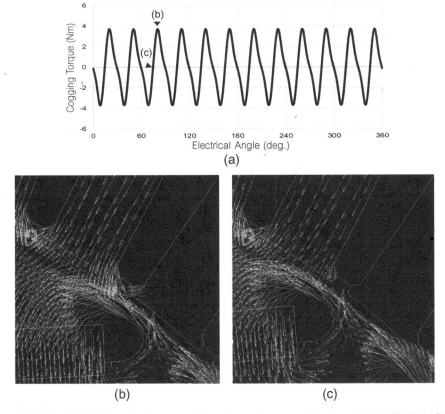

Figure 11.32: FEM simulation results of an 8 pole, 48 slot, 100 kW IPMSM: (a) cogging torque and flux vectors at (b) peak and (c) zero torque.

where P_r and P_α are the radial and tangential force densities (N/m^2), and B_r and B_α are the radial and tangential magnetic field densities, respectively. The normal force deflects the stator yoke inwardly and outwardly, while the tangential force causes the teeth to lean back and forth in the circumferential direction. In the radial flux machines, radial forces are the major power source for the yoke vibration. Usually, the tangential forces have minor effect on noise and vibrations, although it may slightly increase or decrease the amplitude of the first harmonic order radial displacement [25], [45].

Note that the flux direction is normal to the surface at the boundary of a high permeable material. Therefore B_α is relatively small, as is shown in Fig. 11.33. Hence, only the normal component is considered in the following:

$$B_r(\alpha, t) = B_{rs}(\alpha, t) + B_{rr}(\alpha, t) \qquad (11.61)$$

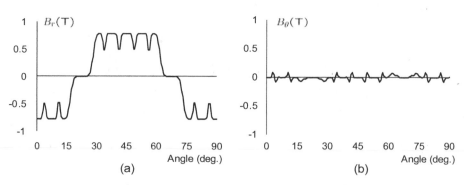

Figure 11.33: Air gap flux density: (a) radial component and (b) tangential component.

The normal field components are modeled as

$$B_{rs}(\alpha, t) = \sum_{\ell=1}^{\infty} B_{s\ell}^m \cos(\ell p \alpha \mp p\omega_r t) \quad \text{for stator} \tag{11.62}$$

$$B_{rr}(\alpha, t) = \sum_{k=1}^{\infty} B_{rk}^m \cos(k p \alpha \mp k p\omega_r t + \phi_k) \quad \text{for rotor.} \tag{11.63}$$

The normal force density is equal to

$$P_r = \frac{1}{2\mu_0} B_r^2 = \frac{1}{2\mu_0} \sum_{\ell=1}^{\infty} \sum_{k=1}^{\infty} \left[B_{rs}^2(\alpha, t) + 2B_{rs}(\alpha, t)B_{rr}(\alpha, t) + B_{rr}^2(\alpha, t) \right], \tag{11.64}$$

where

$$\frac{B_{rs}^2(\alpha, t)}{2\mu_0} = \frac{(B_{s\ell}^m)^2}{4\mu_0} \left[1 + \cos(2\ell p \alpha \mp 2p\omega_r t) \right], \tag{11.65}$$

$$\frac{B_{rr}^2(\alpha, t)}{2\mu_0} = \frac{(B_{rk}^m)^2}{4\mu_0} \left[1 + \cos(2k p \alpha \mp 2k p\omega_r t + 2\phi_k) \right], \tag{11.66}$$

$$\frac{B_{rs}^m B_{rr}^m}{\mu_0} = \frac{B_{s\ell}^m B_{rk}^m}{\mu_0} \left[\cos(p(k - \ell)\alpha \mp (k - 1)p\omega_r t - \phi_k) \right.$$
$$\left. + \cos(p(k + \ell)\alpha \mp (k + 1)p\omega_r t + \phi_k) \right]. \tag{11.67}$$

It shows that the radial (and tangential) stress originates from stator-to-stator, stator-to-rotor, and rotor-to-rotor interactions. Note that constant terms are neglected in the vibration analysis, since they represent uniform static magnetic pressure. Based on (11.64)–(11.67), the general form for the traveling magnetic force ·density is obtained as

$$P_r = \sigma_{n\bar{r}} \cos(\bar{r}\alpha \mp 2\pi n f_r t + \phi_{n\bar{r}}), \tag{11.68}$$

where $\bar{r}$ is the spatial order, $f_r = \omega_r/(2\pi)$, and n is the time-harmonic number. Note that the pressure density varies due to the interaction between the harmonics of stator and rotor fields.

The spatial order is often referred to as *wave number*. Stress waves propagate around the stator bore at the speed of $\omega_r/\bar{r}$. However, since a pair of waves travel in the opposite directions, standing wave takes place for each $\bar{r}$. The circumferential distribution of Maxwell force for $\bar{r} = 0, 1, 2, 3, 4$, is shown in Fig. 11.34.

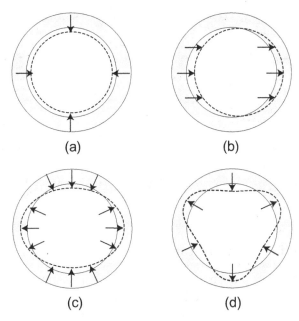

(a)   (b)

(c)   (d)

Figure 11.34: Vibration mode by normal force: (a) $\bar{r} = 0$ (breathing mode), (b) $r = 1$ (pulsating mode), (c) $\bar{r} = 2$ (oval mode), (d) $\bar{r} = 3$.

For $\bar{r} = 0$, the radial force density will be uniform concentric, so the the vibration mode pulsates with $P_r = \sigma_{n0} \cos n\omega_r t$. For $\bar{r} = 1$, the force density appears to be magnetic pull acting on the rotor (shaft) according to $P_r = \sigma_{n1} \cos(\alpha - n\omega_r t)$. It is based on the stator-to-rotor interaction (11.67) and occurs when the stator and rotor pole pair numbers differ by one. The case with $\bar{r} = 2$ is called oval mode: The stator is pressed in one direction, while extended in the other.

Note that the radial displacement of the yoke in response to the $\bar{r}^{th}$ order radial force wave is give by

$$\Delta d \propto F_{r\omega} \frac{l_{st} R_{av}}{E \bar{r}^4 h_{bc}^3}, \qquad r > 0, \tag{11.69}$$

where $F_{r\omega}$ is the magnitude of the r-th order radial force wave at frequency, $\omega/(2\pi)$, l_{st} is the stack length, R_{av} is the yoke average radius, E is the equivalent Young's modulus of yoke, and h_{bc} is the height of back iron (yoke) [26]. Note that the stator deformation is inversely proportional to the fourth power of mode order $\bar{r}$. That is, the magnetic pressure drops drastically as the vibration mode order, r increases. Therefore, low order vibration mode such as $\bar{r} = 1, 2$ should be avoided in the real motor design.

Considering a normal pressure density model, the peak value of radial force in the time and space is plotted for $(\bar{r}, f_r) = (2, 360 \text{ Hz})$ and $(\bar{r}, f_r) = (-2, 360 \text{ Hz})$ as shown in Fig. 11.35.

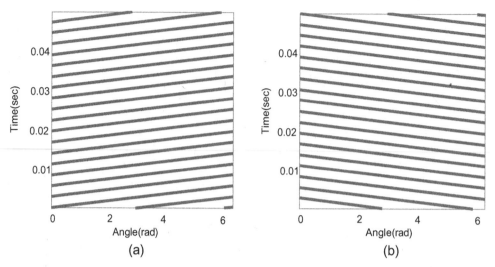

Figure 11.35: Time-space plots: (a) $(\bar{r}, f_r) = (2, 360 \text{ Hz})$ and (b) $(\bar{r}, f_r) = (-2, 360 \text{ Hz})$.

Vibration Modes for Slotted Stator

In the previous section, the normal force resulted from the interaction between the stator and rotor field was considered. In this part, we are focusing at the gap field change due to the stator slots when the rotor PM rotates. In the PM synchronous motor, the PM field is modulated by the stator slots. Such a modulated field component is obtained by multiplying the rotor MMF by the slot permeance [27]:

$$
\begin{aligned}
B_r &= F_r(\alpha, \omega_r t)\Lambda_{sl}(\alpha) \\
&= \left(\sum_{k=1,3,5,}^{\infty} F_{rk} \cos(kp(\alpha \mp \omega_r t) + \phi_k) \right) \left(\Lambda_0 \sum_{m=1,2,3,}^{\infty} A_m \cos(mN_s\alpha) \right) \\
&= \frac{\Lambda_0}{2} \sum_{k=1,3,5,} \sum_{m=1,2,3,} F_{rk} A_m \cos\{(kp \mp mN_s)\alpha \pm kp\omega_r t + \phi_k\}. \quad (11.70)
\end{aligned}
$$

Thus, the $(k, m)^{th}$ component of magnetic pressure is equal to

$$
\begin{aligned}
P_{r(km)} &= \frac{1}{2\mu_0} \frac{\Lambda_0^2}{4} [F_{rk} A_m \cos\{(kp \mp mN_s)\alpha \pm kp\omega_r t + \phi_k\}]^2 \\
&= \frac{1}{8\mu_0} \frac{\Lambda_0^2}{4} F_{rk}^2 A_m^2 [1 + \cos\{2(kp \mp mN_s)\alpha \pm 2kp\omega_r t + 2\phi_k\}].(11.71)
\end{aligned}
$$

It is important to consider low order magnetic pressure $\bar{r} = 2(kp \mp mN_s) = 2, 4, 6,$ and 8 for the stator permeance harmonic number, $m = 1$. Then, it follows that

$$
k = \frac{|0.5\bar{r} \mp N_s|}{p} = 1, 3, 5, \cdots \quad (11.72)
$$

When the motor is spinning in accordance with electrical frequency, f_e, the vibration exciting frequency corresponding to rotor harmonic order, k is obtained as $f_{ex} =$

$2kpf_e$. Note that no even harmonic number in (11.72) due to the half symmetry of the rotor MMF.

Consider 18 slot−8 pole motor running at 600 rpm. Note that the electrical frequency is $f_e = 40$ Hz. For $\bar{r} = 4$, then $k = |0.5 \times 4 + 18|/4 = 5$ is an odd integer. Thus, vibration is excited at frequency $f_{ex} = 2 \times 5 \times 40 = 400$ Hz.

Exercise 11.9   Consider 48 slot−8 pole motor running at 3000 rpm. Find the minimum feasible wave number and the exciting frequency.

Solution.   The electrical frequency is $f = 200$ Hz. For $\bar{r} = 8$, then $k = |0.5 \times 8 + 48|/4 = 11$ or 13. Correspondingly the vibration exciting frequency is $f_{ex} = 2 \times 11 \times 200 = 2200$ Hz or 2600 Hz. ∎

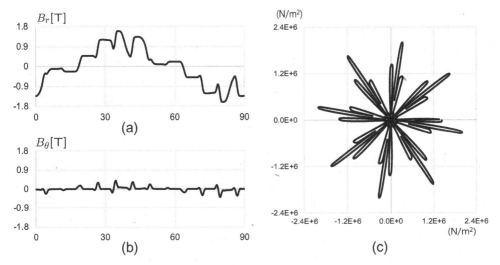

Figure 11.36: (a) B_r, (b) B_α, and (c) force density of an 8 pole, 48 slot, 100 kW IPMSM.

Fig. 11.36 shows FEM simulation results of the radial and tangential flux density of an 8 pole, 48 slot, 100 kW IPMSM, and the normal force density around the stator bore $(0 \sim 360°)$. In Fig. 11.36 (c), the number of spikes is eight which coincides with the vibration mode number $r = 8$ predicted by (11.72).

Table 11.7: Comparison of the motor quality indexes.

$N_s/(2 \cdot p)$	q	LCM($N_s, 2p$)	GCD($N_s, 2p$)
9/8	3/8	72	1
12/10	2/5	60	2
18/12	1/2	36	6
12/14	2/7	84	2
18/16	3/8	144	2
36/32	3/8	288	4

Another criterion is the greatest common divisor (GCD) between the slot and pole numbers. Specifically, $GCD(N_s, 2p)$ is associated with the order of radial force that leads to vibration. The GCD indicates the mode of forcing function. A high GCD means a high order balanced forces, which is less likely to excite the stator vibration. Likewise, a low value of the GCD generates magnetic noises easily. To avoid these negative effects, the GCD has to be an even number and also made as high as possible [28]. Considering all, it is desired to find a slot-pole combination which has high winding factor, high LCM, and high GCD [29]. Table 11.7 summarizes combinations of a number of slots per pole and phase $1/4 < q < 1/2$ for fractional slot concentrated coil motors. The attractive slot-pole combinations are marked by bold faces. A widely used fractional slot machine is 12 slot−10 pole motor. The 9 slot−8 pole combination is a good choice since it has a large winding factor and a high LCM. But, a 9 slot−8 pole machine cannot be used practically since the odd slot number will cause unbalanced magnetic pull and excessive noise and vibration [10]. But, this problem can be overcome by doubling or quadrupling the slot and pole numbers, i.e., 18 slot−16 pole, or 36 slot−32 pole [11], [30]. The duplicated version has the same desirable features as 9 slot−8 pole machine without creating the magnetic pull. Nowadays, fractional slot PM machines are widely used in high torque machines where a high pole number is needed. In-wheel motor is a typical application area [30].

11.5.4   Back Iron Height and Pole Numbers

Under the assumption of linearity, the maximum torque has nothing to do with the pole numbers as far as the rotor volume is the same. Consider, for example, 2 and 4 pole motors that have the same rotor diameter and the same stack length. Assume further that both of them have the same peak PM field densities as shown in Fig. 11.37. Then the sum of positive (or negative) areas are the same, which in turn means that the stored gap field energy is the same. To show the torque identity, assume the following for both cases: number turns per phase $= N_c$; rotor diameter $= D_r$; stack length $= l_{st}$; winding factor $= k_w$; phase current $= I$; shaft speed $= \omega_r$. The phase windings are connected in series in the 4 pole motor. The computation details are shown in Table 11.8. The flux linkage per pole is halved in the 4 pole motor, but the EMF is the same since the phase coils are connected in series. Finally, the gap power and torque are the same independent of the pole number.

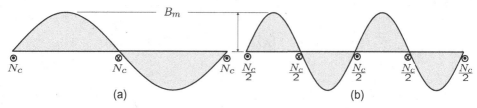

Figure 11.37: Comparison of 2 and 4 pole motors.

Table 11.8: Comparison of 2 and 4 pole motors

	2 Pole	4 Pole	Comparison
Flux linkage	$k_w \frac{\pi D_r}{2} l_{st} \frac{\pi}{2} B_m N_c$	$2 k_w \frac{\pi D_r}{4} l_{st} \frac{\pi}{2} B_m \frac{N_c}{2}$	$\lambda_{2pole} = 2\lambda_{4pole}$
EMF	$\frac{1}{\sqrt{2}} \frac{\pi^2}{4} k_w D_r l_{st} B_m N_c \omega_r$	$2 \times \frac{1}{\sqrt{2}} \frac{\pi^2}{8} k_w D_r l_{st} B_m N_c \omega_r$	$E_{f(2pole)} = E_{f(4pole)}$
Gap power	$3 \frac{1}{\sqrt{2}} \frac{\pi^2}{4} k_w D_r l_{st} B_m N_c \omega_r I$	$3 \frac{2}{\sqrt{2}} \frac{\pi^2}{8} k_w D_r l_{st} B_m N_c \omega_r I$	$P_{e(2pole)} = P_{e(4pole)}$
Torque	$3 \frac{1}{\sqrt{2}} \frac{\pi^2}{4} k_w D_r l_{st} B_m N_c I$	$3 \frac{1}{\sqrt{2}} \frac{\pi^2}{4} k_w D_r l_{st} B_m N_c I$	$T_{e(2pole)} = T_{e(4pole)}$

One clear benefit of high pole motor is the reduced back iron. In particular, the back iron width can be reduced as the pole number increases. Note that stator provides the return path for the rotor field and that the total flux per pole becomes less with a higher pole number. Correspondingly, the back iron width can be reduced as the pole number increases. It is depicted in Fig. 11.38. Basically, the back iron height is about a half of the pole pitch, $h_{bic} \approx \tau_p/2$, since a half of the pole flux passes through the back iron.

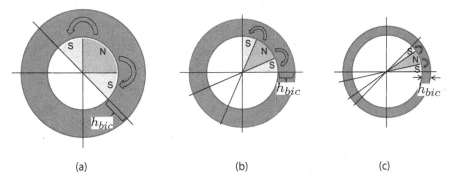

(a)   (b)   (c)

Figure 11.38: Decrease of back iron width with the pole number increase: (a) 4 pole, (b) 8 pole, and (c) 16 pole.

Exercise 11.10 Consider a case where the rated speed is determined by the speed where the open circuit voltage reaches its allowable maximum. Consider the 2 and 4 pole motors that have the same number of turns, the same rotor volume, and the same peak PM field density as shown in Fig. 11.37. Will the two motors have different rated conditions?

Solution. The two motors have the same rated torque at the same rated speed. ∎

11.6   Reluctance Motors

Variable reluctance motors rely solely on the reluctance component in torque generation. They receive attention constantly as an Nd-free solution. There are two types: One is switched reluctance motor (SRM) that has a windmill shape rotor. The other is synchronous reluctance motor (SynRM) which is driven by sinusoidal currents.

11.6.1   Switched Reluctance Motors

In SRMs, typical stator and rotor pole numbers are 6−4 or 12−8. Fig. 11.39 shows a 6−2 SRM at two different rotor positions. The inductance which is represented as the slope of flux versus current, changes with rotor angle. If the rotor pole is not aligned, the inductance is low. However, it increases as the rotor pole aligns with stator teeth.

The operation principle is to inject current when the inductance is low (misaligned state) and remove when the inductance is high (aligned state) [31]. In Fig. 11.39, current is increased in the misaligned state (path A). Thereafter, the rotor will rotate and a rotor pole will be aligned with a stator tooth (path B). Secondly, the current is reduced in the aligned position (path C). The above procedure completes a cycle and a triangle is formed in the flux-current plane. The shaded triangular area represents the energy delivered to the motor, and it is supposedly converted into mechanical energy.

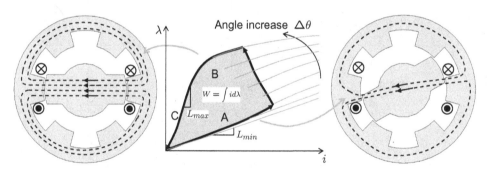

Figure 11.39: Working principle of SRM: current contour in the flux-current plane and the pole alignment states.

Fig. 11.40 shows an ideal case in which current is regulated to be a constant during the pole aligning period. In the second phase ②, the inductance is increased as the pole is aligned. At this time current is fixed, and the corresponding trajectory is depicted as a vertical line in Fig. 11.40 (a). In the third phase ③, current decreases in the high inductance (aligned) state.

Assume that the contour ② in Fig. 11.40 (a) is a vertical straight line. The increase in the co-energy is approximated as

$$W_c = \frac{1}{2}(L_{max} - L_{min})i_a^2. \tag{11.73}$$

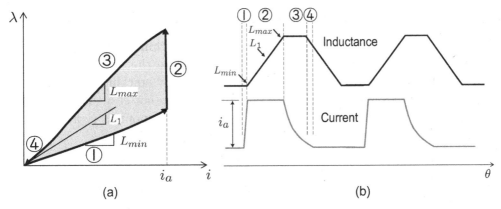

(a)   (b)

Figure 11.40: Working principle of SRM: (a) current contour in the flux-current plane, (b) plots of inductance and current versus rotation angle.

Then the phase voltage also remains high during in the contour ②:

$$v = \frac{d}{dt}L(\theta)i_a = \frac{\partial}{\partial\theta}L(\theta)\frac{d\theta}{dt}i_a = \omega\frac{\partial L(\theta)}{\partial\theta}i_a, \tag{11.74}$$

since $\frac{\partial L(\theta)}{\partial\theta} > 0$. That is, a positive power $vi_a > 0$ is supplied to the motor phase. Thereby, torque is obtained from (11.38) such that

$$T_e = \frac{\partial W_c}{\partial\theta} = \frac{i_a^2}{2}\frac{L_{max} - L_{min}}{\Delta\theta}. \tag{11.75}$$

Fig. 11.41 shows a typical power circuit for SRM inverter. As shown in Fig. 11.40, the currents are unipolar, thereby a normal inverter cannot be used for SRM. Non-sinusoidal current shape is a hurdle in current and torque regulation. In or-

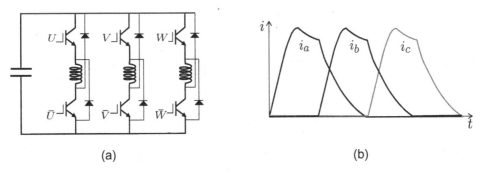

(a)   (b)

Figure 11.41: (a) SRM inverter power circuit and (b) phase currents.

der to make the torque density comparable with other type motors, the inductance variation should be maximized. For example, the saliency ratio needs to be $L_{max} : L_{min} = 6 : 1 \sim 8 : 1$ [32]. To meet such requirement, very narrow air gap (≈ 0.3 mm) is often utilized for small motors. But, it requires precise machining for brackets and precise shaft alignment. This increases manufacturing cost, which may offset the cost advantage resulted from PM elimination. Another limiting factors are poor power factor and severe noise problems.

11.6.2 Synchronous Reluctance Motors

Differently from the SRMs, synchronous reluctance machines (SynRM) are driven by sinusoidal current. There are two types depending on the way of providing the magnetic saliency. One method is to make the rotor by stacking the laminations axially as shown in Fig. 11.42 (a). The manufacturing is not easy since the shapes of laminations are not uniform. Fig. 11.42 (b) show a more popular type of SynRMs in which multiple layers of cavities are utilized to polarize L_d and L_q.

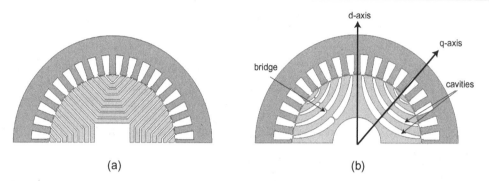

Figure 11.42: SynRM rotors: (a) axial stacking of laminated sheets and (b) multi-layered cavities.

Note that the way of setting magnet axes is different from that of PMSM. In the SynRM, it is common to align the d-axis in parallel with the cavities. Fig. 11.43 (a) and (b) show flux lines when the d and q axis currents are applied. One can see a clear difference in the flux line density: The flux lines are much denser than the d-axis current. Thereby, $L_d > L_q$ and the saliency ratio is normally $L_d/L_q = 6 \sim 7$. Without a PM, torque equation is equal to

$$T_e = \frac{3P}{4}(L_d - L_q)i_d^e i_q^e. \tag{11.76}$$

Since $L_d > L_q$, both d and q-axis currents must be positive to produce positive torque. Then voltage vector resides in the second quadrant according to a current vector in the first one. Fig. 11.43 (c) shows typical voltage and current vectors. It has a relatively large angle, $\phi > 0$ between current and voltage vectors. As a result, SynRMs suffer from a low power factor. For a given current magnitude, the maximum torque is obtained when $i_d^e = i_q^e$. Then the power factor angle is qual to

$$\phi = \tan^{-1}\left(\frac{L_q}{L_d}\right) + \frac{\pi}{4}. \tag{11.77}$$

It shows that the power factor increases as the saliency ratio increases. For example, $\cos\phi = \cos(1/6 + \pi/4) = 0.58$ for $L_d/L_q = 6$, and $\cos\phi = 0.6$ for $L_d/L_q = 7$.

11.6.3 PM Assisted Synchronous Reluctance Machine

A recipe for increasing the power factor is to add PMs in the cavities, as shown in Fig. 11.44 (a). Fig. 11.44 (b) is a corresponding vector diagram showing that

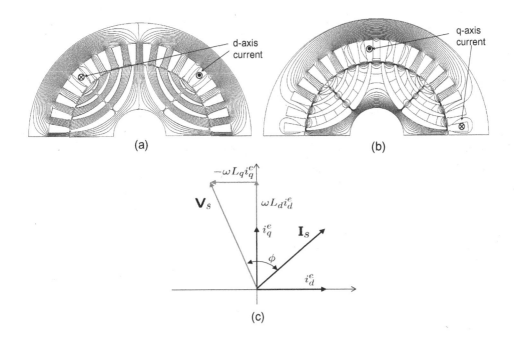

(a)   (b)

(c)

Figure 11.43: Synchronous reluctance machine: (a) d-axis flux lines, (b) q-axis flux lines, and (c) voltage and current vectors.

the voltage vector is shifted into the first quadrant with the help of PM, $\omega\psi_m$. It is commonly called *PM assisted SynRM* or *PMaSynRM*. Basically, PM assisted SynRM is identical to IPMSM. As the PM usage increases, the PM assisted SynRM converges to an IPMSM with a high saliency ratio.

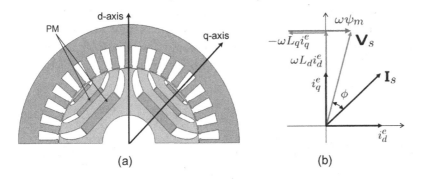

(a)   (b)

Figure 11.44: PM assisted SynRM: (a) stator and rotor core, (b) voltage and current vectors.

Exercise 11.11 Derive a voltage equation for PM assisted SynRM based on the axes set in Fig. 11.44. Find the power factor when $i_d^e = i_q^e$, $\xi = 7$, and $\psi_m = -L_q i_q^e$.

11.7   Motor Types Depending on PM Arrangements

The PMs provide gap field continuously without an external excitation circuit. Due to high energy density, the rare-earth PMs create a motor with high power density, high efficiency, and low moment of inertia. In the case of sintered Nd-Fe-B magnet, the remnant flux density is as high as $B_r = 1.4$ T, and the coercive force is $H_c = 2000$ kW/m. Another invaluable feature is low permeability. As shown in Chapter 6, that attribute is used to exploit the reluctance torque. Fig. 11.45 shows sectional views of 4 pole synchronous motors. The first two motors, PMs mounted on the rotor surface, are categorized as SPMSMs.

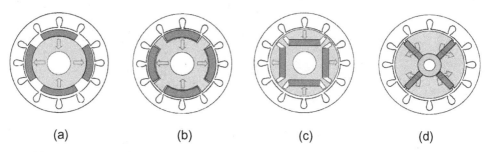

(a)	(b)	(c)	(d)

Figure 11.45: Cross-sectional views of 5 PMSMs (a) surface mounted PMSM, (b) inset PMSM, (c) interior PMSM, and (d) flux concentrating PMSM.

11.7.1   SPMSM and Inset SPMSM

SPMSM, as the name stands for, has magnets attached on the rotor surface. In small power and low speed applications, PMs are usually bonded on the rotor surface by some epoxy adhesives. In some cases, stainless band or glass carbon fiber band is used to fix the PMs, because the bond is not strong enough to endure high centrifugal force. If a stainless band is used for fixing the PMs, then eddy current is created by slot and MMF harmonics. The resulting loss on the band adds heat to the PMs.

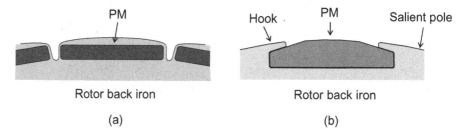

Figure 11.46: PM holding: (a) slight embedding structure and (b) hook design of an inset motor.

Since the PMs are located in the forefront of air gap where high order field harmonics are flourishing, SPMSMs are more susceptible to demagnetization than IPMSMs. To avoid such problem, the PMs are buried slightly in the core as shown in Fig. 11.46 (a). Note that the PM surface is covered with a thin steel strip. Since the steel is highly permeable (2500 ∼ 5000 μ_0), the PMs are partially shielded from harmful air gap harmonics. The covering steel also protects PM mechanically which is typically brittle. Hyundai 2012 Sonata and Honda 2005 Accord utilized such structure. Fig. 11.47 shows two types of rotors: surface mounted PM rotor and embedded PM rotor.

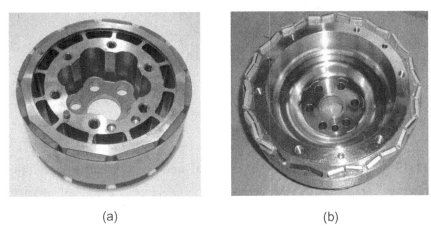

(a)   (b)

Figure 11.47: Rotors of HEV motor: (a) surface mounted and (b) embedded (V-shape) (Hyundai Motor Company).

To decrease the eddy current, PMs are segmented into multiple parts. The eddy current loss decreases inversely proportional to the square of lamination number. Fig. 11.48 shows an example of step skewing along with PM segmentation. The rotor is segmented into 6 pieces axially and assembled with a step skew.

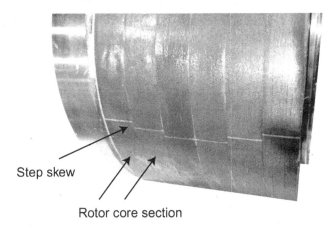

Step skew

Rotor core section

Figure 11.48: Step skewing and PM segmentation in the axial direction.

In some SPMSMs, PMs are inserted into the grooves of the rotor surface, as shown in Fig. 11.46 (b). Those types of motors are called inset PMSMs. The iron bumps between the magnets act as the leads to the q-axis flux. Therefore, the inset motors have the magnetic saliency as $L_q > L_d$. Honda traction motors used an inset structure in the past, and it was reported that the maximum reluctance torque gained by inset configuration amounted to 38% of the magnet torque [33].

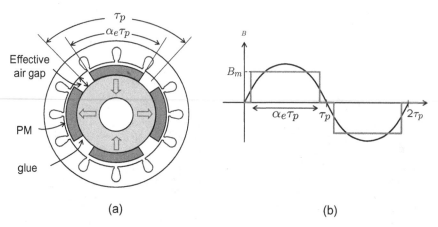

(a)   (b)

Figure 11.49: (a) Schematic view of SPMSM and (b) fundamental component of air gap field.

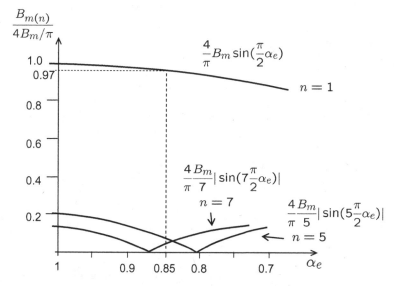

Figure 11.50: Harmonics of air gap field versus PM coverage (α_e): Harmonics comparision versus PM coverage α_e [34].

To make air gap field by PMs sinusoidal, a proper PM coverage should be considered. The PM coverage ratio, α_e is defined as the ratio of PM arc length to the pole pitch. Then, $\alpha_e = 1$ implies that all the rotor surface is covered by PMs. In that case, the fundamental peak will be $\frac{4}{\pi}B_m$ where B_m is the flux density on the surface of PM. Further similarly to (11.2), high order harmonic coefficients are obtained as

$$B_{m(n)} = \frac{4}{\pi}\frac{B_m}{n}\sin(n\alpha_e\frac{\pi}{2}), \qquad n = 1, 3, 5, \cdots . \tag{11.78}$$

Based on (11.78), calculation results of the fundamental, 5th and 7th order harmonics versus α_e are shown in Fig. 11.50. For $\alpha_e = 0.85$, 5th and 7th order harmonics reduce significantly, whereas the fundamental component is decreased only by 3% [34]. Hence, about 15% reduction in the PM coverage is good for reducing the harmonic contents, as well as the PM cost.

11.7.2 IPMSM

Since PMs are inserted in the cavities of the rotor core, no fixation device is necessary for IPMSMs. Since PMs hold a distance from the air gap inside the steel core, PMs are protected from the stator harmonic fields. However, the danger of demagnetization always exists due to the high electric loading, field harmonics, and inefficient cooling.

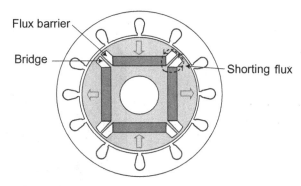

Figure 11.51: Bridge and flux barrier of IPMSM.

The PM flux is meaningful when it crosses the air gap and encircles the stator windings via the back iron of the stator core. But a certain amount of flux does not make such a flux link: Some flux makes a short path inside the rotor, and another flux returns after touching the stator teeth. These are called the leakage flux, and such phenomena are called the flux shorting [35]. Such leakage flux just wastes the PM potential, not contributing to torque production. To minimize the flux shorting, cavities are carved as flux barriers to penalize the leakage flux. However, bridges are required around cavity for mechanical integrity of rotor as shown in Fig. 11.51. The bridge width needs to be optimized between ensuring the mechanical strength and minimizing the leakage flux. The bridge width is 0.7~2 mm, but depends on the thickness the laminated core.

11.7.3 Flux Concentrating PMSM

Flux squeezing and flux focusing are the equivalent terms to flux concentrating. As the name suggests, the PM flux is concentrated in the air gap. It is an extreme case of V-shape arrangement. The idea is to arrange a set of adjacent PMs to face each other and drive all the flux to the air gap as shown in Fig. 11.52. For this purpose, the returning path through the center (shaft) area should be banned.

Thus, a fixture to the shaft is often made with the non-ferromagnetic material such as stainless steel or beryllium-copper.

When the arc of a rotor sector is smaller than twice the PM length, the air gap field density will be higher than that of the PM surface. This type of field concentration is commonly used with weaker magnets such as ferrite. It suits well the high pole machines. However, this topology has a structural drawback against centrifugal force due to the flux barriers around the shaft. Thus, it is suitable for low speed operation. Fig. 11.52 (b) shows a photo of a 10 pole flux concentrating generator being used for BMW i3. Note that the rotor is supported by thin bridges instead of a non-ferromagnetic material. A similar design and fabrication work has been reported in [1]. Additional concern is that the PMs are prone to be demagnetized by large operating current.

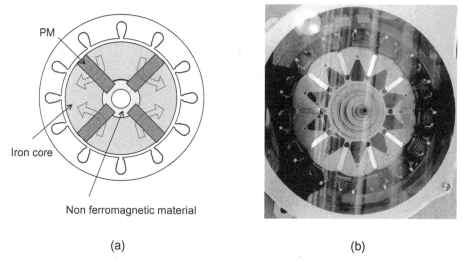

(a)   (b)

Figure 11.52: Flux concentrating PMSM: (a) Schematic view and (b) photo of 10 pole flux concentrating generator used for BMW i3.

11.7.4 Temperature Rise by Copper Loss

Iron and copper losses are the main sources of the temperature increase in the motor. The proportion of two losses changes depending on the supply voltage and the operating speed. In general, the copper loss is more dominant over the iron loss. The copper loss is equal to

$$P_{cu} = 2mPN_c \frac{l_{st} + l_{end}}{\kappa S_{cu}} I^2, \tag{11.79}$$

where l_{end} is the end-turn length, $2mPN_c$ is the total number of conductors in the slots, κ is the conductivity of the conductor, and S_{cu} is the cross sectional area of conductors. The current density of the conductor is defined by $J = \frac{I}{S_{cu}}$. Then,

$$P_{cu} = 2mPN_c \frac{l_{st} + l_{end}}{\kappa} JI = \pi D_r \frac{l_{st} + l_{end}}{\kappa} JA_{rms}, \tag{11.80}$$

where $A_{rms} = A_{max}/\sqrt{2}$ is the rms electric loading.

Note that heat flow, Q (W) causes a temperature gradient, ΔT ($°C$) such that $Q = \alpha_k S_k \Delta T$, where S_k is the area of heat transfer, and ΔT is the temperature difference between the coil surface and motor ambient. In addition, α_k is a heat transfer coefficient of the whole materials laid between the coil and motor outside. In the steady state, a thermal balance is maintained such that $P_{cu} = Q$. Assuming $S_k = \pi D_r(l_{st} + l_{end})$, it follows that

$$\Delta T = \pi D_r \frac{l_{st} + l_{end}}{\kappa \alpha_k S_k} J A_{rms} = \frac{1}{\kappa \alpha_k} J A_{rms}. \tag{11.81}$$

The equation (11.81) states that the temperature increase is proportional to the product of current density and electric loading, $J \times A_{rms}$.

Air cooled motors are designed such that $J = 3.5 \sim 7$ A/mm^2. If the cooling method is more efficient, a higher electric loading can be used. For example, $A_{rms} = 150$ A/cm for a totally enclosed PM servo motor without a fan. But with a water cooling, it can be increased to 700 A/cm. Typical range electric loadings are listed in Table 7.1.

References

[1] Zhu et al., "Instantaneous Magnetic Field Distribution in Brushless Permanent Magnet dc Motors, Part I: Open-circuit Field," *IEEE Trans. on Magn.*, vol. 29, no. 1, pp. 124−135, Jan. 1993.

[2] Z.Q. Zhu, K. Ng, N. Schofield, and D. Howe, "Improved analytical modelling of rotor eddy current loss in brushless machines equipped with surfacemounted permanent magnets," *IEE Proc.-Electr. Power Appl.*, vol. 151, no. 6, pp. 641−650, Nov. 2004.

[3] J. F. Gieras, *Electrical Machines: Fundamentals of Electromechanical Energy Conversion*, CRC Press, 2016.

[4] N. Bianchi, S. Bolognani, and M. D. Pre, "Magnetic Loading of Fractional-Slot Three-Phase PM Motors With Nonoverlapped Coils," *IECON*, pp. 3697−3703, 2011.

[5] N. Bianchi and M. Dai Pre, "Use of the star of slots in designing fractional-slot single-layer synchronous motors," *IEE Proc.-Electr. Power Appl.*, vol. 153, no. 3, pp. 459−466, 2006.

[6] G. Dajaku, S. Spas, X. Dajaku, and D. Gerling, "Comparison of Two FSCW PM Machines for Integrated Traction Motor/Generator," IEMDC, pp. 187−194, 2015.

[7] Y. Yokoi, T. Higuchi, and Y. Miyamoto, "General formulation of winding factor for fractional-slot concentrated winding design," *IET Electr. Power Appl.*, vol. 10, no. 4, pp. 231−239, 2016.

[8] L. Alberti, M. Barcaro, and N. Bianchi, "Design of a Low-Torque-Ripple Fractional-Slot Interior Permanent-Magnet Motor," *IEEE Trans. on Ind. Appl.*, vol. 50, no. 3, pp. 1801–1808, May 2014.

[9] N. Bianchi, S. Bolognani, M. D. Pre, and G. Grezzani, "Design Considerations for Fractional-Slot Winding Configurations of Synchronous Machines," *IEEE Trans. on Ind. Appl.*, vol. 42, no. 4, pp. 997–1006, July 2006.

[10] D. G. Dorrell., M. Popescu, and D. M. Ionel, "Unbalanced Magnetic Pull Due to Asymmetry and Low-Level Static Rotor Eccentricity in Fractional-Slot Brushless Permanent-Magnet Motors With Surface-Magnet and Consequent-Pole Rotors," *IEEE Trans. on Mag.*, vol. 46, no. 7, pp. 2675–2685, 2010.

[11] Z. Q. Zhu, Z.P. Xia, L. J. Wu, and G.W. Jewell, "Influence of Slot and Pole Number Combination on Radial Force and Vibration Modes in Fractional Slot PM Brushless Machines Having Single- and Doublelayer Windings," *IEEE ECCE*, pp. 3443–3450, 2009.

[12] D. Ishak, Z. Q. Zhu, and David Howe, "Permanent-magnet brushless machines with unequal tooth widths and similar slot and pole numbers," *IEEE Trans. on Ind, Appl.*, vol. 41, no. 2, pp. 584 – 590, Mar. 2005.

[13] D. Hanselman, *Brushless Permanent Magnet Motor Design*, 2nd. Ed., The Writers' Collective, Cranston, 2003.

[14] A. M. EL-Refaie, Z. Q. Zhu, T. M. Jahns, and D. Howe, "Winding Inductances of Fractional Slot Surface- Mounted Permanent Magnet Brushless Machines," *IEEE IAS Annual Meeting*, pp. 1–8, 2008.

[15] A.J. Mitcham, G. Antonopoulos, and J.J.A. Cullen, "Favourable slot and pole number combinations for fault-tolerant PM machines," *IEE Proc.-Electr. Power Appl.*, vol. 151, no. 5, pp. 520 – 525, Sep. 2004.

[16] M. Nakano, H. Kometani, M. Kawamura, "A study on eddy-current losses in rotors of surface permanent-magnet synchronous machines," *IEEE Trans. on Magnetics*, vol. 42, no. 2, pp. 429 – 435, Mar. 2006.

[17] D. Ishak, Z.Q. Zhu, D. Howe, "Eddy current loss in the rotor magnets of permanent-magnets brushless machines having a fractional number of slots per pole," *IEEE Trans. on Magnetics*, vol. 41, no. 9, pp. 2462–2469, Sep. 2005.

[18] C. Deak and A. Binder, "Design of compact permanent-magnet synchronous motors with concentrated windings," *Rev. Roum. Sci. Techn. Electrotechn. et Energ.*, vo. 52, no. 2, pp. 183–197, Bucarest, 2007.

[19] B. Koo, M. Lee, and K. Nam, "Groove depth determination based on extended leakage factor in a 12-slot 10-pole machine," *IEEE Int. Conf. on Mech.*, 2015.

[20] J. W. Jiang et al., "Rotor skew pattern design and optimisation for cogging torque reduction," *IET Electr. Syst. Transp.*, vol. 6, no. 2, pp. 126–135, 2016.

[21] Y. Yang, et al., "Design and Comparison of Interior Permanent Magnet Motor Topologies for Traction Applications," *IEEE Trans. on Transp. Electrification*, vol. 3, no. 1, pp. 86−96, Mar. 2017.

[22] D. Zarko, D. Ban, and T. Lipo, "Analytical Solution for Cogging Torque in Surface Permanent-Magnet Motors Using Conformal Mapping," *IEEE Trans. on Magnetics*, vol. 44, no. 1, pp. 52−65, Jan. 2008.

[23] L. Zhu, S. Z. Jiang, Z. Q. Zhu, and C. C. Chan, "Analytical Methods for Minimizing Cogging Torque in Permanent-Magnet Machines," *IEEE Trans. on Magn.*, vol. 45, no. 4, pp. 2023−2031, Apr. 2009.

[24] Y. Park, J. Cho, and D. Kim, "Cogging Torque Reduction of Single-Phase Brushless DC Motor With a Tapered Air-Gap Using Optimizing Notch Size and Position," *IEEE Trans. on Ind. Appl.* vol. 51, no. 6, pp. 4455−4463, Nov. 2015.

[25] S. Jia and R. Qu, "Analysis of FSCW SPM Servo Motor with Static, Dynamic and Mixed Eccentricity in Aspects of Radial Force and Vibration," IEEE ECCE, pp. 1745−1753, 2014.

[26] Shoedel, *Vibrations of Shells and Plates*, 3rd. Ed. Marcel Dekker 2004.

[27] J. F. Gieras and M. Wing, *Permanent Magnet Motor Technonology, Design and Applications*, 3rd. Ed., CRC Press, 2010.

[28] B. Aslan, E. Semail, J. Korecki, and J. Legranger, "Slot/pole Combinations Choice for Concentrated Multiphase Machines dedicated to Mild-Hybrid Applications," *IEEE Trans. on Ind. Appl.*, vol. 44, no. 5, pp. 1513−1521, Sep./Oct. 2008.

[29] E.Carraro, N. Bianchi, S. Zhang, and M. Koch, "Permanent Magnet Volume Minimization of Spoke Type Fractional Slot Synchronous Motors," IEEE ECCE, pp. 4180−4187, 2014.

[30] K. Reis and A. Binder, "Development of a permanent magnet outer rotor direct drive for use in wheel-hub drives," ICEM, pp. 2418−2424, 2014.

[31] *Switched Reluctance Motor*, http://www.intechopen.com/books/torque-control/switchedreluctance-motor, 2011.

[32] J. M. Miller, *Proportion Systems for Hybrid Vehicle*, IEE Power and Energy Series 4, IEE, London, 2004.

[33] H. Satoh, S. Akutsu, T. Miyamura, and H. Shinoki, "Development of traction motor for fuel cell vehicle," *SAE World Congress*, Detroit, no. 2004-01-0567, Mar. 2004.

[34] A. Binder, *Permanent Magnet Synchronous Machine Design*, Motor Design Symposium, Pohang, May 2008.

[35] N. Matsui, Y. Taketa, S. Morimoto, Y. Honda, "Design and Control of IPMSM," Ohmsha Ltd. (in Japanese), 2001.

[36] A. M. EL-Refaie, J. P. Alexander, S. Galioto, P. B. Reddy, K. Huh, P. de Bock, and X. Shen, "Advanced High Power-Density Interior Permanent Magnet Motor for Traction Applications," *IEEE Trans. on Ind. Appl.*, vol. 50, no. 5, pp. 3235–3248, Sep./Oct. 2014.

[37] K. Wang, Z. Q. Zhu, G. Ombach, M. Koch, S. Zhang, and J. Xu, "Electromagnetic performance of an 18-slot/ 10-pole fractional-slot surface-mounted permanent-magnet machine," *IEEE Trans. on Ind., Appl.*, vol. 50, no. 6, pp. 3685–3696, Nov./Dec. 2014.

[38] A. E. Fitzgerlad, C. Kingsley, Jr., and S. D. Umans, *Electric Machinery*, 5th Ed., McGraw Hill, New York, 2003.

[39] N. Bianchi, S. Bolognani, and F. Luise, "Potentials and limits of high-speed PM motors," *IEEE Trans. on Ind. Appl.*, vol. 40, no. 6, pp. 1570–1578, Nov. 2004.

[40] J. Cros and P. Viarouge, "Synthesis of high performance PM motors with concentrated windings," *IEEE Trans. on Energy Conv.*, vol. 17, no. 2, pp. 248 – 253, June 2002.

[41] J.R. Hendershot Jr. and T.J.E Miller, *Design of Brushless Pemanent-Magnet Motors*, Magna Physics Publishing and Clarendon Press, Oxford, 1994.

[42] S. Jurkovic et al., *Design, Optimization and Development of Electric Machine for traction application for GM battery electric vehicle*, IEEE ECCE, pp. 1814–1819, 2015.

[43] K. Rahman et al., "Retrospective of Electric Machines for EV and HEV Traction Applications at General Motors," IEEE ECCE, 2016.

[44] T. Jahns, "Getting rare-earth magnets out of EV traction machines," *IEEE Electrification Magazine*, vol. 5, no. 1, pp. 6–18, Mar. 2017.

[45] H.Y. I. Du, L. Hao, and H. Lin, "Modeling and Analysis of Electromagnetic Vibrations in Fractional Slot PM Machines for Electric Propulsion," IEEE ECCE, pp. 5077–5084, 2013.

[46] C. Hsiao et al., "A Novel Cogging Torque Simulation Method for Permanent-Magnet Synchronous Machines," *Energies* no. 4, pp. 2166–2179, 2011

[47] J. F. Gieras, C. Wang, and J.C. Lai, *Noise of Polyphase Electric Motors*, CRC Press, 2006.

Problems

11.1 Consider the MMF of a concentric winding shown below when $(i_a, i_b, i_c) = (-\frac{1}{2}, 1, -\frac{1}{2})$. Calculate the second order harmonics.

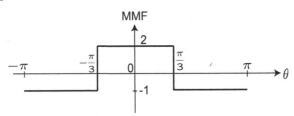

11.2 Consider a C-type magnet circuit with a PM in the center, shown below. Let B_m denote the flux density, H_m the magnetizing force, S_m the cross-sectional area, and ℓ_m the height of the PM. Let S_g be the air-gap area, g the air gap length, and B_g the air gap flux density.

a) Assume that the permeability of the core is infinity. Derive an equation for B_g from Ampere's law.

b) The permeance coefficient is defined as $P_c = \frac{B_m}{\mu_0 H_m}$. Assuming $S_m = S_g$, derive a simple expression for the permeance coefficient.

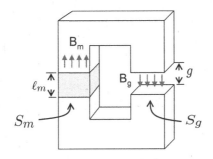

11.3 The stator current is modeled as a current sheet $A_m = \frac{2mNI}{P\tau_p}$ on the stator side, and the PM magnet has a coercive force H_c, and its field directs upward, as shown below. Let g and h_m be the air gap height and PM height, respectively.

a) Determine the rotor moving direction.

b) Applying the Ampere's law along a contour indicated by the loop, determine the air gap field intensity, H_g as a function of x.

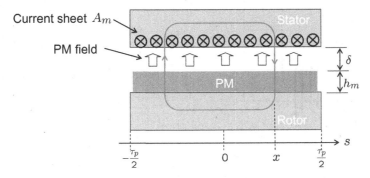

11.4 Consider the magnet circuit shown below. The PM (NdFeB) is assumed to be N40SH ($B_r = 1.2$ T, $H_c = -950$ kA/m). The PM cross-sectional area is $S_m = 10$ cm^2, and its height is 10 mm. The air gap area is $S_m = 10$ cm^2 and the air gap height is 1 mm.
a) Calculate the air gap field density when $i = 0$.
b) Calculate the air gap field density when $Ni = 3000$ A turn.

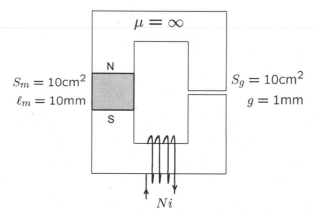

11.5 Consider a double layer 18 slot$-$10 pole motor [37]. Draw a star diagram, coil arrangements, and obtain the pitch factor.

11.6 Consider the demagnetization curve shown below: The basic parameters are $B_r = 1.2$ T, $H_c = 950$ kA/m and $g = 1$ mm.
a) When $H_{knee} = 808$ kA/m, find the corresponding field, B_{knee}.
b) When $NI = 700$ A-turn, find the minimum PM height, ℓ_m to prevent an irreversible demagnetization.
c) When $\ell_m = 10$ mm, what is the maximum feasible NI that does not incur an irreversible demagnetization?

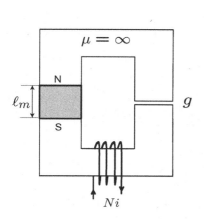
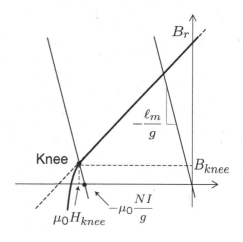

11.7 Consider three types fractional slot machines: (3 slot, 2 pole), (12 slot, 10 pole), and (36 slot, 32 pole).
a) Compare the GCD($N_s, 2p$) and LCM($N_s, 2p$) respectively.
b) Select a machine with the lowest cogging torque and a machine with the lowest noise and vibration.

11.8 Consider an ideal SRM which has 6 stator poles and 2 rotor poles. Assume that the inductance and current of a-phase winding vary with the rotor angle as shown below.
a) Find the maximum and minimum phase inductance from the flux to current contour shown below.
b) Calculate the change in co-energy W_c per a half period (180°) of a-phase winding.
c) Calculate the energy $\oint L(\theta)idi(\theta)$ and show that it is equal to the co-energy change obtained in b).
d) Plot the torque waveform of a-phase coil for a half period.
e) Plot the inductance and current waveform for b and c phases. Also plot the shaft torque.
f) Using the result of e), find the mechanical power at 1200 rpm.

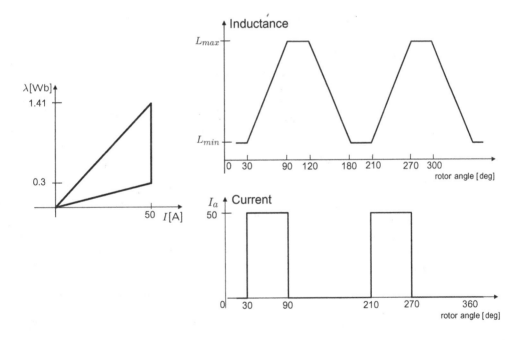

11.9 Assume that a 36 slot 8 pole motor is running at 2400 rpm. Find the minimum feasible wave number and the exciting frequency.

11.10 Assume that the rotor MMF, F_r is given as a step function and the air-gap permeance, Λ_g as a square function as shown below:

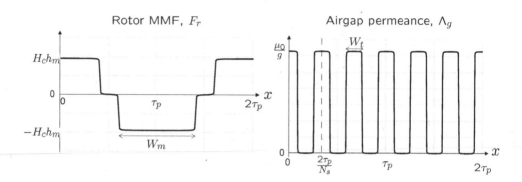

a) Denote by x the absolute position, whereas by x_r a displacement of the rotor. The rotor MMF and the permeance can be expressed as in Fourier series:

$$F_r = \sum_n F_n \cos\left(\frac{n\pi}{\tau_p}(x - x_r)\right)$$

$$\Lambda_g = \Lambda_0 + \sum_m \Lambda_m \cos\left(\frac{mN_s\pi}{\tau_p}x\right).$$

Show that

$$F_n = 2H_c h_m \frac{W_m}{\tau_p} \frac{\sin\left(\frac{n\pi}{2}\frac{W_m}{\tau_p}\right)}{\frac{n\pi}{2}\frac{W_m}{\tau_p}} \qquad \text{for } n = 1, 3, 5, \cdots$$

$$\Lambda_m = \frac{\mu_0}{g} N_s \frac{W_t}{\tau_p} \frac{\sin\left(\frac{mN_s\pi}{2}\frac{W_t}{\tau_p}\right)}{\frac{mN_s\pi}{2}\frac{W_t}{\tau_p}} \qquad \text{for } m = 1, 2, 3, \cdots .$$

$$\Lambda_m = \frac{W_t N_s}{2\tau_p} \frac{\mu_0}{g}.$$

b) Assume that $H_c = 950$ kA/m, $h_m = 10$ mm, $g = 1$ mm, $N_s = 6$, $\frac{W_m}{\tau_p} = 0.8$, $\frac{W_t}{\tau_p} = \frac{1}{6}$, $x_r = 1$, and $\tau_p = 1$. Using MATLAB, calculate the Fourier coefficients and plot the air gap field density, $B_r(x) = \Lambda_g F_r$.

c) When the rotor MMF travels over $x_r = [0, 2\tau_p]$, plot the air gap field variation at the point $x = \tau_p$, i.e., $B_r(\tau_p)$ for $0 \le x_r \le 2\tau_p$.

d) Using the result of c), plot the radial force density $P_r = \left.\frac{B_r^2}{2\mu_0}\right|_{x=\tau_p}$ for $0 \le x_r \le 2\tau_p$ using MATLAB.

Chapter 12

EV Motor Design and Control

EV share is steadily increasing as the battery cost reduces and power density increases. For vehicle drive, the motor should have a high starting torque and produce high power during high speed. Internal combustion engines (ICEs) are limited in the working range in comparison to electric motors. For example, ICEs cannot produce torque at zero speed, and its efficiency drops quickly with speed. These restrictions are solved by gear shift and slip mechanism. Gear shift offers several torque-speed curves as shown in Fig. 12.1. On the other hand, electric motors have a large starting torque, and can work over a wide speed range. EV motors are normally used without a gear shifting mechanism. The EV range (dotted line) is compared with multiple torque-speed curves of ICE in Fig. 12.1.

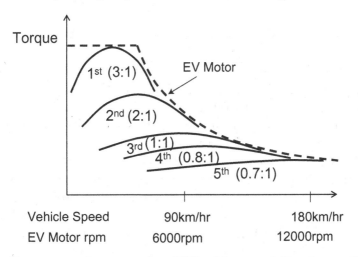

Figure 12.1: Torque-speed curves of an ICE with gear shifts along with an electric motor torque-speed curve (dotted line).

In this chapter, typical rotor topologies of IPMSMs are compared in terms of torque ripple, demagnetization, mechanical stress, radial force distribution, etc. Further, design and manufacturing processes are illustrated. Finally, practical measurement and control methods are explained.

12.1   Requirements of EV Motor

The EV motors need high power density, high short-time overload capability, wide constant power speed range (CPSR), high efficiency, robustness etc. Among the requirements, high power density and high efficiency are the most important properties that a propulsion device should have. The overall efficiency needs to be evaluated based on driving cycles. Thus, an EV motor must be optimized in a most frequently used area. It is not easy though to increase the efficiency while reducing material costs and size. Commonly accepted power densities are 5.7 kW/liter and 2 kW/kg. For high power density, most EV motors adopt water cooling. The motor volume cost is around US$ 500/kW. EV requirements are depicted in Fig. 12.2.

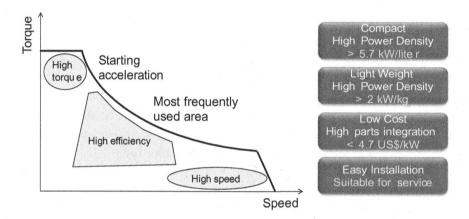

Figure 12.2: EV motor requirements.

Short time high torque capability is required for starting, climbing uphill, and acceleration. The maximum torque is normally twice as large as the rated torque and designed to last for 20~40 seconds. The sustaining time is limited by the coil temperature. In such a high current period, the rotor PM can be demagnetized. The PM should not be demagnetized under all operation conditions.

The CPSR needs to be wide for high speed driving. Usually, it is 3.5~4 times greater than the base speed. PM cavities and flux barriers are designed to polarize the magnetic anisotropy between d and q axes. But, such an effort requires a sacrifice in the mechanical rigidity against centrifugal force. In addition, the back EMF needs to be limited. Otherwise, a high terminal voltage may destroy semiconductor switches (IGBTs) when the inverter looses control. In addition, EV motors are required to have low torque ripple and low noise and vibration. Normally, numerous designs are carried out to optimize the motor under given requirements.

Main specifications of running EV and HEV motors are compared in Table 12.1. Various types of EV motors currently in use are compared in Fig. 12.3. IMs, robust and cost effective, meet most of EV requirements. But the efficiency is lower than that of IPMSM. Therefore, IMs are not popularly used for EV propulsion, although they are used in Tesla EVs and some GM hybrid electric vehicles (HEVs). Fig. 12.4

Table 12.1: Torque and power data of EV and HEV motors.

Vehicle (year)	Leaf 2012	Prius 2010	Lexus 2008	Camry 2007	BMWi3 2016	Volt Gen2 2015
Max. Power (kW)	80	60	110	70	125	87
Pole/slot numbers	8/48	8/48	8/48	8/48	12/72	8/72
Power density (kW/kg)	1.4	1.6	2.5	1.7	2.5	2.25
Max. Torque (Nm)	280	207	300	270	250	280
Max. Speed (rpm)	10,390	13,500	10,230	14,000	11,400	11,000

shows photos of Nissan Leaf motor and rotor of Tesla IM. Table 12.2 lists major parameters of Nissan Leaf 2018 model.

		Structure	PM	Number of Poles	Power Density	Max. Speed	Company
Induction Motor		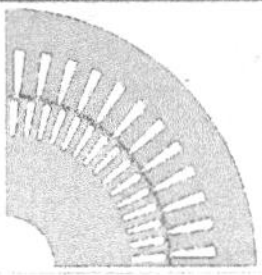	NA	4	2.66 kW/kg	18300 rpm	Tesla Model S
Wound rotor Synchronous Motor		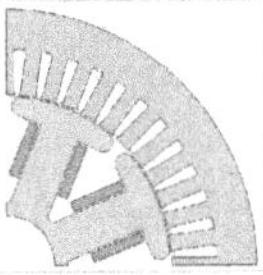	NA	8	1.8 (active) kW/kg	11300 rpm	Renault ZOE
PM Motor	**V Shape**	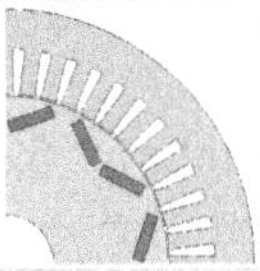	Nd	8	1.6 kW/kg	13500 rpm	Prius 2011
	Double V Shape	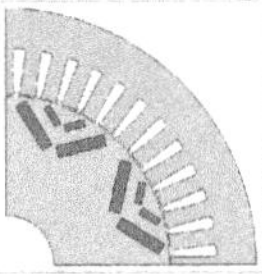	Nd	12	1.97 kW/kg	8810 rpm	Chevy Bolt
	Nabla Shape	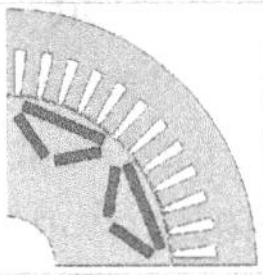	Nd	8	1.4 kW/kg	10400 rpm	Leaf 2012
			Nd	8	2.5 kW/kg	10230 rpm	Lexus LS 600h 2008
	Double I Shape	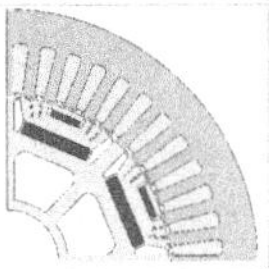	Nd	12	2.5 kW/kg	11400 rpm	BMW i3

Figure 12.3: Comparison of EV motors.

(a)   (b)

Figure 12.4: EV propulsion motors: (a) Nissan Leaf motor and (b) rotor of Tesla motor.

Table 12.2: 2018 Nissan Leaf specification.

Requirement	Target	Condition
Max. speed	10,390 rpm	
Rated speed	3,283 rpm	
Peak output power	110 kW	3283~9795 rpm
Max. torque	320 Nm	0~3283 rpm
Battery capacity	40 kWh	
On board charger	3 or 6 kW	charging time : 16 or 8 hr
Fast charger	50~60 kW	80% charging time : 40 min
Driving range	400 km	JC08
Vehicle gross weight	1765~1795 kg	Curb weight : 1520 kg
Acceleration time	8.5 sec.	0~100 km/hr

12.2   PMSM Design for EVs

PMSM design requires a multivariate optimization process that includes improving efficiency and power density, reducing torque ripple and noise, and enhancing durability. Typical constraints are rigidity and motor cost. Thus, costly PM usage must be minimized while keeping ruggedness against demagnetization. Furthermore, the mechanical rigidity should be assured in the cavity and bridge structure around the PM.

12.2.1   Pole and Slot Numbers

The pole and slot numbers are the most critical numbers that affect the overall motor performances. Concentrated winding offers the advantage of short end winding. But, distributed windings are generally used in EV propulsion motors, because high efficiency is required in a wide speed range with a low PM loss. As shown in Table 12.1, frequently used pole-slot combinations are $(N_s, P) = (48, 8)$ and $(72, 12)$, where N_s and P are the slot and pole numbers. The 48 slots and eight poles are used in EV motors, such as Toyota Prius, Honda Accord, Nissan Leaf, etc. Chevrolet Volt Gen 2 PHEV and BMW i3 motors utilize 72 slots twelve poles. In both cases, the number of slots per pole per phase is $q = 2$. Higher slot number per pole is better to minimize the torque ripple and reduce the air gap harmonics. However, the manufacturing cost increases with decreased slot fill factor.

As shown in Section 11.5.4, high pole machines are advantageous in reducing the stator back iron. Therefore, a high pole number is preferred in order to minimize the motor weight. But, high pole machine suffers a large core loss since it demands high electrical frequencies. Even at such a risk, BMW i3 utilizes twelve pole motors to increase the power density as high as 2.5 kW/kg. Note that the electrical frequency at the maximum speed is $11400/60 \times 6 = 1140$ Hz. Iron loss is normally severe with 1.1 kHz fundamental wave. To mitigate the eddy current loss, thin silicon steel (0.2 mm) is utilized in the i3 stator core. Note that thin silicon steel is expensive due to the difficult rolling work. To minimize the scrap after stamping, stator core is constructed by joining band type segmented pieces having dovetails. Unlike the stator, the rotor is made of a thick standard silicon steel.

12.2.2 PM and Flux Barrier Arrangements

PM imbedded topologies are preferred in traction applications due to the saliency and mechanical robustness. Shape and arrangement of flux barriers determine the reluctance torque while affecting the torque ripple and noise and vibration. Furthermore, the flux barrier should be mechanically robust against centrifugal force. Manufacturing convenience is another important requirement. By and large, the PM arrangements are categorized as V-type, double V-type, ∇-type, and double I-type as shown in Fig. 12.5. The ∇-type is often called delta-type. Single V-type arrangements are used in most Toyota hybrid vehicles like Prius, Camry, Lexus RX450h, etc. Double V-type is used by General Motors in Spark EV and Chevrolet Volt. ∇-type is used in Nissan Leaf and Toyota Lexus LS 600H. Finally, double I-type is used in BMW i3.

Although types are different, all of them can be understood in the context of synchronous reluctance motor (synRM). Note that group of cavities are arranged in rows, so each row acts as a flux barrier. Except for the V-type, the rest have double layer structures. In general, multi-layer structures are better in suppressing noise and torque ripple, since they allow more degrees of freedom in shaping the PM flux density. It was reported in [6] that the influence of armature reaction decreased as the layer number increased. As a result, the cross coupling inductances, L_{dq} and L_{qd} are low with the ∇-type structure.

But the overall performances of four types are almost the same, while the V-type is advantageous in low manufacturing cost and high demagnetization resistance. Using the frozen permeability technique [7], torque can be decomposed into reluctance and magnetic components.

Regarding the torque ripple, it is recommended to choose the number of layers according to $N_s/P \pm 2$ [5]. Thus for a machine with 48 slots and eight poles, the low ripple choices are 4 or 8 layers. Note, however, that manufacturing complexity and

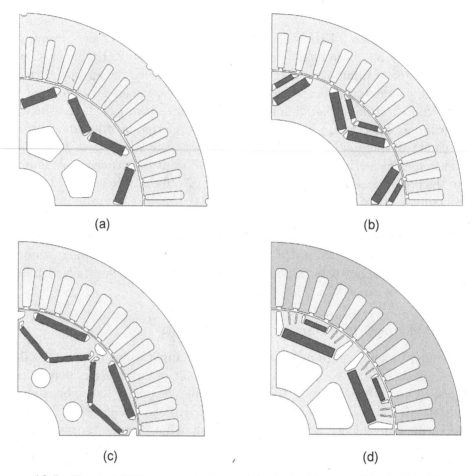

(a)   (b)

(c)   (d)

Figure 12.5: Typical EV motor designs: (a) single V-type, (b) double V-type, (c) ∇-type, and (d) double I-type.

cost increase as the layer number increases. Fig. 12.6 shows a four layer structure of Chevrolet Volt ferrite motor (Motor A) [2].

Figure 12.6: Rotor of Chevrolet Volt ferrite motor that has four layers.

Single V-Shape

Two PMs are placed facing each other with the same polarity in a pole, so that squeezed flux is forced out to the air gap. Cavities for PMs are extended from the center bridge to the rotor surface. As a result, two of them form a long cavity band. It acts like a flux barrier of SynRM. It results in a high saliency ratio, since the flux barrier hinders the passage of the stator d-axis flux. Due to the large saliency ratio, reluctance torque is effectively utilized in the field weakening range. Note that there are two q-axis flux passages inside and outside of the flux barrier. The PM size, position, flux barrier size, and PM coverage ratio determine the overall performance.

The bridges are required for structural integrity. The bridge ends are treated with round corners to avoid stress concentration. A large curve radius is more effective in distributing the stress regularly. However, part of PM flux leaks through the bridge, reducing the PM flux linkage to the stator winding. Therefore, the thinner the better from the magnetic view point. Therefore, the bridge width is determined while compromising the mechanical strength with the loss of PM flux.

The V-type is relatively easier to manufacture than the double V or delta types, because the number of magnets are smaller. Further, the PMs are less vulnerable to demagnetization since the PMs are placed away from the air gap. However, the V shape yields relatively large cross-magnetization. In particular, the symmetric distribution of PM flux is easily distorted by the q-axis current (armature reaction), and the variation is intensified as the current increases.

Double V-Shape

Double V shaped design has two rows of flux barriers at each pole, though the widths are different. The expected benefits are: The saliency increases as the flux barrier doubles. More degrees of freedom are given in locating the PMs and flux barriers, so that the torque can be maximized while reducing the torque ripple and noise-vibration. The double V design is more resistant to the armature reaction due to the increased number of flux barriers. As a consequence, the inductances, L_d and L_q of double V type machine do not change as much as a single V type machine.

Delta and Double I-Shape

The delta-shaped design can be seen as a mixture of V and I shapes, or a variation of double V shape. The I magnet increases the peak field density, while making the EMF closer to a sinusoidal shape. The same benefits can be expected with the double I machine. It utilizes fewer magnet numbers than the double V. The forefront cavity for I magnet blocks the penetration q-axis field, thereby alleviates the armature reaction. Thus, the cross coupled saturation effect is reduced. However, the I magnet tends to be demagnetized by the field harmonics. The slotting effects are more detrimental in the normal high inductance machines. Several exquisite cavities around the I magnet can reduce EMF harmonics and demagnetization possibilities.

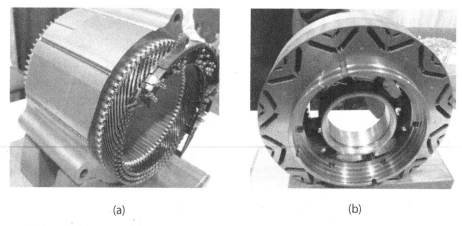

(a)   (b)

Figure 12.7: GM Volt Gen 2: (a) stator and (b) rotor.

12.3   PMSM Design for EV Based on FEA

Major design processes are illustrated with an example PMSM developed for a *C*-class passenger vehicle. The major parameters are listed in Table 12.3. Fig. 12.8 shows a photo of a developed motor and its coil design layout, and Table 12.4 lists coil data.

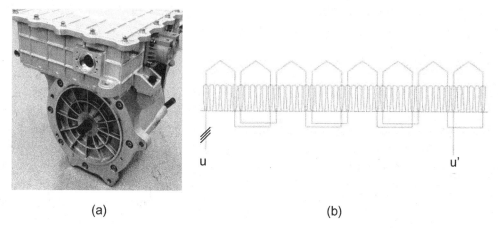

(a)   (b)

Figure 12.8: (a) Photo of a developed motor and (b) coil design layout.

12.3.1   Flux Density and Back EMF Simulation

Fig. 12.9 shows a mesh model and flux density simulations under $(I, \beta) = (150\,\mathrm{A_{rms}}, 30°)$ and $(250\,\mathrm{A_{rms}}, 43°)$. The flux density reaches $1.8 \sim 1.9$ T at the teeth of major flux passage under high load conditions. It indicates that local saturation is developed by high q-axis current. In general, motors are designed so that the maximum flux density does not exceed 1.6 T. Nevertheless, high saturation values are tolerated in EV motors, since more emphasis is put on the power density rather than the linearity.

Table 12.3: Major parameters for an IPMSM for EV propulsion.

No. of poles	8
Power (max)	100 kW
Max. torque (60 sec.)	320 Nm
Max. current	293 A_{rms}
Rated torque ($\geq$ 30 min)	160 Nm
Rated current	150 A_{rms}
DC link voltage	260~360 V
Base speed	3600 rpm
Max. speed	12000 rpm
Inductance (L_d/L_q)	0.234/0.562 mH
Coil resistance (r_s)	13 mΩ
Cooling	Water (12 liter/min)
Max. current density (r_s)	17.4 A/mm^2
Max. electric loading	854 A_{rms}/cm
PM flux (ψ_m)	0.0927 Wb

Table 12.4: Coil specifications

Insulation class	N	Coil diameter	1.0 mm
Connection	Y	Coil area/slot	51.8 mm^2
No. of turns per phase	24	Slot fill factor	47.4% (bare copper)
No. of conductor/slot	3	Coil length/phase	10.9 m
No. of strands/phase winding	22	Resistance/phase	0.015 Ω (100°C)

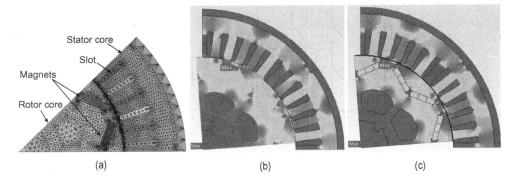

Figure 12.9: (a) Mesh model, (b) flux density under (150 A_{rms}, 30°), and (c) (250 A_{rms}, 43°).

Fig. 12.10 (a) and (b) show a back EMF simulation result and its harmonic spectrum at the rated speed when the PM temperature is 140°C. From the fundamental component, the back EMF coefficient is calculated as $\psi_m = 143.8/(60 \times 4 \times 2\pi) = 0.095$ Wb. Note that the flux density reduces as the temperature increases, i.e.,

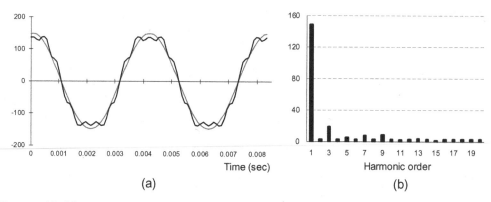

Figure 12.10: Back EMF simulation at 3600 rpm: (a) plot versus time and (b) harmonic analysis.

the temperature coefficient of remanence flux density of Nd magnet is -0.12 %/K. Thus, 0.095 Wb is a smallest EMF coefficient, since motor PM is regulated below $140°$C in all normal conditions.

12.3.2 Voltage Vector Calculation

Fig. 12.11 shows a moment of current flow when the full inverter DC link voltage is applied to the motor $u - w$ terminal pair. Note that the peak line-to-line voltage is equal to the DC link voltage. Therefore, in this example the maximum phase voltage is $V_{dc}/\sqrt{3} = 360/1.732 = 207.9$ V.

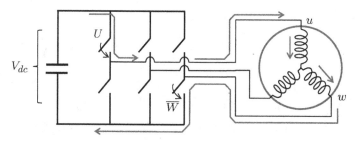

Figure 12.11: Inverter schematic diagram showing that the peak line-to-line voltage is equal to V_{dc}.

Exercise 12.1   Determine the motor terminal voltages at 2500, 7200, and 12000 rpm when the current magnitudes and angles are $(I, \beta) = (453, 43°)$, $(453, 75.25°)$, and $(453, 80.75°)$, respectively. Note here that $I = \sqrt{i_d^{e\,2} + i_q^{e\,2}}$. Using the data in Table 12.3, draw voltage and current vectors for the three cases. Check that the phase voltages are less than 207.9 V.

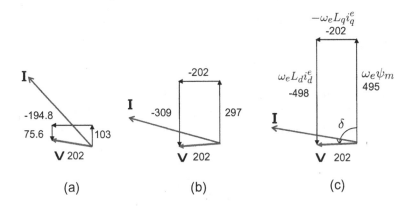

Figure 12.12: Voltage and current vectors for (a) 2500, (b) 7200, and (c) 12000 rpm (Solution to Exercise 12.1).

Solution. The electrical speed is $\frac{2500}{60} \times 2\pi \times 4 = 1047.2$ rad/sec. The back EMF is equal to $\omega_e \psi_m = 1047.2 \times 0.0984 = 103$ V.

$$\omega_e L_d i_d^e = 1047.2 \times 0.000234 \times (-308.6) = -75.6 \text{ V}$$
$$\omega_e L_q i_q^e = 1047.2 \times 0.000562 \times 331 = 194.8 \text{ V}.$$

On the other hand, the ohmic drop is equal to $r_s i_d^e = 0.0147 \times (-308.6) = -4.5$ V and $r_s i_q^e = 0.0147 \times 331 = 4.9$ V. The terminal voltage is

$$
\begin{aligned}
V_m &= \sqrt{(\psi_m \omega_e + \omega_e L_d i_d^e + r_s i_q^e)^2 + (\omega_e L_q i_q^e + r_s i_d^e)^2} \\
&= \sqrt{(103 - 75.6 + 4.9)^2 + (-194.8 - 4.5)^2} = 202 \text{ V}.
\end{aligned}
$$

Therefore, the voltage angle is equal to

$$\delta = \tan^{-1}\left(\frac{-v_d^e}{v_q^e}\right) = \tan^{-1}\left(\frac{199.3}{32.3}\right) = 80.8°$$

Since $\beta = 43°$, PF $= \cos(\beta - \delta) = \cos(-37.8°) = 0.79$. Repeating the same for 7200 and 12000 rpm, we have voltage and current vectors as shown in Fig. 12.12, in which the relative magnitudes and angles are respected. ∎

12.3.3 Flux Linkage Simulation and Inductance Calculation

Fig. 12.13 shows a result of time stepping analysis. The flux linkages, λ_a, λ_b, and λ_c are obtained for one pole period $0 \le \theta_e \le \pi$ with the phase currents, $(i_a, i_b, i_c) = (I \cos \omega_e t, I \cos(\omega_e t - \frac{2\pi}{3}), I \cos(\omega_e t - \frac{4\pi}{3}))$, where $\theta_e = \omega_e t$. The d- and q-axis flux linkages are calculated by the transformation map:

$$
\begin{bmatrix} \lambda_d^e(\theta_e) \\ \lambda_q^e(\theta_e) \end{bmatrix} = \frac{2}{3} \begin{bmatrix} \cos\theta_e & \sin\theta_e \\ -\sin\theta_e & \cos\theta_e \end{bmatrix} \begin{bmatrix} 1 & -\frac{1}{2} & -\frac{1}{2} \\ 0 & \frac{\sqrt{3}}{2} & -\frac{\sqrt{3}}{2} \end{bmatrix} \begin{bmatrix} \lambda_a(\theta_e) \\ \lambda_b(\theta_e) \\ \lambda_c(\theta_e) \end{bmatrix} \tag{12.1}
$$

Fig. 12.13 (a) and (b) show $(\lambda_a, \lambda_b, \lambda_c)$ and the corresponding $(\lambda_d^e, \lambda_q^e)$ [8], [9]. As discussed in Section 11.5, the ripples in λ_d^e and λ_q^e are caused by the stator slotting, rotor permeance harmonics, and the MMF harmonics. Fig. 12.14 shows three dimensional plots of flux linkages over the whole current range. It shows a decoupling between the d and q axes, though not perfect. That is, λ_d^e depends only on i_d^e, whereas λ_q^e only on i_q^e. Note from Fig. 12.14 (a) that λ_d^e has a big offset when $i_d^e = i_q^e = 0$. It is due to the rotor PM flux in the d-axis. However, $\lambda_q^e = 0$ at zero current. It justifies

$$\lambda_d^e(i_d^e, i_q^e) = \psi_m + L_d(i_d^e, i_q^e)i_d^e \tag{12.2}$$
$$\lambda_q^e(i_d^e, i_q^e) = L_q(i_d^e, i_q^e)i_q^e, \tag{12.3}$$

where L_d and L_q are d and q-axis inductances, respectively. It should be noted here that L_d and L_q are not constants due to core saturation. Therefore, they are represented as functions of currents.

Fig. 12.15 (a) and (b) show three dimensional plots of flux linkages of the IPMSM. Inductances are calculated from (12.2) and (12.3) such that

$$L_d(i_d, i_q) = \frac{\lambda_d(i_d, i_q) - \psi_m}{i_d} \tag{12.4}$$

$$L_q(i_d, i_q) = \frac{\lambda_q(i_d, i_q)}{i_q}. \tag{12.5}$$

Fig. 12.15 shows inductances calculated by an FEM tool over a wide current range. The L_d variation is minor, whereas the L_q variation is noticeable. Fig. 12.15 (c) and (d) show inductances versus current angle, β when the current magnitude is fixed at 150 A$_{rms}$ and 250 A$_{rms}$. Note that L_q increases as β increases. The reason is that the q-axis current decreases along with β. Furthermore, nullification of PM flux in the air gap becomes intense, thereby the core is released from saturation.

12.3.4   Method of Drawing Torque-Speed Curve

It is complex to find the maximum torque curve under voltage and current limit, since the solutions are obtained from the conditions of MTPA, maximum power, and

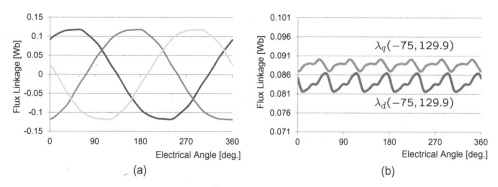

Figure 12.13: Time-stepping analysis: (a) flux linkage of phase coils, (b) flux linkage of d and q axes in the synchronous frame.

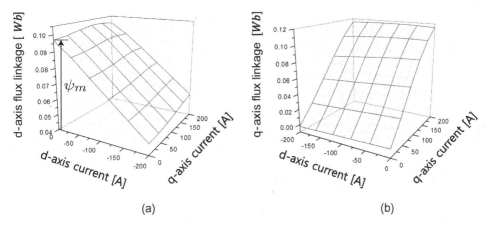

(a)   (b)

Figure 12.14: Flux linkage of IPMSM: (a) λ_d^e and (b) λ_q^e.

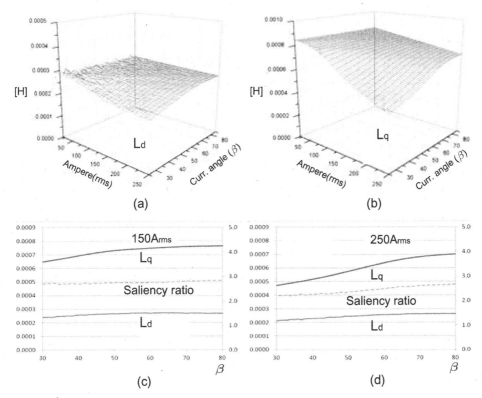

Figure 12.15: Inductance plots: (a) L_d, (b) L_q, (c) and (d) inductance versus current angle when the current magnitude is fixed at 150 and 250 A_{rms}.

MTPV depending on the speed range. It was illustrated previously in Section 7.4.3 and 7.4.4.

An easy and practical method is to utilize Excel spreadsheet. In the spreadsheet, set P, ψ_m, r_s, L_d and L_q as fixed parameters. On the other hand, let I_s and β the input variables to be adjusted manually. A decision criterion is the motor terminal voltage. That is, the maximum power is tracked below the voltage limit at each speed, while adjusting the current magnitude and angle. Furthermore, an additional voltage margin (about $7 \sim 10$ V) is necessary for current regulation. The voltage margin should be reserved to accommodate the voltage reference that comes out of the PI controller. Inductances are assumed to be constant for convenience in the first place. Later, the case will be treated with the inductance variation.

Fig. 12.16 shows a calculation method of torque-speed curve using the Excel sheet. The procedure for using the Excel program is:

1. Input the motor parameters P, r_s, L_d, L_q, ψ_m in the first column.
2. Put the motor motor speeds in rpm in column 'C' and put the corresponding electrical speeds in column 'D'.
3. Put a (maximum) current magnitude, $I_s = \sqrt{i_d^{e\,2} + i_d^{e\,2}}$ in column 'E' and put current angle, β in column 'F'. The current angle must be

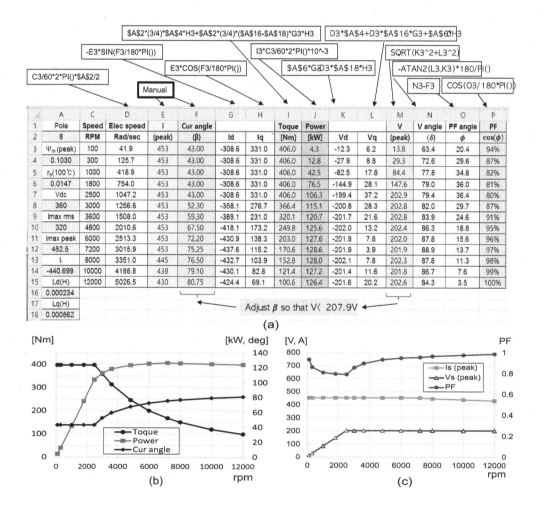

Figure 12.16: (a) Excel spreadsheet for torque-speed curve, (b) torque, power, current angle, and (c) current, voltage, power factor.

increased with speed. Put $i_d^e = -I_s \sin \beta$ and $i_q^e = I_s \cos \beta$ in columns 'G' and 'H'.

4. Based on (i_d^e, i_q^e), write torque and power equations such that $T_e = 3P/4(\psi_m i_q^e + (L_d - L_q)i_d^e i_q^e)$ and $P_e = T_e \times 2\omega_e/P$. Put them in column 'I' and 'J'.

5. Based on (i_d^e, i_q^e), write voltage equations such that $v_d^e = r_s i_d^e - \omega_e L_q i_q^e$, $v_q^e = r_s i_q^e + \omega_e L_d i_d^e + \omega_e \psi_m$, and $V_s = \sqrt{v_d^{e\,2} + v_q^{e\,2}}$. Put them in columns 'K', 'L', and 'M'.

6. Select an initial current angle, β which maximizes the torque. The MTPA current angle can be found by changing β manually. In this example, the initial current angle is $\beta = 43°$. Put the same β in the following column until the motor voltage, $V_s + \Delta V$ reaches nearly the maximum, $V_{dc}/\sqrt{3}$, where $\Delta V \approx 7 \sim 10$ V is a voltage margin provided for current controller. The speed at which $V_s + \Delta V = V_{dc}/\sqrt{3}$ determines the end of the constant torque region. In this example, the constant torque range is $(0, 2500)$ rpm.

7. In the CPSR (from row 8), increase β manually. Maximize the torque, but check if the voltage condition, $V_s + \Delta V < V_{dc}/\sqrt{3}$ is satisfied.

8. Repeat the process of increasing both speed and current angle. If β is close to 75°, the voltage limit condition cannot be satisfied by adjusting β only. We have to reduce the current magnitude, I_s. It is the range of MTPV. In such a high speed range (from row 13), change both I_s and β.

Fig. 12.16 (b) and (c) show the plots of torque, power, current angle, current, voltage, and power factor based on the data in Fig. 12.16 (a). Note that the constant power speed region (CPSR) is extended to 12000 rpm. According to the Fig. 12.16 (b), the starting torque is as large as 406 Nm. It is not a realistic value since the core saturation is not reflected.

In Step **6**, the torque is maximized in the low speed region. The voltage limit does not put any limit in the low speed region. In Step **7**, the speed increase results in the voltage increase. To meet the voltage condition, the current angle must be increased. However, a large increase will cause a big power drop. So, it is necessary to increase β slightly at each step. In this example, $V_{dc}/\sqrt{3} = 360/\sqrt{3} = 207.9$ V and $\Delta V = 5.9$ V. It means that V_s should be less than 202 V. In Step **8**, I is reduced to 445 A at 7200 rpm to make $V_s = 202$ V with $\beta \leq 81°$.

Exercise 12.2   Determine the current magnitude and angle so that the motor produces the maximum power in the field weakening region up to 10000 rpm. The motor parameters are $\psi_m = 0.103$ Wb, $r_s = 0.0147$ Ω, $V_{dc} = 360$ V, $L_d = 234$ μH, $L_q = 562$ μH. The maximum current is $I_s = 280$ A (peak), and set the voltage margin, $\Delta V = 6$ V. Calculate the maximum torque and power using Excel spreadsheet. Plot torque, power, and current versus speed.

Excel Spreadsheet with Inductance Change

The coupling terms, $-\omega_e L_q i_q^e$ and $\omega_e L_d i_d^e$ are large constituents in the phase voltage when the speed is high. Thus, the inductance variation affects a lot voltage and torque equations. Fig. 12.17 shows an extended version of Excel sheet that incorporates the change of L_d and L_q. Two columns are provided to write different inductance values depending on current magnitude and angle. For this purpose, an inductance look-up table as shown in Fig. 12.15 (a) and (b) is needed. Inductances

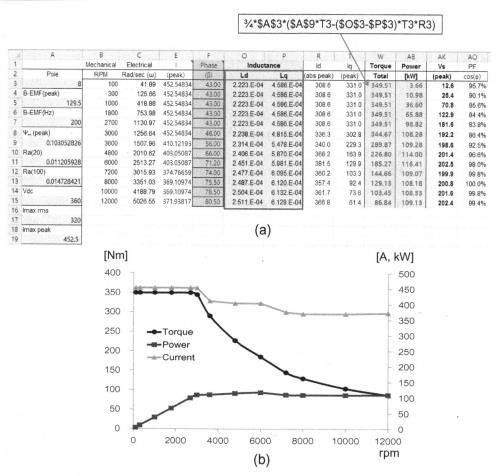

¾*A3*(A9*T3-(O3-P3)*T3*R3)

	A	B	C	E	F	O	P	R	T	W	AB	AK	AO
		Mechanical	Electrical		Phase	Inductance		Id	Iq	Torque	Power	Vs	PF
	Pole	RPM	Rad/sec (ω)	(peak)	(β)	Ld	Lq	(abs peak)	(peak)	Total	[kW]	(peak)	cos(φ)
3	8	100	41.89	452.54834	43.00	2.223.E-04	4.586.E-04	308.6	331.0	349.51	3.66	12.6	95.7%
4	B-EMF(peak)	300	125.66	452.54834	43.00	2.223.E-04	4.586.E-04	308.6	331.0	349.51	10.98	25.4	90.1%
5	129.5	1000	418.88	452.54834	43.00	2.223.E-04	4.586.E-04	308.6	331.0	349.51	36.60	70.8	85.6%
6	B-EMF(Hz)	1800	753.98	452.54834	43.00	2.223.E-04	4.586.E-04	308.6	331.0	349.51	65.88	122.9	84.4%
7	200	2700	1130.97	452.54834	43.00	2.223.E-04	4.586.E-04	308.6	331.0	349.51	98.82	181.6	83.8%
8	Ψm (peak)	3000	1256.64	452.54834	48.00	2.238.E-04	4.815.E-04	336.3	302.8	344.67	108.28	192.2	86.4%
9	0.103052826	3600	1507.96	410.12193	56.00	2.314.E-04	5.478.E-04	340.0	229.3	289.87	109.28	198.6	92.5%
10	Ra(20)	4800	2010.62	403.05087	66.00	2.406.E-04	5.870.E-04	368.2	163.9	226.80	114.00	201.4	96.6%
11	0.011205928	6000	2513.27	403.05087	71.20	2.451.E-04	5.981.E-04	381.5	129.9	185.27	116.41	202.5	98.0%
12	Ra(100)	7200	3015.93	374.76659	74.00	2.477.E-04	6.095.E-04	360.2	103.3	144.66	109.07	199.9	99.8%
13	0.014728421	8000	3351.03	369.10974	75.50	2.487.E-04	6.120.E-04	357.4	92.4	129.13	108.18	200.8	100.0%
14	Vdc	10000	4188.79	369.10974	78.50	2.504.E-04	6.132.E-04	361.7	73.6	103.45	108.33	201.6	99.8%
15	360	12000	5026.55	371.93817	80.50	2.511.E-04	6.129.E-04	366.8	61.4	86.84	109.13	202.4	99.4%
16	Imax rms												
17	320												
18	Imax peak												
19	452.5												

(a)

(b)

Figure 12.17: (a) Excel spreadsheet for power and torque calculation with the inductance look up table and (b) the corresponding torque and power curves.

are either computed by a FEM tool, or obtained by experiments. It requires an iterative process between (I_s, β) and (L_d, L_q). As we change (I_s, β) to find a maximum torque below a voltage limit, the inductance also changes. Then, it requires us to change (I_s, β) again to satisfy the voltage limit. While doing such a minor tuning, some interpolation methods can be used to find L_d and L_q that are not listed in the look-up table.

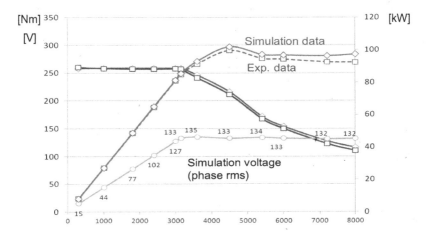

Figure 12.18: Comparison between Excel spreadsheet calculation result and experimental data.

Comparison with Experimental Data

Experimental and computed results are compared in Fig. 12.18. The computation was done with the Excel spreadsheet. The inductance look up table was obtained by using an FEM tool. Generally, the computed results match well with the experimental ones in overall range. The maximum error is less than 7%, and the errors are resulted from the inductance error and the uncounted effects such as friction and windage loss.

Exercise 12.3   Consider the motor with the following parameters: number of poles, $P = 8$, rotor flux linkage, $\psi_m = 0.0984$ Wb, and phase resistance, $r_s = 0.015$ Ω. In all cases, the terminal line-to-line (peak) voltage should be less than $325/\sqrt{3}$. The maximum current magnitude is $I_s = 212$ A. Determine the angle so that the motor produces power over 60 kW in the field weakening region up to 8000 rpm. Fill the blanks in the following Excel spreadsheet, and plot the curves of torque, power, terminal voltage, and power factor.

12.4   Finite Element Analysis

Nowadays, commercial finite element analysis (FEA) tools are very powerful covering a wide range of electro-magnetic, thermal, and mechanical stress analyses. Through time-stepping analysis, torque ripple and core loss can be computed. With a 3D-FEA model, the PM loss can be calculated.

12.4.1   Torque Simulation

Fig. 12.20 shows toque simulation results for different current conditions before and after skewing at 3600 rpm. The amount of ripple reaches 28.5%, but it reduces to

| Mechanical | Electrical | I | Phase | Inductance | | Id | Iq | Torque | Power | Vs | V. angle | PF angle | PF |
RPM	Rad/sec	(peak)	(β)	Ld	Lq	(peak)	(peak)	Total	[kW]	(peak)			
100	41.89	212.13	36.00	2.219.E-04	6.414.E-04								
300	125.66	212.13	36.00	2.219.E-04	6.414.E-04								
1000	418.88	212.13	36.00	2.219.E-04	6.414.E-04								
1800	753.98	212.13	36.00	2.219.E-04	6.414.E-04								
2700	1130.97	212.13	36.00	2.219.E-04	6.414.E-04								
3000	1256.64	212.13	36.00	2.219.E-04	6.414.E-04								
3600	1507.96	212.13		2.278.E-04	6.535.E-04								
4800	2010.62	212.13		2.443.E-04	6.660.E-04								
6000	2513.27	212.13		2.502.E-04	6.638.E-04								
7200	3015.93	229.1		2.524.E-04	6.563.E-04								
8000	3351.03	240.42		2.529.E-04	6.522.E-04								

Figure 12.19: (Exercise 12.3) Excel spreadsheet for power and torque calculation with variable inductances.

10% after skewing. It shows that the torque ripple is low with the low currents. Since the motor pole number is eight, one electrical cycle is 90° mechanically. It can be easily observed that the dominant ripple component is the 6th order. The skew angle is usually chosen to be a half slot pitch. In this 48 slot machine, the slot pitch is equal to 3.75°, which corresponds to 15° in electrical angle.

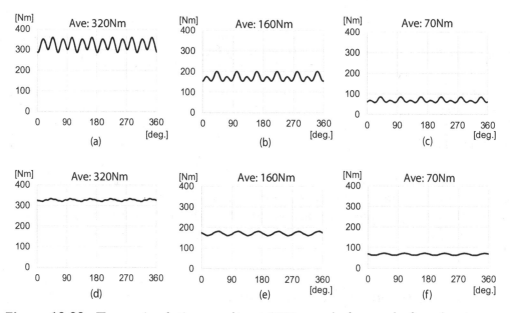

Figure 12.20: Toque simulation results at 3600 rpm before and after skewing.

12.4.2 Loss Analysis

An FEM tool is utilized to calculate the core loss. Fig. 12.21 shows the stator core loss density plots under various current and speed conditions. As for an electrical steel, the loss data of 27PNF1500 was utilized: The specific core loss is $W_{15/50} = 2.03$ W/kg and $W_{10/400} = 13.3$ W/kg, and the electrical resistivity is 56.5×10^{-8} $\Omega \cdot$

m. Most loss takes place around the boundary of stator teeth. It shows a dramatic increase at 12000 rpm (800 Hz). Note that the losses in the rotor core and PMs are relatively small, since the fundamental component is seen as a DC.

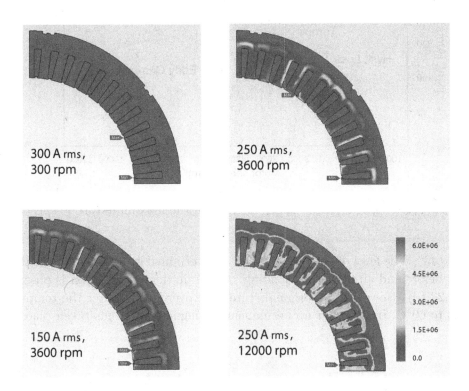

Figure 12.21: Iron loss distribution: (a) 300 A_{rms}, 300 rpm (b) 250 A_{rms}, 3600 rpm, and (c) 250 A_{rms}, 1200 rpm.

Fig. 12.22 shows copper and iron losses along with the maximum torque-speed curve. The copper loss is calculated according to the ohmic law. The iron loss is separated into hysteresis and eddy current losses. In the low speed (constant torque) range, the copper loss is dominant over the iron loss. With the speed increase, the current is reduced to meet the voltage limit ($V_{dc} = 360$ V). Correspondingly, copper loss decreases. In the low speed region, the iron loss is minor. But in the high speed region, it is comparable with the copper loss. Among the loss components, the proportion of eddy current loss expands with the speed, since it is proportional to the square of speed. At 12000 rpm, the electrical frequency is $12000/60 \times 4 = 800$ Hz, with which the eddy loss itself amounts to 2 kW. Note on the other hand that the hysteresis loss is almost steady.

12.4.3 PM Demagnetization Analysis

Fig. 12.23 shows the level of PM demagnetization, when the PM goes through a temperature cycle of 60°C, 180°C, and 60°C. It is under a current loading; 320 A_{rms}

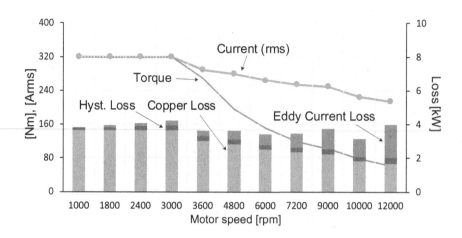

Figure 12.22: Copper and iron losses along with the maximum torque-speed curve.

and $\beta = 43°$. The level of demagnetization is determined by a change in the PM flux density before and after thermal loading. A certain demagnetization is observed in Fig. 12.23 (d). However, the original state is recovered fully after the temperature returns to 60°C. It says that no permanent demagnetization has taken place.

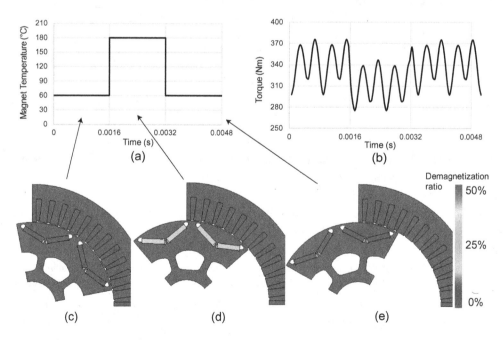

Figure 12.23: Demagnetization analysis with $I = 320$ A$_{rms}$ and $\beta = 45°$: (a) temperature profile, (b) torque responses, and (c), (d), (e) percentage of demagnetization.

12.4.4 Mechanical Stress Analysis

The PMSM rotors have many cut-outs to insert PM and to block out flux passage. However, narrow bridges are necessary to hold mechanical integrity. As the rotor speed increases, the centrifugal force is concentrated on the edge of flux barriers and bridges. Under the maximum speed condition, the mechanical stress should not exceed the yield stress of the steel. If a high stress point is found, the bridge design should be modified to disperse the stress concentration: The design change includes

1) Widen bridge width;
2) Increase the radius of curvature in the cavity corner;
3) Add fillets at the corner of bridges.

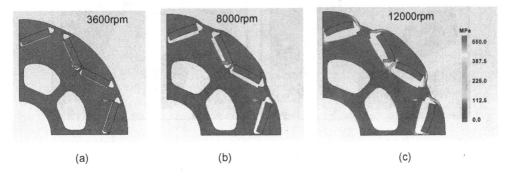

Figure 12.24: Mises stress and displacement ($\times 400$) simulation: (a) 3600 rpm, (b) 8000 rpm, and (c) 12000 rpm.

Widening bridge often increases the PM leakage flux, so that it reduces torque or increases the PM usage. Additionally, many magnetically unnecessary parts are cut out to reduce the mass of rotor. It may reduce the stress in the bridge, or help stress distribution.

Fig. 12.24 shows Mises stress and displacement at different speeds. The displacement was 400 times exaggerated. A rule of sum is to keep the maximum displacement less than 10 μm at 20% over speed. The maximum displacement does not have a direct relationship with the maximum stress. The maximum displacement occurs in the center part of a pole, whereas the maximum stress occurs in the bridges.

Table 12.5 shows the steel and resin (magnet adhesive) data required for mechanical FEA. In most EV motors, the cavities are filled with some resin, which, once solidified, disperses the stress on the larger area. It minimizes deformation and reduces stress peak. Fig. 12.25 shows effects of resin at 12000 rpm. One can see a contrast before and after filling the resin in the PM cavities.

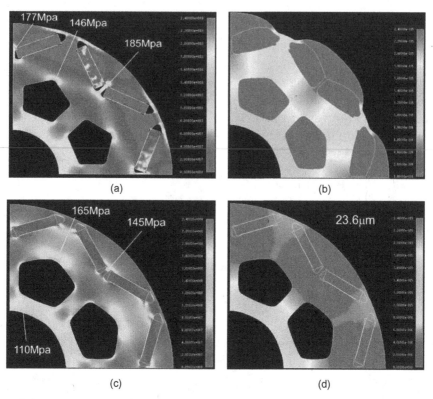

Figure 12.25: Rotor distribution of Mises stress and deformation at 12000 rpm without resin (a), (b) and with resin (c), (d): (a) (c) Misses stress and (b) (d) deformation.

Table 12.5: Mechanical properties of electrical steel and magnet adhesive.

Material	27PNF1500 (POSCO Elec. Steel)	2214 (3M Magnet Adhesive)
Tensile strength	410 Mpa	
Poisson ratio	0.3	0.35
Young's modulus	165 Gpa	5.17 Gpa
Density	7600 kg/m^3	1438 kg/m^3

12.5 PMSM Fabrication

The stator of PMSM is basically the same as that of induction motor. But, the rotor fabrication is different in many ways due to the PM. The general assembly process is illustrated with pictures.

12.5.1 Stator Assembly

Fig. 12.26 shows the states of stator core stacking, welding, and coil insertion. The stack is made of multiple sub-stacks. For example, Fig. 12.26 (b) shows the

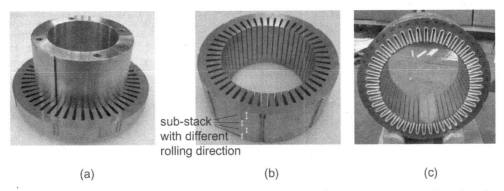

(a)   (b)   (c)

Figure 12.26: Stator assembly: (a) core stacking, (b) welding, and (c) slot liner insertion.

whole stator core is made of 4 sub-stacks. Even isotropic steel sheet has different permeabilities depending on the directions. The permeability is greater along the rolling direction than the transversal direction. Thus, not all steel sheets should be stacked in a uniform direction. It is necessary to randomize the rolling direction in the stack. A normal practice is to stack sub-stacks with 90° shifts as shown in Fig. 12.26 (b). The sides of stack are welded while being pressed by a 10 ton press.

Both sides of silicon steel sheet is coated with enamel of thickness 0.5 ~ 0.8 μm. Including the enamel thickness, the stacking factor is in the range of 97.5 ~ 98%. The slot liner has a high dielectric strength (18 ~ 34 kV/mm), thus it is used as an isolator from the stator core. It has a smooth surface to facilitate coil insertion. In addition, the required properties are mechanical toughness, thermal stability, flexibility, and resilience against chemicals.

The coil is classified depending on the thermal capability as shown in Table 12.6. In EV motor, 'N'-class coils are used which are coated with poly-amideimide insulation material. In the example motor, 22 strands of AWG 18 coils are used in the phase winding. Further, 3 coil bundles are inserted in each slot, whose area is 109 mm^2. Since the diameter of AWG 18 coil is 1.02 mm, the sum of coil sectional area in each slot is equal to $\pi \times 0.5^2 \times 22 \times 3 = 54.3$ mm^2. Therefore, the slot fill factor is $54.3/109 \times 100 = 49.8\%$ on the basis of bare copper. On the other hand, the maximum current density is equal to $J = \frac{320}{\pi \times 0.5^2 \times 22} = 18.52$ A/mm^2.

Table 12.6: Thermal class of coils

Thermal classes	Y	A	E	B	F	H	N
Endurance temp. (°C)	90	105	120	130	155	180	200

Fig. 12.27 shows a model for coil end turn calculation and the length per loop is also estimated according to

$$l_{turn} = 2l_{st} + 2l_b, \tag{12.6}$$
$$l_b = 1.3\tau_p + 3h_q + 2l_a, \tag{12.7}$$

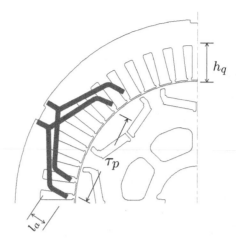

Figure 12.27: Coil end turn length calculation model.

where l_b is the end turn length, τ_p is pole pitch, h_q is the slot height, and l_a is the length of coil neck over the stator core. In the example motor, $l_b = 1.3 \times 67.15 + 3 \times 24.45 + 2 \times 3 = 166.7$ mm. Thus, the total length is equal to $2 \times (166.7 + 120) \times 24 \times 10^{-3} = 13.8$ m. Therefore, the coil resistance consisting of 22 strands is $0.021/22 \times 13.8 = 0.013$ Ω.

Fig. 12.28 shows a second part of stator core assembly. The neutral point is isolated with a special tape made of silicon rubber and glass cloth. It is an adhesive tape having self fusing property, as well as the heat and water resistance. The end turn coil temperature is measured using an NTC sensor made of a semiconductor material such as sintered metal oxides. Its resistance decreases as the temperature rises. Varnish impregnation is a common process that

> increases the overall dielectric strength,
> improves the structural integrity of the coils,
> reduces or eliminate winding noise,
> improves heat transfer,
> reduces partial discharge.

Vacuum impregnation is normally used to remove the trapped air bubbles from the slot and fill the empty space with the resin that has a high dielectric strength. The resin, once solidified, facilitates heat transfer from coils and bonds the wires tightly. Thus, it increases cooling efficiency and reduces noise during operation.

Fig. 12.29 shows stator housing, core insertion, and isolation testing. The body part of stator housing consists of two pieces of cast aluminum (AC4C-T6). The inner part has spiral cooling water channel. The core assembly is inserted into the housing via shrink-fitting.

A typical isolation test is dielectric strength testing (Hipot test) that measures the ability to withstand a medium-duration voltage surge without sparkover. It may be destructive in the event of fault. Another non-destructive method is to measure

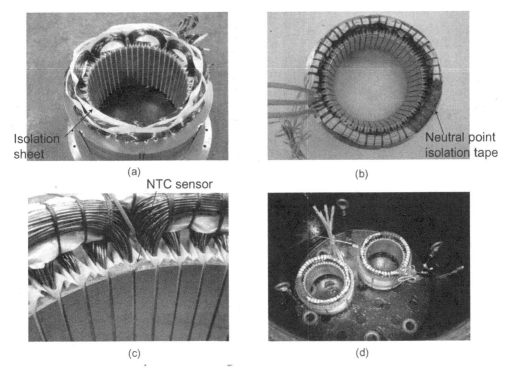

Figure 12.28: Second part of stator core assembly: (a) end-turn isolation between phase coils, (b) neutral point isolation, (c) temperature sensor attachment, and (d) vanish impregnation.

the insulation resistance. Testing standards for low voltage motors are summarized in Table 12.7

Table 12.7: Low voltage motor testing standards.

	Condition	Standard
Dielectric withstand test	1 kV(1 min)	< 5 mA
Insulation resistance test	500 V_{dc}	> 100 MΩ

12.5.2 Rotor Assembly

To give a step skew, the rotor is segmented axially into many sub-stacks. In general, a larger number of segments yield a less torque ripple. Four to six step skews are widely used for EV motors. For example, see Fig. 11.48. In some cases, a magnet in a cavity is segmented into multiple pieces to reduce more the eddy current loss. Fig. 12.30 shows an example. PMs are coated with Nickel plating to prevent corrosion. Specific magnet data are listed in Table 12.8.

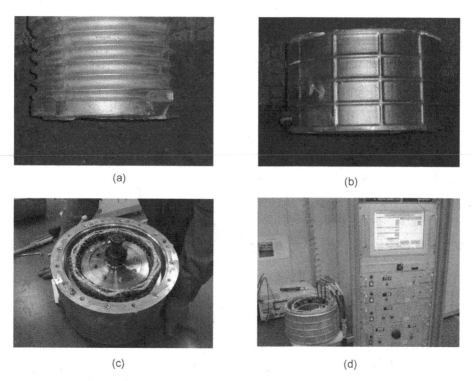

(a) (b)

(c) (d)

Figure 12.29: Stator housing assembly and isolation test: (a) aluminum housing inner piece, (b) housing outer piece, (c) core insertion via shrink fitting, and (d) coil isolation test.

Table 12.8: Nd magnet data

Name	N39UH	Mass Density	7600 kg/m^3
B_r	1.22~1.28 T	$(BH)_{max}$	286~318 kJ/m^3
$_iH_c$	1989 kA/m	$_bH_c$	915 kA/m
Poisson Ratio	0.3	Young's Modulus	150 GPa

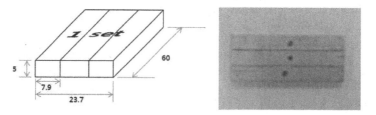

Figure 12.30: PM segmentation.

Fig. 12.31 shows rotor core assembly and balancing. Cores are stacked via a stacking jig. Segmented magnets are inserted with glue. The glue Poisson ratio is 0.35, which is similar to that (0.3) of magnet and steel core. The rotor stack is tightened with two end plates, then it goes through balancing process. Unbalanced

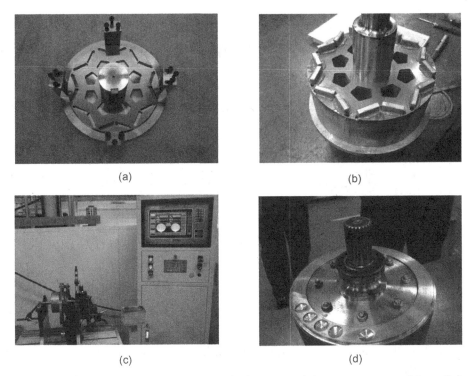

Figure 12.31: Rotor core assembly and balancing: (a) rotor core stacking, (b) magnet insertion with glue, (c) balancing process, and (d) result of end plate trimming out for balancing.

state is corrected by trimming the end plate. Fig. 12.32 shows the shaft components consisting of spline coupler, motor shaft, bearing and wave spring washer, and resolver rotor and stator. However, the resolver is attached after housing assembly. The shaft material is a Cr-Mo alloy steel (SCM440) and its surface is treated by abrasive machining. Fig. 12.33 shows the last assembly step. Resolver components are installed after the rear bracket is assembled.

12.6 PMSM Control in Practice

The real machine shows more diverse feature than can be described with an ideal model. The core saturation is a typical nonlinearity that is hardly handled with a linear model. Further, core losses and torque ripples are not rendered in the dynamic model. The coil resistance, inductance, and the back EMF constant are the three important motor parameters. In this section, motor parameter measurements and practical control methods are considered.

12.6.1 Coil Resistance Measurement

The coil resistance can be measured simply by an Ohm-meter. But, the resistance changes depending on the coil temperature. Concerning the temperature influence,

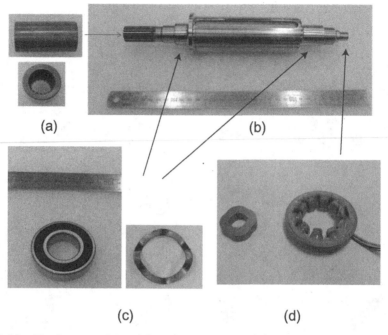

Figure 12.32: Shaft assembly: (a) spline coupler, (b) motor shaft, (c) bearing and wave spring washer, and (d) resolver rotor and stator.

the following two empirical equations are widely used.

$$\rho_\vartheta = \rho_{20}(1 + \alpha(\vartheta - 20)), \tag{12.8}$$

where $\rho_{20} = 1.72 \times 10^{-8}$ $\Omega \cdot$m is the resistivity of copper, $\alpha = 0.0039$ K^{-1}, and ϑ is the temperature in $°C$. Alternatively, the following interpolation method can be used:

$$\frac{\rho_2}{\rho_1} = \frac{234.5 + \vartheta_2}{234.5 + \vartheta_1}. \tag{12.9}$$

For example, if the coil temperature is $150°C$, then the resistance will increase by 51% from that at $20°C$.

12.6.2 Back EMF Constant Measurement

The rotor flux constant is measured simply by detecting the open circuit line-to-line voltage of the motor terminal. The measurement procedure is

    a) Set the motor under test on a dynamometer;
    b) Rotate the shaft of the motor at a fixed speed by the dynamo (load) motor;
    c) Measure the line-to-line voltage for 5 ~ 10 periods;
    d) Analyze the measured voltage using the Fourier series expansion, and read the fundamental component, $V_{ll(1)}$;

Figure 12.33: Last assembly step: (a) rotor, bearing insertion, and a front bracket assembly, (b) resolver insertion to the rear bracket, (c) housing assembly, and (d) coil end termination.

e) Calculate the rotor flux according to

$$\psi_m = \frac{V_{ll(1)}/\sqrt{3}}{\omega_r \times P/2}.$$ (12.10)

Exercise 12.4   Determine the back EMF constant of an 8 pole motor based on the experimental result shown in Fig. 12.34. The open circuit line-to-line voltage was measured at 1000rpm.

Solution. Since the fundamental component of line-to-line rms voltage is equal to 51.8 V, the phase voltage peak is $51.8 \times \sqrt{2}/\sqrt{3} = 42.3$ V. Note also that $\omega_e = (1000/60) \times 2\pi \times 4 = 418.9$ rad/s. Since $v_q^e = \omega_e \psi_m$ when $i_d^e = i_q^e = 0$, we have $\psi_m = 42.3/418.9 = 0.101$ V · s.   ∎

12.6.3   Inductance Measurement

The inductances are measured in the steady state utilizing the voltage current relation in the synchronous frame. An inverter and a dynamometer should be prepared before the measurement. The following is the measuring procedure when the current is (i_d^e, i_q^e):

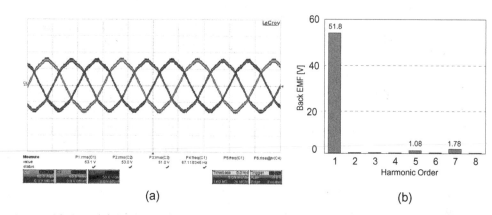

(a)   (b)

Figure 12.34: (a) Line-to-line open circuit voltage at 1000 rpm and (b) its spectrum.

a) Set the motor under test on a dynamometer;

b) Set the inverter in the current control mode and fix the current command at the given value, (i_d^e, i_q^e);

c) Rotate the shaft of the motor at a fixed speed by the dynamo (load) motor;

d) Send the inverter internal data of voltage, (v_d^e, v_q^e) and current, (i_d^e, i_q^e) to a measuring PC via a communication port;

e) Calculate

$$L_d = \frac{v_q^e - r_s i_q^e - \omega_e \psi_m}{\omega_e i_d^e} \qquad (12.11)$$

$$L_q = \frac{r_s i_d^e - v_d^e}{\omega_e i_q^e}. \qquad (12.12)$$

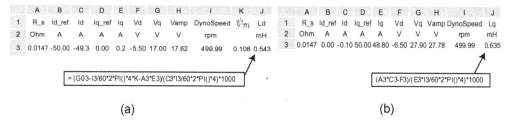

(a)   (b)

Figure 12.35: Excel spreadsheet for inductance measurements: (a) L_d and (b) L_q.

Fig. 12.35 shows a calculation example. A low speed 500 rpm is selected with $(i_d^{e*}, i_q^{e*}) = (-50, 0)$ and $(i_d^{e*}, i_q^{e*}) = (0, 50)$. Fig. 12.36 (a) and (b) shows plots of L_d and L_q versus current. Due to core saturation a general trend is that the inductance decreases as the current magnitude increases. But L_q increases initially and then decreases along with i_q^e. The rotor flux seems to be responsible for such a nonlinear behavior. Fig. 12.37 shows 3D plots of inductance measurements versus (i_d^e, i_q^e).

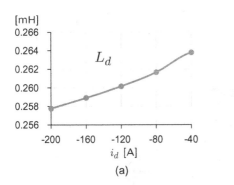
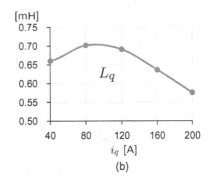

Figure 12.36: Inductance plot versus current: (a) L_d when $i_q^e = 0$ and (b) L_q when $i_d^e = 0$.

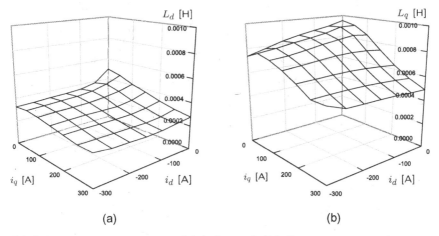

Figure 12.37: 3D measurements of (a) L_d and (b) L_q.

12.6.4 Look-up Table for Optimal Current Commands

As discussed in Section 7.4.1, the torque depends on both i_d^e and i_q^e in IPMSMs. An optimal current command, (i_d^{e*}, i_q^{e*}), should be selected for each torque and speed, (T_e, ω_r) to maximize the efficiency. In practice, the optimal current commands are found experimentally, and stored as a current command look-up table. The experimental method searches (i_d^{e*}, i_q^{e*}) of the minimum magnitude for each (T_e, ω_r) using a dynamo system. Therefore, it does not require an inductance table. The nonlinearity or current dependency is already reflected in the experimental data. Therefore, the experimental method provides a more accurate table than the model based one. The (optimal current) look-up table is used for the control in practice.

Experimental Method for a Reference Current Table

The optimal current commands are found by dynamo experiments. Figs. 12.38 and 12.39 show a photo of a dynamo system and its connection diagram. A bidirectional DC power supply acts as a battery simulator, and is connected to the inverter DC link. The power analyzer monitors motor current and voltage at the motor terminal, from which electric power to/from the motor can be calculated. A torque sensor monitors the shaft torque and speed. During the test, most of the energy circulates in the loop consisting of mechanical link and AC/DC/AC electrical devices.

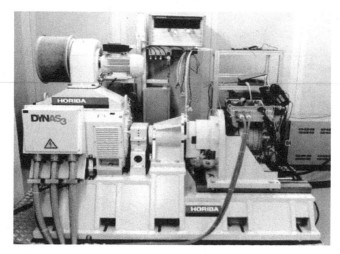

Figure 12.38: Photo of a dynamometer.

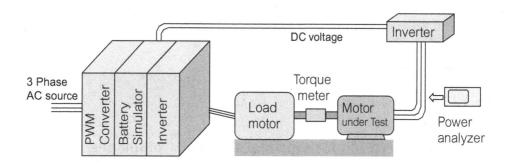

Figure 12.39: Dynamo connection diagram.

The experimental procedure is listed in the following. The basic idea is to find the current of minimum magnitude under a voltage limit among multiple current sets that yield the same torque at a fixed speed. As shown in Chapter 7, the desired solution is found at an MTPA point or an intersection between torque and voltage curves. Due to the voltage limit, experiments start with a large value of current angle.

a) Set an inverter for an IPMSM under test at the current control mode, connect the inverter DC link to the bidirectional DC power supply, and set the DC link voltage at V_{dc};

b) Let $i = 0$;

c) Rotate the motor shaft at a constant speed, ω_r^i by the dynamo motor;

d) Let $j = 0$;

e) Set current magnitude, I^{ij}. Choose a sufficiently large current angle, β^{ijk} for $k = 0$. The current angle needs to be large so that the terminal voltage does not exceed its limit. That is, $\sqrt{v_d^{e\,2} + v_q^{e\,2}} + \Delta V \leq \frac{V_{dc}}{\sqrt{3}}$ for the current set, $(i_d^e, i_q^e) = (-I^{ij}\cos\beta^{ijk}, I^{ij}\sin\beta^{ijk})$, where $\Delta V > 0$ is a voltage margin for current control.

f) Operate the inverter with the current command, $(i_d^{e*}, i_q^{e*}) = (-I^{ij}\cos\beta^{ijk}, I^{ij}\sin\beta^{ijk})$, and measure torque, T_e^{ijk};

g) Increase the index $k \leftarrow k+1$ and let $\beta^{ij(k+1)} = \beta^{ijk} - \Delta\beta^{ik}$, where $\Delta\beta^{ik} > 0$. If $\sqrt{v_d^{e\,2} + v_q^{e\,2}} + \Delta V \leq \frac{V_{dc}}{\sqrt{3}}$ for $\beta^{ij(k+1)}$, goto f);

h) Else, find $\max\{T_e^{ij0}, T_e^{ij1}, \cdots, T_e^{ijk}\}$, and increase the index $j \leftarrow j+1$ and let $I^{i(j+1)} = I^{ij} + \Delta I^i$. If $I^{i(j+1)} < I_s$, goto e);

i) Else $\omega_r^{i+1} = \omega_r^i + \Delta\omega_r$ for some $\Delta\omega_r > 0$, and goto c).

Fig. 12.40 shows a flow chart for experimental search. It consists of three loops. The k-loop represents a repeated torque testing procedure for a given current magnitude while decreasing β. On the other hand, the current magnitude is increased in the j-loop which includes the k-loop. Examples at 2750 rpm and 10000 rpm are shown in Fig. 12.41. In the case of low speed (2750 rpm), the maximum torque is found at the MTPA points. However in the high speed (10000 rpm), the maximum torque is found at the intersection with the voltage limit.

The final search results are shown in Fig. 12.42, in which optimal current values are indexed by torque and speed. Note that each curve in Fig. 12.42 represents a group of optimal current commands for different torques at a fixed speed. That is, the coordinate of a point is an optimal current command, (i_d^{e*}, i_q^{e*}). In other words, it is a current of the minimum magnitude that produces a marked torque. For example, $(i_d^e, i_q^e) = (-195, 230)$ A is the optimal current to produce 216 Nm at 2750 rpm.

The look-up table maps from (T_e, ω_r) to (i_d^e, i_q^e). Fig. 12.43 shows an example look-up table for $V_{dc} = 360$ V. Current command is read based on the percentage torque throttle and flux.

Speed Calibration Based on DC-link Voltage

In most EVs, the main battery is directly connected to the DC link and the battery voltage changes depending on the state of charge (SOC). It will be onerous if we were to make as many current command tables as different V_{dc}'s. The key question here is 'Is there a way to use a single reference table independently of the DC link

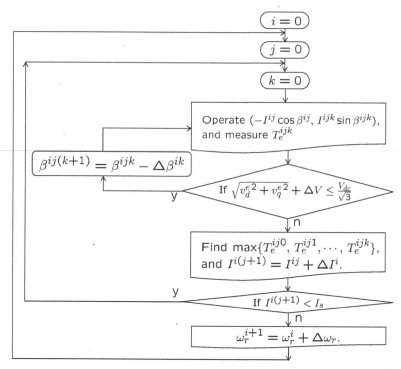

Figure 12.40: Experimental procedure searching for the optimal current map.

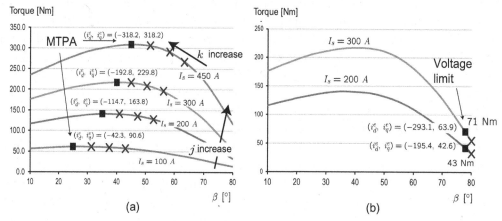

Figure 12.41: Measured torque data: (a) 2750 rpm and (b) 10000 rpm.

voltage?' The answer is 'yes', and the method is to use a *speed scaling strategy* depending on the DC link voltage. In the following, fictitious speeds are utilized. The basic idea is to reflect the voltage change as a change of speed in reading the reference current (look up) table.

Note that the motor speed have a fixed linearity with the flux when the DC link voltage is fixed. Therefore, the table can be read based on flux, instead of speed. The speed is scaled via flux. Note that $\sqrt{\lambda_d^{e\,2} + \lambda_q^{e\,2}} = V_{dc}/(\sqrt{3}\omega_e)$ in the steady state if ohmic drop is neglected. The detail is illustrated with the following

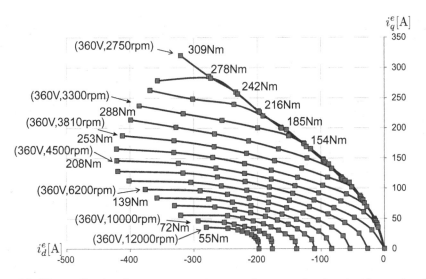

Figure 12.42: Optimal current command set obtained through dynamo-experiments when $V_{dc} = 360$ V.

		Speed	Torque throttle [%]										
Step	Flux	at 360V [rpm]	0	10	20	30	40	50	60	70	80	90	100
0	0.180435	2750	(0.0 ,0.0)	(-16.2 ,49.8)	(-39.7 ,89.1)	(-69.3 ,120.0)	(-94.4 ,151.1)	(-128.8 ,177.3)	(-163.8 ,202.2)	(-200.4 ,230.5)	(-234.7 ,260.6)	(-278.0 ,287.9)	(-320.0 ,320.0)
1	0.170397	2913	(0.0 ,0.0)	(-16.2 ,49.8)	(-39.4 ,88.5)	(-68.6 ,118.8)	(-96.3 ,148.3)	(-127.2 ,175.1)	(-162.9 ,201.1)	(-197.6 ,227.3)	(-231.8 ,257.5)	(-274.6 ,284.3)	(-346.7 ,290.9)
2	0.160359	3095	(0.0 ,0.0)	(-16.2 ,49.8)	(-38.5 ,86.6)	(-65.1 ,117.5)	(-94.0 ,144.7)	(-124.7 ,171.6)	(-155.0 ,198.4)	(-190.1 ,222.6)	(-239.6 ,240.4)	(-302.7 ,248.6)	(-370.7 ,259.6)
3	0.150321	3301	(0.0 ,0.0)	(-15.7 ,48.4)	(-37.4 ,84.0)	(-63.1 ,113.8)	(-88.4 ,141.5)	(-116.8 ,166.8)	(-151.5 ,188.5)	(-201.0 ,201.0)	(-254.2 ,211.8)	(-315.9 ,222.0)	(-387.9 ,233.1)
4	0.140283	3538	(0.0 ,0.0)	(-14.1 ,46.0)	(-34.3 ,80.7)	(-56.5 ,110.9)	(-82.3 ,137.0)	(-124.2 ,152.3)	(-168.2 ,165.8)	(-215.0 ,177.2)	(-268.4 ,188.7)	(-329.4 ,198.7)	(-401.4 ,209.0)
5	0.130244	3810	(0.0 ,0.0)	(-12.5 ,43.5)	(-32.0 ,75.5)	(-52.6 ,103.3)	(-91.5 ,120.5)	(-132.8 ,133.2)	(-176.8 ,145.7)	(-221.3 ,157.3)	(-272.2 ,167.5)	(-332.5 ,176.0)	(-413.4 ,184.1)
6	0.120206	4128	(0.0 ,0.0)	(-11.3 ,39.4)	(-28.6 ,70.8)	(-60.6 ,90.5)	(-98.6 ,105.3)	(-139.5 ,117.5)	(-181.2 ,128.3)	(-223.6 ,138.6)	(-272.4 ,147.3)	(-333.2 ,155.4)	(-423.1 ,160.7)
7	0.110168	4504	(0.0 ,0.0)	(-10.5 ,36.7)	(-33.5 ,62.3)	(-68.9 ,78.7)	(-105.3 ,92.2)	(-143.4 ,103.4)	(-180.7 ,113.8)	(-224.3 ,123.3)	(-269.7 ,131.0)	(-328.8 ,137.5)	(-430.2 ,140.6)
8	0.10013	4956	(0.0 ,0.0)	(-20.0 ,30.8)	(-48.1 ,51.8)	(-78.6 ,66.9)	(-111.8 ,79.5)	(-145.3 ,90.4)	(-180.6 ,100.1)	(-219.3 ,108.4)	(-261.4 ,115.3)	(-317.1 ,121.1)	(-420.8 ,123.0)
9	0.090092	5508	(-28.3 ,0.0)	(-44.4 ,24.8)	(-65.0 ,42.7)	(-92.7 ,57.1)	(-121.9 ,68.9)	(-151.8 ,79.0)	(-182.2 ,87.7)	(-217.6 ,95.5)	(-256.2 ,102.0)	(-305.6 ,107.3)	(-395.2 ,109.6)
10	0.080054	6199	(-55.2 ,0.0)	(-65.9 ,21.4)	(-83.2 ,35.7)	(-104.1 ,47.9)	(-128.8 ,58.4)	(-155.7 ,67.4)	(-184.5 ,75.7)	(-215.2 ,82.6)	(-250.6 ,88.7)	(-295.2 ,93.6)	(-377.0 ,95.4)
11	0.070016	7087	(-82.0 ,0.0)	(-93.2 ,16.9)	(-106.4 ,28.9)	(-121.0 ,39.3)	(-141.7 ,48.8)	(-164.4 ,56.9)	(-187.4 ,63.8)	(-215.1 ,70.3)	(-244.5 ,75.9)	(-283.0 ,80.1)	(-365.9 ,81.1)
12	0.059978	8273	(-111.7 ,0.0)	(-119.3 ,14.9)	(-127.7 ,24.8)	(-139.1 ,32.6)	(-154.6 ,40.6)	(-173.2 ,47.4)	(-193.5 ,53.7)	(-216.9 ,59.3)	(-243.5 ,63.9)	(-278.9 ,68.0)	(-346.6 ,69.9)
13	0.04994	9936	(-138.6 ,0.0)	(-147.9 ,12.7)	(-157.2 ,19.3)	(-164.8 ,26.1)	(-178.1 ,32.4)	(-191.4 ,38.2)	(-207.7 ,43.0)	(-226.9 ,47.8)	(-247.6 ,52.6)	(-275.8 ,56.1)	(-327.2 ,58.3)
14	0.039902	12436	(-173.9 ,0.0)	(-177.8 ,11.5)	(-178.8 ,16.6)	(-186.8 ,21.6)	(-192.0 ,26.3)	(-199.9 ,30.6)	(-213.5 ,35.0)	(-230.1 ,38.5)	(-246.8 ,41.7)	(-270.7 ,44.6)	(-307.6 ,46.5)
15	0.029864	16616	(-198.0 ,0.0)	(-199.1 ,11.8)	(-203.1 ,15.3)	(-205.7 ,18.4)	(-212.5 ,21.6)	(-219.3 ,24.2)	(-228.9 ,27.3)	(-237.2 ,29.5)	(-248.2 ,32.2)	(-263.6 ,34.7)	(-294.4 ,35.7)

Figure 12.43: Example (i_d^e, i_q^e) look-up table for $V_{dc} = 360$ V.

example: First, consider the maximum flux which is utilized at a rated speed. The rated speed of example motor is $\omega_r = 2750$ rpm, and the nominal battery voltage is $V_{dc} = 360$ V. Thus, the maximum flux is calculated as

$$\lambda_{max} = \frac{360}{\sqrt{3} \times 2750/60 \times 2\pi \times 4} = 0.18043 \text{ Wb.} \qquad (12.13)$$

Second, consider a case of the minimum flux. It will be found at the maximum speed operation under the minimum voltage. Assume that the battery voltage is reduced to 260V in a complete depletion state. Then, the minimum flux is calculated at the maximum speed, 12000 rpm such that

$$\lambda_{min} = \frac{260}{\sqrt{3} \times 12000/60 \times 2\pi \times 4} = 0.02986 \text{ Wb.} \qquad (12.14)$$

Table 12.9: (a) Reference table between flux and speed at the nominal voltage $V_{dc} = 360$ V and (b) speed calibration depending on different DC link voltages.

	DC Link Volt. Vdc=360V	
	RPM	Flux
0	2750	0.18043
1	2913	0.17040
2	3095	0.16036
3	3301	0.15032
4	3538	0.14028
5	3810	0.13024
6	4128	0.12021
7	4504	0.11017
8	4956	0.10013
9	5508	0.09009
10	6199	0.08005
11	7087	0.07001
12	8273	0.05998
13	9936	0.04994
14	12436	0.03990
15	16616	0.02986

(a)

	Flux	Calibrated RPM for Vdc=260V	Calibrated RPM for Vdc=430V
0	0.18043	1987	3285
1	0.17040	2104	3479
2	0.16036	2235	3696
3	0.15032	2384	3943
4	0.14028	2555	4225
5	0.13024	2752	4551
6	0.12021	2982	4931
7	0.11017	3253	5380
8	0.10013	3579	5920
9	0.09009	3978	6579
10	0.08005	4477	7404
11	0.07001	5119	8465
12	0.05998	5975	9882
13	0.04994	7176	11868
14	0.03990	8982	14854
15	0.02986	12000	19847

(b)

The full flux range is $[\lambda_{max}, \lambda_{min}]$. Make $N+1$ flux levels by dividing $\lambda_{max} - \lambda_{min}$ equally. The next procedure is to determine the speed, namely the *fictitious speed*, corresponding to each flux level such that

$$\omega_{rk} = \frac{V_{dc}}{\sqrt{3} \times \lambda_k}, \qquad \text{for } 1 \leq k \leq N. \tag{12.15}$$

Table 12.9 (a) shows a reference speed-flux relation. Note that the speed was extended to 16616 rpm with the lowest flux level.

This reference table can be used to recalculate the speeds for different DC link voltages. Table 12.9 (b) shows the fictitious speeds for $V_{dc} = 260$ V and $V_{dc} = 430$ V. The speed is recalculated to account for the DC link voltage change, i.e., the fictitious speed is made according to (12.15) for different V_{dc}'s. It tells us that if $V_{dc} = 260$ V, 0.13 Wb is the maximum tolerable flux at 2752 rpm (See row 5 of Table 12.9 (b)). On the other hand, if $V_{dc} = 430$ V, the speed can be increased to 4551 rpm for the same flux.

Reading Optimal Current Commands

Fig. 12.44 shows an illustration how an optimal current command is read from the reference table when the DC link voltage is different. Recall that Fig. 12.42 was developed for $V_{dc} = 360$ V. Here, it is desired to use the reference table when the DC-link voltage is different.

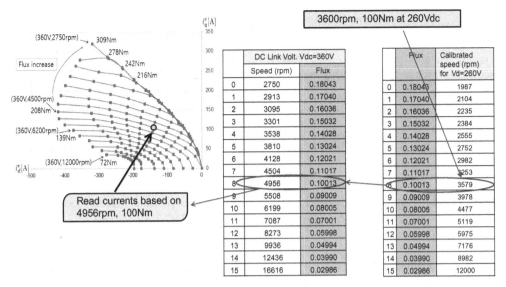

Figure 12.44: Procedure for reading current command based on the fictitious speeds.

For example, assume that it is desired to find an optimal current command to produce 100 Nm at 3600 rpm, when the battery voltage is $V_{dc} = 260$ V. A first step is to read a feasible flux for 3600 rpm from the 260 V column in Fig. 12.45. It is read as 0.1 Wb. Then, find a row of the nearest flux value from the reference (360 V) column. The eighth row of 360 V column has the same flux level, and the corresponding speed is 4596 rpm. It is a fictitious speed that was translated into the 360 V system. Finally, read the current commands from the current look-up table corresponding to 4596 rpm and 100 Nm. In this example, the desired current command is read $(i_d^{e*}, i_q^{e*}) = (-140, 105)$ A from the chart shown in Fig. 12.42.

Exercise 12.5 Suppose that the battery voltage is reduced to 300 V when the motor runs at 4200 rpm. It is desired to generate 110 Nm. Find an approximate (i_d^{e*}, i_q^{e*}) using the chart shown in Fig. 12.44.

Fig. 12.45 shows a command block which yields the current commands based on torque, speed, and the DC-link voltage. Note that the table is indexed by quantized torque and flux, and that the table readings are interpolated before outputting.

Motor Efficiency Plot

Fig. 12.46 shows an efficiency map of the motor obtained via experiments. For each operating point, the output torque and speed are measured to calculate the shaft mechanical power. Via a power analyzer, the input power to the motor is measured. The efficiency is calculated by dividing the mechanical power by the electrical input power. In general, efficiency is the lowest in the high torque-low speed region, where the copper loss is dominant. The other low efficiency region is the low torque-high

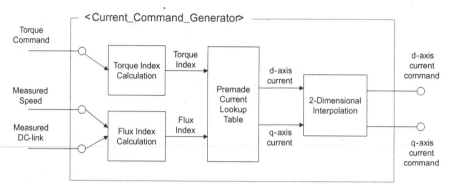

Figure 12.45: Current command block including an optimal current look up table and a speed calibration algorithm depending on the DC link voltage.

speed region, where the mechanical loss and core eddy current loss are high. The highest efficiency reaches 97.5%, and is found in the middle of the 45 degree line.

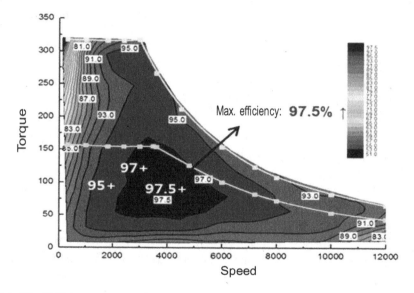

Figure 12.46: Efficiency map obtained by experiments.

12.6.5 Torque Control with Voltage Anti-Windup

Fig. 12.47 shows a typical block diagram for an EV motor torque control. The torque command T^* is read from a vehicle control unit (VCU) to the current command block. Then, an optimal current command is read from the current command look-up table based on torque command, speed, and DC link voltage. The inductance look-up table is used to calculate more precisely the decoupling compensation terms. It also contains a voltage anti-windup scheme. It computes the reference voltage (line-to-line voltage peak, $\sqrt{3}\sqrt{v_d^{e2} + v_q^{e2}}$) and compares it with the DC link voltage measurement, V_{dc}. If the DC link voltage is lower than the voltage reference, then it should decrease the flux level. In such a case, the voltage anti-windup feeds back a negative value so that a reduced flux is fed to the current command block.

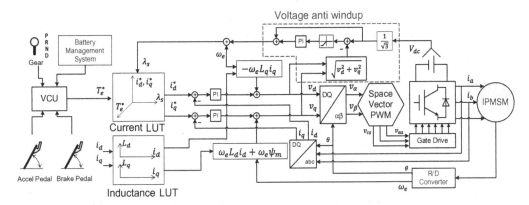

Figure 12.47: EV control block diagram containing a current look-up table and a voltage anti-windup.

References

[1] A. M. EL-Refaie et al., "Advanced High Power-Density Interior Permanent Magnet Motor for Traction Applications," *IEEE IEMDC*, pp. 581−587, 2013.

[2] S. Jurkovic et al., "Retrospective of Electric Machines for EV and HEV Traction Applications at General Motors," *IEEE ECCE*, pp. 5219−5226, 2015.

[3] T. Jahns, "Getting rare-earth magnets out of EV traction machines," *IEEE Electrification mag.*, vol. 5, no. 1, pp. 6-18, Mar. 2017.

[4] C. Hsiao et al., "Next Generation Chevy Volt Electric Machines; Design, Optimization and Control for Performance and Rare-Earth Mitigation," *Energies* no. 4, pp. 2166−2179, 2011.

[5] G. Pellegrino, A. Vagati, B. Boazzo, and P. Guglielmi, "Comparison of Induction and PM Synchronous Motor Drives for EV Application Including Design Examples" *IEEE Trans. on. Ind. Appl.*, vol. 48, no. 6, pp. 2322−2332, Dec. 2012.

[6] K. Yamazaki and M. Kumagai, "Torque Analysis of Interior Permanent-Magnet Synchronous Motors by Considering Cross-Magnetization: Variation in Torque Components With Permanent-Magnet Configurations," *IEEE Trans. on Ind. Appl.*, vol. 61, no. 7, pp. 3192 − 3200, July 2014.

[7] J. A. Walker, D. G. Dorrell, and C. Cossar, Flux-linkage calculation in permanent-magnet motors using the frozen permeabilities method, *IEEE Trans. on Magn.*, vol. 41, no. 10, pp. 3946 − 3948, Oct. 2005.

[8] S. Jung, J. Hong, and K. Nam, "Current Minimizing Torque Control of the IPMSM Using Ferrari's Method," *IEEE Trans. Power Elec.*, vol. 28, no.12, pp. 5603–5617, Dec. 2013.

[9] I. Jeong and K. Nam, "Analytic Expressions of Torque and Inductances via Polynomial Approximations of Flux Linkages," *IEEE Trans. on Magn.* vol. 51, no. 7, article sequence no. 7209009, 2015.

Problems

12.1 Consider an eight pole, 48 slot IPMSM. The air gap bore radius is 78 mm, and has a triple layered winding. The peak phase current is 350 A_{rms}. Find the electric loading.

12.2 Using the data in Table 12.3, determine the motor terminal voltages at 10000 rpm when the current magnitude and angle are $(I_s, \beta) = (350, 55°)$. Find the minimum β to make the magnitude of line-to-line voltage less than 202 V. Find the PF.

12.3 Consider an IPMSM with $P = 8$, $\psi_m = 0.09$ Wb, $r_s = 0.01$ Ω, $L_d = 280$ μH, and $L_q = 500$ μH. The inverter DC link voltage is $V_{dc} = 325$ V. Determine the current magnitude and angle so that the motor produces the maximum power up to 10000 rpm. The maximum current is $I_s = 280$ A (peak), and set the voltage margin, $\Delta V = 6$ V. Calculate the maximum power curve using Excel spreadsheet. Plot torque, power, and current.

12.4 Determine the back EMF constant of a six pole motor when the measured line to line fundamental peak voltage is 100 V at 3600 rpm.

12.5 Consider an eight pole IPMSM with the following parameters: $\psi_m = 0.09$ Wb, $r_s = 0.01$ Ω, $L_d = 280$ μH, and $L_q = 500$ μH. Using the method shown in Fig. 12.40, obtain a set of torque maximizing current commands, (i_d^{e*}, i_q^{e*}) for $I_s = 50, 100, 150, 200, 250$ and 300 A when the motor speed is 4500 rpm and the battery voltage is 320 V. Plot (i_d^{e*}, i_q^{e*}) in the current plane.

12.6 Consider an IPMSM that yields the optimal current sets shown in Fig. 12.42.
a) Assume that the IPMSM is producing 185 Nm at 4500 rpm. Read the optimal current command from Fig. 12.42.   b) Assume that the battery voltage is reduced to 320 V from 360 V. Using Figs. 12.42 and 12.44, find an optimal current set, (i_d^{e*}, i_q^{e*}) that generates 185 Nm at 4500 rpm under $V_{dc} = 320$ V.
c) Suppose that the IPMSM parameters are $L_d = 0.234$ mH, $L_q = 0.562$ mH, $r_s = 13$ mΩ, and $\psi_m = 0.0927$ Wb. Using Excel spreadsheet, find an optimal current solution (i_d^{e*}, i_q^{e*}) that generates 185 Nm at 4500 rpm under $V_{dc} = 320$ V. Compare the result with the one of b).

12.7 Assume that a 3 phase eight pole IPMSM is rotating at 3000 rpm by an external force. Before $t = 0.0175$ s, no current was injected. Neglect the coil resistance.

a) Find the back EMF coefficient, ψ_m from the open circuit voltage.

b) Suppose that current flows after $t = 0.0175$ s. Then the voltage is reduced to $V_s = 86$ V, and the voltage angle, measured from the q-axis, is $45°$. Assume that that current magnitude is $I_s = 86$ A, and that the current is delayed by $20°$ from the voltage, i.e., the current angle is $20°$. Find v_d^e, v_q^e, i_d^e, and i_q^e using the voltage vector diagram. Find L_d and L_q.

c) Construct an IPMSM model with L_d and L_q obtained in b) with MATLAB Simulink. Construct d and q-axis current controllers, and set the current commands by the currents obtained in b). Show that the a-phase voltage waveform is identical to the one shown in the following figure (c).

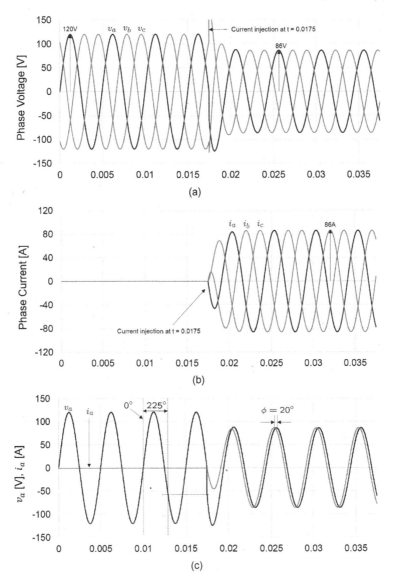

12.8   Consider a three-phase twelve pole PMSM for the EV bus which has following inductances. The rated current is $I_s = 420$ A and the battery voltage is 600 V_{DC}.

a) The fundamental component of a phase voltage is 67.482 V when the shaft speed is 300 rpm. Find the back EMF coefficient, ψ_m.

b) Using the following inductance table, draw the torque curve versus current angle β from 30° to 80° when the rated current flows. Determine the current angle that maximizes the torque.

Beta (°)	L_d (mH)	L_q (mH)
30	0.69	1.59
35	0.70	1.63
40	0.71	1.67
45	0.72	1.72
50	0.73	1.77
55	0.74	1.82
60	0.75	1.88
65	0.77	1.95
70	0.79	2.02
75	0.83	2.11
80	0.88	2.27

c) Using Excel spreadsheet, calculate the torque and power for $I_s = 420$ A at different speeds: 300, 600, 900, 1000, 1180, 1300, 1500, 1800, 2300, 3200 rpm, Draw torque and power curves versus speed.

12.9   Consider a three phase 48 slot eight pole motor which has following motor parameters: number of turns per phase $N_{ph} = 16$, stack length $l_{st} = 150$ mm, stator inner diameter (air gap radius) $D_{si} = 160$ mm, and slot height $h_q = 25$ mm, slot width $w_q = 8$ mm.

a) The coil diameter is 1.0 mm. Calculate the number of strands so that the fill factor is in $[0.43 \sim 0.45]$

b) Calculate the coil length per phase using (12.6) and (12.7), where l_a is 3 mm. (Hint: Calculate the pole pitch τ_p using D_{si} and use $l_{ph} = N_{ph} \cdot l_{turn}$.)

c) Calculate the resistances of the phase winding when the coil temperature is 20°C and 140°C. Use the resistivity $\rho_{20} = 1.724 \times 10^{-5}$ $\Omega \cdot$ mm at 20°C. At 140°C, use formula (12.9).

d) Calculate the copper loss when the current is 160 A_{rms} at 140°C.

12.10   A typical BLDC motor has 120° coil span and 180° PM pole coverage arc in electric angle. Therefore, the ideal back EMF exhibits a trapezoidal waveform with 120° flat top (bottom) regions as shown in the following:

a) Using T and E, expand the back EMF in Fourier series.

b) Note that the EMF is plotted versus mechanical angle. Let $T = \frac{\pi}{2}$. Determine the number of poles.

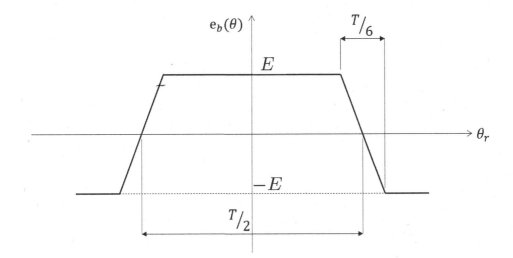

c) Assume that $E = 100$ V. Calculate the Fourier coefficients for $n = 1, 2, 3, \cdots, 15$ and plot the harmonic spectrum.

d) When the rotating speed is 3000 rpm, calculate the back EMF coefficient ψ_m.

Chapter 13

Vehicle Dynamics

Since the land vehicles are designed to move primarily in one direction, only single dimensional lateral dynamics are considered. There are five lateral force components: inertial force, longitudinal traction force, air drag, tire rolling resistance, and gravity when the vehicle is moving up or down a hill.

13.1 Longitudinal Vehicle Dynamics

Fig. 13.1 shows the lateral force components for a vehicle moving on an inclined road. Air drag, F_{aero}, rolling resistance, F_{roll}, and gravity constitute the road load. Traction force, F_x, is provided via the slip between tire and road, and the engine or electric motor is the real power source for the slip generation. The difference between the sum of road loads and the traction force is used for acceleration or deceleration:

$$m_v \frac{d}{dt} V_x = F_x - F_{aero} - F_{roll} - m_v g \sin \alpha \qquad (13.1)$$

where V_x is the vehicle velocity along the longitudinal x-direction, m_v is the vehicle mass including passenger loads, g is the acceleration of gravity, and α is the incline angle of the road.

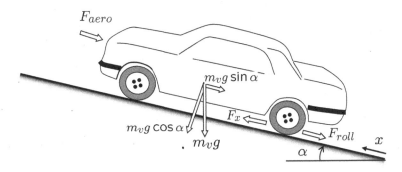

Figure 13.1: Lateral force components of a vehicle.

13.1.1   Aerodynamic Drag Force

When a headwind blows at a speed of V_{wind} to a moving vehicle, the aerodynamic drag force exerted on a vehicle is calculated as

$$F_{aero} = \frac{\rho C_d A_F}{2}(V_x + V_{wind})^2 \tag{13.2}$$

where ρ is the mass density of air, C_d is the aerodynamic drag coefficient, and A_F is the equivalent frontal area of the vehicle. The mass density of air is equal to $\rho = 1.225$ kg/m^3 at the commonly used standard set of conditions (15°C and a 101.32 kPa). The frontal area A_F is in the range of $79-84\%$ of the area calculated from the vehicle width and height for passenger cars [5]. The aerodynamic drag coefficient is typically in the range $0.2 < C_d < 0.4$: $C_d = 0.3$ for common passenger cars and $C_d = 0.4$ for common sports utility vehicles.

To investigate the aerodynamic effect separately, we assume in the following that a vehicle is traveling a horizontal road when the headwind is equal to zero. Neglecting the rolling resistance, (13.1) reduces to

$$m_v \frac{dV_x}{dt} = -\frac{\rho A_F C_d}{2}V_x^2 + F_x. \tag{13.3}$$

Let $K_1 = \sqrt{\frac{\rho A_F C_d}{2m_v}}$ and $K_2 = \sqrt{\frac{F_x}{m_v}}$. Then it follows from (13.3) that

$$\frac{dV_x}{dt} = -K_1^2 V_x^2 + K_2^2.$$

Separating variables, we obtain

$$\frac{dV_x}{V_x - K_2/K_1} - \frac{dV_x}{V_x + K_2/K_1} = -2K_1 K_2 dt.$$

Thus, integrating over $[0, t]$, it follows that

$$\ln \frac{|V_x - K_2/K_1|}{|V_x + K_2/K_1|} = -2K_1 K_2 t.$$

Note however $V_x < K_2/K_1$. Thus, we have

$$-V_x + K_2/K_1 = (V_x + K_2/K_1)e^{-2K_1 K_2 t}.$$

Therefore, the velocity is given as [3]

$$V_x(t) = \frac{K_2}{K_1}\frac{e^{K_1 K_2 t} - e^{-K_1 K_2 t}}{e^{K_1 K_2 t} + e^{-K_1 K_2 t}} = \frac{K_2}{K_1}\tanh\left(K_1 K_2 t\right). \tag{13.4}$$

Since the hyperbolic tangent function is monotonically increasing, the maximum velocity is obtained as

$$\max V_x(t) = \lim_{t \to \infty} \frac{K_2}{K_1}\tanh\left(K_1 K_2 t\right) = \frac{K_2}{K_1} = \sqrt{\frac{2F_x}{\rho A_F C_d}}. \tag{13.5}$$

This is a velocity limit caused by the aerodynamic drag force. That is, the maximum velocity is determined mostly by the thrust and aerodynamic coefficients.

Exercise 13.1 Consider a vehicle with parameters: $m_v = 1000$ kg, $A_F = 2.5$ m^2, $\rho = 1.225$ kg/m^3, and $C_d = 0.3$. Assume that the traction force is equal to 2 kN. Considering only the aerodynamic resistance, calculate the maximum velocity. Repeat the calculation when $A_F = 1.5$m^2.

Solution. $K_1 = \sqrt{\rho A_F C_d/(2m_v)} = 0.0214$ and $K_2 = 1.414$. The maximum velocity is equal to 66.07 m/s $= 238$ km/h. When $A_F = 1.5$ m^2, the maximum velocity is 307 km/h. ■

13.1.2 Rolling Resistance

As the tire rotates, a part of the tire is continuously depressed at the bottom, and then released back to its original shape after it leaves the contact region. Fig. 13.2 shows a tire depression. These depressing and releasing processes are not totally elastic. That is, due to the damping action, energy is consumed during the deforming and recovering processes. Such a loss of energy in the tire is reflected as a rolling resistance that opposes the motion of the vehicle.

Obviously, the amount of deformation depends on the vehicle's weight. Typically, the rolling resistance is modeled to be proportional to the normal force, F_z, on the tire, i.e., the sum of rolling resistance is

$$F_{roll} = f_r F_z = f_r m_v g \cos \alpha, \tag{13.6}$$

where f_r is the rolling resistance coefficient. Typical values for radial tires are in the range of $0.009 \sim 0.015$. The rolling resistance is affected by the vehicle speed, but the speed dependent term is very small so that it is usually neglected [1]. If the weight distribution between the front and rear wheels is not even, the rolling resistance should be calculated separately.

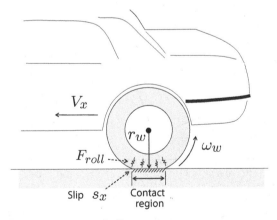

Figure 13.2: Rolling resistance and slip.

Exercise 13.2 Suppose that the weight distribution of front and rear wheels is 6:4, and the vehicle mass is 1200 kg. Find the rolling resistance force, when

$f_r = 0.015$. Assume that the vehicle is running at the speed of 80 km/h, while the traction power is equal to $P_x = 30$ kW. Find the percentage of the rolling resistance out of the total traction force.

Solution.

$$\text{Front wheels} \quad : \quad F_{roll} \quad = \quad 0.015 \times 1200 \times 0.6 \times 9.8 = 105.8 \text{ (N)},$$
$$\text{Rear wheels} \quad : \quad F_{roll} \quad = \quad 0.015 \times 1200 \times 0.4 \times 9.8 = 70.6 \text{ (N)}.$$

The total tire traction force is equal to $F_x = P_x/V_x = 30000/(80000/3600) = 1350$ N. Thus, the percentage of the rolling resistance is equal to $(105.8{+}70.6)/1350 \times 100 = 13\%$. ∎

13.1.3   Longitudinal Traction Force

The longitudinal traction force F_x is based on a friction that is proportional to the slip between the tire and road surfaces. The slip speed is defined as the difference between the circumferential speed of the tire, $r_w\omega_w$ and vehicle velocity, V_x, where r_w is the wheel dynamic rolling radius and ω_w is the angular speed of the wheel. The normalized slip is defined as [2]

$$s_x \quad = \quad \frac{r_w\omega_w - V_x}{r_w\omega_w}. \tag{13.7}$$

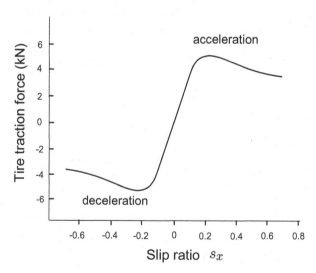

Figure 13.3: Longitudinal traction force as a function of wheel slip.

Experimental results have shown that the longitudinal traction force increases in proportion to the slip in the low slip range. The force grows almost linearly with the slip but decays after a certain slip. Fig. 13.3 shows a typical longitudinal friction coefficient versus slip and it is modeled with a nonlinear function, $\mu(\cdot)$, as

$$F_x = \mu(s_x) F_{z(dr)}, \tag{13.8}$$

where $F_{z(dr)}$ is the normal force on the drive axle. However, it depends on the road and tire conditions. On dry concrete or asphalt, the traction force increases until the slip reaches 20% [1]. Fig. 13.3 also indicates that the slip always takes place whenever traction force is generated. In the linear region, the traction force is modeled as

$$F_x = \mu_{s0} s_x m_v g \cos \alpha. \tag{13.9}$$

where μ_{s0} is the longitudinal friction coefficient.

Exercise 13.3 Assume that a vehicle of $m_v = 1200$ kg is on the horizontal plane. According to the curve shown in Fig. 13.3, the traction force is equal to 4kN at $s_x = 0.1$. Determine μ_{s0}.

Solution.
$$\mu_{s0} = \frac{4000}{0.1 \times 1200 \times 9.8} = 3.4. \qquad\blacksquare$$

13.1.4   Grade

The grade is defined as a percentage, i.e., % grade $= \frac{\Delta h}{d} \times 100$, where d and Δh are horizontal and vertical lengths of a slope, respectively. Then the grade angle, α is equal to

$$\alpha = \arctan \frac{\%\text{grade}}{100}. \tag{13.10}$$

Typical grade specifications are 7.2% for normal driving and 33% for vehicle launch [1].

Exercise 13.4 Assume that a vehicle of mass 2000 kg is moving on a uphill of grade 20%. Calculate the force due to gravity.

Solution. Note that $\alpha = \arctan(20/100) = 11.3°$.
Thus, $mg \sin \alpha = 2000 \times 9.8 \times \sin 11.3° = 3841$ N. $\qquad\blacksquare$

13.2   Acceleration Performance and Vehicle Power

Vehicle dynamics (13.1) is rewritten as

$$\frac{dV_x}{dt} = \frac{1}{m_v} \left(F_x - \frac{\rho C_d A_F}{2} V_x^2 - f_r m_v g \cos \alpha - m_v g \sin \alpha \right). \tag{13.11}$$

The difference of tractive force, F_x, from the sum of road loads (gravity, rolling, and aerodynamic resistances) is used for accelerating the vehicle.

Acceleration is an important feature in the vehicle performance evaluation. Most passenger cars have $4 \sim 12$ second acceleration time from zero to 100 km/h. Consider, for example, a car that has 10 seconds acceleration time for 0–100 km/h. Then, an average acceleration is 0.28 g. Fig. 13.4 shows the inertial force components for 0.28 g acceleration. Note that the inertial force takes the largest percentage (See Fig. 13.4(a)). The next largest component is the gravity when the grade is 7.5%. The aerodynamic force becomes noticeable when the vehicle speed exceeds 100 km/h.

Exercise 13.5   Consider a vehicle with parameters: $m_v = 1500$ kg, $A_F = 2$ m^2, $\rho = 1.225$ kg/m^3, $C_d = 0.3$, and $f_r = 0.01$. Suppose that the maximum acceleration is 0.28 g.

a) Calculate the lateral force components; $\frac{\rho}{2}C_d A_F V_x^2$, $m_v \frac{dV_x}{dt}$, $f_r m_v g \cos \alpha$, and $m_v g \sin \alpha$, when the grade is 7.5%.

b) Repeat the same calculation when the vehicle runs at constant speeds when the grade is 20%.

Solution. Fig. 13.4 shows the magnitudes of aerodynamic, gravity, tire rolling, and inertial forces.   ■

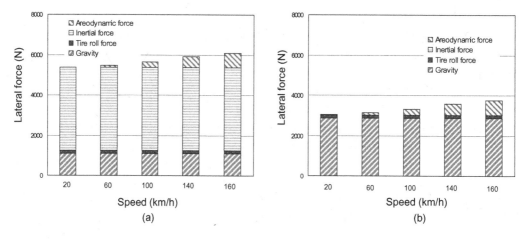

Figure 13.4: Lateral force components versus speeds: (a) acceleration=0.28 g and grade = 7.5% and (b) zero inertial force with grade = 20% (Solution to Exercise 13.5).

As indicated by (13.8), the vehicle thrust is generated via the slip between the driving wheels and the load. Note however that the wheel slip is a function of the vehicle speed, V_x, Therefore, the traction force cannot be determined, unless vehicle speed is known, i.e., the complete vehicle dynamics are described by a closed loop:

Substituting (13.9) for F_x, we obtain from (13.11) that

$$\frac{dV_x}{dt} = \frac{1}{m_v}\left(\mu_{s0}\frac{r_w\omega_w - V_x}{r_w\omega_w}m_v g\cos\alpha - \frac{\rho C_d A_F}{2}V_x^2 - f_r m_v g\cos\alpha - m_v g\sin\alpha\right).$$

$$(13.12)$$

Exercise 13.6 [6] Simplified longitudinal vehicle dynamics are:

$$J_w\dot\omega_w = T_w - r_w F_x \qquad (13.13)$$

$$m_v\dot V_x = F_x - F_{rl} \qquad (13.14)$$

$$F_x = m_v g \times \mu_{s0}\frac{r_w\omega_w - V_x}{r_w\omega_w}, \qquad (13.15)$$

where J_w is the wheel inertia, T_w is the wheel shaft torque, and F_{rl} is the road load representing the sum of the rolling resistance, gradient, and aerodynamic drag forces. Draw a block diagram of the longitudinal dynamics based on (13.13)–(13.15) regarding F_{rl} as an external disturbance.

13.2.1 Final Drive

Consider an EV drive line model, shown in Fig. 13.5. Denote by g_{dr} the whole gear ratio from the motor shaft to the wheel axle. Also, we denote by η_{dr} the drive line efficiency, i.e., the efficiency of the drive train between the motor and the axle.

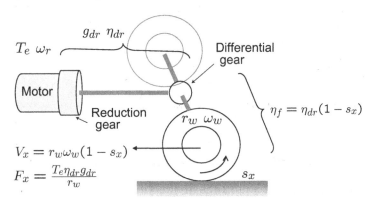

Figure 13.5: EV drive line model.

Note that

$$V_x = r_w\omega_w(1 - s_x) = r_w\left(\frac{\omega_r}{g_{dr}}\right)(1 - s_x), \qquad (13.16)$$

where r_w is the effective radius of the driving wheels. Therefore, the vehicle traction power is equal to

$$P_x = T_w\frac{V_x}{r_w} = T_w\left(\frac{\omega_r}{g_{dr}}\right)(1 - s_x)$$

$$= T_e\omega_r\eta_{dr}(1 - s_x), \qquad (13.17)$$

where T_w is the torque of the wheel axle. In the second equality of (13.17), the loss in the drive line is reflected by multiplying η_{dr}, i.e., the motor shaft torque is reduced such that $T_e\eta_{dr} = \frac{T_w}{g_{dr}}$.

We also denote by η_f the whole efficiency from motor shaft power, P_e, to vehicle traction power. Since $P_e = T_e\omega_r$, it follows from (13.17) that

$$\eta_f \equiv \frac{P_x}{P_e} = \eta_{dr}(1 - s_x). \tag{13.18}$$

Thus, the whole efficiency is comprised of the drive line efficiency and the efficiency in the force conversion between the wheel surface and the road. Note that there is a significant loss mechanism associated with the wheel sleep. Specifically, the power loss caused by the slip is equal to $P_e\eta_{dr}s_x$ or $F_xV_x\frac{s_x}{1-s_x}$. The Sankey diagram for the motor power is depicted in Fig. 13.6.

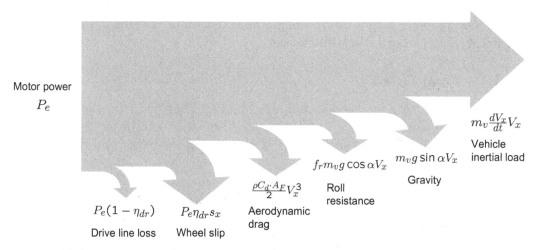

Figure 13.6: Sankey diagram for the EV motor power.

13.2.2 Speed Calculation with Torque Profile

The maximum torque curve of a motor is divided into two segments: constant torque and constant power regions. The two regions are separated by the base speed, ω_b, as shown in Fig. 13.7. Therefore, different values of the tractive effort should be used for the maximum acceleration.

Utilizing (13.17) and (13.18), it follows that

$$F_x = \frac{P_x}{V_x} = \frac{T_e\eta_{dr}\omega_r(1 - s_x)}{r_w\omega_w(1 - s_x)} = \frac{T_e\eta_{dr}g_{dr}}{r_w}. \tag{13.19}$$

Using the base speed, ω_b, of the traction motor as a pivot value, the maximum tractive effort of EV is set as

$$F_x = \begin{cases} \frac{T_e g_{dr}\eta_{dr}}{r_w}, & \omega_r \le \omega_b \\ \frac{P_e\eta_f}{V_x}, & \omega_r > \omega_b. \end{cases}$$

Table 13.1: Example EV parameters

Vehicle		Motor	
Curb weight	1313 kg	Max. torque	400 Nm
Drag Coeff., C_d	0.3	Base speed	1200 rpm
Frontal area, A_F	1.746 m^2	Max. speed	6400 rpm
Rolling resistance, f_r	0.009	Max. power	50 kW
Dynamic tire radius, r_w	0.29 m		

Then, the governing equations for the maximum acceleration are

$$\frac{dV_x}{dt} = \frac{1}{m_v}\left(\frac{T_e g_{dr} \eta_{dr}}{r_w} - f_r m_v g - \frac{\rho C_d A_F}{2}V_x^2\right) \quad \text{for constant torque region.}$$

$$(13.20)$$

$$\frac{dV_x}{dt} = \frac{1}{m_v}\left(\frac{P_e \eta_f}{V_x} - f_r m_v g - \frac{\rho C_d A_F}{2}V_x^2\right) \quad \text{for constant power region.}$$

$$(13.21)$$

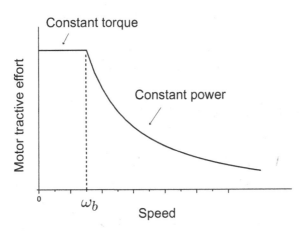

Figure 13.7: An EV motor tractive effort.

Exercise 13.7   Consider an EV with the parameter listed in Table 13.1.

a) Assume that $g_{dr} = 4.1$ and $\eta_{dr} = 0.95$. Calculate the maximum tractive effort. Draw the tractive effort versus speed for 50 kW constant power.

b) Assume that the vehicle mass with a load is 1500 kg. Compute road loads according to

$$F_{rl} = \frac{\rho C_d A_F}{2}V_x^2 + f_r m_v g \cos\alpha + m_v g \sin\alpha$$

when the grades are 0, 10, 20, 30, and 35%. Draw the load lines versus speed.

Solution.

$$F_x = \frac{T_e g_{dr} \eta_{dr}}{r_w} = \frac{400 \times 4.1 \times 0.95}{0.29} = 5372 \text{ N.}$$

The vehicle speed corresponding to the motor base speed (with zero slip) is

$$\frac{1200}{60} \times 2\pi \times \frac{1}{4.1} \times 0.29 \times \frac{3600}{1000} = 32 \text{ km/h.}$$

Calculation results are shown in Fig. 13.8.   ■

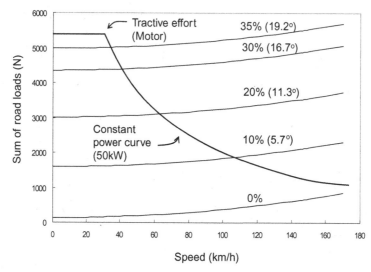

Figure 13.8: Motor tractive force and the sum of lateral road loads of an EV listed in Table 13.1. (Solution to Exercise 13.7).

Exercise 13.8   Consider an EV with parameters: $m_v = 1598$ kg, $A_F = 2$ m^2, $\rho = 1.225$ kg/m^3, $C_d = 0.3$, $f_r = 0.015$, $r_w = 0.284$ m, $g_{dr} = 9$, $\eta_{dr} = 0.9$, $\eta_f = 0.8$, and $\alpha = 0$. The motor has the constant maximum motor torque of 270 Nm under a base speed. Assume that the vehicle speed corresponding to the motor base speed is $V_{xb} = 30$ km/h.

a) Calculate the time elapsed before the vehicle speed reaches 30 km/h. Neglect the aerodynamic force, since it is small in a low speed range.

b) Calculate the maximum motor power at the time vehicle speed reaches $V_x = 30$ km/h.

c) Compute the acceleration performance utilizing Runge-Kutta 4th method based on (13.20) and (13.21).

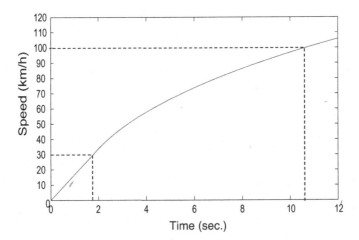

Figure 13.9: Acceleration performance (Solution to Exercise 13.8).

Solution.
a) Neglecting the aerodynamic force, we have

$$\frac{dV_x}{dt} = \frac{1}{1598}\left(\frac{270 \times 9 \times 0.9}{0.284} - 0.015 \times 1598 \times 9.8\right) = 4.67 \text{ m/s}^2.$$

Therefore, the elapsed time for $V_x = 30$ km/h is equal to

$$\frac{30 \times 1000}{3600} \times \frac{1}{4.67} = 1.78 \text{ s.}$$

b) The motor power is equal to

$$P_e = T_e \frac{g_{dr} V_x}{\eta_f r_w} = 270 \times \frac{9}{0.8} \times \frac{30000}{3600 \times 0.284} = 80.2 \text{ kW.}$$

c) Fig. 13.9 shows a simulation result based on (13.20) and (13.21) according to the maximum torque and constant power curves shown in Fig. 13.7. Note that 0–100 km/h acceleration time is 10.7 s. The simulation result shows that the acceleration time up to 30km/h is 1.8 s, which is quite close to the approximate solution (1.78 s) obtained in a). ∎

Exercise 13.9 Consider an EV with $m_v = 1500$ kg and $f_r = 0.015$. Assume that the drive line efficiency is $\eta_f = 0.82$. Neglecting the aerodynamic force, calculate the road load, F_x when the road grade is 7.5%. Calculate the required motor power to maintain a vehicle speed at 36 km/h.

Solution. Note that $\alpha = \tan^{-1}(0.075) = 4.3°$. The required force at steady state is equal to

$$
\begin{aligned}
F_x &= f_r m_v g \cos\alpha + m_v g \sin\alpha \\
&= 0.015 \times 1500 \times 9.8 \times \cos 4.3° + 1500 \times 9.8 \times \sin 4.3° = 1319.5 \text{ N.}
\end{aligned}
$$

Thus,

$$P_e = F_x V_x \times \frac{1}{\eta_f} = 1319.5 \times \frac{36000}{3600} \times \frac{1}{0.82} = 16091 \text{ W}. \qquad \blacksquare$$

13.3   Driving Cycle

A driving cycle is a standardized driving pattern designed to test the efficiency of vehicle engines or drive trains. The pattern is a speed-time table which stands for urban stop-and-go trips or relatively smooth highway cycles. The US Environmental Protection Agency (EPA) developed the Federal Test Procedure (FTP75) to assess the performance of vehicles, such as fuel consumption and polluting emissions. The urban portion of the FTP75 is taken as the urban dynamometer driving schedule (UDDS). The UDDS, often called LA-4 or FTP72 cycle, has the high percentage of stop and acceleration/deceleration, and the maximum speed is less than 100 km/h. A hybrid power train can show a significant gain in fuel economy with the UDDS.

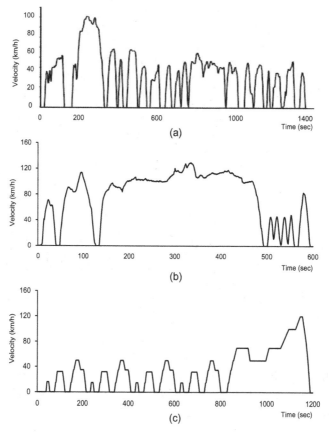

Figure 13.10: (a) EPA urban dynamometer driving schedule (UDDS), (b) highway driving cycle (US06 Hwy), and (c) NEDC.

NYCC is the most representative of urban driving that includes signals and congestion, with an average speed of only 11.4 km/h. US 06 cycle is an EPA highway cycle representing a high speed traffic flow. The average speed is 77.2 km/h, but the maximum acceleration is 3.8 m/s². New European driving cycle (NEDC) consists of four repeated ECE-15 driving cycles and an Extra-Urban driving cycle. But, the maximum velocity is lower than the real vehicles speed on European highway. Fig. 13.10 shows the speed plots of EPA UDDS, US06 Hwy, and NEDC, whose average speeds are 34, 77.2, and 32.2 km/h.

With the vehicle dynamics, it is possible to calculate the energy consumed for a given driving cycle. The calculation process consists of two parts: one for the mechanical part and the other for the electrical part. The first mechanical part covers from the speed profile to motor shaft torque. Based on the vehicle dynamic model, the required thrust can be calculated. It is depicted in the left column of Fig. 13.11. The electric power calculation starts from the motor to the battery. This process is described in the right column Fig 13.11.

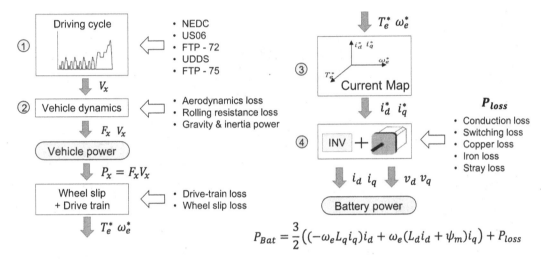

Figure 13.11: Loss calculation based on driving cycle.

13.3.1 Mechanical Power Calculation

In order to calculate the traction force, acceleration is calculated numerically from the speed data such that $\frac{dV_x}{dt} = (V_x(t+T) - V_x(t))/T$, where V_x is the vehicle speed. Once $\frac{dV_x}{dt}$ is given, the vehicle thrust or vehicle traction force can be calculated according to

$$F_x = \underbrace{m_v \frac{dV_x}{dt}}_{\text{inertial}} + \underbrace{\frac{\rho C_d A_F}{2} V_x^2}_{\text{aero}} + \underbrace{f_r m_v g \cos\alpha}_{\text{roll}} + \underbrace{m_v g \sin\alpha}_{\text{gravity}}. \tag{13.22}$$

Note that the wheel circumferential force is equal to the vehicle traction force, F_x. In other words, the wheel torque must be equal to $T_w = r_w F_x$, where r_w is the effective wheel radius. Note on the other hand that vehicle thrust is made by the wheel slip, meaning that the circumferential speed and vehicle speed are different.

Recall that the normalized wheel slip is given by $s_x = (r_w\omega_w - V_x)/r_w\omega_w$, where ω_w is the wheel speed. Then the wheel power is divided into the two components:

$$
\begin{aligned}
T_w\omega_w &= T_w\omega_w(1 - s_x) + T_w\omega_w s_x \\
&= \underbrace{\frac{T_w}{r_w}}_{F_x} \underbrace{r_w\omega_w(1 - s_x)}_{V_x} + \underbrace{T_w\omega_w s_x}_{\text{road loss}}.
\end{aligned} \tag{13.23}
$$

To summarize, the wheel torque and speed are related to the vehicle counterparts by

$$
T_w = r_w F_x \tag{13.24}
$$

$$
\omega_w = \frac{V_x}{r_w(1 - s_x)}. \tag{13.25}
$$

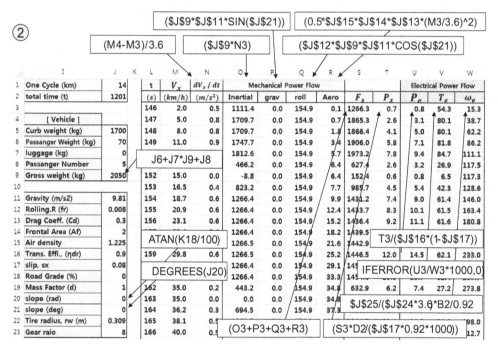

Callout formulas (top, left to right):

- (M4-M3)/3.6
- (J9*N3)
- (J9*J11*SIN(J21))
- (J12*J9*J11*COS(J21))
- (0.5*J15*J14*J13*(M3/3.6)^2)

	I	J	K (t, s)	L (V_x, km/h)	M (dV_x/dt, m/s²)	N Inertial	O grav	P roll	Q Aero	R F_x	S P_x	T P_e	U T_e	V w_e
1	One Cycle (km)	14				Mechanical Power Flow						Electrical Power Flow		
2	total time (t)	1201	(s)	(km/h)	(m/s²)	Inertial	grav	roll	Aero	F_x	P_x	P_e	T_e	w_e
3			146	2.0	0.5	1111.4	0.0	154.9	0.1	1266.3	0.7	0.8	54.3	15.3
4	[Vehicle]		147	5.0	0.8	1709.7	0.0	154.9	0.7	1865.3	2.6	3.1	80.1	38.7
5	Curb weight (kg)	1700	148	8.0	0.8	1709.7	0.0	154.9	1.8	1866.4	4.1	5.0	80.1	62.2
6	Passanger Weight (kg)	70	149	11.0	0.9	1747.7	0.0	154.9	3.4	1906.0	5.8	7.1	81.8	86.2
7	luggage (kg)	0	J6+J7*J9+J8			1812.6	0.0	154.9	5.7	1973.2	7.8	9.4	84.7	111.1
8	Passanger Number	5				466.2	0.0	154.9	6.4	627.4	2.6	3.2	26.9	117.5
9	Gross weight (kg)	2050	152	15.0	0.0	-8.8	0.0	154.9	6.4	152.4	0.6	0.8	6.5	117.3
10			153	16.5	0.4	823.2	0.0	154.9	7.7	985.7	4.5	5.4	42.3	128.6
11	Gravity (m/s2)	9.81	154	18.7	0.6	1266.4	0.0	154.9	9.9	1431.2	7.4	9.0	61.4	146.0
12	Rolling.R (fr)	0.008	155	20.9	0.6	1266.4	0.0	154.9	12.4	1433.7	8.3	10.1	61.5	163.4
13	Drag Coeff. (Cd)	0.3	156	23.1	0.6	1266.4	0.0	154.9	15.2	1436.4	9.2	11.1	61.6	180.8
14	Frontal Area (Af)	2	ATAN(K18/100)			1266.4	0.0	154.9	18.2	1439.5				
15	Air density	1.225				1266.5	0.0	154.9	21.6	1442.9	T3/(J16*(1-J17))			
16	Trans. Effi., (ηdr)	0.9	159	29.8	0.6	1266.5	0.0	154.9	25.2	1446.5	12.0	14.5	62.1	233.0
17	slip, sx	0.08	DEGREES(J20)			1266.4	0.0	154.9	29.1	145..	IFERROR(U3/W3*1000,0)			
18	Road Grade (%)	0				1266.4	0.0	154.9	33.3	145..				
19	Mass Factor (d)	1	162	35.0	0.2	1266.4	0.0	154.9	34.8	632.9	6.2	7.4	27.2	273.8
20	slope (rad)	0	163	35.0	0.0	0.0	0.0	154.9	34.8	J25/(J24*3.6*B2/0.92)				
21	slope (deg)	0	164	36.2	0.3	694.5	0.0	154.9	37.3					
22	Tire radius, rw (m)	0.309	165	38.1	0.5	(O3+P3+Q3+R3)				(S3*D2/(J17*0.92*1000))				98.0
23	Gear raio	8	166	40.0	0.5									12.7

Figure 13.12: Spread sheet for vehicle parameters and traction force calculation.

Between the wheel and the motor shaft, a drive line exists consisting of the reduction gear, differential gear, and drive shaft. Normally there are two step speed reductions: The first one is in the reduction gear box installed in front of the motor, and the second one is in the differential gear box. The total reduction ratio is denoted by $1/g_{dr}$. The loss also takes place in the drive train mainly due to the gear friction. The corresponding efficiency drop is denoted by η_{dr}. The total reduction ratio and efficiency of drive line is shown in Fig. 13.5. Thereby, torque and speed are related to motor and wheel shafts such that

$$\omega_e = g_{dr}\omega_r, \tag{13.26}$$

$$T_e = \frac{T_w}{\eta_{dr}g_{dr}}. \tag{13.27}$$

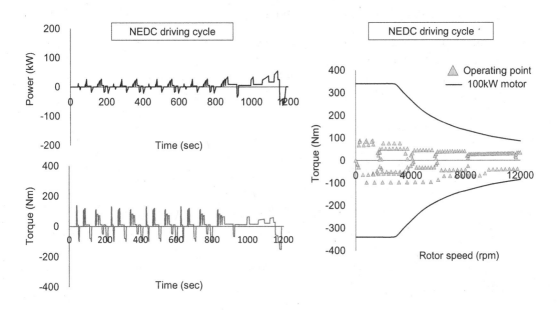

Figure 13.13: Torque and power for NEDC.

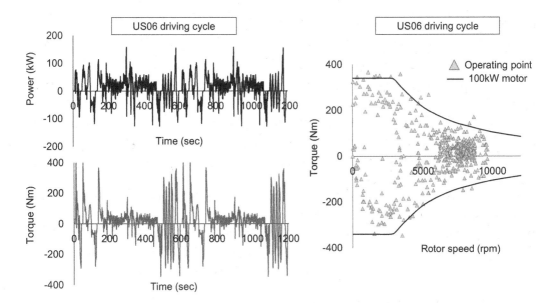

Figure 13.14: Torque and power for US06 Highway.

An example of spread sheet calculation is shown in Fig. 13.12. Computed results for NEDC and US06 Highway are shown in Fig. 13.13 and 13.14. The negative torque and negative power are generated during deceleration. Note also that the operation range is wide and wild with US06. In the case of US06, two cycles were simulated.

13.3.2   Electrical Power Calculation

Most commonly, torque and speed commands are transmitted from the vehicle control unit (VCU) to the inverter. Then the inverter finds an optimal current command, (i_d^{e*}, i_q^{e*}) from the look up table (LUT) based on torque and speed. The method of creating the LUT was discussed in detail in Section 12.6.4.   It is assumed

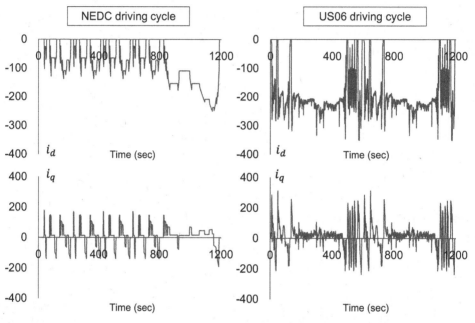

Figure 13.15: Current plots for driving cycle simulation.

here that the inverter current controller is perfect so that the real current follows the command exactly. In brief, the optimal current set is utilized at each step to generate the required torque of a driving cycle. Fig. 13.15 and Fig. 13.16 show d and q axis current commands during over the driving cycles. Note that the d-axis current is always negative and its profile looks similar to the speed profile of the driving cycle. It is because there is a tendency that the d-axis current magnitude is proportional to the speed.

13.3.3   Motor and Inverter Loss Calculation

Then, the copper loss can be calculated such that $P_{cu} = \frac{3}{2} r_s (i_d^{e\,2} + i_q^{e\,2})$. At the same time, the motor terminal voltage is calculated from the motor model such that $v_d^e = r_s i_d^e - \omega_e L_q i_q^e$ and $v_q^e = r_s i_q^e + \omega_e L_d i_d^e + \omega_e \psi_m$. Neglecting the ohmic

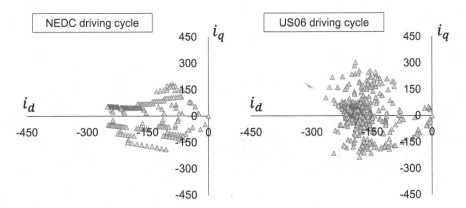

Figure 13.16: Current data in the dq current plane applied for driving cycle simulation.

drop, $\|\boldsymbol{\lambda}\|^2 = \lambda_d^2 + \lambda_q^2 \approx (v_d^{e\,2} + v_q^{e\,2})/\omega_e^2$. Thus, the iron loss can be calculated: $P_{fe} = c_{fe}\omega_e^{\gamma}(\lambda_d^2 + \lambda_q^2) = c_{fe}\omega_e^{\gamma}\left((L_d i_d + \psi_m)^2 + (L_q i_q)^2\right)$, where $c_{fe} = 0.7$ and $\gamma = 1.6$ were chosen based on the FEA loss calculation results.

On the other hand, the inverter loss is calculated based on the IGBT switching loss, on-drop loss, and diode reverse recovery loss:

$$P_{inv} = 3 \times \left[\frac{2V_{dc}}{\pi} I(t_r + t_f) + V_{on}\frac{2I}{\pi}t_{on} + \frac{V_{dc}}{2}I_{rr}t_{rr} \right] f_{sw}, \qquad (13.28)$$

where I, t_r, V_{on}, I_{rr}, t_{rr} and f_{sw} are current magnitude, rise time, fall time, collector-emitter on-state voltage, reverse recovery current, reverse recovery time and switching frequency, respectively. In the calculation, typical automotive IGBT data are applied: $t_r = 0.09$ μs, $t_f = 0.16$ μs, and $t_{rr} = 0.06$ μs. The diode reverse recovery current is estimated as $I_{rr} = \frac{1}{3}I$. Computed results of P_{cu}, P_{fe}, and P_{inv} are shown in Fig. 13.17. In the inverter loss calculation, $V_{dc} = 360$ V and $f_{sw} = 8$ kHz were utilized. The losses were calculated over all torque–speed range using the optimal current command (i_d^{e*}, i_q^{e*}) for each $(T_e, \frac{P}{2}\omega_e)$.

13.3.4 Efficiency over Driving Cycle

The the required power to and from the battery is equal to

$$P_{bat} = \frac{3}{2}(v_d i_d^{e*} + v_q i_q^{e*}) + \underbrace{P_{cu} + P_{fe} + P_{str}}_{=P_{motor}} + P_{inv}$$

$$= \frac{3}{2}\left((-\omega_e L_q i_q^{e*})i_d^{e*} + (\omega_e L_d i_d^{e*} + \omega_e\psi_m)i_q^{e*}\right) + P_{motor} + P_{inv}. \quad (13.29)$$

Note again that (i_d^{e*}, i_q^{e*}) is read from the optimal LUT based on the required torque which was resulted from the driving cycle simulation. Total energy needed to complete a driving cycle is equal to

$$E_{DC} = \int_0^t P_{bat}dt. \qquad (13.30)$$

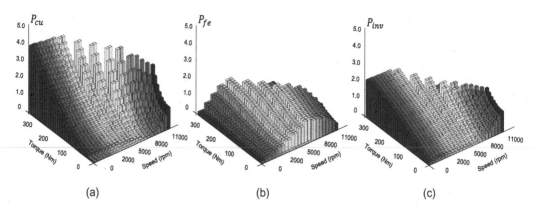

Figure 13.17: (a) Copper, (b)iron, and (c) inverter losses using the optimal current command sets.

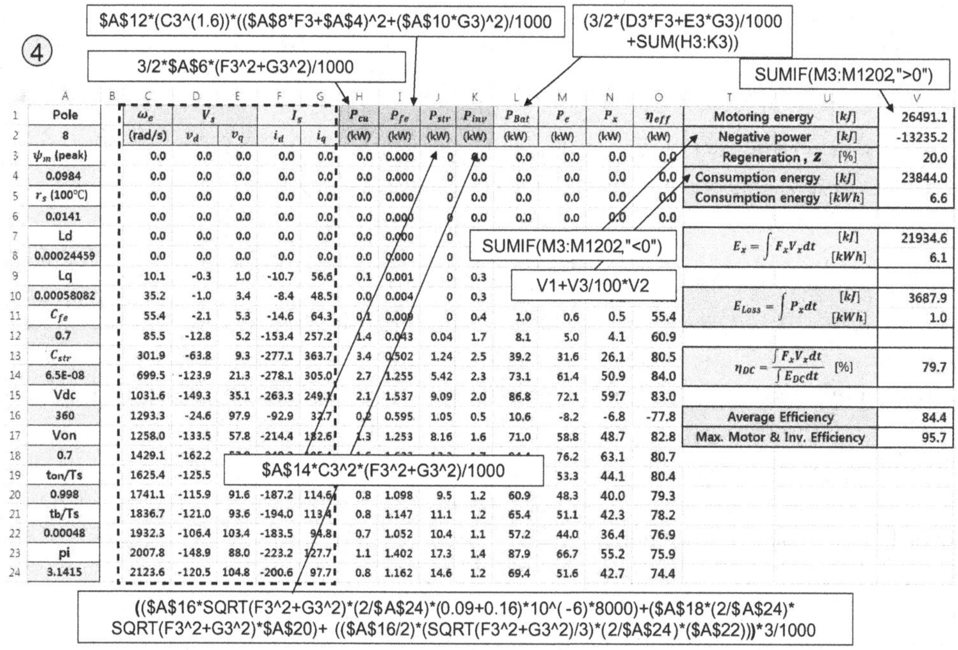

Figure 13.18: Chart of loss calculation and driving cycle efficiency calculation.

During the deceleration periods, negative power can be generated. If the vehicle is running down the hill, negative power is created continuously from the wheel. However, all the negative power cannot be recuperated into the battery. If the battery is fully charged, the battery cannot receive the braking power. Another reason not recovering power is that mechanical brake is more reliable than the electrical braking. On the average, $20 \sim 30\%$ of negative power is recovered in EVs. As a result, the total consumed energy will be different under the braking scenario.

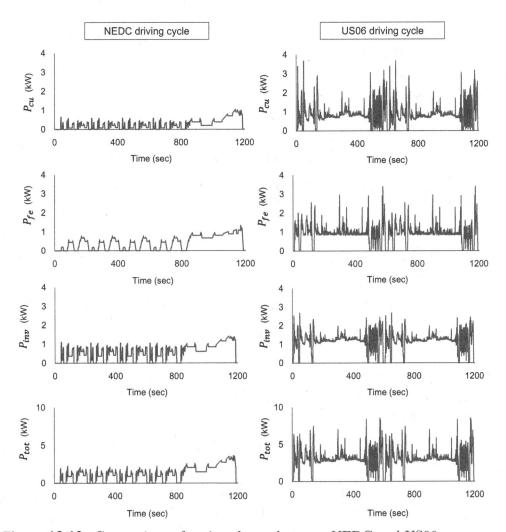

Figure 13.19: Comparison of various losses between NEDC and US06.

For computational convenience, the consumed energy is redefined as

$$E_{DC(z)} = \int_0^t P_{bat} \vee (-zP_{max}) \, dt, \qquad (13.31)$$

where $0 < z < 1$ is a constant, $P_{max} = \max_{0 \leq \tau \leq t} \{|P_{bat}(\tau)|\}$, and

$$f \vee g = \begin{cases} f & \text{if } f \geq g \\ g & \text{if } g > f. \end{cases} \qquad (13.32)$$

It implies that the braking power larger than zP_{max} is not recovered. With such an energy recovery scheme, the driving cycle efficiency is calculated as

$$\eta_{DC(z)} = \frac{\int_0^t F_x V_x \, dt}{E_{DC(z)}}. \qquad (13.33)$$

Fig. 13.18 shows a chart of calculating the loss and the efficiency. Fig. 13.19 shows plots of various losses during NEDC and US06 simulations.

References

[1] J. M. Miller, *Propulsion Systems for Hybrid Vehicles,* IEE, London, 2004.

[2] R. Rajamani, *Vehicle Dynamics and Control,* Springer, 2006.

[3] I. Husain, *Electric and Hybrid Vehicles, Design Fundamentals,* CRC Press, 2003.

[4] W. Gao, "Performance comparison of a fuel cell-battery hybrid powertrain and a fuel cell-ultracapacitor hybrid powertrain," *IEEE Trans. on Vehicular Techonology,* vol. 54, no. 3, pp. 846–855, May 2005.

[5] J. Y. Wong, *Theory of Ground Vehicles,* Wiley-Interscience, 3rd. Ed., 2001.

[6] D. Yin and Yoichi Hori, "A novel traction control of EV based on maximum effective torque estimation," *IEEE VPPC,* Harbin, Sep. 2008.

[7] K. Muta, M. Yamazaki, and J. Tokieda, "Development of new-generation hybrid system THS II - Drastic improvement of power performance and fuel economy," *Proc. of SAE World Congress,* Detroit, Mar., 2004.

[8] J. Miller, " Propulsion Systems for Hybrid Vehicles," *IET Power and Energy Series,* vol. 45, Stevenage, IET, 2004.

[9] C. C. Ming, and S. J. Cin, "Performance analysis of EV powertrain system with/without transmission," *World Electric Vehicle Journal,* vol. 4, pp. 2032–6653, 2010.

[10] A. Sorniotti, et al., "A Novel Seamless 2-Speed Transmission System for Electric Vehicles: Principles and Simulation Results," *SAE Int. J. Engines,* vol. 4, no. 2, pp. 2671–2685, 2011.

[11] T.Holdstock, A. Sorniotti, M. Everitt, M. Fracchia er al., "Energy consumption analysis of a novel four-speed dual motor drivetrain for electric vehicles," *IEEE VPPC,* pp. 295–300, 2012.

[12] H. Grotstollen, "Optimal design of motor and gear for drives with high acceleration by consideration of torque-speed and torque-acceleration product," *IEEE Trans. Ind. Appl.,* vol. 47, no. 1, pp. 144–152, Apr. 2011.

[13] S. Pinto, et al., "Torque-fill control and energy management for a four-wheel-drive electric vehicle layout with two-speed transmission," *IEEE Trans. Ind. Appl.,* vol. 53, no. 1, pp. 447–458, Jan. 2017.

[14] P. J. Kollmeyer, J. D. McFarland, and T. M. Jahns, "Comparison of class 2a truck electric vehicle drivetrain losses for single- and two-speed gearbox systems with IPM traction machines," *IEEE IEMDC*, pp. 1501–1507, 2015.

Problems

13.1 Consider a vehicle with parameters: $m_v = 1200$ kg, $A_F = 1.5$ m^2, $\rho = 1.225$ kg/m^3, and $C_d = 0.35$. Assume that the traction force is equal to 4 kN.

a) Calculate the acceleration time for $0-100$ km/h considering only the aerodynamic drag and inertial force.

b) Calculate the theoretical speed limit.

c) Calculate the acceleration time again for $0-100$ km/h when the wind blows at 10 m/s in the opposite direction.

13.2 Obtain the solution of the following in an explicit form:

$$s(t) = \int_0^t V_x(\tau)d\tau = \int_0^t \frac{K_2}{K_1}\tanh(K_1 K_2 \tau)d\tau.$$

Calculate the distance traveled during $0-100$ km/h acceleration under the conditions of Problem 13.1.

13.3 Consider an EV listed in Table 13.1. Assume $s_x = 0.1$ and $g_{dr} = 4.1$. Utilizing $V_x = \frac{\omega_r r_w}{g_{dr}}(1-s_x)$, determine the maximum vehicle speed when the motor runs at 6400 rpm.

13.4 Consider a driving pattern show below.

a) Calculate the rolling resistance, gravity, and inertial force using the data listed in Table 13.1 when the vehicle mass with a load is 1350 kg and the grade is 10%.

b) Draw the corresponding power plot.

c) Calculate the motor torque and speed for $7 \leq t < 15$ sec when the wheel slip is $s_x = 0.1$, the drive line gear ratio is $g_{dr} = 4.1$, and the drive line efficiency is $\eta_{dr} = 0.95$.

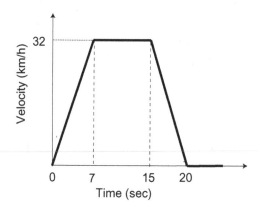

13.5 Consider an EV with parameters: the vehicle mass is $m_v = 1500$ kg, the maximum motor torque is $T_e = 250$ Nm, the drive line gear ratio is $g_{dr} = 4.1$, the drive line efficiency is $\eta_{dr} = 0.95$, and the wheel radius is $r_w = 0.29$ m

a) Calculate the maximum grade of an EV at launching. Neglect the roll force.

b) Calculate the base speed of the motor when the rated power is equal to 55 kW and the rated torque 450 Nm.

c) Assuming $s_x = 0.1$, calculate the vehicle speed when the motor operates at the base speed.

13.6 Consider an electric mini-bus with parameters: $m_v = 4500$ kg, $r_w = 0.35$ m, and $f_r = 0.01$.

a) Assuming the slip, $s_x = 0.05$, calculate wheel speed when the bus runs at 60 km/h.

b) The motor speed is desired to be 7500 rpm when the bus runs at 60 km/h with the slip 0.05. Determine the gear ratio.

c) The required thrust is $F_x = 4000$ N at $V_x = 60$ km/h. Determine the motor power assuming that the drive line efficiency with slip is 0.95.

d) Calculate the roll force according to $F_{roll} = f_r m g$.

e) Suppose that the motor starting torque is 330 Nm. Determine the maximum grade at start.

13.7 Consider EV parameters listed in Table 13.1 and refer to the motor tractive effort shown in Fig. 13.8. Calculate the vehicle $0 - 100$ km/h acceleration time with the following instructions:

In $(0, 32)$ km/h range, the maximum torque is applied. Neglect the aerodynamic drag in calculating the road load.
In $(32, 100)$ km/h range, let $\eta_f = \eta_{dr}(1 - s_x) = 0.8$.

Obtain a solution by solving (13.20) and (13.21) with the Rounge-Kutta method.

13.8 Consider a one ton electric truck with the parameters: $m_v = 2736$ kg, $r_w = 0.343$ m, $f_r = 0.014$, $C_d = 0.32$, $A_F = 4.317$, $\rho = 1.225$, $\eta_{dr} = 0.96$ and $s_x = 0.04$. Using Excel spreadsheet, solve the following problems.

a) Obtain NEDC data from www.google.com with the sampling period, $T = 1$ sec. Find V_x and $(V_x(t + T) - V_x(t))/T$ as shown in Fig. 13.12.

b) Plot thrust force, F_x and mechanical power, P_x versus t.

c) The motor shaft speed is calculated according to $\omega_r = g_{dr} \frac{V_x}{r_w(1-s_x)}$. Plot the required motor torque and electrical power for the NEDC when the gear ratio, g_{dr} is 7.767.

d) Assume that the maximum motor torque is 320 Nm, the base speed is 2750 rpm, and the maximum speed is 12000 rpm. Plot the frequency map with the motor TN curve as shown in Fig. 13.13.

e) The efficiencies of motor and inverter are $\eta_m = 0.93$ and $\eta_{inv} = 0.97$, respectively. Find the battery power over one cycle of NEDC, taking into account the loss of the motor and the inverter. (Hint: $P_{bat} = P_e/(\eta_m \eta_{inv})$.)

f) Find the consumed energy in kWh required for NEDC 10 cycles, while neglecting the regeneration power. (Hint. $E_{bat} = \int P_{bat} \vee 0 dt$)

g) Find the average consumed energy per km driving in the NEDC mode when 20% of regeneration power is recuperated. A battery is full capacity is 30 kWh. Its SOC window is 90%–10%. Find the total driving mileage per charge in km based on the average NEDC power consumption.

Chapter 14

Hybrid Electric Vehicles

Hybrid electric vehicles (HEVs), as the name stands for, have two types of power sources. An electric power source is added to the conventional internal combustion engine (ICE), which helps to improve the fuel economy through load sharing and regenerative braking. Note that ICEs have the optimum fuel efficiency in the middle speed and high torque range. However, the efficiency is low in the low speed/low torque region. Therefore, the ICEs have relatively poor performance in urban driving where there are many stop-go situations. On the other hand, electric motors generate high torque naturally in the low speed range and have a pretty high efficiency in overall operating region. When the vehicle is driven by a motor in a low speed region, the overall fuel economy will be improved significantly. In addition, the motor can force the engine to operate in an optimal condition independently of the road load. The motor also enables the regenerative braking, i.e., vehicle's kinetic energy can be recovered in the battery when the vehicle is decelerating. But, the ICE's power is sustainable over a long driving period and have the advantage of short refueling times.

14.1 HEV Basics

Typical driving patterns include start, acceleration/incline, cruise, decline, and deceleration/stop, as shown in Fig. 14.1. Direct benefits of HEVs can be summarized as follows:

Idle Off: Idle off means that the engine shuts down during even brief stops. Its functionality is designed so as not to be noticeable by the driver. As soon as a driver releases the brake pedal, the vehicle is initially propelled by the motor, and engine cranking follows soon afterwards. Idle off is particularly advantageous for urban traffic situations; taxis and buses are ideal candidates for idle off application. According to the tests undertaken by Tokyo Metropolitan authorities, idling-stopping improved fuel economy as much as 14%.

Regenerative Braking: Regenerative braking is an important feature of the hybrid systems. The vehicle's kinetic energy can be retrieved in the form of electrical energy during the period of deceleration or braking. That is, the regenerative braking saves the energy that would be lost as waste heat. In the LA-4 driving mode, about 25% efficiency improvement is obtained by hybridization [15] But the regeneration capacity is limited mainly by the battery rating: It is limited by the maximum permissible charging rate of the battery and also by the energy capacity. For example, the maximum power rating of the Prius II battery is 21 kW, whereas the required braking power is greater than 50 kW. Also, when the battery is already charged fully regenerative braking is not possible. To improve the fuel economy through regeneration, the battery capacity should be increased.

Power Assist or Power Split: Conventional ICEs are relatively inefficient at low speeds (efficiency: $5 - 10\%$). Hence, it is better to use the electric power at low speeds, and change to the ICE at high speeds. In the series power train architecture, the ICE, being dedicated to electric power generation, is decoupled from the wheel power demand. The ICE can operate at an optimal operating point, or stops. In the parallel power architecture, the motor assists the ICE while coping with surges and power deficiencies on the wheels. Of course, excess engine power is stored in the battery, and used later at high power requirements. The separating the ICE operation from the direct wheeling requirement is referred to as "power split". Efficiency enhancement of HEVs is mostly attributed to this power split.

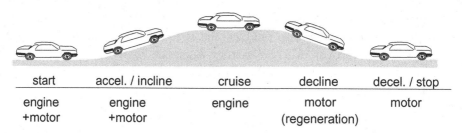

start	accel. / incline	cruise	decline	decel. / stop
engine +motor	engine +motor	engine	motor (regeneration)	motor

Figure 14.1: A driving cycle.

14.1.1 Types of Hybrids

HEVs are typically grouped into five categories according to the following criteria [1],[2].

1) Idle-off capability
2) Regenerative braking capacity (up to step 2 : "micro" hybrid)
3) Power assist and engine downsizing (up to step 3 : "mild" hybrid)
4) Electric-only drive (up to step step 4 : "full" hybrid)
5) Extended battery-electric range (up to step 5 : "plug-in" hybrid)

i) **Micro Hybrid**: Micro hybrid systems have only the idle off function. A conventional ICE vehicle can be converted easily into a micro hybrid car by replacing alternator by an integrated starter−generator (ISG). Mostly the existing ISGs are powered by 48 V battery. The ISG pulley is connected to the crank via a 3:1 belt and the maximum torque is 35 ∼ 45 Nm. Thus, the trice torque is applied to the crank shaft during starting. Usually, the maximum power rating is 10 kW for 12 seconds. Fuel efficiency gain is about 5 − 15%. Photos of ISG motor and inverter are shown in Fig. 14.2.

ii) **Mild Hybrid**: An electric motor is incorporated in the drive train, but the ICE is ·a dominant power source. Mild hybrid systems provide important HEV features such as idle off, regenerative braking, and power assisting. Drive motor often replaces torque converter in conventional automatic transmission. However, the engine power split is imperfect and the efficiency gain is about 30%−50%. Honda 'Civic Hybrid' and Hyundai 'Sonata Hybrid' are typical mild hybrid vehicles.

iii) **Full Hybrid**: Full hybrids have the power split feature, i.e., the electrical power train enables the vehicle to run its engine more frequently at its most efficient operating point. Further, the rated electrical power is large so that the vehicle can be driven solely by the electric motor (EV mode), though the range and speed are limited. In a low speed range, the electric motor drives the vehicle, but at a high speed the engine cuts in. Efficiency gain is about 50%−100%. Toyota 'Prius', Chevrolet 'Volt' and Ford "Escape" are the full hybrid vehicles.

iv) **Plug-in Hybrid**: The battery capacity is increased in plug-in hybrid electric vehicles (PHEVs) so that the vehicles can drive a long range in the EV mode. Since the vehicle battery can be charged from the power grid, a PHEV may not consume a drop of gasoline when used daily for commuting. Depending on the all electric range (AER) in miles, they are sorted as PHEV 10, PHEV 40, and PHEV 60. For example, PHEVs with the energy capacity of 10 mile driving are called PHEV10. The PHEV is a viable solution to reduce the emission of CO_2 and toxic gases in urban areas. Plug-in hybrids are favored by potential customers due to low operation costs.

Fig. 14.3 shows various types of EVs ranked according to the electrification of the power train. The conventional ICE vehicle is at the left end, while the battery electric vehicle (BEV) is at the right end. HEVs and PHEVs are in between, but PHEVs rely more on electrical power. The battery size of a PHEV determines the driving range in the all electric mode. In some PHEVs, the ICE behaves as a charging power source when the battery is depleted below a predetermined level. Hence, the ICE with a generator is often called a range extender. Further PHEVs carry on−board chargers so that household electricity can be utilized for battery charging. Characteristics of hybrid systems are summarized in Table 14.1.

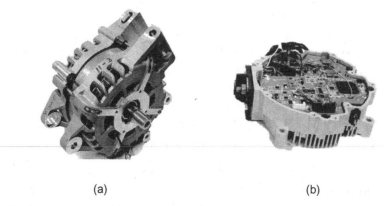

(a)   (b)

Figure 14.2: Photos of (a) ISG and (b) inverter.

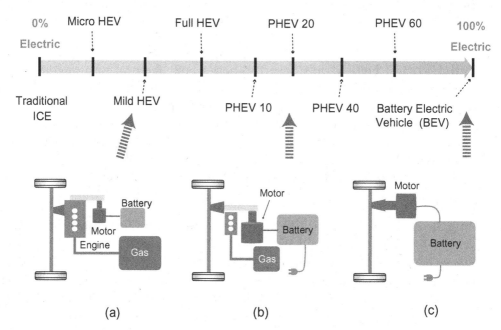

(a)   (b)   (c)

Figure 14.3: Degrees of electrification: (a) HEV (b) PHEV (c) BEV.

14.1.2   HEV Power Train Components

HEV power trains consist of an ICE, a motor, a generator, a battery, and a power control unit. The battery stores and supplies energy from and to the motor and generator. Another key component is the power split device, which splits the engine power to the motor, generator, and wheels.

a) ICE:

The efficiency of an engine is based on the fuel flow rate per useful power output, and brake specific fuel consumption (BSFC) is a measure of fuel efficiency: The measured power at the crankshaft is denoted by P_o and the fuel (mass) flow rate

Table 14.1: Hybrid systems characteristics.

	Micro Hybrid	Mild Hybrid	Full Hybrid	Plug-in Hybrid
Elec. power	ISG	Motor	Motor-Gen.	Motor-Gen.
Motor power	5 kW	16 kW	60 kW	100~120 kW
Engine power split	no	no	yes	yes
Batt. voltage	48V	200-400 V	200-400 V	>300 V
Batt. capacity		1 kWh	2 kWh	4~16 kWh
Fuel eff. gain	5~15%	30~50%	50~100%	50~100%
AER	no	no	< 1 km	16~64 km
Gas emission	yes	yes	yes	no/yes

denoted by $\dot{m}_f$. Then the BSFC is defined as

$$BSFC = \frac{\dot{m}_f}{P_o}. \qquad (14.1)$$

The BSFC map shows the group of contours indicating equi-fuel consumption rate

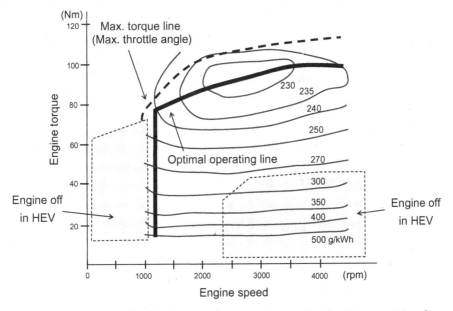

Figure 14.4: Brake specific fuel map of an engine and a basic operating line.

per shaft power production. Fig. 14.4 shows an example BSFC map, in which the units of BSFC is given by gram/kWh. Therefore, the smaller number represents the higher efficiency. Note that the highest efficiency (2800 rpm, 100 Nm) is achieved at a medium speed and near the maximum torque line. However, the efficiency is almost halved in the low torque region (under 20 Nm). The optimal operating line represents the trajectory of the most efficient operating points with increasing power and speed. To achieve maximum efficiency, it is better to narrow down the ICE operation area near (2800 rpm, 100 Nm), or stop the operation. A (dual) hybrid power train allows the ICE to operate at maximum efficiency, while the wheels can roll at independent speeds. The top dotted line represents the maximum torque of the engine.

b) Motor and Generator:
The motor and generator are functionally identical. However, depending on the power flow, they operate in either motoring or generating mode. They are named after the main use. As for the motor and the generator, permanent magnet synchronous machines are normally used for high efficiency. The motor power rating is normally larger than that of the generator, because the motor has to take care of peak power demands while the generator responds to more or less the average power.

c) Power Control Unit:
The power control unit refers to an electronic circuit system that includes two inverters for motor and generator, and DC-DC converter for battery voltage boost. An inverter controls the motor shaft torque, and the other inverter regulates the generator current, so that proper battery charging or discharging takes place. The DC-DC converter boosts up the battery voltage depending on the motor speed.

d) Battery:
The battery is the most typical energy storage system for EVs, and should be evaluated according to energy density, power density, life time, safety, and cost. Nowadays, lithium ion batteries are most popularly used.

e) Power Split Device:
The power split device is required in the series/parallel hybridization. In a wise sense, the motor and generator are also elements of power split device. But in strict sense, it refers to planetary gear and clutch. The engine power is split into a direct mechanical path and an electrical path.

14.2   HEV Power Train Configurations

The major design targets of HEVs are

  . maximum fuel economy,
  . minimum emissions,
  . minimum system costs,
  . good driving performance.

The power control strategies for HEVs involve the following considerations:

Optimal engine operation: Limiting the engine operation to the optimum range not only brings fuel savings, but also emission reduction. In the low efficiency regions, it is better to turn off the engine and drive the vehicle only with the motor. That is, a best way to achieve maximum fuel efficiency is either to operate engine in the optimal region, or to stop. However, the engine should not be turned on and off frequently, and the engine speed should not be changed quickly. Otherwise, additional fuel consumption and emissions result.

Safe battery operation: The battery should be protected from overcharging and excessive dissipation. Overcharging can lead to battery exploding and excessive charge depletion shortens battery life. Thus, there are upper and lower limits in the state of charge. During battery operation, the current should be limited so that the maximum power rating is not exceeded.

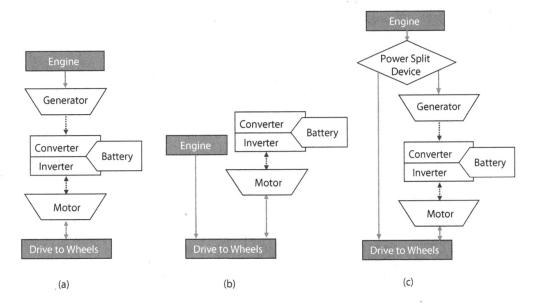

(a) (b) (c)

Figure 14.5: HEV power train topologies. (solid line : mechanical power, dotted line : electric power) : (a) series, (b) parallel, (c) series/parallel.

According to the power flow, HEVs are divided into three categories: parallel hybrids, series hybrids, and series/parallel hybrids, as shown in Fig. 14.5 [5], [1]. Dotted lines indicate electrical power flow, while solid lines indicate mechanical power flow. The series HEV topology is described in Fig. 14.5 (a): The engine, operating in the on-off modes, is devoted to generating electricity. The battery charging current is regulated by the converter. The motor delivers power to the wheel, utilizing the stored energy in the battery. Since the engine operation is independent of the vehicle speed and road load, it can work in its optimal condition. Since the series power train is simple and enables the engine to operate optimally, it is recently deployed in PHEVs.

In the parallel configuration, shown in Fig. 14.5 (b), both mechanical and electrical paths coexist. Therefore, the engine and motor can drive the wheels individually or collaboratively. It does not employ a separate generator. But, the motor, behaving as a generator, retrieves excess engine power when the engine operates in the vicinity of an optimum operation point. The parallel configuration is not so efficient, since the engine power split is imperfect.

The series/parallel hybrid employs a power split device such that the engine power can be delivered via two paths, as shown in Fig. 14.5 (c). Various power flow control modes are possible, and the configuration varieties are tuned for optimal engine operations.

14.3   Series Drive Train

Series hybrids are also referred to as extended-range electric vehicles (EREV). The series power train topology is shown in Fig. 14.6. The drive train is the simplest since it does not require a clutch, a multi-speed gearbox, or a power split device. The mechanical transmission is replaced by a generator and electric traction motors.

The engine is dedicated to the generation of electric power by turning the generator, and only the motor is responsible for satisfying varying power demands of driving. The engine/generator only operates if the battery is depleted, or to charge the batteries. Since the engine is decoupled from the varying power demands of wheels, it can operate at the optimum efficiency. The engine is controlled to perform optimally or stop.

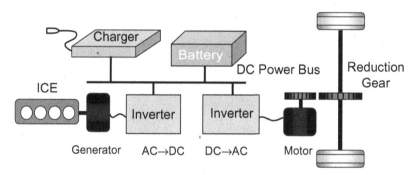

Figure 14.6: Series power train topology.

One disadvantage of the series connection is that the drive train efficiency is low due to the cascaded structure: Suppose that the motor and generator have 92% efficiency and that the inverter and converter have 97% efficiency. Then, the total efficiency, being the product of all efficiencies, is merely 80%, which is much lower compared with that of the direct mechanical path. Furthermore, since the power units in the series path have the full power rating, the serial hybrid systems are relatively heavy, expensive, and bulky. Thus, this configuration was not used in passenger cars, but is commonly used in diesel-electric locomotives and ships.

Note, on the other hand, that the engine size is typically smaller in HEVs because it has to meet only the average driving power demand. Another big advantage of the series HEV is that it can be easily converted into a PHEV by increasing the battery capacity and attaching an on-board charger. In such PHEVs, the ICE acts as a range extender. Note that 2017 Prius and Chevrolet Gen 2 Volt can be run in HEV mode.

Operation modes of the series power train are classified as the engine on-off mode operation and the blended mode operation. In the engine on-off mode operation, the motor is driven solely by the battery power if the battery charge level is sufficiently high. It is called *EV mode* or em charge depleting mode. But the engine turns on when the SOC reaches a prescribed low limit. At this time, the engine operates at the most efficient operating point.

However, in the blended mode operation the engine turns on whenever the power demand from the drive line exceeds the battery discharging capacity. That is, the engine cooperates with the battery to cover a peak load.

14.3.1   Simulation Results of Series Hybrids

A driving cycle simulation of a series HEV was performed using the simulation tool, "ADVISOR". In the simulation scenario, the vehicle mass with cargo was 1197kg, and GEO 1l (41 kW) SI engine model was used. A 58 kW PMSM motor model was adopted as a propulsion motor, and a lead acid battery model consisting of 25 modules with 307V nominal voltage, 3.5 kWh capacity, 30 kW peak power, and 120kg weight was used.

Fig. 14.7 shows a characteristic feature of the series hybrid power train. It is a magnified view of an initial part of the UDDS. As the speed increases, the power requirement grows. Note that the maximum power rating of the battery is 30 kW. However, the required peak power by motor is 40 kW. Therefore, to make up the deficiency in power, the engine starts to operate at $t = 190$ sec. Fig. 14.7 (d) shows the corresponding operating point, (3300 rpm, 58 Nm). Note that the engine operates at a point of maximum efficiency (0.33 g/kWh). At this time, the engine power (20 kW) is more than sufficient, so that the excess power is used for battery charging. Negative battery power shown in Fig. 14.7 (b) indicates the battery charging.

Fig. 14.8 (b) shows a SOC swing over two UDDS cycles. Note that SOC window is between 0.4 and 0.65, and charging/discharging repeats every 1400 seconds (23 minutes). One can see from Fig. 14.8 that the engine operates when either power assist is required, or when the battery SOC low limit is reached. Power assist is needed at 190 seconds, and the battery SOC low limit is hit at 1220 seconds. But since the battery SOC high limit was hit at 1480 seconds, the engine stopped.

14.4   Parallel Drive Train

The parallel HEV is basically an ICE vehicle equipped with an electric assisting device. In the parallel hybrid architecture, both an ICE and an electric motor are coupled jointly to the drive shaft, thereby each or both of them can drive the car. The motor can add up or subtract torque while turning at equal speeds. The electric motor can be used as a generator to charge the battery during regenerative braking or when recovering excess ICE power to the motor. This makes them more efficient in urban 'stop-and-go' conditions.

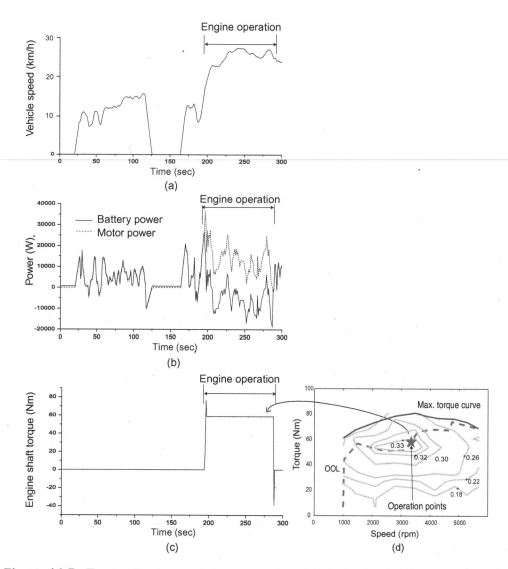

Figure 14.7: Engine involvement in a series HEV: (a) the UDDS driving cycle, (b) motor power and battery power, (c) engine shaft torque, and (d) engine operation point (GEO 1l, 41 kW).

The ICE is usually dominant, while the vehicle can run on the electric system alone. Usually the motor power rating is a fraction of the ICE power. For example, the Honda Insight has a 70 kW engine, but the motor power rating is only 10 kW. In this regard, the size and cost of parallel HEVs is less than those of series HEVs. However since the speed ratio between motor and engine is fixed, the roles of the motor are limited.

ICE alone and motor alone operations are feasible. To separate the ICE during the motor alone operation (all electric powering mode), the ICE is isolated by a clutch. Fig. 14.9 shows the parallel hybrid power train topology that is utilized in the Hyundai Avante.

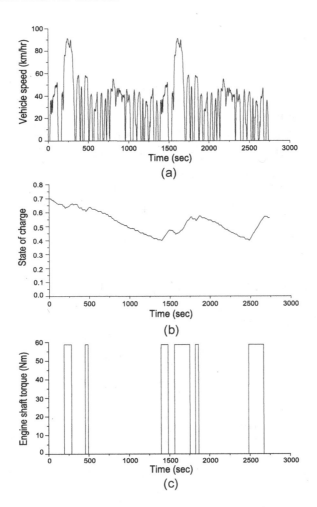

Figure 14.8: SOC swing and engine operation in a series HEV: (a) the UDDS driving cycle, (b) SOC change, and (c) engine shaft torque.

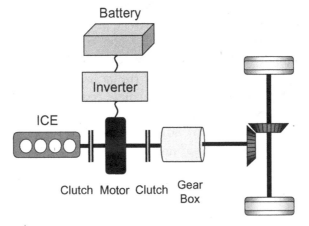

Figure 14.9: Parallel power train architecture.

The parallel assist strategy can be summarized as:

1) The motor alone propels the vehicle with the engine shut off in a low speed region, or in the region where the engine would run inefficiently.

2) The motor assists the engine if the required torque is larger than the maximum produceable by the engine.

3) The motor charges the battery during regenerative braking.

4) When the battery SOC is low, the engine will provide excess torque to charge the battery. Fig. 14.10 shows a view of the parallel power train of Hyndai "Avante Hybrid". In Hyundai 'Sonata' HEV, a 32 kW, 170 Nm PMSM is attached to front end of 6 speed automatic transmission.

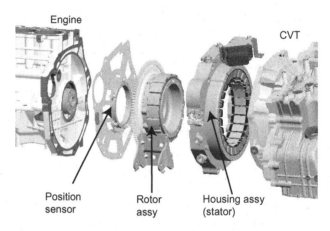

Figure 14.10: Parallel power train of Hyndai "Avante Hybrid".

Fig. 14.11 shows an overall control block diagram of a parallel hybrid vehicle. The battery management system (BMS) monitors the cell voltages, temperature, and state of charge, and communicate with the power control unit via CAN. It also regulates the cell voltages evenly to prevent excessive voltage unbalance. Through the use of a DC-DC converter, the main battery power is transferred to the 14 volt system where conventional electric loads are connected. The isolation is necessary for the safety of human and battery. A typical topology is the full bridge zero voltage switching (ZVS) circuit. The power rating is 1.6~2 kW. The power control unit (PCU) receives all necessary sensor signals required for PMSM control and outputs PWM gate signals. It also communicates with the engine controller via CAN. Engine power is controlled by adjusting the throttle angle, and the throttle position sensor (TPS) signal is also monitored by the PCU. Engine/motor shaft power is delivered to the drive wheels via the CVT. Engine control, motor control, and CVT control are coordinated in the PCU to optimize the overall fuel efficiency.

Fig. 14.12 shows the "ADVISOR" [18] simulation results for UDDS and US06 Hwy driving cycles. Both engine and motor operation points are shown in a single map. The motor operates mostly in a low torque region. Note that regenerations took place frequently in the urbane driving cycle. Obviously, engine operation points are widely spread compared with the cases of series and series/parallel hybrids.

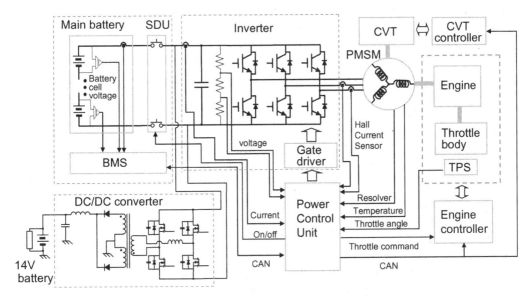

Figure 14.11: Overall control block diagram of a parallel hybrid vehicle.

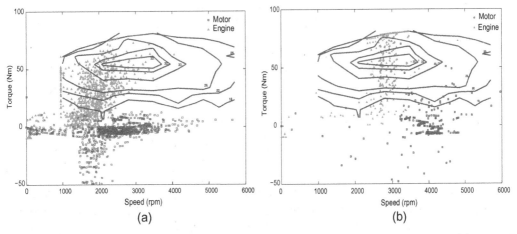

Figure 14.12: Engine and motor operating points of a parallel HEV for (a) UDDS and (b) US06 Hwy driving cycles.

14.4.1 Electrical Continuous Variable Transmission

Continuously variable transmission (CVT) is a transmission that can change the speed ratio seamlessly. It is usually constructed with a V-belt, tapered pulleys, and a hydraulic actuator. A CVT enables the ICE to run at a fixed speed when the wheel speed changes. Since it can restrict the engine speed in its most efficient range, it improves fuel economy. However, mechanical CVT is a friction based device so that the efficiency is low (about 88%), which offsets the efficiency gain in the engine. Besides, a CVT alone does not suffice all the necessary requirements for optimum ICE operation. The power balance should also be satisfied.

The electrical CVT (e-CVT) is a power splitting mechanism between the mechanical and electrical drive trains. As shown in Fig. 14.5 (c), the electrical drive train consists of two motors, two inverters, and a battery. Since the motors can act as generators, they are noted as 'M/G' in the following. It was first developed by TRW around 1970, and is being used in the Toyota, Ford, and GM hybrid transmissions. The e-CVT not only facilitates continuous speed change, but achieves the power balance, allowing the ICE to operate optimally. During low torque operation, excess power of the ICE is saved via a generator in the battery. But in a high torque mode, the generator, acting as a motor, adds more torque to the wheel.

14.4.2 Planetary Gear

A planetary gear together with a motor and a generator plays a key role in the engine power splitting in the series/parallel hybrid configuration. The planetary gear consists of sun gear, planetary gear, and ring gear, as shown in Fig. 14.13. The sun gear is located in the center, and planetary gears connected to a common carrier plate rotate around it while also meshing with an outer ring that has inward facing gear teeth. Based on the schematic shown in Fig. 14.13 (a), it follows that

$$R_r \omega_r - R_c \omega_c = R_c \omega_c - R_s \omega_s, \tag{14.2}$$

where R_s, R_p, and R_r are the radii of sun gear, carrier, and ring gear, and ω_s, ω_c, and ω_r are the speeds of sun gear, carrier, and ring gear, respectively. Let the gear ratio be defined by $R_r/R_s = k_p$. Then, dividing the right hand side by the left hand side of (14.2), it follows that

$$\frac{R_r \omega_r - \frac{R_r+R_s}{2}\omega_c}{\frac{R_r+R_s}{2}\omega_c - R_s \omega_s} = \frac{2\omega_r k_p - (k_p+1)\omega_c}{(k_p+1)\omega_c - 2\omega_s} = 1. \tag{14.3}$$

This is the fundamental equation of the planetary gear (i.e., epicyclic gearing), indicating that the gear rotation must maintain a fixed ratio, k_p, of angular velocity relative to the carrier body. After rearranging, (14.3) reduces to the more general form [5]:

$$\omega_s + k_p \omega_r - (k_p+1)\omega_c = 0. \tag{14.4}$$

Note that there is one constraint equation, (14.4) among three independent variables, ω_s, ω_c, and ω_r. Therefore, the planetary gear has two degrees of freedom, i.e., it takes up two independent shaft speeds. This illustrates the inherent speed summing nature of the planetary gear and the reason why it is used as a key power split device. Equation (14.4) of planetary gear is rearranged such that

$$\omega_c = \frac{1}{k_p+1}\omega_s + \frac{k_p}{k_p+1}\omega_r. \tag{14.5}$$

Geometrically, (14.5) appears to be a straight line as shown in Fig. 14.14, and it is called a lever diagram. The planetary gear of the Prius II has the following individual gears:

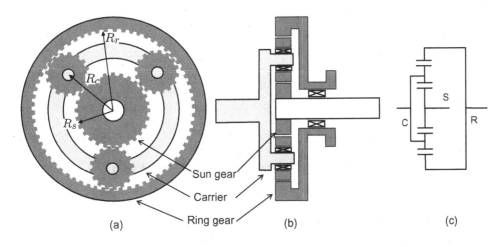

Figure 14.13: Planetary gear: (a) front view, (b) side view, (c) symbol.

sun gear : number of teeth, $N_s = 30$,
ring gear : number of teeth, $N_r = 78$.

Therefore, $k_p = R_r/R_s = N_r/N_s = 78/30 = 2.6$.

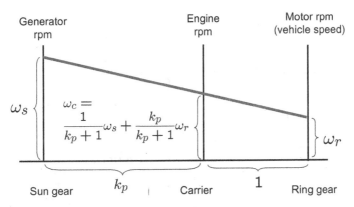

Figure 14.14: Lever diagram for the planetary gear system.

The torque balance and power balance conditions in the steady state are

$$T_c = T_r + T_s, \tag{14.6}$$
$$T_c \omega_c = T_r \omega_r + T_s \omega_s, \tag{14.7}$$

where T_c, T_r, and T_s are the shaft torques of carrier, ring, and sun gears, respectively. Since (14.6) and (14.7) are the steady state equations, inertial forces of gears are omitted. Further, the gear efficiency is assumed to be unity. It follows from (14.5) and (14.7) that [6]−[8]

$$T_c = (1 + k_p)T_s, \tag{14.8}$$
$$T_c = \left(1 + \frac{1}{k_p}\right)T_r. \tag{14.9}$$

Exercise 14.1
Consider a planetary gear shown in Fig. 14.14. Show that

$$-k_p = -\frac{R_r}{R_s} = \frac{\omega_s - \omega_c}{\omega_r - \omega_c}. \tag{14.10}$$

Exercise 14.2
Consider a planetary gear system, shown in Fig. 14.14. Assume $k_p = 2.6$, $\omega_c = 2500$ rpm, and $\omega_s = 3600$ rpm.
a) Calculate the speed of ring gear.
b) Assume that the engine produces 60 kW and that the generator applies 70 Nm load torque to the engine. Calculate the engine power directed to the ring gear.

Solution.
a)

$$\omega_r = \left(1 + \frac{1}{k_p}\right)\omega_c - \frac{\omega_s}{k_p} = 1.385 \times 2500 - 0.385 \times 3600 = 2078 \text{ rpm}$$

b)

$$P_g = \frac{3600}{60} \times 2\pi \times 70 = 26,389 \text{ W}.$$

Thus, $60 - 26.389 = 33.611$ (kW).   ■

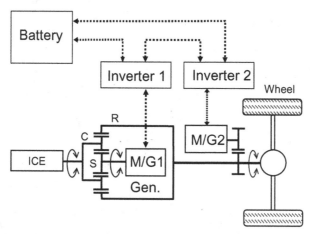

Figure 14.15: Schematic drawing of THS II employing a planetary gear as the input power split device (dotted line : electrical power train).

14.4.3   Power Split with Speeder and Torquer

A planetary gear is used as a power split device in the series/parallel HEVs. Fig. 14.15 shows a schematic diagram of the e-CVT of Toyota Hybrid System (THS) in Prius II

Table 14.2: Specifications of Prius II, III, and IV [15], [16], [19].

		Prius IV	Prius III	Prius II
Engine	Power	72 kW	73 kW	57 kW
	Torque	142 Nm	142 Nm	115 Nm
Motor	Type		IPM	IPM
	Max. power	53 kW	60 kW	50 kW
	Max. torque	163 Nm	207 Nm	400 Nm
	Max. speed		17000 rpm	6400 rpm
Generator	Type			IPM	IPM
	Max. power			22 kW
	Max. speed			11000 rpm
Battery	Type	NiMH	NiMH	NiMH
	Power		21 kW	21 kW
	Voltage	202 V	201.6 V	201.6 V
	Weight			45 kg
	Capacity	1 kWh	1.3 kWh	1.3 kWh
System voltage			650 Vmax	500 Vmax
Motor gear ratio			2.636	
Differential gear ratio		4.11	3.267	4.113

[15]. In the Prius III, the maximum motor speed was increased to 13500 rpm, reducing the mass by about 35% [16]. Major power train specifications are listed in Table 14.2. Since the power split device is positioned at the input (upstream) side, it is classified as input split. Since it contains a single planetary gear, it is called single mode.

The engine crank shaft is connected to the carrier, The M/G1, labeled 'generator', is connected to the sun gear. On the other hand, the M/G2, labeled 'motor', is connected to the ring gear. Note that the ring gear is coupled with the drive axle via the chain, final gear, and differential gear. Thus, the motor has a fixed gear ratio with the axle. Therefore, a part of the engine power is transmitted directly to wheels through the ring gear. The rest of the engine power is taken by the generator. The generated electric power is transmitted to the motor or stored in the battery and may be used later for EV mode or acceleration. The motor not only provides an additional torque to the drive line, but retrieves a part of vehicle kinetic energy into the battery by applying regenerative braking when the vehicle decelerates. A THS schematic diagram is shown in Fig. 14.15. Electrical paths are denoted by dotted lines, whereas mechanical paths are denoted by solid lines.

A more detailed diagram and power circuit are shown in Fig. 14.16. Two inverters are connected back to back with a common DC link bus. Note however that the battery voltage is around 200 V, but the DC link voltage is boosted to 500 V or 650 V. Thus, it requires the use of a bi-directional buck-boost converter. The reason for this voltage boosting is to reduce the current rating of cable and

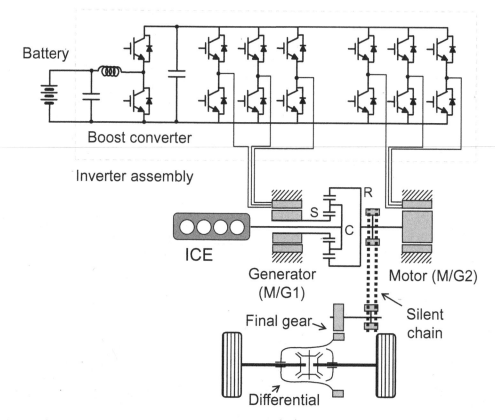

Figure 14.16: THS schematic diagram and power circuit.

M/G coils and to minimize M/G volumes. The chain in the Prius II was replaced by another planetary gear in the Prius III. Neither launch device, such as a torque converter in an automatic transmission, nor a clutch in a manual transmission, is used. The engine can remain directly connected to the transmission at all speeds.

14.4.4   Motor/Generator Operation Principle

It is assumed that the vehicle speed and the load torque do not change during a short time interval due to the large inertia. Here, both T_{dr} and ω_r are treated as fixed numbers. A power determining procedure consists of the following steps:

i)   For a given engine torque, T_c^*, determine ω_c^* such that the engine power is equal to the drive line power requirement, i.e., $T_c^*\omega_c^* = P_{dr}$.

ii)   Choose ω_s so that $\omega_s = (k_p + 1)\omega_c^* - k_p\omega_r$.

The engine speed is selected such that the engine power meets the demand required by the drive line. Since the ring gear speed, ω_r is fixed by the vehicle speed, the engine speed, ω_c, is passively determined by the sun gear speed, ω_s. In other words,

ω_c is determined by ω_s according to the linear equation, (14.4). For this purpose, the generator (M/G1) is controlled in the speed control mode; thereby it is called *the speeder* [3].

Once engine torque, T_c, is fixed, the torques of sun and ring gear shafts are determined according to (14.8) and (14.9). That is, the engine power is split according to

$$T_c\omega_c = \underbrace{\frac{T_c\omega_s}{k_p + 1}}_{=P_{elec}} + \underbrace{\frac{k_p T_c\omega_r}{k_p + 1}}_{=P_{mech}}. \qquad (14.11)$$

Note that the electrical path consists of sun gear–generator (M/G1)–inverter 1–inverter 2–motor-ring gear, whereas the mechanical path is made by the direct gear coupling – the carrier and ring gear. Note however that the split powers are summed at the drive line, as shown in Fig. 14.17. The power summation is possible since the motor (M/G2) is controlled in a torque control mode. Specifically, the motor can add/subtract torque at an arbitrary drive line speed. For this reason, the motor is called *the torquer* [3].

The ratio between the two power flows is equal to

$$\frac{P_{elec}}{P_{mech}} = \frac{\omega_s}{k_p\omega_r}. \qquad (14.12)$$

Therefore, as ω_s increase, the electrical path proportion increases.

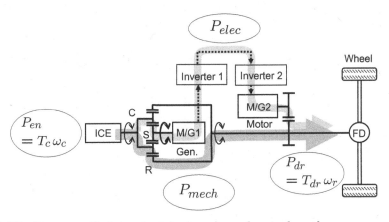

Figure 14.17: Power split into electrical and mechanical paths.

As mentioned in the above, the power transfer efficiency of the electrical path is low compared with that of the mechanical path. This implies that delivering 100% power over the mechanical path would be better. Then, the question is: Why is it necessary to split the power and sum the splits at the drive line? The benefit of the power split is illustrated by the following example: Assume that the drive line operates at 280 rad/sec with a load of 55.7 Nm. Since $P_{dr} = 15.595$ kW, the same power has to be delivered from the engine. Suppose that the generator speed is forced to down from 208 rad/sec, to 28 and to -30 rad/sec by a speed control action, while the throttle angle is controlled to yield a constant power, 15.6 kW: Let the operation points be marked such that 'A: ω_s = 280 rad/sec, B: ω_s = 28 rad/sec, and C: ω_s = −30 rad/sec. The corresponding engine speeds and torque are

A: (ω_c, T_c) = (2482 rpm, 60 Nm),
B: (ω_c, T_c) = (2005 rpm, 74 Nm),
C: (ω_c, T_c) = (1853 rpm, 80.4 Nm).

The engine speed, ω_c, decreases in proportion to ω_s, as shown in Fig. 14.18 (b). Since the engine maintains a constant power mode, the engine operation points migrate from "A" to "B" and "C" along a constant power line as shown in Fig. 14.18 (a). It should be noted that "C" is closer to the OOL than "A" and "B". Specifically, the fuel efficiency is increased to 235 g/kWh from 250 g/kWh.

The power split device enables the engine speed to move to the OOL while satisfying the wheel power requirements. Specifically, the speeder determines the engine speed at a point with high efficiency, independently of the wheel speed. In this way, the engine speed is decoupled from the wheel speed. In summary, the power split improves the system efficiency by allowing the engine to operate in the vicinity of the OOL. A minor loss takes places along with electrical power conversion.

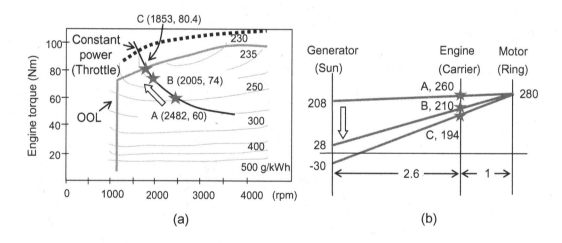

Figure 14.18: Operating point migration toward the OOL as the generator speed is controlled down to −30 rad/sec. (a) operating points in the engine torque-speed plane and (b) lever diagram.

This section is summarized with the following remarks:

1.   Since the generator operates usually in the speed control mode; it is called "speeder". The generator speed is determined to limit the engine operation within an optimal operation range. Recall that the engine is efficient when it is heavily loaded, i.e., when the operation point is near to the maximum torque line. Thus, the generator generally applies an additional load to the engine so that the operation point is close to the OOL when the vehicle road

load is low. The generated electric power is transmitted to the motor for power boosting, or used to charge the battery for later EV mode operation.

2.   Since the motor operates typically in the torque control mode, it is called "torquer". The motor supplies supplementary torque to the drive shaft utilizing the electric power. If the power from the generator is insufficient during a sudden acceleration, the battery power is also utilized.

14.5   Series/Parallel Drive Train

The THS is a typical series/parallel hybrid, in which both the series and parallel features coexist. With a planetary gear, operation is much more flexible compared with the parallel hybrid system. For example, it accommodates "electric mode driving" without utilizing a clutch: The vehicle can be driven by the motor while the engine is stationary. The series/parallel vehicle operates like a series vehicle in a low speed region, whereas it acts like a parallel HEV in the high speed region. Extensive researches are being performed regarding the power train modeling and control optimization [9]−[14].

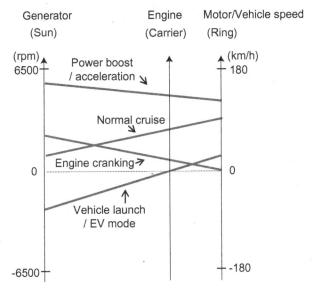

Figure 14.19: Lever diagrams.

Various operation modes include:

i) Vehicle launch / EV mode operation

ii) Engine cranking

iii) Normal Cruise with battery charging

iv) Power boosting for acceleration

v) Regenerative braking during deceleration

The corresponding lever diagrams are shown in Fig. 14.19. Recall that the motor has a fixed ratio to the vehicle speed (if the wheel slip is constant). During engine cranking, the vehicle speed (ring gear speed) is zero. In normal mode operation, the engine power is split and transported to the wheels via mechanical and electrical paths. For sudden acceleration, the battery power is used to increase the motor power. Both M/G1 and M/G2 serve as either a motor or a generator, depending on the driving conditions. For example, the generator (M/G1) normally generates electric power, but acts as a motor during engine cranking. On the other hand, the motor (M/G2) normally generates torque, but acts as a generator during regenerative braking.

1) Vehicle Launch and Battery EV Mode

For vehicle start and a low speed operation, the vehicle is powered only by the battery power. Thus, this mode is called EV mode. Note that the engine is stopped without a clutch mechanism while the vehicle is moving. It is clearly illustrated with the lever diagram: For a given positive ring gear (motor) speed, the sun gear (generator) speed is determined at a negative speed such that the carrier (engine) speed is zero. This is possible because the generator is controlled in a speed control mode. Fig. 14.20 shows an example of the battery EV mode.

Exercise 14.3
Planetary gear ratio is equal to $k_p = 2.6$, and the gear ratio of the drive line is $g_{dr} = 4.1$. The effective tire rolling radius is $r_w = 0.29$m. The vehicle operates in the EV mode with the engine shut off.

a) Determine the ring gear speed in the EV mode when the generator speed is 4500 rpm.

b) Assume that the engine cuts in when generator speed reaches 4500 rpm. Determine vehicle speed at the time of engine cut-in. Neglect the wheel slip.

Solution.
a) Using (14.10) with $w_c = 0$, it follows that

$$w_r = -\frac{w_s}{k_p} = -\frac{(-4500)}{60} \times 2\pi \times \frac{1}{2.6} = 181.2 \text{ rad/sec.}$$

b)

$$\frac{w_r}{g_{dr}} \times r_w = \frac{181.2}{4.1} \times 0.29 = 12.8 \text{m/sec} = 46 \text{ km/h.} \qquad \blacksquare$$

Note that the engine cuts in at 45 km/h in Prius II.

Exercise 14.4   (Continued from Exercise 14.3)
Suppose that the vehicle runs in an EV mode at 40 km/h against a 600 N road load. Assume that the motor and inverter efficiencies are 0.92 and 0.97, respectively. Assume further that efficiency of the final drive is $\eta_f = 0.82$ including the wheel slip.

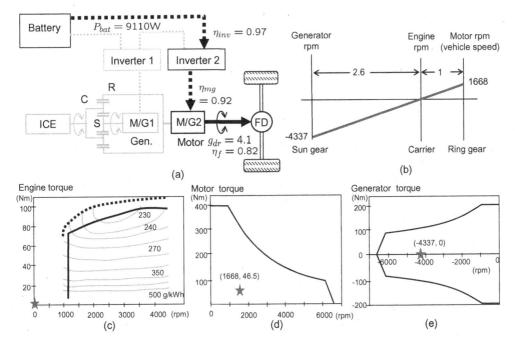

Figure 14.20: Battery EV mode: (a) power flow, (b) lever diagram, (c) engine fuel map, (d) motor torque-speed curve, and (e) generator torque-speed curve (Exercise 14.4).

a) Calculate speed, torque, and power of the motor when the average slip is $s_x = 0.1$.

b) Calculate the battery output power.

Solution.

a) Wheel power is equal to $P_x = -40000/3600 \times 600 = -6667$ W. Reflecting the final drive efficiency, motor power is $6667/0.82 = 8130$ W. Angular speed of wheel is

$$\omega_w = \frac{V_x}{r_w(1 - s_x)} = \frac{40000}{3600} \times \frac{1}{0.29 \times 0.9} = 42.6 \text{ rad/sec.}$$

Motor speed is $42.6 \times 4.1 = 174.7$ rad/sec (1668 rpm). Motor torque is equal to $8130/174.7 = 46.5$ Nm.

b) Battery power is equal to $P_{bat} = 8130/(0.92 \times 0.97) = 9110$(W). The result is depicted in Fig. 14.20. ∎

2) Normal Driving (Cruise) Mode

When the vehicle is cruising at a moderately high speed, the power demand is not so high. The engine power is split by the planetary gear: Through the carrier−ring gears, a portion of engine power is transmitted to wheels mechanically. The other portion is converted into an electrical power and transferred to the motor for torque assist. Fig. 14.21 shows a normal driving mode.

Exercise 14.5
Assume that the vehicle runs at a constant speed, 64.8 km/h, bearing a road load, $F_{rl} = 666.7$ N. Assume that the M/G and inverter efficiencies are 0.92 and 0.97, respectively, and that the efficiency of the final drive is $\eta_f = 0.82$ including the wheel slip.

a) Calculate the required power at the drive line, P_{dr}.

b) The effective tire rolling radius is $r_w = 0.29$ m and the final drive gear ratio is $g_{dr} = 4.1$. Calculate the motor angular speed, ω_r when the wheel slip is $s_x = 0.1$.

c) Assuming a steady state, determine the sun gear (generator) speed such that the engine speed is 220 rad/sec (2101 rpm).

d) Suppose that the engine output torque is regulated at 67.5 Nm. Calculate the power directed to the ring gear, P_r.

e) Calculate the required motor shaft power, $P_e = P_{dr} - P_r$.

f) Calculate the total efficiency of the electrical power train consisting of generator, inverter 1, inverter 2 and motor, when the inverter efficiency is $\eta_{inv} = 0.97$ and the efficiency of motor and generator is $\eta_{mg} = 0.92$.

g) Calculate the motor shaft power driven by the electric power transferred from the generator.

Solution.
a) At constant speeds, there is no inertial force. Therefore, $F_{rl} = F_x$. The vehicle propulsion power is $P_x = -F_x V_x = -666.7 \times 64800/3600 = -12001$ W. Since the efficiency of the final drive is $\eta_f = 0.82$, the power at the drive line should be $P_{dr} = 12001/0.82 = 14636$ W.
b)

$$\omega_r = \frac{V_x}{r_w(1 - s_x)} g_{dr} = \frac{64800}{3600 \times 0.29 \times 0.9} \times 4.1 = 282.8 \text{ rad/sec.}$$

c) Using (14.5), the generator speed is $\omega_s = 3.6 \times 220 - 2.6 \times 282.8 = 56.7$ rad/sec.
d) According to (14.8), generator torque is equal to $67.5/3.6 = 18.75$ Nm.
Generator power is $P_s = 56.7 \times 18.75 = 1063$ W. Therefore, power of the ring gear shaft is $P_r = P_{en} - P_s = 220 \times 67.5 - 1063 = 13787$ W.
e) Amount of power that should be assisted by the motor is
$P_{dr} - P_r = 14636 - 13787 = 849$ W.
f) Since there are two M/Gs and two inverters, the efficiency of the electrical path is $(0.97 \times 0.92)^2 = 0.796$.
g) Therefore, the motor power is $P_e = P_s \times \eta_{inv}^2 \eta_{mg}^2 = 1063 \times 0.796 = 847$ W.   ∎

Note from e) and g) that the drive line lacks 849 W, whereas the motor power received from the electric path is 847 W. That is, they are almost equal, so that there is no need for battery discharging. The solutions for the above normal mode are summarized in Fig. 14.21.

Fig. 14.22 shows power flow, lever diagram, and operating points of engine, generator, and motor at the normal load of Exercise 14.5. Note that the motor

(M/G2) operation point lies in the first quadrant of a torque-speed map. Since the generator applies a load torque to the engine, it produces a negative torque, i.e., the operation point locates in a regeneration region (4th quadrant).

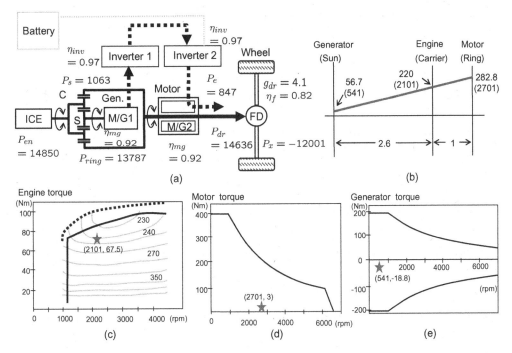

Figure 14.21: Normal driving mode: (a) power flow, (b) lever diagram, (c) engine fuel map, (d) motor torque-speed, and (e) generator torque-speed (Exercise 14.5).

The corresponding Sankey diagram is shown in Fig. 14.22: The engine power, P_{en}, is split into P_s and P_r. The generator power, P_s is used for powering the motor. In the electric power train, losses of generator, motor, and inverters are considered. Two powers, P_e and P_r, are summed at the drive line, P_{dr}. Finally, traction power, P_x is obtained bearing the mechanical loss of the final drive.

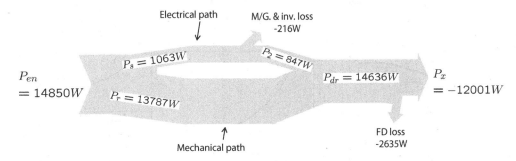

Figure 14.22: Sankey diagram for a normal driving mode (Exercise 14.5).

3) Power Boost Mode at Sudden Acceleration

When a required road load torque is higher than the full throttle torque of the engine, the motor acts as a power booster utilizing the battery power. That is, in addition to the generator power, more power is drawn from the battery into the motor.

Exercise 14.6

Assume that the vehicle runs at a constant speed, 80 km/h bearing a road load of $F_{rl} = 1400$ N. The engine speed and torque are regulated at $(T_{en}, \omega_c) = (97$ Nm, 314.2 rad/sec). Assume that the M/G and inverter efficiencies are 0.92 and 0.97, respectively, and that the efficiency of the final drive is $\eta_f = 0.82$ including the wheel slip.

a) The effective tire rolling radius is $r_w = 0.29$ m and the final drive gear ratio is $g_{dr} = 4.1$. Calculate the motor angular speed, ω_r, assuming that the average wheel slip is $s_x = 0.15$.

b) Assuming a steady state, determine the generator speed, ω_s using the lever diagram of $k_p = 2.6$.

c) Determine the generator torque and calculate the split powers, i.e., P_s and P_r.

d) Calculate the drive line power, P_{dr} and the battery (discharging) power, P_{bat}.

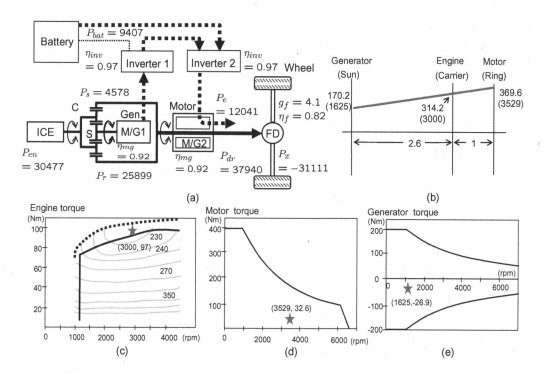

Figure 14.23: Power boost mode: (a) power flow, (b) lever diagram, (c) engine fuel map, (d) motor torque-speed, and (e) generator torque-speed (Exercise 14.6).

Solution.

a) The motor speed is

$$\omega_r = \frac{V_x}{r_w(1 - s_x)} \times g_{dr} = \frac{80000}{3600 \times 0.29 \times 0.85} \times 4.1 = 369.6 \text{ rad/sec.}$$

b) Since the engine speed is 314.2rad/sec, it follows from the level diagram that $\omega_s = 3.6\omega_c - 2.6\omega_r = 3.6 \times 314.2 - 2.6 \times 369.6 = 170.2$ rad/sec (1625 rpm).

c) The engine power is $P_{en} = 97 \times 314.2 = 30477$(W). According to (14.8), the generator torque is equal to $97/3.6 = 26.9$ Nm. Thus, the generator power is $P_s = 170.2 \times 26.9 = 4578$ W. Therefore, $P_r = P_{en} - P_s = 30477 - 4578 = 25899$ W.

d) At constant speeds, $F_x = F_{rl}$. The vehicle power is $P_x = -80000/3600 \times 1400 = -31111$ W. Considering the final drive efficiency, the power at the drive line is $P_{dr} = 31111/0.82 = 37940$ W. Thus, the required motor power is $P_e = P_{dr} - P_r = 37940 - 25899 = 12041$ W. A portion of motor power transmitted from the generator is $P_s \times \eta_{inv}^2 \eta_{mg}^2 = 4578 \times (0.92 \times 0.97)^2 = 3646$ W. Additional motor power that should come from the battery is $12041 - 3646 = 8395$ W. Thus, the battery (discharging) power is $P_{bat} = \frac{8395}{0.92 \times 0.97} = 9407$ W. ∎

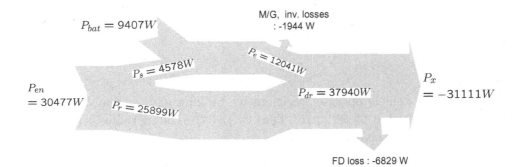

Figure 14.24: Sankey diagram for power boost mode (Exercise 14.6).

The corresponding Sankey diagram is shown in Fig. 14.24: It is shown that the battery (discharging) power, P_{bat} is added to the electrical power train so as to make up the lack of engine power, when the drive line power, P_{dr} exceeds the engine power, P_{en}.

4) Regenerative Braking

While the vehicle is decelerating or declining a hill, the vehicle's kinetic energy can be recovered by operating the motor in the generator mode. At this time, the motor is rotating in the forward direction. But by changing the q-axis current polarity to be negative, the braking power is transformed into electric power and used for recharging the battery. The engine is shut off during the regenerative braking. The regenerative braking is basically the same as the EV mode, but the power flow is reversed.

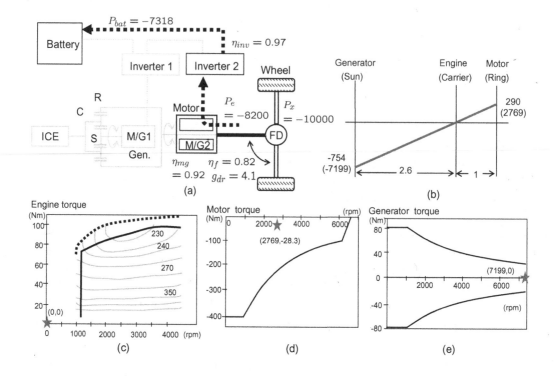

Figure 14.25: Regenerative braking (Exercise 14.7).

Exercise 14.7
Assume that the vehicle operates regenerative braking with engine shut off while the speed is regulated at 66.42 km/h. The effective tire rolling radius is $r_w = 0.29\,\text{m}$ and the gear ratio of the drive train is $g_{dr} = 4.1$.

a) Assuming that the wheel slip is $s_x = -0.1$. Calculate the motor speed, ω_r.

b) Assume that the drive line efficiency is $\eta_f = 0.82$. Determine the motor power, P_e, and the battery recharging power, P_{bat}, when the road load of $-542\,\text{N}$.

Solution.
a)
$$\omega_r = \frac{V_x}{r_w(1 - s_x)} \times g_{dr} = \frac{66420}{3600 \times 0.29 \times 1.1} \times 4.1 = 290\,\text{rad/sec.}$$

b) The wheel power is equal to $P_x = 66420/(3600) \times (-542) = -10000$ (W). Therefore, the drive line power is equal to $P_{dr} = -8200$ W. Also, $P_{dr} = P_e = -8200$ W. Considering the motor efficiency, the battery recharging power is $P_{bat} = -8200 \times 0.92 \times 0.97 = -7318$ (W). ∎

5) Engine Cranking
During the engine cranking, the vehicle is stationary, so that the motor speed is equal to zero. The generator forces the engine to spin, so that it operates in a motoring mode. The lever diagram for engine cranking is shown in Fig. 14.26. Note

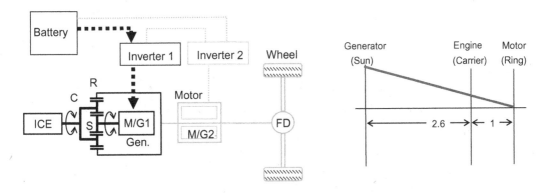

Figure 14.26: Engine cranking mode.

also that the lever diagram for stationary battery charging is the same as that of engine cranking. But, the power flow is reversed.

14.5.1   Prius Driving Cycle Simulation

The simulation was performed with the Prius model using a drive simulation tool, ADVISOR. The used vehicle simulation parameters are:

Curb mass with cargo (m_v) = 1398 kg,
CD (C_d) = 0.3,
Frontal area (A_F) = 1.746 m^2
Rolling coefficient (f_r) = 0.009.

Fig. 14.27 (b) the engine, motor, and generator operation points in the torque-speed plane under simple driving cycles shown in Fig. 14.27 (a). Four driving cycles were repeated to obtain dense operation points. It is shown that the motor generated a high torque during the acceleration period, especially in a low speed area. Then, a negative torque was applied in the deceleration period, whereby kinetic energy was recovered. It should be noted that engine operation points are not spread wide from the OOL.

The maximum motor speed is read 2400 rpm from Fig. 14.27 (b). Assume that the gear ratio between the wheel and motor shafts is 4.1 and that the effective tire rolling radius is 0.29 m. When we neglect the wheel slip, the peak velocity is estimated such that

$$V_x = 2400 \times \frac{2\pi}{60} \times \frac{1}{4.1} \times 0.29 \times \frac{3600}{1000} = 64 \text{ km/h}.$$

Note that the estimation is close to the peak speed (64 km/h), as shown in Fig. 14.27 (a).

Fig. 14.28 shows the loci of the engine operation points along a highway driving cycle (US06 Hwy). The operation points were aggregated to the optimum efficiency line, and the engine torque reached the maximum values. Fig. 14.29 shows the SOC pattern of the Prius II for urbane driving cycle (EPA UDDS). Note the the SOC window of Prius is very narrow (50~55%), i.e, the battery is cautiously utilized.

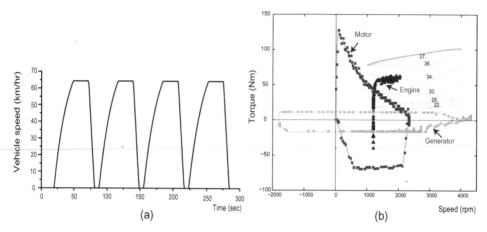

Figure 14.27: Operating points loci of Prius II engine, motor, and generator for a simple driving cycle.

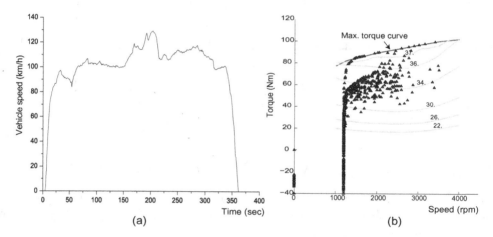

Figure 14.28: Engine operating points for a highway driving cycle (US06 HWY).

14.5.2   2017 Prius Plug-in Two-Mode Transmission

Recently Prius adopted a new design that used two clutches and two planetary gear sets as shown in Fig. 14.30 [24]. The intension is to enhance a dual motor EV mode to improve efficiency at higher vehicle speeds. Then, it looks similar to the Chevrolet Volt transmission which appears in the following. Each sun gear is connected to each motor, while two ring gears are connected together to the wheels. The first carrier is connected to the ICE. Two carrier gears are connected via Clutch 1 and then to the chassis via Clutch 2.

14.5.3   Gen 2 Volt Powertrain

Like Prius, 2nd generation Volt transmission utilizes two motors as shown in Fig. 14.31: The smaller motor, M/G A is 48 kW and 140 Nm, whereas the larger one, M/G B 87 kW and 280 Nm. In M/G A, ferrite magnets are utilized while maximizing the rotor saliency. On the other hand, M/G B utilizes NdFeB magnets, but the

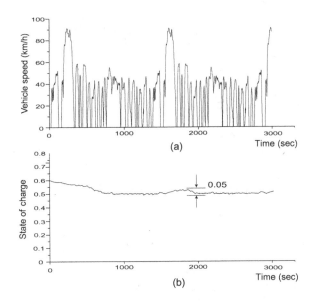

(a)

(b)

Figure 14.29: The SOC level change of Prius II during UDDS driving cycles: (a) EPA UDDS, (b) SOC.

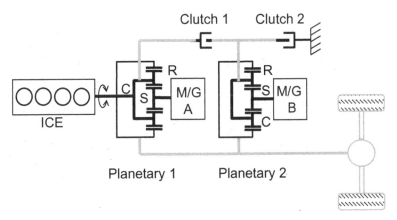

Figure 14.30: 2017 Toyota Prius power train architecture.

use of dysprosium is minimized by applying the grain boundary diffusion method. It also contains three clutches and two planetary gears as shown in Fig. 14.31 [23]. The ICE is connected to the ring of the first planetary gear set and M/G A is connected to the sun gear. M/G B is connected to the sun gear of the second planetary gear set. The carriers of two planetary gear sets are connected to the wheel shafts. The clutches reconfigure the power flow optimally at different vehicle speeds and torque requirements. Clutch 3 is a one-way clutch that only allows rotation in one direction. It does not engage in the mode change, but acts when it is necessary to hold the ICE shaft. The concept of extended range EV (EREV) is reflected in Gen 2 Volt design [20]−[22]. At the same time, it widens the use of a direct power path

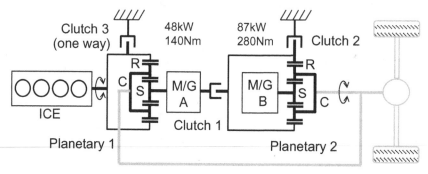

Figure 14.31: Chevrolet Volt power train architecture.

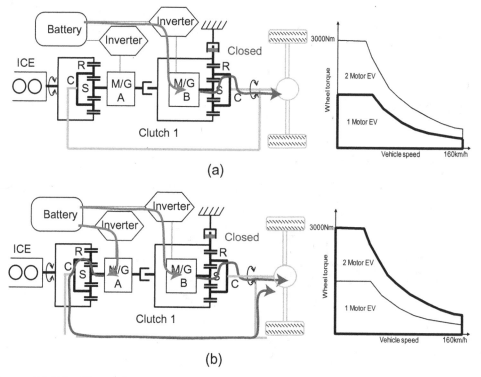

Figure 14.32: Chevrolet Volt EV modes: (a) one motor EV and (b) two motor EV.

from ICE to wheels, thereby the energy loss resulting from electric power conversion is minimized. In the following, 5 driving modes are illustrated in detail.

One Motor EV Mode

It is a low torque EV mode in which M/G A and ICE are off and just a M/G B delivers the battery power to the wheels. The M/G B design is optimized for low torque operation over the whole speed range.

Two Motor EV Mode

It is a high torque EV mode in which both M/G A and M/G B work in parallel, while ICE is off.

Low Extended Range Mode

The operation region is restricted high torque low speed range with ICE on. It is an input split mode, where ICE power is split between wheels and M/G A. The M/G A power is delivered to the battery or to drive M/G B. The two motors adjust their power to make the ICE operate at its most efficient operating point. At low torque/low speed, it goes into 'One Motor EV Mode' with ICE off.

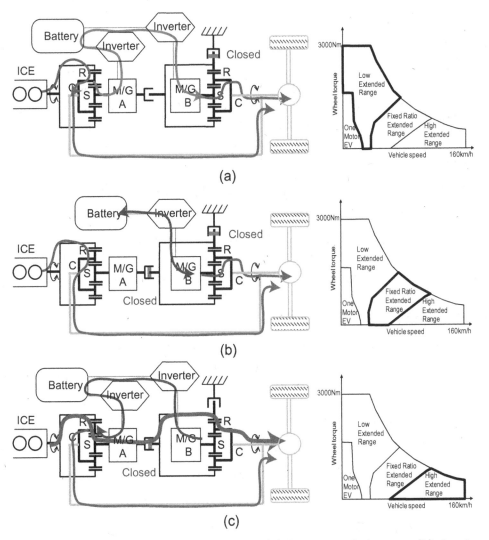

(a)

(b)

(c)

Figure 14.33: Chevrolet Volt EV modes: (a) low extended range (b) fixed ratio extended range, and (c) high extended range.

CS2: Fixed Ratio Extended Range Mode

Both Clutch 1 and 2 are closed, so that M/G A is locked. The ICE power is delivered directly to wheels, and M/G B also participates in to cover the peak load. However, excess ICE power is recovered by M/G B to the battery in the low torque region. It is the most efficient mode since the electrical power flow is minimized.

CS3: High Extended Range Mode

Clutch 1 is closed while Clutch 2 is open. The ICE power is split into two mechanical paths: One goes directly to wheels, while the others via planetary gears. By turning M/G B slowly, the gear ratio can be changed along with ICE rpm. At this time, small power is by-passed via M/G B to M/G A or battery.

Summary

- THS and Volt power trains are two typical hybrid architectures. Initially, THS did not use clutch and used a single planetary gear. However, two clutches and two planetary gears used in 2017 Prius. Thus, the recent THS becomes similar to Gen 2 Volt power train. A major impetus for change was to extend the EV mode to a high speed range. With the battery size upgrade, the HEV can be easily converted into a PHEV.

- Clutches offer a greater variety in power flow paths, thereby power split becomes more flexible.

- Both motor and generator can be used as torquer in the EV mode. On the other hand, they press the engine to operate around the OOL by applying more load, whereby the excess energy is stored in the battery.

References

[1] K.T. Chau and Y.S. Wong, "Overview of power management in hybrid electric vehicles," *Energy Conversion and Management*, vol. 43, pp. 1953−1968, 2002.

[2] C. C. Chan, "The state of the art of electric, hybrid, and fuel cell vehicles," *Proceedings of IEEE*, vol. 95, no. 4, pp. 704−718, Apr. 2007.

[3] J. M. Miller, *Proportion Systems for Hybrid Vehicle,* IEE Power & Energy Series 4, IEEE, London, 2004.

[4] Toyota, Toyota Hybrid System THS II, *Brochure.*

[5] J. M. Miller, "Hybrid electric vehicle propulsion system architectures of the e-CVT type," *IEEE Trans. on Power Electronics*, vol. 21, no. 3, pp. 756−767, May 2006.

[6] J. Liu and H. Peng, "Control optimization for a power-split hybrid vehicle," in Proceedings of the 2006 American Control Conference, Minneapolis, June, 2006.

[7] Y. Yu, H. Peng, Y. Gao, and Q. Wang, "Parametric design of power-split HEV drive train," IEEE VPPC, pp. 1058–1063, Dearborn, Michigan, Sep. 2009.

[8] Z. Chen and C. C. Mi, "An adaptive online energy management controller for power-split HEV based on dynamic programming and fuzzy logic," IEEE VPPC, pp. 335–339, Dearborn, Michigan, Sep. 2009.

[9] J. Liu, H. Peng and Z. Filipi, "Modeling and Analysis of the Toyota Hybrid System," Proceedings of the 2005 IEEE/ASME International Conference on Advanced Intelligent Mechatronics, Monterey, California, pp. 134–139, Jul. 2005.

[10] Y. Cheng, S. Cui, and C.C.Chan, "Control strategies for an electric variable transmission based hybrid electric vehicle," IEEE VPPC, pp. 1296–1300, Dearborn, Michigan, Sep. 2009.

[11] Y. Du, J. Gao, L. Yu, J. Song, and F. Zhao, "HEV system based on electric variable transmission," IEEE VPPC, pp. 578–583, Dearborn, Michigan, Sep. 2009.

[12] R. Zanasi and F. Grossi, "The POG technique for modeling planetary gears and hybrid automotive systems," IEEE VPPC, pp. 1301–1307, Dearborn, Michigan, Sep. 2009.

[13] J. Liu and H. Peng "Modeling and control of a power-split hybrid vehicle," *IEEE Trans. on Control Systems Technology*, vol. 16, no. 6, pp. 1242–1251, Nov. 2008.

[14] K. Chen, et al., "Comparison of two series-parallel hybrid electric vehicles focusing on control structures and operation modes," IEEE VPPC, pp. 1308–1315, Dearborn, Michigan, Sep. 2009.

[15] K. Muta, M. Yamazaki, and J. Tokieda, "Development of new-generation hybrid system THS II - Drastic improvement of power performance and fuel economy," Proc. of SAE World Congress, paper no. 2004-01-0064, Mar. 2004.

[16] Y. Mizuno et al., "Development of new hybrid transmission for compact-class vehicles," Proc. of SAE World Congress, paper no. 2009-01-0726, Apr. 2009.

[17] M. Ehsani, Y. Gao, S.E. Gay, and A. Emadi, *Modern Electric, Hybrid Electric, and Fuel Cell Vehicle, Fundamentals, Theory, and Design*, CRC Press, London, 2004.

[18] AVL ADVISOR (Advanced Vehicle Simulator), www.avl.com.

[19] M. Taniguchi et al., "Development of New Hybrid Transaxle for Compact-Class Vehicles," Proc. of SAE World Congress, paper no. 2016-01-1163, Apr. 2016.

[20] S. Jurkovic et al., "Design, Optimization and Development of Electric Machine for traction application for GM battery electric vehicle," IEEE IEMDC, pp. 1814–1819, 2015.

[21] K. M. Rahman et al., "Design and Performance of Electrical Propulsion System of Extended Range Electric Vehicle (EREV) Chevrolet Volt," *IEEE Trans. on Ind. Appl.*, vol. 51, no. 3, pp. 2479–2488, 2015.

[22] K. Rahman et al., "Retrospective of Electric Machines for EV and HEV Traction Applications at General Motors," IEEE ECCE, 2016.

[23] http://gm-volt.com/2015/02/20/gen-2-volt-transmission-operating-modes-explained/ .

[24] http://www.hybridcars.com/revenge-of-the-two-mode-hybrid/ .

Problems

14.1 Consider a planetary gear. Let $k_n = R_s/R_c$. Show that

$$k_n \omega_s + (2 + k_n)\omega_r - 2(1 + k_n)\omega_c = 0.$$

14.2 Consider a planetary gear system, shown below. For the carrier, the dynamics is represented by

$$T_{en} - (J_{carr} + J_{en})\dot{\omega}_c = T_{sun} + T_{ring}.$$

Derive the rest of equations for the sun and ring.

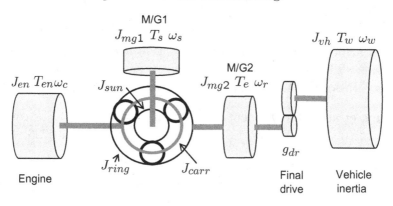

14.3 Consider the engine torque-speed chart shown in Fig. 14.18. The motor speed is fixed at 280 rad/sec.

a) Obtain the generator speed, ω_s so that the engine operation point moves to (1750 rpm, 82 Nm).

b) Calculate the generator power.

14.4 (Normal driving mode: Low load) Assume that a series/parall HEV of the THS type runs at 80 km/h against load road of 270 N. The gear ratio of the drive line is $g_{dr} = 3.9$ and the effective tire rolling radius is $r_w = 0.29$. The engine speed and torque are regulated at $(T_{en}, \omega_{en}) = (84 \text{ Nm}, 150 \text{ rad/sec})$ while the generator torque is 24 Nm. Assume that the M/G and inverter efficiencies are 0.92 and 0.97, respectively, and that the efficiency of the final drive is $\eta_f = 0.82$.

a) Calculate the motor speed when the wheel slip is $s_x = 0.1$.

b) Determine the generator speed and draw the lever diagram.

c) Calculate the power that should be provided by the motor.

d) Calculate the battery charging power.

14.5 Consider a series HEV drive train. Assume that 15 kW power is being used by the drive line while the engine operates with the full throttle at (60 Nm, 3400 rpm). The battery voltage is 307 V. Calculate the battery charging current when the efficiencies of motor and generator are 0.92 and those of their inverters are 0.97.

Index

Printed in the United States
by Baker & Taylor Publisher Services